Extracellular Polysaccharides

Tanmay Sarkar • Dibyajit Lahiri • Moupriya Nag
Editors

Extracellular Polysaccharides

Springer

Editors
Tanmay Sarkar
Department of Food Processing Technology
Malda Polytechnic, West Bengal State
Council of Technical Education,
Government of West Bengal
Malda, West Bengal, India

Dibyajit Lahiri
Department of Biotechnology
University of Engineering and Management
Kolkata, West Bengal, India

Moupriya Nag
Department of Biotechnology
University of Engineering and Management
Kolkata, West Bengal, India

ISBN 978-3-031-94471-0 ISBN 978-3-031-94472-7 (eBook)
https://doi.org/10.1007/978-3-031-94472-7

This Springer imprint is published by the registered company Springer Nature Switzerland AG
The registered company address is: Gewerbestrasse 11, 6330 Cham, Switzerland

Preface

Extracellular Polysaccharides: A Novel Frontier in Alternative Food Strategies represents a pioneering venture into the uncharted territory of extracellular polysaccharides (EPS) and their transformative potential in alternative food solutions. This groundbreaking work transcends the conventional discourse on food production challenges by unveiling the multifaceted applications of EPS sourced from diverse organisms like bacteria, fungi, algae, and plants. The book meticulously navigates through sustainable extraction techniques, nutritional insights, and cutting-edge technologies, offering a comprehensive roadmap for revolutionizing the global food industry.

Readers embark on a journey that goes beyond the theory, focusing on practical applications of EPS in various food products, from enhancing textures to serving as natural preservatives. Addressing current concerns in the food industry, this book serves as a catalyst for change, providing a holistic perspective on EPS that is both scientifically rigorous and accessible to a wide audience.

This book aims to provide a comprehensive overview of EPS, exploring their diverse applications, cutting-edge research, and future trends. It is unique in its holistic exploration of EPS, offering a more comprehensive understanding of their potential in alternative food strategies compared to titles that focus on specific sources or applications. A key focus is placed on the functional applications of EPS and explores their role as natural preservatives, contributing to food safety and shelf life. The book also provides global perspectives on EPS applications, showcasing international case studies and analyzing cultural influences on their adoption. Furthermore, it highlights specific food applications, such as their use in confectionery, snacks, and fermented foods, demonstrating their versatility.

Gut health is a major theme throughout the book, with dedicated chapters exploring the power of exopolysaccharides for improved gut health, their prebiotic functions, and their immunomodulatory role in the food and dairy industry. The impact of EPS on the quality of baked goods, including gluten-free and healthier alternatives, is also examined.

Looking ahead, the book explores future trends in EPS research and the development and futuristic applications of genetically modified exopolysaccharides. Finally, it addresses the crucial aspect of sustainability, examining the economic feasibility of microbial exopolysaccharide production. Our book places a strong emphasis on sustainability, addressing the ecological impact of EPS production and its role in sustainable food practices, aligning with current global trends toward environmentally friendly food solutions. While some titles touch upon food applications, our book extensively explores practical implementations of EPS in various food products, from beverages to baked goods, offering actionable insights for industry professionals and researchers. Addressing a gap in existing literature, our book focuses on the complex regulatory landscape surrounding EPS in food products. This coverage is crucial for readers navigating the intersection of scientific advancements and regulatory compliance.

Researchers, industry professionals, and policymakers will find invaluable insights, discovering novel strategies for addressing food security, improving nutritional profiles, and navigating the complex landscape of regulatory considerations. The book promotes an interdisciplinary perspective by fostering collaboration between these groups, ensuring a well-rounded exploration of EPS.

Extracellular Polysaccharides thus stands out as a comprehensive and forward-looking resource, providing a unique blend of scientific depth, practical applications, and a sustainability-focused perspective in the context of alternative food strategies. It is not merely a book; it's a guide to a sustainable and resilient food future, empowering its readers to be at the forefront of innovation and positive change. We hope this book will serve as a valuable resource and inspire further exploration into this exciting area.

Malda, West Bengal, India — Tanmay Sarkar
Kolkata, West Bengal, India — Dibyajit Lahiri
Kolkata, West Bengal, India — Moupriya Nag

Contents

About the Editors

Tanmay Sarkar is currently working as a Lecturer in the Government of West Bengal. He has received his Bachelor of Technology, Master of Technology, and PhD in Engineering from Jadavpur University, Kolkata, India. With 9 years of experience in both teaching and research, his areas of interest include bioactive components from natural sources, ethnic foods, mathematical modeling, and phytochemicals. He is an editor for several SCI, SCIE, and Scopus-indexed journals and has edited 12 books with Elsevier, Springer, and Wiley. Dr. Sarkar has been classified among the top 2% of global highly cited researchers published by Stanford University, USA.

Dibyajit Lahiri is at present working as Associate Professor in the Department of Biotechnology, University of Engineering and Management, Kolkata. His research interest includes biofilm isolated from human prosthesis, antimicrobial resistance, enzyme technology, and alternative therapeutics. He is in editorial boards of various journals like *Applied Biotechnology and Biochemistry*, *BMC Microbiology*, *Discover Bacteria*, *Biocatalysis*, and *Agricultural Biotechnology*. At present Dr. Lahiri has more than 130 research/review papers, 80 book chapters, and 10 books in the account of his publication. He has been accounted as World's 2% scientist by Stanford University and Scopus.

Moupriya Nag is presently working as Associate Professor in the Department of Biotechnology, University of Engineering & Management, Kolkata. She pursued her doctoral studies from Chemical Sciences Division, Saha Institute of nuclear Physics, Kolkata, India, and completed her postdoctoral research work at the Department of Biophysics, Bose Institute, Kolkata, India. Her research interests include single molecule Biophysics of protein unfolding/refolding and protein aggregation kinetics and its mechanism of action, especially the amyloid proteins in neurodegenerative disorders. Recently, her research focuses on functional amyloids in bacteria and their relation in forming biofilms. Her work deals with developing novel antimicrobial and anti-biofilm compounds from natural sources including plants, microbes, and green synthesized nanoparticles. She is serving as an editorial board member in *Applied Biochemistry & Biotechnology* (Springer) and *Biocatalysis & Agricultural Biotechnology* (Elsevier), *PLOS One*, *Biotechnology and Applied Biochemistry* (Wiley), *Natural Products* (Sage). She has about 14 years of teaching experience in Biophysics, Bioinformatics, Bioprocess and Biotechnology, Enzyme Technology, Computing, Molecular Modeling, and Drug Design.

Role of Extracellular Polysaccharides (EPSs) in Sustainable Food Chain Supply

Muhammad Usman, Muhammad Kasib Khan, Muhammad Shahid Mahmood, Zaheer Abbas, Sultan Ali, Riffat Maqsood, and Abdullah Azeem

1 Introduction

The interest and need for research in the food industry are increasing day by day to meet the nutritional requirements of the growing population. Polysaccharides are gaining great attention from researchers in the food sector. Polysaccharides are macromolecular polymers which are composed of hundreds to thousands of monosaccharides (Mohammed et al., 2021; Guo et al., 2017). Polysaccharides are supported by glycosidic bonds (Muhamad et al., 2019). There are many types of polysaccharides which are secreted by microorganisms like bacteria, fungi, algae, and protozoa (Darge et al., 2019). Some common types of polysaccharides are exopolysaccharides, lipopolysaccharides, and capsular polysaccharides (Yildiz & Karatas, 2018). Exopolysaccharide or extracellular polysaccharide is the main focus corner in this study. Extracellular polysaccharides are natural biopolymers which are synthesized outside or within microorganisms and then secreted (Hundschell & Wagemans, 2019). They are divided into two categories: homopolysaccharides and heteropolysaccharides (Moradi et al., 2021). Homopolysaccharides are made up of a single type of monosaccharide, but in contrast, heteropolysaccharides consist of repeating units of more than two types of monosaccharides. Only one enzyme, named glycosyltransferase, is widely used in the formation of homopolysaccharides, while a complex process involving various enzymes plays its role in the production of heteropolysaccharides. Microorganisms that secrete polysaccharides are

M. Usman · R. Maqsood
Faculty of Veterinary Science, University of Agriculture, Faisalabad, Pakistan

M. K. Khan (✉) · Z. Abbas · A. Azeem
Department of Parasitology, University of Agriculture, Faisalabad, Pakistan

M. S. Mahmood · S. Ali
Institute of Microbiology, University of Agriculture, Faisalabad, Pakistan

T. Sarkar et al. (eds.), *Extracellular Polysaccharides*,
https://doi.org/10.1007/978-3-031-94472-7_1

useful for the formation of their aggregates, adhesion of microorganisms to surfaces, and their protection (Dragoš & Kovács, 2017). EPS acts as a storage for energy and carbon for microorganisms. They play an enormous role in enhancing the quality of food. In addition, they are used in food packaging, fermented meat, dairy, and bakery products (Korcz & Varga, 2021). They are also enhancing the rheological properties of food. Extracellular polysaccharides play a role in various biological functions within an individual. They regulate the immune system and are antitumor agents (González-Garcinuño et al., 2018). They facilitate cholesterol lowering and antibacterial functioning (Korcz et al., 2018). EPS produced by different microorganisms has specific characteristics.

2 Sources of Extracellular Polysaccharides

Various microorganisms like algae, fungi, and bacteria are involved in the secretion of exopolysaccharides. Specific properties are assigned to exopolysaccharides produced by different microorganisms. Here, we will discuss these EPSs in detail.

3 Microalgal-Derived EPS

Microalgae, the photosynthetic microorganisms, are cultivated in water to get valuable and precious compounds like polysaccharides and proteins (Bezerra et al., 2022). Natural biopolymers named extracellular polysaccharides are produced and secreted by microalgae. The production of EPS from microalgae depends upon several factors like the availability of nutrients, light, temperature, and salinity. They have a significant scope in the food industry. Additionally, they also facilitate the survival of algae in various ways. Microalgae usually prefer to live in the form of groups like granules rather than as single cells. Microalgal-derived EPS facilitates the cohesion of microalgae in the form of groups to protect them from harmful substances (Xiao & Zheng, 2016). Under stress conditions, they facilitate the survival of microalgae by acting as a reservoir for energy and various nonmetals like carbon (Xiao & Zheng, 2016; Zhou et al., 2018). The functioning of EPS varies with the presence of sugars like glucose and galactose and chemical groups like sulfate. Some common types of microalgae like red algae, green algae, and diatoms usually secrete EPS. The composition of microalgal-derived EPS varies with species (Xiao & Zheng, 2016). The composition of EPS produced from some microalgae is discussed below.

(a) *From Green Microalgae*

The composition of EPS varies among different species of green algae. Some microalgal species like *Ankistrodesmus angustus*, *Chlorella* sp., *Botryococcus braunii*, and *Dunaliella salina* produce extracellular polysaccharides. They are usually composed of monosaccharides such as hexose, glucose, uronic acid, pentose, galactose, mannose, fucose, and sulfates, while EPS produced from some species such as *Kirchneriella aperta*, *Scenedesmus acuminatus*, and *Ankistrodesmus densus* primarily contain mannose (Xiao & Zheng, 2016).

(b) *From Red Microalgae*

The composition of EPS remains almost the same across different red algae species. Commonly, they are composed of xylose, glucuronic acid, galactose, and glucose (Gaignard et al., 2019; Delattre et al., 2016). But the ratio of different monosaccharides may vary among its different species. But commonly, these EPSs are composed of alternating 1,4- and 1,3-linked galactose chains. Methyl and sulfate groups are usually attached to these galactose chains (Borjas Esqueda et al., 2022).

(c) *From Diatoms*

Diatoms are key contributors to carbon cycling which are found in moist environments (Bohórquez et al., 2017). Composition of EPS varies with the growth phase and its type of species. Usually, diatom-derived EPSs are composed of sugars like mannose and a small amount of proteins (Xiao & Zheng, 2016).

4 Applications of Microalgae-Produced EPS in the Food Industry

1. *Improving the Texture of Food*

   Various microalgal species like *D. salina*, *Cosmarium turpinii*, and *Palmella texensis* play their functional role in the food industry by producing valuable EPS. EPS produced by *D. salina* alters various characteristics of food. They enhance the stability of food in various ways. They act as renewable bioemulsifiers to increase the stability of food. The bioemulsifying activity of EPS produced by *D. salina* and commercial bioemulsifiers is about 85% and 65%, respectively (Morais et al., 2022). So they are potential alternatives to commercial bioemulsifiers. In addition, they are thermostable at high temperatures up to 270 °C.

   EPS produced from *Cosmarium turpinii* increases the thickness due to its gel-like structure and pseudoplastic properties. So it acts as a thickening agent in the food industry to stabilize food. Meanwhile, *Palmella texensis*–derived EPS acts as a stabilizer in food due to its low viscosity. It plays a role in maintaining the texture of food even at undesirable pH and temperature (Xiao & Zheng, 2016).

2. *Used as Additives in Food to Improve Health*

   Natural additives are admired by people because of their crucial importance in improving health. Due to the increasing demand for natural additives, microalgal-derived EPSs are used in food. EPS from some microalgal *Porphyridium* species has diverse uses as a natural additive in food. Firstly, it is linked with a reduction in inflammation and stimulation of the immune system.

It enhances macrophage activity to support the immune system. Secondly, it is used in food to lower the cholesterol level to prevent heart diseases in humans (Laurienzo, 2010). These EPSs have potential uses in chicken feed to enhance the quality of eggs by lowering cholesterol levels in blood and eggs. Thirdly, these EPSs are used in food to maintain its stable rheological properties like viscosity and thickness across a wide range of temperatures, pH, and salinity levels (Laurienzo, 2010).

5 Fungal-Derived EPS

Due to structural complexity, fungi are considered a rich source of EPS. EPS derived from these fungi may be homopolysaccharides (contain only glucose) or hetero-polysaccharides (made up of different sugars). Ascomycota and Basidiomycota are two fungal groups involved in EPS production which are commonly studied. Some fungal species like *Phellinus linteus*, *Ganoderma lucidum*, and *Inonotus obliquus* belong to the phylum Basidiomycota and are commonly reported for EPS production, while *Fusarium* and *Pleurotus* spp. are included in the phylum Ascomycota and are involved in EPS production (Guo et al., 2013). The composition of EPS produced from Ascomycota and Basidiomycota is studied by using various techniques like chromatography and spectroscopy. Different types of linkages are present in these EPSs, but the $1 \rightarrow 6$ linkage type is very common in them.

6 Applications of Fungal-Derived EPS in the Food Industry

Some homopolysaccharides having nutritional significance like levan, pullulan, and scleroglucan are produced by fungi. These homopolysaccharides are described below.

1. *Pullulan*

   Pullulan produced from *Aureobasidium pullulans* is a nontoxic polysaccharide used in the food industry due to its unique characteristics, structure, and functions (Osemwegie et al., 2020; Prajapati et al., 2013). It consists of repeating units of maltotriose, providing a flexible and stable structure. It is used in packaging film due to its strength, flexibility, and ability to resist pH changes. As a carrier of antimicrobial substances and flavors, edible films of pullulan are widely used in the food industry. For instance, a film of extract of sweet basil extract and meadowsweet flower mixed with pullulan is applied to an apple. It is done to protect apples from the attack of fungi like *Rhizopus arrhizus* during storage (Gniewosz et al., 2014; Synowiec et al., 2014).
2. *Scleroglucan*

   Scleroglucans are homopolysaccharides which are derived from *Porodisculus pendulus* (Tiwari et al., 2020).

The following are applications of scleroglucan in the food industry:

(a) They are used in food products due to their high viscosity and thermostable properties to enhance the stability of food.
(b) Due to pseudoplastic behavior, they improve the quality of frozen foods.
(c) As an emulsifier, they are used in various food items like ice creams (Alghmadi, 2016).

3. *Levan*

Levan is a brittle homopolymer usually secreted by some plants, fungi, and bacteria (Öner et al., 2016). It is used as an encapsulating agent for film packaging in the food industry. Due to its brittle nature, it is mixed with various plasticizers like glycerol in the correct percentage during the manufacturing of film (Bahroudi et al., 2020). This is done to increase the flexibility of the film. In addition, it may be used as a carrier for flavors and thickeners in the food industry (Basiri, 2021).

7 Bacterial-Derived EPS

Various groups of bacteria are involved in the production of EPS, among which bifidobacteria and lactic acid bacteria are significant. EPS covers the surface of bifidobacteria and lactic acid bacteria in the form of slime and capsule (Mazzoli et al., 2014; Prasanna et al., 2014). They are necessary for the protection of bacteria (Leroy & De Vuyst, 2016). Usually, they are composed of repeating units of glucose and galactose. They are divided into homopolysaccharides (dextran, ß-glucan, and alternan) and heteropolysaccharides (Xu et al., 2019). There are some variations in homopolysaccharides and heteropolysaccharides. The variations in homopolysaccharides are mainly due to the involvement of specific enzymes, different sizes of polymers, and types of glycosidic bonds. The structural diversity in heteropolysaccharides is due to the complexity of the interaction between substrate and enzyme (Zhou et al., 2019).

8 Applications of Bacterial-Derived EPS in the Food Industry

1. *In Dairy Products*

EPSs from bacteria (levan, dextran, and inulin) play a vital role in manufacturing dairy items.

- Levan is commonly used in cheese production to improve its moisture. It is also used as a stabilizer and emulsifier in food (Tomulescu et al., 2016).

- Dextran is used in bakery and dairy items for the retention of moisture. In addition, it is usually used in the manufacturing of fermented drinks and ice cream.
- Inulin as a natural additive and stabilizer improves the texture and nutritional balance of food without affecting taste. As a food texturizer, it is the best alternative to fats in manufacturing cheese and butter. For example, in yogurt production, they prevent syneresis and provide a creamy and smooth texture to yogurt (Zannini et al., 2016).

2. *In Fermented Meat*

   Lactic acid bacteria especially lactobacillus and pediococcus are used in the manufacturing of fermented meat. The production of EPS is controlled in order to consistently ensure the good quality of food (Korcz & Varga, 2021). They have a wide range of applications in fermented meat. Firstly, in order to improve food safety, they are used to lower the levels of harmful substances. Secondly, they also provide color and moisture to meat (Ryan et al., 2015). Thirdly, in food manufacturing, EPSs protect them during harsh conditions like high temperature and low pH.
3. *In Bread and Sausages*

   Bacterial-derived EPSs are widely used in the manufacturing of bread. They improve the quality of bread by making it softer, increasing loaf volume, and improving shelf life and retention of moisture of bread (Wang et al., 2019). They are used to enhance the quality of sausages by improving their toughness and firmness (Korcz & Varga, 2021). Figure 1 shows the sources and applications of EPS in the food industry.

9 Addressing Challenges to the Food Industry

The food industry is playing a central role in meeting the nutritional requirements of the world. But there are several challenges which are alarming for the food industry. Among them, some important challenges are described below.

1. *Food-borne Parasites*

   In order to maintain health, the food industry is playing an essential role in meeting the increasing dietary demands. There are many parasites playing a leading role in disrupting health by causing various diseases. Some important parasites which are destroying human health and are spreading form food are listed below.

   (a) ***Toxoplasma gondii***

   T. gondii is a protozoan related to the phylum *Apicomplexa* which causes toxoplasmosis in humans. It is usually present as tachyzoites, bradyzoites, and sporozoites. Tachyzoites proliferate in the host and destroy infected host cells. Bradyzoites are formed from cysts in tissues and remain dormant. On

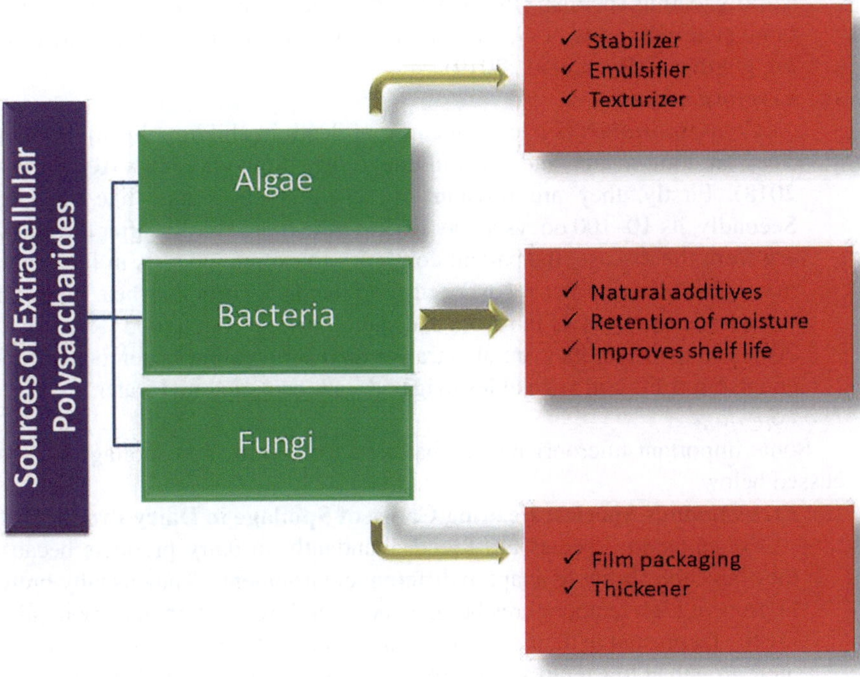

Fig. 1 Sources and applications of EPS in the food industry

the other hand, sporozoites are present within oocysts in cat faeces. They remain dormant under harsh conditions and form spores. After studying their form of presence, the entrance of *T. gondii* is also crucial. *T. gondii* enters the human body through food contaminated with its spores. These spores enter the human body mainly through partially cooked food, mainly meat. After ingestion of spores, their cell wall rupture and releases bradyzoites (Bintsis, 2017). After release, they are transported across the intestinal epithelium and move across the whole body. They are transformed into tachyzoites and result in the death of the infected host cell. The immune system converts these tachyzoites into oocytes, and they invade adjacent cells. In this way, they remain inside the host throughout life and cause the death of host cells. In this way, these food-borne parasites are alarming for health (Bintsis, 2017).

(b) ***Giardia duodenalis***

G. duodenalis can transfer to humans from food or water contaminated with it and can cause giardiasis. They may be in the form of dormant or active (disease-causing) cysts. There are various ways through which food is contaminated with them. Firstly, direct contact with infected workers involved in food harvesting and packaging may result in giardiasis (Greig et al., 2007). Secondly, the use of salad or herbs irrigated with contaminated water may result in the transmission of disease (Semenza et al., 2012).

Thirdly, eating shellfish (filters water by accumulating *Giardia cysts)* contaminated with cysts of *G. duodenalis may damage human health by causing* giardiasis (Ryan et al., 2019).

(c) ***Cryptosporidium***

Cryptosporidium is a leading cause of death in children due to diarrhea. They are long-lasting and harmful due to their following traits (Ryan et al., 2018). Firstly, they are resistant to several disinfectants like chlorine. Secondly, its 10–100 oocysts may lead to infection. Thirdly, after the recovery from the disease, the patient continues to excrete oocysts in faeces and may spread them. After recovery of food handlers from diarrhea, they continue to secrete oocysts in faeces and spread them through food (Ryan et al., 2018). In addition, they are also transmitted from eating undercooked food and using fruits and vegetables irrigated with contaminated water.

2. *Food Spoilage*

Some important microorganisms that are involved in food spoilage are discussed below.

(a) ***Pseudomonas* Species: Leading Cause of Spoilage in Dairy Products**

Pseudomonas species are found abundantly in dairy products because they have the ability to adapt to different environments. They usually prefer to grow at high temperatures but may adapt to low temperatures by regulating the fat present in their cell membrane in liquid form. They may stimulate their growth at low temperatures by producing some enzymes and pigments. They cause food spoilage by producing various enzymes in raw milk, butter, and cheese.

In raw milk, various *Pseudomonas* species like *P. gessardii*, *P. fragi*, and *P. putida* are found. Ultra-heat treatment (UHT) is given to milk to prevent it from enzymes like protease and lipase produced by these bacterial species. But due to the heat-resistant property of these enzymes, they tolerate UHT (Stuknytė et al., 2016). An enzyme named protease leads to the breakdown of proteins, especially k-casein, over time and affects the process of milk curdling with rennet (D'Incecco et al., 2019). This leads to a change in flavor, texture, and yield of cheese. On the other hand, lipase is released from *P. fragi*, which leads to the breakdown of triglycerides into glycerol and free fatty acids and affects the flavor of milk. The alteration of the flavor of the milk is directly linked to the length of the carbon chain. Short-chain fatty acids (C4–C8) can bring rancid odor and flavor, While medium-chain fatty acids (C10–C12) can result in soapy taste.

If cheese like mozzarella is stored in a liquid medium, it may allow *Pseudomonas* species, especially *P. fluorescens*, to grow without any hindrance. *P. fluorescens* also produces a protease enzyme, resulting in the breakdown of α- and β-casein in mozzarella. This breakdown of protein leads to the release of more amino acids. This decrease in protein causes weakening of its structure and starts collapsing slowly over time (Baruzzi et al., 2012; Caputo et al., 2015).

In butter, the presence of various psychrotrophic pseudomonads like *P. fragi*, *P. putrefaciens*, *P. nigrificans*, and *P. fluorescens* can affect the qual-

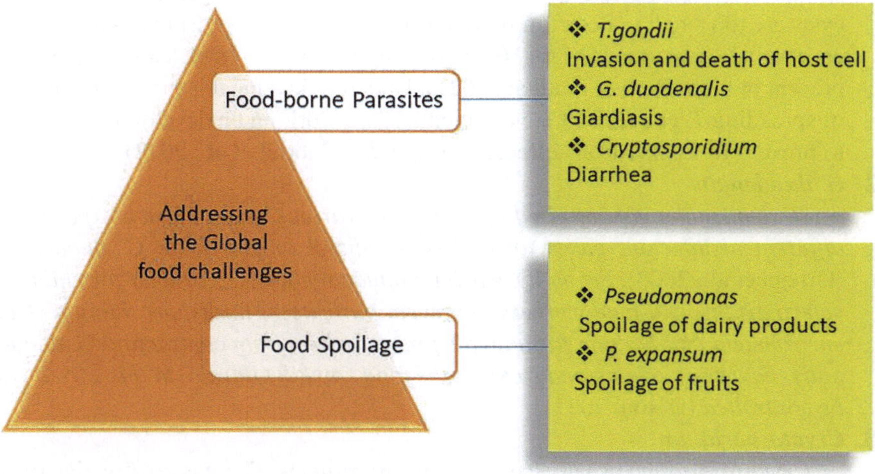

Fig. 2 Global food challenges

ity of butter in various ways (Ledenbach & Marshall, 2009). *P. putrefaciens* and *P. nigrificans* have the ability to grow on the surface of bacteria and lead to black discoloration. If butter is stored for 7–10 days at low temperature, it may develop a putrid odor due to *P. putrefaciens* and *P. nigrificans*. *P. fragi* and, less commonly, *P. fluorescens* can spoil butter by making it rancid (Quintieri et al., 2021).

(b) ***Penicillium expansum*: Responsible for Fruit Spoilage**

P. expansum is a fungus found in soil, vegetables, dairy products, and seafood. It is a leading cause of fruit spoilage and can come in contact with them during their handling. It releases toxins like patulin even inside seeds. It is produced in different parts of the mold cell, and a series of enzymes and genes are involved in its production. Acetyl-CoA and malonyl-CoA are the basic building blocks of the patulin biosynthesis pathway (Tannous et al., 2014). It starts to accumulate within fruits depending on acidity, climate, and storage. As the fungi grow and produce patulin, it leads to softening of the fruit and changes its color to brown. This leads to the spoilage of fruits, and eating these fruits may lead to various disorders in people (Yu et al., 2020). Figure 2 shows the global food challenges.

10 Suitable Solutions for Challenges in the Food Industry

1. ***T. gondii***

T. gondii is a food-borne parasite causing various disorders in humans. However, vegetables and meat contaminated with it are a leading cause of its spread in humans. There is no standardized method for its prevention. Its tissue cysts are too small to detect in food during food inspection. But cooking meat at an internal temperature of 64 °C or freezing it below −18 °C can kill parasites present in food (Rani & Pradhan, 2021). Undercooked meat is increasing the risk of spreading *T. gondii*. Moreover, scientists are working on developing a vaccine to prevent from disastrous effects of *T. gondii* (Gabriël et al., 2022).

2. ***G. duodenalis***

   G. duodenalis is controlled effectively in various ways. Firstly, exposure of different disinfectant gases to food is beneficial to get rid of G. duodenalis (Ortega et al., 2008). *Secondly, washing hands regularly is crucial for children, veterinarians, and food workers to prevent parasites. Thirdly, providing a clean environment and bathing for animals prevents them from contracting G. duodenalis. Fourthly, cooking and freezing the food can kill Giardia. In this way, it can be controlled* (Dixon, 2021).

3. ***Cryptosporidium***

   Cryptosporidium usually causes cryptosporidiosis. There is no effective treatment for cryptosporidiosis, so various preventive measures are used to get rid of *Cryptosporidium.* The following are its preventive measures:

   (a) Exposure to a temperature of about 64 °C for 5 min or 72 °C for 1 min can kill the oocysts of *Cryptosporidium.*
   (b) Moreover, being subjected to a temperature of about −70 °C or lower could be helpful for its elimination.
   (c) Ultraviolet rays are helpful in destroying oocysts (Ryan et al., 2018).
   (d) Some disinfectants like ammonia, chlorine dioxide, ozone, and hydrogen peroxide are effective against *Cryptosporidium.*
   (e) Predatory protozoa and rotifers are helpful for controlling oocysts of *Cryptosporidium* in water bodies by eating them (Ryan et al., 2018).

4. **Control of *Pseudomonas* Species in Dairy Products**

   In dairy products, various *Pseudomonas* species are responsible for food spoilage. Different tactics are used to control them. Due to their antibiotic resistance, antimicrobial peptides (AMPs) and vegetable-derived substances like phenols are used to control them. AMP is responsible for slowing the growth of *Pseudomonas* species by producing bovine lactoferrin (BLF). They are widely used in human food supplements due to their antimicrobial effects. These BFLs are used in dairy products like cheese and butter to prevent them from discoloration and reduce the growth of *Pseudomonas* species (Quintieri et al., 2019).

5. **Postharvest Control of *P. expansum***

   Various physical and chemical methods are used in the postharvest control of *P. expansum.* Physical methods like heat and UV rays are used to control it. Before freezing the fruits, they are soaked in hot water (40–45 °C) for a short period (5–10 min). This leads to inhibition of fungal spore growth and enhances the proficiency of the fruit's natural defense, while the exposure of UV-C waves of about 190–280 nm to the fruit's surface may result in the death of pathogens

(Romanazzi et al., 2016). Its effectiveness depends upon the texture of the fruit and the duration of exposure to UV light.

On the other hand, various chemical tactics are used to control the *P. expansum*. The use of fungicides like benomyl, pyrimethanil, and thiabendazole is restricted due to their devastating effects on the environment and human health. Some antifungal agents like food additives and plant hormones are commonly used as an alternative for fungicides (Zhang et al., 2018). Food additives like sodium bicarbonate and calcium chloride are applied to control the rots due to *P. expansum*. The growth and activity of fungal pectinase enzymes (responsible for the breakdown of pectin in the cell wall) are disturbed due to their presence. Similarly, plant hormones like salicylic acid are used to control *P. expansum*. Salicylic acid is a phenolic compound safe for both humans and the environment. It is used to enhance the activity of the natural defense system of plants by inducing the activity of enzymes like lipoxygenase, phenylalanine ammonia-lyase, and peroxidase (Yu et al., 2020).

11 Conclusion

Extracellular polysaccharides from various sources like bacteria, fungi, and algae are widely used in the food industry as a thickener, emulsifier, and stabilizer, while the food industry is facing many serious challenges like food spoilage and food-borne parasite diseases. There are certain preventive measures to protect the food industry from these challenges.

References

Alghmadi, A. Y. (2016). Rheological characteristics of Ganoderma applanatum exopolysaccharides. *African Journal of Microbiology Research, 10*(6), 147–155.

Bahroudi, S., et al. (2020). Levan exerts health benefit effect through alteration in bifidobacteria population. *Iranian Biomedical Journal, 24*(1), 54.

Baruzzi, F., et al. (2012). Occurrence of non-lactic acid bacteria populations involved in protein hydrolysis of cold-stored high moisture Mozzarella cheese. *Food Microbiology, 30*(1), 37–44.

Basiri, S. (2021). Applications of microbial exopolysaccharides in the food industry. *Avicenna Journal of Medical Biochemistry, 9*(2), 107–120.

Bezerra, P., et al. (2022). Innovative application of brackish groundwater without the addition of nutrients in the cultivation of Spirulina and Chlorella for carbohydrate and lipid production. *Bioresource Technology, 345*, 126543.

Bintsis, T. (2017). Foodborne pathogens. *AIMS Microbiology, 3*(3), 529.

Bohórquez, J., et al. (2017). Different types of diatom-derived extracellular polymeric substances drive changes in heterotrophic bacterial communities from intertidal sediments. *Frontiers in Microbiology, 8*, 245.

Borjas Esqueda, A., Gardarin, C., & Laroche, C. (2022). Exploring the diversity of red microalgae for exopolysaccharide production. *Marine Drugs, 20*(4), 246.

Caputo, L., et al. (2015). Pepsin-digested bovine lactoferrin prevents Mozzarella cheese blue discoloration caused by Pseudomonas fluorescens. *Food Microbiology, 46*, 15–24.

Darge, H. F., et al. (2019). Polysaccharide and polypeptide based injectable thermo-sensitive hydrogels for local biomedical applications. *International Journal of Biological Macromolecules, 133*, 545–563.

Delattre, C., et al. (2016). Production, extraction and characterization of microalgal and cyanobacterial exopolysaccharides. *Biotechnology Advances, 34*(7), 1159–1179.

D'Incecco, P., et al. (2019). Bacterial proteolysis of casein leading to UHT milk gelation: An applicative study. *Food Chemistry, 292*, 217–226.

Dixon, B. R. (2021). Giardia duodenalis in humans and animals–transmission and disease. *Research in Veterinary Science, 135*, 283–289.

Dragoš, A., & Kovács, Á. T. (2017). The peculiar functions of the bacterial extracellular matrix. *Trends in Microbiology, 25*(4), 257–266.

Gabriël, S., et al. (2022). Foodborne parasites and their complex life cycles challenging food safety in different food chains. *Food, 12*(1), 142.

Gaignard, C., et al. (2019). Screening of marine microalgae: Investigation of new exopolysaccharide producers. *Algal Research, 44*, 101711.

Gniewosz, M., et al. (2014). The antimicrobial activity of pullulan film incorporated with meadowsweet flower extracts (Filipendulae ulmariae flos) on postharvest quality of apples. *Food Control, 37*, 351–361.

González-Garcinuño, Á., et al. (2018). Levan and levansucrases: Polymer, enzyme, microorganisms and biomedical applications. *Biocatalysis and Biotransformation, 36*(3), 233–244.

Greig, J. D., et al. (2007). Outbreaks where food workers have been implicated in the spread of foodborne disease. Part 1. Description of the problem, methods, and agents involved. *Journal of Food Protection, 70*(7), 1752–1761.

Guo, S., et al. (2013). Structural elucidation of the exopolysaccharide produced by fungus Fusarium oxysporum Y24-2. *Carbohydrate Research, 365*, 9–13.

Guo, M. Q., et al. (2017). Polysaccharides: Structure and solubility. *Solubility of Polysaccharides, 2*, 8–21.

Hundschell, C. S., & Wagemans, A. M. (2019). Rheology of common uncharged exopolysaccharides for food applications. *Current Opinion in Food Science, 27*, 1–7.

Korcz, E., & Varga, L. (2021). Exopolysaccharides from lactic acid bacteria: Techno-functional application in the food industry. *Trends in Food Science & Technology, 110*, 375–384.

Korcz, E., Kerényi, Z., & Varga, L. (2018). Dietary fibers, prebiotics, and exopolysaccharides produced by lactic acid bacteria: Potential health benefits with special regard to cholesterol-lowering effects. *Food & Function, 9*(6), 3057–3068.

Laurienzo, P. (2010). Marine polysaccharides in pharmaceutical applications: An overview. *Marine Drugs, 8*(9), 2435–2465.

Ledenbach, L. H., & Marshall, R. T. (2009). Microbiological spoilage of dairy products. In *Compendium of the microbiological spoilage of foods and beverages* (pp. 41–67). Springer.

Leroy, F., & De Vuyst, L. (2016). Advances in production and simplified methods for recovery and quantification of exopolysaccharides for applications in food and health. *Journal of Dairy Science, 99*(4), 3229–3238.

Mazzoli, R., et al. (2014). Towards lactic acid bacteria-based biorefineries. *Biotechnology Advances, 32*(7), 1216–1236.

Mohammed, A. S. A., Naveed, M., & Jost, N. (2021). Polysaccharides; classification, chemical properties, and future perspective applications in fields of pharmacology and biological medicine (a review of current applications and upcoming potentialities). *Journal of Polymers and the Environment, 29*, 2359–2371.

Moradi, M., Guimarães, J. T., & Sahin, S. (2021). Current applications of exopolysaccharides from lactic acid bacteria in the development of food active edible packaging. *Current Opinion in Food Science, 40*, 33–39.

Morais, M., et al. (2022). Exopolysaccharides from microalgae: Production in a biorefinery framework and potential applications. *Bioresource Technology Reports, 18*, 101006.

Muhamad, I. I., Lazim, N. A. M., & Selvakumaran, S. (2019). Natural polysaccharide-based composites for drug delivery and biomedical applications. In *Natural polysaccharides in drug delivery and biomedical applications* (pp. 419–440). Elsevier.

Öner, E. T., Hernández, L., & Combie, J. (2016). Review of Levan polysaccharide: From a century of past experiences to future prospects. *Biotechnology Advances, 34*(5), 827–844.

Ortega, Y. R., et al. (2008). Efficacy of gaseous chlorine dioxide as a sanitizer against Cryptosporidium parvum, Cyclospora cayetanensis, and Encephalitozoon intestinalis on produce. *Journal of Food Protection, 71*(12), 2410–2414.

Osemwegie, O. O., et al. (2020). Exopolysaccharides from bacteria and fungi: Current status and perspectives in Africa. *Heliyon, 6*(6), e04205.

Prajapati, V. D., Jani, G. K., & Khanda, S. M. (2013). Pullulan: An exopolysaccharide and its various applications. *Carbohydrate Polymers, 95*(1), 540–549.

Prasanna, P., Grandison, A. S., & Charalampopoulos, D. (2014). Bifidobacteria in milk products: An overview of physiological and biochemical properties, exopolysaccharide production, selection criteria of milk products and health benefits. *Food Research International, 55*, 247–262.

Quintieri, L., Fanelli, F., & Caputo, L. (2019). Antibiotic resistant Pseudomonas spp. spoilers in fresh dairy products: An underestimated risk and the control strategies. *Food, 8*(9), 372.

Quintieri, L., et al. (2021). Recent advances in the mechanisms and regulation of QS in dairy spoilage by Pseudomonas spp. *Food, 10*(12), 3088.

Rani, S., & Pradhan, A. K. (2021). Evaluating uncertainty and variability associated with Toxoplasma gondii survival during cooking and low temperature storage of fresh cut meats. *International Journal of Food Microbiology, 341*, 109031.

Romanazzi, G., et al. (2016). Integrated management of postharvest gray mold on fruit crops. *Postharvest Biology and Technology, 113*, 69–76.

Ryan, P., et al. (2015). Sugar-coated: Exopolysaccharide producing lactic acid bacteria for food and human health applications. *Food & Function, 6*(3), 679–693.

Ryan, U., Hijjawi, N., & Xiao, L. (2018). Foodborne cryptosporidiosis. *International Journal for Parasitology, 48*(1), 1–12.

Ryan, U., et al. (2019). Giardia: An under-reported foodborne parasite. *International Journal for Parasitology, 49*(1), 1–11.

Semenza, J. C., et al. (2012). Climate change impact assessment of food- and waterborne diseases. *Critical Reviews in Environmental Science and Technology, 42*(8), 857–890.

Stuknytė, M., et al. (2016). Extracellular thermostable proteolytic activity of the milk spoilage bacterium Pseudomonas fluorescens PS19 on bovine caseins. *Journal of Dairy Science, 99*(6), 4188–4195.

Synowiec, A., et al. (2014). Antimicrobial and antioxidant properties of pullulan film containing sweet basil extract and an evaluation of coating effectiveness in the prolongation of the shelf life of apples stored in refrigeration conditions. *Innovative Food Science & Emerging Technologies, 23*, 171–181.

Tannous, J., et al. (2014). Sequencing, physical organization and kinetic expression of the patulin biosynthetic gene cluster from Penicillium expansum. *International Journal of Food Microbiology, 189*, 51–60.

Tiwari, O. N., et al. (2020). Application of microbial extracellular carbohydrate polymeric substances in food and allied industries. *3 Biotech, 10*, 1–17.

Tomulescu, C., et al. (2016). Levan-a mini review. *Scientific Bulletin Series F. Biotechnologies, 20*, 309–317.

Wang, Y., et al. (2019). Influence of dextran synthesized in situ on the rheological, technological and nutritional properties of whole grain pearl millet bread. *Food Chemistry, 285*, 221–230.

Xiao, R., & Zheng, Y. (2016). Overview of microalgal extracellular polymeric substances (EPS) and their applications. *Biotechnology Advances, 34*(7), 1225–1244.

Xu, Y., et al. (2019). Exopolysaccharides produced by lactic acid bacteria and Bifidobacteria: Structures, physiochemical functions and applications in the food industry. *Food Hydrocolloids, 94*, 475–499.

Yildiz, H., & Karatas, N. (2018). Microbial exopolysaccharides: Resources and bioactive properties. *Process Biochemistry, 72*, 41–46.

Yu, L., et al. (2020). Postharvest control of Penicillium expansum in fruits: A review. *Food Bioscience, 36*, 100633.

Zannini, E., et al. (2016). Production, properties, and industrial food application of lactic acid bacteria-derived exopolysaccharides. *Applied Microbiology and Biotechnology, 100*, 1121–1135.

Zhang, H., et al. (2018). Recent developments in the enhancement of some postharvest biocontrol agents with unconventional chemicals compounds. *Trends in Food Science & Technology, 78*, 180–187.

Zhou, Y., et al. (2018). Quantification of heterotrophic bacteria during the growth of Synechocystis sp. PCC 6803 using fluorescence activated cell sorting and microscopy. *Algal Research, 30*, 94–100.

Zhou, Y., Cui, Y., & Qu, X. (2019). Exopolysaccharides of lactic acid bacteria: Structure, bioactivity and associations: A review. *Carbohydrate Polymers, 207*, 317–332.

EPS: Nature's Hidden Treasures

Devanshi Patel, Vivek Goswami, Hiral G. Chaudhari, Viral Surati, Saklain Mustak Saiyad, Tarun Patel, and Siddhi D. Shah

1 Introduction

Exopolysaccharides (EPS) are biopolymers synthesized thematically in bacteria, fungi, algae, and plants. These endogenous compounds play essential functions in the ecosystem as fertilizers and, first and foremost, as factors that influence the water regime and plant–microbe interactions. Crop rotation, which was employed in the old world, for example, through the farming of legumes, unknown to the ancients, supported EPS benefits by cultivating EPS-producing bacteria such as *Rhizobium* in the soil. These EPS have, in production, developed from appurtenant instruments to specialized prime agriculture into indispensable items in the conservation of the environment and sustainable agriculture. EPS has taken a significant surge in the last decades, and many researchers have tried to utilize it in climate change mitigation, pollution control, and other industrial areas. Due to its biopolymeric characteristics, the EPS has specific functional properties (as shown in Fig. 1) that provide it a high perspective for usage in different branches; therefore, it can be considered as a perspective material to contribute to the development of sustainable technologies and preserving ecosystems (Costa et al., 2018; Saha et al., 2020).

D. Patel · V. Goswami · S. D. Shah (✉)
Vidhyadeep Institute of Science, Vidhyadeep University, Surat, Gujarat, India

H. G. Chaudhari · V. Surati
Shri Alpesh N. Patel Post Graduate Institute of Science & Research, Anand, Gujarat, India

S. M. Saiyad
Department of Biotechnology, Institute of Science & Technology for Advanced studies & Research (ISTAR), The Charutar Vidya Mandal (CVM) University, Vallabh Vidyanagar, Gujarat, India

T. Patel
V.S. Patel College of Arts & Science, Billimora, Gujarat, India

T. Sarkar et al. (eds.), *Extracellular Polysaccharides*,
https://doi.org/10.1007/978-3-031-94472-7_2

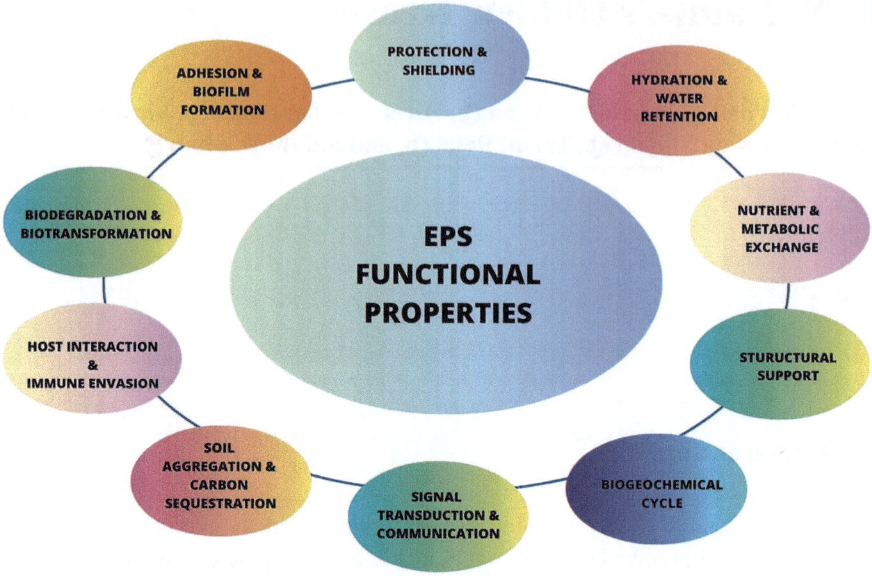

Fig. 1 General functional properties of EPS

1.1 History of EPS

Exopolysaccharides, or EPS, arose as important biopolymers with numerous applications synthesized by microorganisms, plants, and algae that perform different roles in the ecosystem. Early societies also benefitted from the advantages of EPS-producing organisms in their agronomic activities, though they were unaware of the fact. Ancient Indian, Egyptian, and Mesopotamian farmers noticed that chickpeas and peas experienced improved soil fertility when grown in an alternating manner with other crops, albeit they did not comprehend the role of the EPS-producing nitrogen-fixing *Rhizobium* bacteria. As early as the Lean Dough Era of China more than 8000 years ago, water-logged rice fields selectivity husbanded warm water EPS producing algae and bacteria overnight and endowed the soil with favorable structure and water management capacity for sustainable agriculture..

Investigation into the EPS-generating microorganisms was initiated at the end of the nineteenth century with the works of a few researchers and scientists, including Louis Pasteur and Robert Koch. For instance, in the early twentieth century, the EPS-producing bacteria *Xanthomonas campestris* was discovered, which led to the production of xanthan gum, which is used to thicken food and cosmetics. This led to the commercial use of EPS, which inaugurated the industrial application of the microorganisms responsible for its production (Nadda et al., 2021).

Subsequently, starting in the middle of the twentieth century, the use of EPS in soil microbiology and agriculture was widely accepted. Research conducted by researchers established that EPS-producing bacteria played a crucial role in improving soil structure, moisture content, and nutrient content. In India, EPS-based biofertilizers, notably Rhizobium, were popularized during the Green Revolution in the 1960s and 1970s. In addition to boosting yields and lessening reliance on chemical fertilizers, this strategy provided sustainable agriculture a better prospect in arid regions with water scarcity. Later, the Indian Council of Agricultural Research (ICAR) also encouraged biofertilizers as the concept of sustainable agriculture emerged (Babiak & Krzemińska, 2021).

Starting in the mid-twentieth century, EPS has been used in environmental fields with an emphasis on bioremediation and pollution management. As shown by *Aspergillus* and *Penicillium*, EPS has the desired characteristic of binding heavy metals an essential component of dealing with industrial pollutants. The EPS-producing organisms were used in the industrial areas of India, which demonstrated a positive outcome of EPS for bioremediation by immobilizing the pollutants and thus restricting their spread in the groundwater (El-Sayed, 2022).

In the twenty-first century, EPS research covered such areas as climate change mitigation practices and soil conservation. EPS from the algal community provides the capability to absorb carbon dioxide in the seawater but EPS from both cyanobacteria and algae is helpful in the stabilization of soil in sensitive coastal biogeographical regions, such as the Sundarbans of Bangladesh. In the Indian deserts, EPS-based biofertilizers are employed to prevent soil erosion and retain moisture.

At the moment, EPS research is moving on to bioplastics and encapsulation applications, in line with Indian environment-friendly programs, including the Swachh Bharat mission of reducing plastic waste. Biopolymers based on EPS are used instead of traditional plastic, and micro- and macroencapsulation make it possible to have the sustained release of biologically active substances in fertilizers or pharmaceuticals. Further innovations in respective EPS applications go on targeting environmental and agriculturally related issues in the future, which gives accentuating importance to sustainability and climate change adaption (Patil & Kamble, 2011).

1.2 Importance of EPS

1.2.1 Enhancing Agricultural Productivity through Soil Health

Currently, EPS research is extending to bioplastics and encapsulation use with an aim to reduce plastic use. As mentioned earlier, carbon-neutral EPS biopolymer is employed instead of conventional thermally derived plastics, whereas micro- and macroencapsulation mechanisms allow the availability of biologically active substances- long activity release that is unique with fertilizer or drug. Future enhancements in respective EPS applications continue to address environmental- and agriculture-oriented problems, which underlines the accentuating significance of sustainability and climate change adaptation (Kumar et al., 2024).

1.2.2 Biofilms and Ecosystem Services in Water Purification

EPS is used in the production of biofilms—complex associations of microbial cells surrounded by a matrix of EPS that fulfills important roles in ecosystems, such as water treatment, nutrient recycling, and bioremediation. In India, the natural Brooklyn is applied in the Ganga River to improve water quality as it facilitates the breakdown of biofilms of organic pollutants. EPS manufacturing bacteria and algae are attached to the river surfaces and are capable of immobilizing heavy metals and inhibiting the transport of contaminants. Moreover, treatment wetlands of Tamil Nadu and Karnataka make use of EPS-enriched biofilms to enhance nutrient and pollutant removal from the wastewater. It is integrated to control water pollution from agriculture and urban areas, thus protecting Indian freshwater sources (Sundaramoorthy et al., 2021).

1.2.3 Carbon and Nitrogen Cycling for Climate Change Mitigation

EPS bears significant input to the carbon and nitrogen cycles of the world, which is crucial for combating climate change. In marine environments, algal EPS forms “marine snow’ ‘by entrapment of organic carbon that ascends to the marine bed and thereby reduces the level of atmospheric CO_2. For instance, in the Sundarbans mangrove forest, there are EPS-producing algae and cyanobacteria that immobilize sediments and thus fix organic carbon in an effective carbon sink. Furthermore, in Maharashtra and Madhya Pradesh, nitrogen-fixing bacteria, that is, *Rhizobium*, was utilized to release EPS that promotes nitrogen fixation in legume crops, thereby decreasing the risk of synthetic fertilizer and thus greenhouse gas emission from agriculture (Rahman et al., 2015; Kumar et al., 2018; Barik et al., 2021).

1.2.4 Soil Stabilization and Erosion Control in Arid Regions

EPS plays an important role in the binding of soils and has been used in systematic eradicating measures, particularly in regions such as Rajasthan and Gujrat, which face the critical problem of soil erosion and desert land formation. EPS-producing bacteria such as *Bacillus* and *Pseudomonas* fix soil particles and improve water-carrying capacities, which also leads to the formation of structures that will not erode. EPS bacteria in biofertilizers promote the structure of soil, and thus drought-endurable crops, and mitigate wind erosion of sand and water erosion in the Thar Desert. This sustainable approach assists in reducing desertification and enhances agriculture and food security in the Iraq-like conditions that prevail in these regions of India (Vardharajula & Ali, 2015; Choudhary & Varma, 2016; Kumar et al., 2018).

1.2.5 Protection Against Environmental Stress in Extreme Conditions

EPS usually offers critical shielding to microorganisms in saline soils, deserts, and high UV regions. Cyanobacteria present in the Rann of Kutch in Gujarat, such as Anabaena and *Spirulina* produce EPS in high-saline conditions; these EPS form a boundary layer that prevents loss of moisture and helps survive during droughts. These cyanobacteria also have the potential for the management of saline soils through the interaction of sodium ions to minimize toxicity impact and improve the growth of salt-tolerant crops. This sustainable measure restores the efficiency of agricultural output in salt-soaked zones across Gujarat and Haryana (Singh & Kumar, 2016).

1.2.6 Bioremediation and Pollution Control

EPS is particularly important for bioremediation processes because it can adsorb heavy metals and contaminants throughout the industry zones. The contaminated area at Vapi, Gujarat, India is taken as a case study, where EPS-producing fungi such as *Aspergillus* and *Penicillium* are used to stabilize toxic contaminants. These fungi produce an EPS matrix biofilm that immobilizes heavy metals and keeps them from polluting water and soil. It is a cheap bioremediation technique that is natural since it seeks to lessen industrial pollution, hence contributing to the restoration of the natural resource base and the efficient management of pollution (Patil & Kamble, 2011; Shah et al., 2013).

1.3 Classification of EPS

1.3.1 Classification by Origin

1.3.1.1 Bacterial EPS

Bacteria produce tremendous amounts of EPS, and it is used in biofilm formation, offers stress protection, and encourages nutrient acquisition. One of the most important inputs in the case of the Indian agricultural sector is *Rhizobium*, a nitrogen-fixing bacterium. It synthesizes compounds that enable the formation of root nodules in legumes such as chickpeas (*Cicer arietinum*) and pigeon peas (*Cajanus cajan*), both very important food crops in India. These EPS enhance soil nitrogen-fixing capability hence across the states of Madhya Pradesh, Maharashtra, etc. (Kumar & Verma, 2018).

Another important bacterial EPS producer is *Bacillus*, which is being used for rice biocontrol in West Bengal and Odisha. *Bacillus* produces EPS, which helps it to find a place in plant roots and form a biofilm that protects crops from diseases and improves sustainable farming (Vardharajula & Ali, 2015; Choudhary & Varma, 2016).

1.3.1.2 Fungal EPS

EPS synthesized by fungi provides a robust ecological role, which includes enhancement of soil health, nutrient cycling, and bioremediation. Fungal EPS is involved in a direct way in reducing organic material to nutrient-rich humus by a process known as humification. Among the species, *Aspergillus* and *Penicillium* are given much importance for their bioremediation characteristics, where much research is conducted, particularly in the areas contaminated with pollutants in different regions of India. These fungi also secrete EPS that chelates with the heavy metals, hence minimizing the chances of water aquifer pollution. This application is very helpful in the reduction of pollution in areas that are densely industrialized and in practicing sustainable environments (Gadd, 2004; Singh & Sharma, 2020; Rana & Rana, 2021).

1.3.1.3 Algal EPS

EPS production is mainly evident in microalgae and cyanobacteria, which play critical roles in aquatic biomass production and carbon sink. In coastal Indian regions such as the Sundarbans, EPS-providing algae and cyanobacteria are very important in contributing to sediment stabilization, which reduces coastal erosions. *Spirulina* and *Anabaena* species, for instance, release very adhesive EPS, with capabilities of capturing settling sediments, involved in the formation of marine biofilms, which play the role of sequestering carbon, hence controlling climate change (Zhou et al., 2024).

In freshwater bodies, *Chlorella*, a green microalga, synthesizes EPS, and the net result is enhanced water quality due to pollution control. Particularly it becomes helpful in the case of watering reservoirs in states such as Karnataka, where the algal EPS plays a role in water treatment and can be beneficial to sustainable water supply programs (Manna et al., 2010; Zhang et al., 2021).

1.3.1.4 Plant-Derived EPS

It is well understood that EPS is synthesized and secreted in plants by the root as part of the composition of the soil structure and water. In India, *Oryza sativa* and *Triticum aestivum* are two crops that benefit the soil since they secrete EPS, which contributes to the improvement of aggregation and water involvement in soils. Whereas, EPS contribute to soil fertility in Indo-Gangetic Plains, India which is one of the most important agricultural soils. Additionally, chickpeas, pigeon peas, and other leguminous plants release EPS, which soil microbes feed on, thereby improving nutrient acquisition and maintaining favorable structure and productivity of soil, especially in rain-fed environments, where water management is important (Chenu, 1993; Alami et al., 2000; Sandhya & Ali, 2015).

1.3.2 Classification by Composition

1.3.2.1 Homopolysaccharides

According to their monosaccharide units, homopolysaccharides are made of one type of monosaccharide subunits. Such polymers as cellulose, curdlan, and dextran are applied in industries. In India, bacterial cellulose prepared from *Gluconacetobacter xylinus* has been investigated for its application in nonwoven fabric in the form of wound dressings because of its purity and biocompatibility. Moreover, dextran, synthesized by *Leuconostoc* species, is apparently used in the pharmaceutical industry for plasma expansion in hospitals (Patel & Prajapati, 2013; Shah et al., 2013; Pandey et al., 2024). Some common structures of EPS are depicted in Fig. 2.

1.3.2.2 Heteropolysaccharides

Heteropolysaccharide molecules contain nonsimilar monosaccharide units and are more complex in structure and roles. A few well-known examples include Xanthan gum, pectins, and agar. Xanthan gum is a heteropolysaccharide produced commercially by *Xanthomonas campestris* and is employed in the food industries of India as a thickening agent and stabilizer. In agriculture, it is employed to enhance water retention within the soil in regions such as Rajasthan, which is well known for its inadequate water resources (Vardharajula & Ali, 2015; Choudhary & Varma, 2016; Kumar & Verma, 2018).

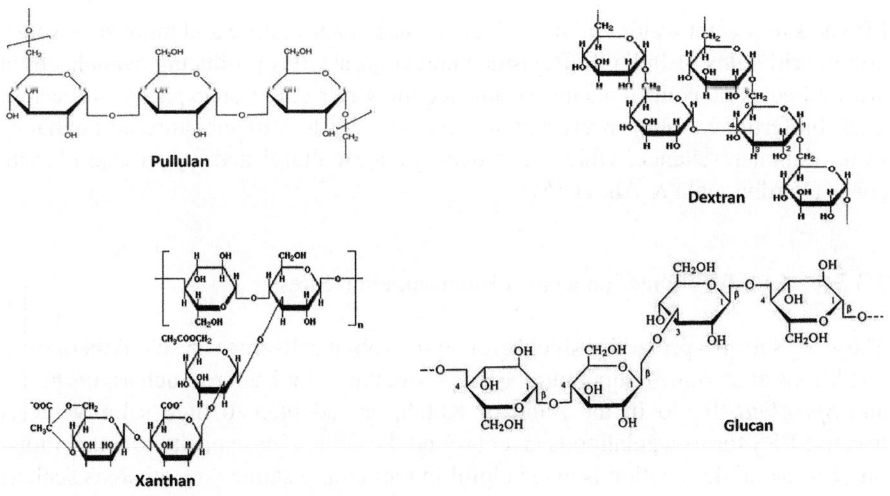

Fig. 2 A representative structure of EPS

1.3.3 Classification by Function

1.3.3.1 EPS for Biofilm Formation

EPS has a crucial role in biofilm development and functions as a nutrient sequestering center, sink for pollutant, and promoter of environmental stability. Freshwater biofilms in the Ganga River comprise EPS-forming bacteria and algae that play a role in water purification by trapping pollutants and organic material particles. These biofilms are essential for establishing water quality within natural water systems and are used in the technological systems of constructed wetlands for wastewater treatment across Tamil Nadu and Karnataka (Vardharajula & Ali, 2015; Choudhary & Varma, 2016; Kumar & Verma, 2018).

1.3.3.2 EPS for Soil Structure and Fertility

Valuable functions of EPS include mechanisms where EPS binds soil particles to enhance soil aeration and water infiltration, as well as to promote soil fertility. This is especially important in agriculture, particularly in India, because soils are fundamental to supporting yields. Benefits of EPS such as increase in the availability of nutritens and improving soil fertility since it can fix nitrogen from the atmosphere with the help of Rhizobium, bacteria. In the dry region of the Thar Desert, for soil management and protection from erosion in areas prone to drought, EPS-producing bacteria incorporated into biofertilizers are used (Choudhary & Varma, 2016).

1.3.3.3 EPS for Water Retention

EPS acts as a great water retainer, which is vital in agriculture and more so in semi-arid to arid fields of India. In Rajasthan and Gujarat, EPS-producing cyanobacteria are applied as biocontrol agents to enhance the water retention capacity of the soil. Such biofertilizers help to prevent soil erosion and loss of moisture and enhance water deficit resistance, which is critical for agricultural activity in arid climate zones (Vardharajula & Ali, 2015).

1.3.3.4 EPS for Protection against Environmental Stress

These EPS form a protective shield around microbial cells that can shield them from conditions of extreme temperature. EPS-producing cyanobacteria such as *Anabaena* and *Spirulina*, found in the Rann of Kutch, are adapted to high salinity levels because EPS forms a gelatinous layer around the cells. This capacity to retain moisture and avoid desiccation is also helpful in reclaiming saline soils in areas such as Gujarat and Haryana. This is because EPS-producing organisms lower the toxicity of the soil and boost the production of saline-tolerant crops (Vardharajula & Ali, 2015).

2 EPS Sources: A Diverse Kingdom

2.1 Bacterial EPS

EPS is important for bacteria as it provides the basis for biofilm construction and protection from unfavorable conditions and from the negative influences of antibiotics and toxins. These are specific EPS-producing bacteria: *Pseudomonas*, *Bacillus*, *Rhizobium*, and *Lactobacillus*. *Rhizobium* is widely used in legume production systems in India, especially in rain-fed areas, to improve nitrogen-fixing capabilities in legumes such as chickpeas and pigeon peas (Table 1). Also, *Bacillus* species support the planet as biocontrol since these bacteria develop a layer around the root systems of plants, which helps to keep undesirable microbes away and increase the defence mechanisms. These applications indicate the importance of bacterial EPS in yet, another adaption of agriculture and the environment (Choudhary & Varma, 2016).

2.2 Fungal EPS

Several EPS produced by fungi have important functions in the soil, such as nutrient uptake, decomposition of organic materials, and the capacity to form mycorrhizal linkages with plants. *Aspergillus*, *Penicillium*, and *Ganoderma* are vital fungi in the production of EPS. In India, Aspergillus species are quite useful as they help in improving the structure and moisture retention capability of the soil for better decomposition of organic matter, which is favorable for agriculture (Table 1). Fungal EPS is also important in bioremediation, such as in the industrial areas in Vapi, where it helps bind heavy metals and thus minimizes the possibilities of soil pollution. Furthermore, mycorrhizal associations with fungal EPS help enhance nutrient acquisition of water-stressed crops and are critical for sustainable agriculture in the driest environments (Paria & Chakraborty, 2019).

2.3 Algal EPS

Exopolysaccharides (EPS) from microalgae and cyanobacteria have essential functional uses in maintaining aquatic environments' stability. These EPS provide a capsule-like covering around cells, aiding in nutrient transport and as insulation against factors such as UV radiation. In India, *Spirulina* and *Anabaena* form biofilms of freshwater bodies and help the water bodies by entrapment of particles and

Table 1 EPS-producing microorganisms with their biological properties

Producer organism	Name of EPS	Biological properties	References
Lactobacillus plantarum	LpF1	Antioxidant, immunomodulatory, and anticancer properties	Yang et al. (2023
Rhizobium spp.	Glucuronan	Enhances soil fertility and nitrogen fixation	El-Sayed (2022)
Bacillus subtilis	Levan	Acts as a biocontrol agent in agriculture	
Aspergillus spp.	Glucan	Facilitates bioremediation of heavy metals	
Spirulina spp.	Calcium spirulan	Stabilizes aquatic ecosystems	
Chlorella spp.	Chlorella derived EPS	Used in biofuel and bioplastic production	
Anabaena spp.	Anabaena EPS	Aids in saline soil reclamation and enhances drought resistance	
Penicillium spp.	Penicillium EPS	Assists in pollutant immobilization in industrial zones	
Streptococcus thermophilus	Thermophilic EPS	Improves texture and viscosity in dairy products	Salazar et al. (2009)
Xanthomonas campestris	Xanthan gum	Used as a thickening and stabilizing agent in food and industrial applications	Jansson et al. (1975)
Leuconostoc mesenteroides	Dextran	Employed in medical applications and as a food additive	Purama et al. (2009)
Aureobasidium pullulans	Pullulan	Utilized in edible films, coatings, and pharmaceutical capsules	Singh et al. (2008)
Pseudomonas aeruginosa	Alginate	Contributes to biofilm formation and pathogenicity in clinical settings	Ghafoor et al. (2011)
Rhizobium spp.	Rhizobial EPS	Enhances nitrogen fixation and improves soil fertility in legume cultivation	Singh & Saini (2017)
Bacillus spp.	Levan	Used in biocontrol and aids biofilm formation for crop protection	Borriss (2011)
Aspergillus spp.	Glucans	Effective in bioremediation and binds heavy metals in polluted soils	Gadd (2010)
Spirulina spp.	Spirulan	Provides salinity tolerance in arid environments and has potential in biofuel production	Chaudhuri & Balasubramanian 2023
Xanthomonas campestris	Xanthan	Widely used as a thickening agent in food, cosmetics, and industrial applications	Laskin et al. (2014)

stabilization of sediments. In coastal ecosystems, such as the Sundarbans, EPS derived from algae and cyanobacteria offers protection from erosion and species habitat. *Chlorella* EPS is examined in India for its biodiesel capacities and for being an organic substance in biopolymers for furthering green industrial usages (Table 1) (Paria & Chakraborty, 2019; Babiak & Krzemińska, 2021).

3 Applications

3.1 *Agriculture: Enhancing Soil Health and Crop Productivity*

EPS finds useful application in agriculture due to its ability to enhance the physical properties of the soil, such as structure, water holding capacity, and nutrient supply. Bacteria that form EPS arrays, such as *Rhizobium*, are often involved in modulation with legume plants, including chickpeas and pigeon peas. In that way, *Rhizobium* contributes to the natural increase in fertility of the soil and at the same time decreases the dependence on chemical fertilizers. In rain-fed farming, EPS plays major role in holding soil together to form aggregates, which enhance water infilteration and soil aeration. It is sustainable farming because the use of EPS-based biofertilizers encourages healthier soil and improved crop yields for farmers in Indo- Gangetic plains of India (Saha et al., 2020).

3.2 *Water Purification and Ecosystem Services*

EPS plays a crucial role in water clarification through the adoption of a biofilm that assists capture and breakdown of toxins. In the natural water systems, microbial communitites including bacteria and microalgae form biofilms on submerged surfaces in the Ganga River for instance, and microscopically degrade organic pollutants, heavy metals, and contaminants. Also, the biofilms made from EPS in the constructed wetlands of Tamil Nadu and Karnataka states are used to discharge nutrients and pollutants from wastewater. This biofilm-driven water purification makes use of the most efficient approach to deal with water pollution by minimizing the contaminants in the sources of freshwater supply and promoting sustainable health of aquatic life (Salama et al., 2016).

3.3 *Climate Change Mitigation Through Carbon and Nitrogen Cycling*

EPS-producing organisms have an efficient contribution toward the global carbon and nitrogen cycle, whereby they aid fight climate change. Algal EPS in marine environments promotes carbon sequestration by a process called "marine snow," whereby the EPS structure harbors organic carbon and descends to the ocean bed, thus lowering the amount of CO_2 in the atmosphere. Similarly, ephemeral particles such as alga and cyanobacteria among the Sundarbans mangroves are responsible for the immobilization of sediment and trapping of carbon particles so that they don't get released into the atmosphere. For example, EPS produced by the biofilm of nitrogen-fixing bacteria such as Rhizobium, holds soil nutrients especially nitrogen resulting enhance soil fertility. It also cut greenhouse gas emissions linked to chemical fertilizer use (Mota et al., 2022).

3.4 Soil Stabilization and Erosion Control in Arid Regions

EPS has found the most utility in deserts and other areas that receive relatively low rainfall because it reduces soil erosion. EPS-forming bacteria, including *Bacillus* and *Pseudomonas*, form soil structures that offer resistance to erosion by the wind and water. In the Thar Desert area, use of biofertilizers with EPS producers promotes the physical condition of the soil and helps in moisture conservation and vegetation growth. These properties facilitate practices such as growing crops that require little water and hence can withstand instances of drought. This leads to the sustainable agriculture and suitable environment for the deserts- (Costa et al., 2018).

3.5 Bioremediation and Pollution Control

These EPS-forming organisms can bioremediate pollution by capturing and immobilizing pollutants in water in highly polluted areas. *Aspergillus* and *Penicillium* are fungi that can be used in industrial belts because the EPS that they release immobilizes heavy metals and other pollutants from spreading. Such EPS-based biofilms encapsulate those contaminants in a material that defeats their purpose by lowering their solubility and therefore their environmental effects. This is an abiotic, low-cost method that plays an important role in the process of rehabilitating ecosystems degraded by industrial activities (Kumawat et al., 2021).

3.6 Medical and Pharmaceutical Applications

EPS has been used to a great effect in medicine because it is biocompatible and forms layers of protection. EPS-based biofilms find applications in drug delivery systems, wound dressing, and in tissue engineering. For example, bacterial cellulose, an EPS produced by *Gluconacetobacter xylinus*, is used in the formation of wound dressing for keeping moisture and enhancing the growth rate of new tissues. Also, EPSs with antioxidant, immunomodulatory, and anticancer activities are explored for therapeutic purposes, indicating their prospect as natural and efficient bioactive compounds (Liang, 2023).

3.7 *Bioplastics and Industrial Uses*

The companies that used to rely on synthetic materials are now turning into EPS based approach for their products. For example, proteins and lipids from *Chlorella* and other algae are proposed for bioplastic manufacturing in response to increased efforts globally in combating plastic pollution. In the petroleum sector, they assist with more improvement within the structure for oil to be mobilized. Because of biodegradability and functionality, EPS is used as a biodegradable substitute for bioplastics, emulsifying agents, and thickeners, making industries greener (Mogany et al., 2024).

4 Future Aspects

4.1 *Bioplastics and Sustainable Materials*

EPS is being considered as a sustainable option for conventional plastics. Owing to increasing awareness about the negative impact of plastic wastes on the environment, studies are being conducted on the conversion of EPS into bioplastic, in particular those produced from algae like *Chlorella*. EPS-based bioplastics have a superior property in that they are biodegradable, thus eliminating environmental threats. In future, EPS based bioplastics can be applied for the packaging, insulation and protective materials in the various sectors. In India, the drive to decrease plastic waste includes programs such as Swachh Bharat, where EPS-based bioplastics are marketed as an essential solution (Mogany et al., 2024).

4.2 *Agricultural Applications: Climate-Resilient Farming*

Policies have estimated that EPS-based solutions will also drive sustainable agriculture since climate change affects farming around the world. EPS-forming microorganisms benefit the soil by enhancing the physical appearance of the soil structure, water management, and nutrient recycling, which make such microorganisms appropriate to be used in dry/eroded regions. More studies involve improving the EPS-based biofertilizers in the coming years and also for the greater ability of the crop to grow in extreme temperatures, salinity stresses, and water-deficient environments. EPS from cyanobacteria and bacteria may be useful for reclaiming lost soil quality, enhancing crop productivity, and diminishing the usage of chemical fertilizers, consequently contributing worldwide for the sustainable agriculture practices (Tang et al., 2024).

4.3 Advanced Bioremediation and Environmental Restoration

The capacity of EPS to precipitate with heavy metals and pollutants regards EPS as an asset in controlling pollution. Future applications could involve isolating and applying particular species of EPS producers for the elimination of particular pollutants in industries, mines, and water resources. For this reason, it could be designed to optimize pollutant uptake and degradation, which makes it perfect for environmental remedy. Rising pollution levels in industrial regions worldwide make EPS-based bioremediation solutions more viable and efficient (Syed et al., 2023).

4.4 Medical and Pharmaceutical Innovations

EPS based polymers are good source of biocompatible material in the medical fields. Future work involves their application in drug delivery vehicles, wound healing patches, and tissue engineering support structures. Because EPS is nontoxic and protective, it can package drugs for slow release, which would enhance the effectiveness of this treatment without the negative effects. It is also being explored for its part in the process of tissue regeneration for reconstructing bioscaffolds for the repair of tissues that have been destroyed or injured. Furthermore, EPS with immunomodulatory, antioxidant, and antimicrobial potentials are the new frontiers for natural application as a supplement or drug in the pharmaceutical industry (Gomaa, 2022).

4.5 Bioenergy and Carbon Sequestration

Carbon cycling through the ecosystem and sequestration by animals has been put under consideration by EPS of climate change. The EPS-producing algae and cyanobacteria play an important role in marine biological pumps by capturing organic carbon and concentrating it to form marine snow, which settles on the ocean floor. The next studies may exploit this process in large-scale bioenergy systems, utilizing algae to sequester and store CO_2 from the atmosphere efficiently. Algal EPS could be utilized in biofuel production systems, where algal EPS plays the role of both capturing carbon and generating renewable energy, thereby contributing to energy and climate change goals (Ali et al., 2024).

4.6 *Nanotechnology and Material Engineering*

Some of the physical and chemical characteristics of EPS qualify it for use in nanotechnology and material science. In the future, EPS could be developed in a highly controlled manner at the nanoscale, where certain properties, such as strength, flexibility, or biodegradability, can be produced as needed. Such components could be used in diverse fields with an aim of environmental conservation. EPS-based nanomaterials could also have applications in the case of environmental sensing, where it is used to identify and remove pollutants at the nano level (Wani et al., 2023).

4.7 *Water Resource Management*

EPS biofilms have potential in water treatment due to their ability to trap and decompose potential contaminants in the water. Potential future uses may involve methods for using EPS to design filters for water purification, which could be employed in rural areas and in urban centers to supply safe water. This technology could be of particular help in areas where water pollution is observed due to secretion from agricultural fields or industries. EPS could also improve constructed wetlands for wastewater treatment, which are cost-effective, efficient, and eco-friendly solutions for water management in the future (Kumawat et al., 2021).

4.8 *Genetic Engineering for Enhanced EPS Production*

Future improvements in genetics engineering can enable researchers to isolate and enhance EPS biosynthesis in distinct microorganisms to suit their certain applications. As such, the yield, structural properties, and pollutant-binding abilities of EPS may be improved by genetic modifications in the future to improve the efficiency and economic feasibility of future applications. Modifying EPS-synthesizing organisms to produce EPS involving specific functionalities could improve their applicability in bioremediation, agriculture, and material science, thus ensuring early adoption of EPS-based solutions in various areas (Kumawat et al., 2021).

5 Limitations

5.1 *High Production Costs*

The possibility of manufacturing EPS at the industrial level is still expensive. For industrial production of EPS, some special growth requirements, such as effective pH, growth temperature, and nutrient availability, are needed to entail costs.

Furthermore, the steps for isolation and extraction of EPS are normally time-consuming and costly, hence reducing the possibilities of large-scale production and usage. Thus, unless economical approaches are developed, the application of EPS in such spheres as agriculture, bioremediation, or others might be too expensive (Hasan et al., 2024).

5.2 *Variability in Yield and Composition*

The parameters of EPS production can show large fluctuations depending on the species of microorganism, the conditions in which it grows, and the substrate on which it feeds. There are differences in composition and characteristics of the EPS synthesized by different microorganisms; therefore, it is difficult for there to be production standardization between batches. This variation influences the characteristics of EPS, such as its quality, functionality, and performance in diverse uses such as biofertilizers, bioplastics, and medical. Sustaining consistency in the basic formulation of EPS and yield always poses a major test to scale up the production process (Atmakuri et al., 2024).

5.3 *Limited Stability in Extreme Conditions*

EPS is generally highly crystalline and suffers from variations in its stability under extreme conditions in terms of heat, ultraviolet radiation, and high or low pH. However, in many applications, such as that of desert agriculture or marine systems, the instability of EPS means that the applications are restricted. The emission trading system of the EPS under these conditions may diminish their application as a soil conditioner, and water insulator or a (Costa et al., 2018).

5.4 *Complex Regulatory and Safety Approvals*

As EPS application is in agriculture, food, or medicinal products, EPS products must pass through various tests to gain approval for usage. The increased use of genetically modified organisms (GMOs) for the improvement of EPS production makes the analysis more complicated because their usage is highly regulated in many countries. Lengthy and costly approval procedures for EPS-based products such as biofertilizers for crops, food additives, and medical material have limited the use of EPS. Also, any accidental effects of the product on the environment or ecosystems need further research and testing (Hasan et al., 2024).

5.5 Challenges in Scaling for Industrial Applications

EPS production from a laboratory scale to an industrial scale offers technical difficulties. To ensure that EPS is produced consistently, it can be quite expensive and technical for the equipment needed, microbial cultures, and environmental conditions all out of consistent EPS production. For certain industries such as bioplastics and bioremediation, no other problem poses as much of a challenge as the ability to upscale processes and deliver the same level of EPS standard. In addition, contamination hazards are possible during high production volume, which affects yield and quality and therefore potentially more variable production (Dave et al., 2020).

5.6 Short Shelf Life and Storage Issues

EPS tend to delaminate and become mechanically damaged or lose their utility and create an issue of storage and shelf life. For EPS to remain stable and maintain its value during storage, conditions such as low temperatures or use of preservatives are necessary and hence the cost. The stability of these compounds may be affected by their shelf life, which may further reduce their application in biofertilizers and the production of some pharmaceutical products. Encapsulation betterments are required to increase the durability of EPS-based products (Atmakuri et al., 2024).

5.7 Environmental Concerns with Certain EPS

It is well established that EPS is biodegradable; however, the EPS that is altered in some way to meet certain end uses degrades at a slower rate. This can lead to accumulation in the environment, like in the case of the synthetic polymers that we already know. Overall, large-scale bioplastics demand could be problematic when EPS remains a slow degrading material, especially in delicate and disturbed ecosystems. Further studies are called for to improve the biodegradability of these EPS types to avoid any negative environmental effects (Baria et al., 2022).

5.8 Limited Knowledge of Long-Term Effects

The consequences of EPS production on the health of soil, plant development, and the functioning of ecosystems have been studied only in the long run. As EPS has revealed positive means in the short run, more researches are required to explore their constant effects, especially for agriculture and environmental usage. For example, the continual impact of EPS-based biofertilizers on the supply, nature, and variety of microbial communities, nutrients, and ecological balance of the soil needs to be researched to ascertain they do not unbalance natural systems (Costa et al., 2018).

6 Concluding Remarks

EPS can be regarded as nature's diverse gift in the form of biomaterials that can be used in environmental problem-solving, agriculture, industry, and climate issues. Getting back to the components of the ecosystem, EPS provides important ecosystem services in the form of nutrient cycling, soil stabilization, and bioremediation. EPS application potential increases with time and is expected to rise in the upcoming years as research continues to develop, addressing various problems affecting sustainability and conservation. Nonetheless, some problems, for example, elevated production costs and regulatory issues, entail innovative approaches to reach the top level of industrial applicability. As the exploration proceeds further, EPS has the potential to be the foundation for sustainable consumption and ecological harmony.

References

Alami, Y., Achouak, W., Marol, C., & Heulin, T. (2000). Rhizosphere soil aggregation and plant growth promotion of sunflowers by an exopolysaccharide-producing Rhizobium sp. strain isolated from sunflower roots. *Applied and Environmental Microbiology, 66*(8), 3393–3398. https://doi.org/10.1128/AEM.66.8.3393-3398.2000

Ali, S. S., Hassan, L. H., & El-Sheekh, M. (2024). Microalgae-mediated bioremediation: Current trends and opportunities—A review. *Archives of Microbiology, 206*, Article 343. https://doi.org/10.1007/s00203-024-04052-x

Atmakuri, A., Yadav, B., Tiwari, B., et al. (2024). Nature's architects: A comprehensive review of extracellular polymeric substances and their diverse applications. *Waste Disposal and Sustainable Energy., 6*, 529. https://doi.org/10.1007/s42768-024-00205-2

Awasthi, R., Tewari, R., & Nayyar, H. (2011). Synergy between plants and *Pseudomonas putida* improves tolerance to drought stress: Involvement of indole acetic acid, antioxidant defense and osmolyte accumulation. *Acta Physiologiae Plantarum, 33*(6), 2133–2144. https://doi.org/10.1007/s11738-011-0743-0

Babiak, W., & Krzemińska, I. (2021). Extracellular polymeric substances (EPS) as microalgal bioproducts: A review of factors affecting EPS synthesis and application in flocculation processes. *Energies, 14*(13), 4007. https://doi.org/10.3390/en14134007

Baria, D. M., Patel, N. Y., Yagnik, S. M., et al. (2022). Exopolysaccharides from marine microbes with prowess for environmental cleanup. *Environmental Science and Pollution Research, 29*(52), 76611–76625. https://doi.org/10.1007/s11356-022-23198-z

Barik, J., Sanyal, P., Ghosh, T., & Mukhopadhyay, S. K. (2021). Carbon stock and storage pattern in the Sundarbans mangrove forest, NE coast of India. *Tropical Ecology, 62*(1), 95–106. https://doi.org/10.1007/s42965-020-00130-5

Borriss, R. (2011). Use of plant-associated *Bacillus* strains as biofertilizers and biocontrol agents in agriculture. In D. K. Maheshwari (Ed.), *Bacteria in agrobiology: Plant growth responses* (pp. 41–76). Springer. https://doi.org/10.1007/978-3-642-20332-9_3

Chaudhuri, R., & Balasubramanian, P. (2023). Evaluating the potential of exopolysaccharide extracted from the spent cultivation media of Spirulina sp. as plant biostimulant. *Biomass Conversion and Biorefinery*, 1–10. http://dx.doi.org/10.1007/s13399-023-04865-8

Chenu, C. (1993). Clay- or sand-polysaccharide associations as models for the interface between micro-organisms and soil: Water-related properties and microstructure. *Geoderma, 56*(1–4), 143–156. https://doi.org/10.1016/0016-7061(93)90106-U

Choudhary, D. K., & Varma, A. (Eds.). (2016). *Microbial-mediated induced systemic resistance in plants*. Springer. https://doi.org/10.1007/978-981-10-0388-2

Costa, O. Y. A., Raaijmakers, J. M., & Kuramae, E. E. (2018). Microbial extracellular polymeric substances: Ecological function and impact on soil aggregation. *Frontiers in Microbiology, 9*, Article 1636. https://doi.org/10.3389/fmicb.2018.01636

Dave, S. R., Upadhyay, K. H., Vaishnav, A. M., et al. (2020). Exopolysaccharides from marine bacteria: Production, recovery, and applications. *Environmental Sustainability, 3*(2), 139–154. https://doi.org/10.1007/s42398-020-00101-5

Decho, A. W., & Gutierrez, T. (2017). Microbial extracellular polymeric substances (EPSs) in ocean systems. *Frontiers in Microbiology, 8*, 922. https://doi.org/10.3389/fmicb.2017.00922

El-Sayed, A. S. A. (2022). Potential functions and applications of diverse microbial exopolysaccharides in the environment and biomedicine: A mini-review. *Journal of Genetic Engineering and Biotechnology, 20*, 151. https://doi.org/10.1186/s43141-022-00432-2

Gadd, G. M. (2004). Mycotransformation: Fungal biotransformation of toxic metals and metalloids. *Environmental Microbiology, 6*(9), 925–940. https://doi.org/10.1111/j.1462-2920.2004.00671.x

Gadd, G. M. (2010). Metals, minerals, and microbes: Geomicrobiology and bioremediation. *Microbiology, 156*(3), 609–643. https://doi.org/10.1099/mic.0.037143-0

Garcia-Ochoa, F., Santos, V. E., Casas, J. A., & Gomez, E. (2000). Xanthan gum: Production, recovery, and properties. *Biotechnology Advances, 18*(7), 549–579. https://doi.org/10.1016/S0734-9750(00)00050-1

Ghafoor, A., Hay, I. D., & Rehm, B. H. (2011). Role of exopolysaccharides in Pseudomonas aeruginosa biofilm formation and architecture. *Applied and Environmental Microbiology, 77*(15), 5238–5246. https://doi.org/10.1128/AEM.00637-11

Gomaa, M. (2022). Biodegradable plastics based on algal polymers: Recent advances and appli cations. In G. A. M. Ali & A. S. H. Makhlouf (Eds.), *Handbook of biodegradable materials*. Springer. https://doi.org/10.1007/978-3-030-83783-9_20-1

Gupta, S., & Pandey, S. (2019). Exopolysaccharide producing rhizobacteria and their impact on growth and drought tolerance of wheat seedlings. *Biocatalysis and Agricultural Biotechnology, 20*, 101207. https://doi.org/10.1016/j.bcab.2019.101207

Hasan, H. A., Rahim, N. F. M., Alias, J., Ahmad, J., Said, N. S. M., Ramli, N. N., et al. (2024). A review on the roles of extracellular polymeric substances (EPSs) in wastewater treatment. *Water, 16*(19), 2812. https://doi.org/10.3390/w16192812

Jansson, P. E., Kenne, L., & Lindberg, B. (1975). Structure of the extracellular polysaccharide from *Xanthomonas campestris*. *Carbohydrate Research, 45*(1), 275–282.

Kaushik, A., & Kaushik, C. P. (2018). Bioremediation of salt-affected soils using cyanobacteria in terms of physical structure, nutrient status, and microbial activity. *International journal of Environmental Science and Technology, 15*, 571–580. https://doi.org/10.1007/s13762-017-1419-7

Kumar, A., & Verma, J. P. (2018). Does plant—microbe interaction confer stress tolerance in plants: a review?. *Microbiological Research, 207*, 41–52. https://doi.org/10.1016/j.micres.2017.11.004

Kumar, D., Kaštánek, P., & Adhikary, S. P. (2018). Exopolysaccharides from cyanobacteria and microalgae and their commercial application. *Current Science, 115*(2), 234–241. https://doi.org/10.18520/cs/v115/i2/234-241

Kumar, M., Soni, R., Maniyar, R. S., Saran, P., & Chaudhary, D. R. (2024). Characterization of extracellular polymeric substances from biofilm-forming marine bacteria from the Arabian Sea, India. *Indian Journal of Microbiology*, 1–13. https://dx.doi.org/10.1007/s12088-024-01351-1

Kumawat, T. K., Kumawat, V., Sharma, S., Kandwani, N., & Biyani, M. (2021). Applications of EPS in environmental bioremediations. In A. K. Nadda, Sajna K. V., & S. Sharma (Eds.), Microbial exopolysaccharides as novel and significant biomaterials (pp. 289–307). Springer, . https://doi.org/10.1007/978-3-030-75289-7_11

Laskin, A. I., Gadd, G. M., & Sariaslani, S. (Eds.). (2014). *Advances in applied microbiology* (Vol. 87). Academic.

Liang, S. (2023). Advances in drug delivery applications of modified bacterial cellulose-based materials. *Frontiers in Bioengineering and Biotechnology, 11*, 1252706. https://doi.org/10.3389/fbioe.2023.1252706

Manna, S., Chaudhuri, K., Bhattacharyya, S., & Bhattacharyya, M. (2010). Dynamics of Sundarban estuarine ecosystem: Eutrophication induced threat to mangroves. *Aquatic Biosystems, 6*(8), 1–11. https://doi.org/10.1186/1746-1448-6-8

Mogany, T., Bhola, V., & Bux, F. (2024). Algal-based bioplastics: Global trends in applied research, technologies, and commercialization. *Environmental Science and Pollution Research, 31*(36), 38022–38044. https://doi.org/10.1007/s11356-024-33644-9

Mota, R., Flores, C., & Tamagnini, P. (2022). Cyanobacterial extracellular polymeric substances (EPS). In *Polysaccharides of microbial origin: Biomedical applications* (pp. 139–165). Springer. https://doi.org/10.1007/978-3-030-42215-8_11

Nadda, A. K., Sajna, K. V., & Sharma, S. (Eds.). (2021). *Microbial exopolysaccharides: Biochemistry and functions*. Springer. https://doi.org/10.1007/978-3-030-75289-7

Nisha, R., Kiran, B., Kaushik, A., & Kaushik, C. P. (2018). Bioremediation of salt-affected soils using cyanobacteria in terms of physical structure, nutrient status, and microbial activity. *International journal of Environmental Science and Technology, 15*, 571–580. https://doi.org/10.1007/s13762-017-1419-7

Pandey, A., Singh, M. K., & Singh, A. (2024). Bacterial cellulose: A smart biomaterial for biomedical applications. *Journal of Materials Research, 39*, 2–18. https://doi.org/10.1557/s43578-023-01116-4

Paria, K., & Chakraborty, S. K. (2019). Eco-potential of *Aspergillus penicillioides* (F12): Bioremediation and antibacterial activity. *SN Applied Sciences, 1*, 1515. https://doi.org/10.1007/s42452-019-1545-6

Patel, S., & Prajapati, J. B. (2013). Food and health applications of exopolysaccharides produced by lactic acid bacteria. *Advances in Dairy Research, 1*(2), 1–7. https://doi.org/10.4172/2329-888X.1000107

Patil, S. V., & Kamble, S. S. (2011). Prospective of microbial exopolysaccharide for heavy metal exclusion. *Applied Biochemistry and Biotechnology, 165*(7–8), 2222–2234. https://doi.org/10.1007/s12010-011-9440-2

Purama, R. K., Goyal, A., & Tiwari, K. L. (2009). Dextran and food application. In V. K. Gupta & M. G. Tuohy (Eds.), *Biofuel crops: Production, physiology and genetics* (pp. 353–370). CABI. https://doi.org/10.1007/978-3-319-03751-6_66-1

Rahman, M. M., Khan, M. N. I., Hoque, A. K. F., & Ahmed, I. (2015). Carbon stock in the Sundarbans mangrove forest: Spatial variations in vegetation types and salinity zones. *Wetlands Ecology and Management, 23*(2), 269–283. https://doi.org/10.1007/s11273-014-9379-x

Rana, A., & Rana, K. L. (2021). Fungal exopolysaccharides: Properties and applications. In K. L. Rana & A. K. Sharma (Eds.), *Microbial exopolysaccharides: Biochemistry and functions* (pp. 1–24). Springer. https://doi.org/10.1007/978-981-16-9310-6_1

Saha, I., Datta, S., & Biswas, D. (2020). Exploring the role of bacterial extracellular polymeric substances for sustainable development in agriculture. *Current Microbiology, 77*(11), 3224–3239. https://doi.org/10.1007/s00284-020-02169-y

Salama, Y., Chennaoui, M., Sylla, A., Mountadar, M., Rihani, M., & Assobhei, O. (2016). Characterization, structure, and function of extracellular polymeric substances (EPS) of microbial biofilm in biological wastewater treatment systems: A review. *Desalination and Water Treatment, 57*(35), 16220–16237. https://doi.org/10.1080/19443994.2015.1077739

Salazar, N., Ruas-Madiedo, P., Kolida, S., Collins, M., Rastall, R., Gibson, G., & de los Reyes-Gavilán, C. G. (2009). Exopolysaccharides produced by intestinal *Bifidobacterium* strains act as fermentable substrates for human intestinal bacteria. *Applied and Environmental Microbiology, 75*(20), 6534–6542. https://doi.org/10.1128/aem.00325-08

Sandhya, V., & Ali, S. Z. (2015). The role of exopolysaccharides in the environmental adaptation of bacteria and their implications for developing biological soil crusts to combat desertification. *Journal of Arid Environments, 112*, 4–17. https://doi.org/10.1016/j.jaridenv.2014.07.004

Shah, M. P., Patel, K. A., Nair, S. S., & Darji, A. M. (2013). Environmental bioremediation of dyes by *Pseudomonas aeruginosa* ETL-1 isolated from the final effluent treatment plant of Ankleshwar. *American Journal of Microbiological Research, 1*(4), 105–112. https://doi.org/10.12691/ajmr-1-4-3

Singh, P., & Saini, P. (2017). Food and health potentials of exopolysaccharides derived from Lactobacilli. *Microbiology Research Journal International, 22*, 1–14. http://dx.doi.org/10.9734/MRJI/2017/36935

Singh, H., & Sharma, P. (2020). Role of fungi in bioremediation and environmental sustainability. In A. Kumar, S. Sharma, & P. Singh (Eds.), *Microbial bioremediation and biodegradation* (pp. 153–170). Springer. https://doi.org/10.1007/978-3-030-54422-5_8

Singh, J. S., & Kumar, A. (2016). Cyanobacteria: A precious bio-resource in agriculture, ecosystem, and environmental sustainability. *Frontiers in Microbiology, 7*, 529. https://doi.org/10.3389/fmicb.2016.00529

Singh, R. S., Saini, G. K., & Kennedy, J. F. (2008). Pullulan: Microbial sources, production, and applications. *Carbohydrate Polymers, 73*(4), 515–531. https://doi.org/10.1016/j.carbpol.2008.01.003

Sundaramoorthy, N. S., Thothathri, S., Bhaskaran, M., GaneshPrasad, A., & Nagarajan, S. (2021). Phages from Ganges River curtail in vitro biofilms and planktonic growth of drug resistant Klebsiella pneumoniae in a zebrafish infection model. *Amb Express, 11*, 1–11. https://doi.org/10.1186/s13568-021-01181-0

Syed, Z., Sogani, M., Rajvanshi, J., et al. (2023). Microbial biofilms for environmental bioremediation of heavy metals: A review. *Applied Biochemistry and Biotechnology, 195*, 5693–5711. https://doi.org/10.1007/s12010-022-04276-x

Tang, H., Xiang, G., Xiao, W., Yang, Z., & Zhao, B. (2024). Microbial mediated remediation of heavy metals toxicity: Mechanisms and future prospects. *Frontiers in Plant Science, 15*, Article 1420408. https://doi.org/10.3389/fpls.2024.1420408

Vardharajula, S., & Ali, S. Z. (2015). The role of exopolysaccharides in the environmental adaptation of bacteria and their implications for developing biological soil crusts to combat desertification. *Journal of Arid Environments, 112*, 4–17. https://doi.org/10.1016/j.jaridenv.2014.07.004

Wani, N. R., Dar, A. H., Dash, K. K., Pandey, V. K., Srivastava, S., Jan, S. Y., et al. (2023). Recent advances in the production of bionanomaterials for development of sustainable food packaging: A comprehensive review. *Environmental Research, 237*, 116948. https://doi.org/10.1016/j.envres.2023.116948

Yang, Y., Jiang, G., & Tian, Y. (2023). Biological activities and applications of exopolysaccharides produced by lactic acid bacteria: A mini-review. *World Journal of Microbiology and Biotechnology, 39*, 155. https://doi.org/10.1007/s11274-023-03610-7

Zhang, Y., Li, X., Du, J., Pu, L., & Chen, S. (2021). Effects of *chlorella* extracellular polymeric substances on the aggregation and stability of TiO_2 nanoparticles as electrolytes. *Desalination and Water Treatment, 209*, 334–341. https://doi.org/10.5004/dwt.2021.26619

Zhou, Y., Cui, X., Wu, B., Wang, Z., Liu, Y., Ren, T., ... & Rittmann, B. E. (2024). Microalgal extracellular polymeric substances (EPS) and their roles in cultivation, biomass harvesting, and bioproducts extraction. *Bioresource Technology*, 131054. https://doi.org/10.1016/j.biortech.2024.131054

Extraction Techniques: Past, Present, and Future

Saikat Mazumder, Debasmita Bhattacharya, Dibyajit Lahiri, and Moupriya Nag

1 Introduction

Extracellular polymeric substances are a group of high molecular weight biopolymers that are secreted by various microorganisms, including fungi and bacteria, as byproducts of their metabolic processes into the surrounding environment (Asianezhad et al., 2023). Extracellular polymeric substances are macromolecules that have variable chemical composition and structure, which depend on species and growth conditions. Polysaccharides and proteins are the major components. In addition, fatty acids, nucleic acids, humic acids, amino acids, and other compounds are present in smaller amounts (Shi et al., 2017a). The term extracellular polysaccharides was first referred to in 1972 by Sutherland to describe a broad range of microbial polysaccharides located outside cell walls that may firmly or loosely connect based on their structural and functional interaction with the cell. EPS includes a protective shield against desiccation, osmotic pressure, extreme temperature, and salt among bacterial cells (Cirrincione et al., 2018). Bacteria are mainly used in the production of EPSs, mainly because of their ability to grow rapidly with loosely

S. Mazumder
Department of Biotechnology, Institute of Engineering and Management, Kolkata University of Engineering and Management, Kolkata, West Bengal, India

Department of Food Technology, Guru Nanak Institute of Technology, Kolkata, West Bengal, India

D. Bhattacharya
Department of Basic Science and Humanities, Institute of Engineering and Management, Kolkata, University of Engineering and Management, Kolkata, West Bengal, India

D. Lahiri (✉) · M. Nag (✉)
Department of Biotechnology, Institute of Engineering and Management, Kolkata University of Engineering and Management, Kolkata, West Bengal, India
e-mail: dibyajit.lahiri@uem.edu.in; moupriya.nag@uem.edu.in

T. Sarkar et al. (eds.), *Extracellular Polysaccharides*,
https://doi.org/10.1007/978-3-031-94472-7_3

bound EPS having a mucoid surface, which can be isolated easily from the cells by applying different isolation techniques (Angelin & Kavitha, 2020). Exopolysaccharides have been further divided into two classes, homo-exopolysaccharides (HoEPS) and hetero-exopolysaccharides (HeEPS), depending on the monosaccharide compositions (Zhang et al., 2024). Due to its various physicochemical characteristics, EPS finds use in numerous industries, such as food, pharmaceuticals, paper, and textiles (Casillo et al., 2018; Xiao & Zheng, 2016). The medicinal properties of exopolysaccharides involve antibacterial, antitumor, anti-inflammatory, and antioxidative properties (Delattre et al., 2016; Xiao & Zheng, 2016). Depending on the strain, growth media composition, and environmental stresses, the extracellular polymeric compounds from microalgae have varying properties (Delattre et al., 2016; Xiao & Zheng, 2016).

EPS can be isolated from diverse sources, such as extremophiles, thermophiles, halophiles, psychrophiles, and acidophiles (Nicolaus et al., 2010), which influence their physicochemical properties. However, lactic acid bacteria (LAB) is a potent significant source and extensively studied. The EPS produced by LAB has applications in the fields of food, pharmaceuticals, textiles, and agriculture (Shukla et al., 2019). Bacterial exopolysaccharides have outstanding bioactive potential and immunological modulation, prebiotic effects, and anticancer, antioxidant, antidiabetic, antiviral, antiulcer, and anti-gastritis properties (Angelin & Kavitha, 2020). Over the past years, new product formulations based on bacteria have caused several bacterial biopolymers to be isolated and characterized. This is due to the medical utility of the probiotic bacteria EPS due to its nontoxicity and metabolic compatibility. New advances in genetic alteration and metabolic engineering enable interesting therapeutic applications for EPS. Bacterial EPS is preferable because bacteria are manageable compared to plant EPS. Biofilms play a significant role in chronic human infections involving the colonization of microbiota, shifting health and illness; hence, there is much work being done studying bacterial EPS function and bioactivity for therapeutic application.

2 Characteristics of Extracellular Polymeric Substances

Macromolecular compounds that are released from cells can accumulate and form a viscous layer that surrounds the cells and protects them from the adverse effects of some environmental stress factors (Delattre et al., 2016). EPS from microalgae has a complex structure in nature that developed due to many environmental stress factors (Gaignard et al., 2019). Different types of EPS have been reported in the scientific literature, depending on their properties and ability to bind to cells. Those loosely associated with cells and released into the media are referred to as soluble EPS (SL-EPS). Thus, immobilized EPS represents a form that is connected with cells (Naveed et al., 2019). It consists of loosely bound EPS (LB-EPS) and tightly bound EPS (TB-EPS). LB-EPS and TB-EPS are different in terms of composition and efficiency Y. Li et al. (2021) reported that based on the adsorption capabilities of Pb(II), LB-EPS have higher efficiency compared to both TB-EPS and SB-EPS.

LB-EPS seems to be more critical for responding to deleterious environmental changes compared to other fractions of EPS; therefore, they have a protective function in algal metabolism (Pan et al., 2016). EPS can have a role in nutrient distribution because the extracellular matrix contains substances that are produced by algal cells, such as nucleic acids, enzymes, lipopolysaccharides, and phospholipids. Studies showed that proteins translocated from TB-EPS might reach the extracellular environment to the outside LB-EPS or SB-EPS. The protein content of TB-EPS decreased toward the death phase, but the protein amount in all EPS fractions didn't change within the logarithmic phase (Wang et al., 2019). The nomenclature of EPS is inconsistent; hence, several words for total EPS are included in the literature, such as exopolysaccharides, extracellular polysaccharides, extracellular proteoglycan (EPG), exopolymeric compounds, or extracellular polymers. Both fractions have distinct features and compositions (Pierre et al., 2019; Shi et al., 2017a, b). Rossi and De Philippis (2016) assert that no substantial differences exist. This review examines a released fraction known as EPS. Extracellular polymeric substances (EPSs) are primarily composed of polysaccharides, proteins, lipids, and nucleic acids (Shi et al., 2017a, b). Extracellular polymeric entities can appear as polysaccharides (primarily consisting of sugar components), proteoglycans (which contain proteins within sugar chains), or glycoproteins (where proteins are the primary constituents of the polymer), or they may be an alternative polymer form (Halaj et al., 2019). EPS has a high molecular weight; however, there is a percentage of low molecular weight polymeric compounds (Villacorte et al., 2015). The molecular weight of a polymer extracted from Chlorella vulgaris JSC-7 was 9.86×10^3 g mol^{-1} (Alam et al., 2014). Besides that, the EPS from Dictyosphacrium strains had a high molecular weight fraction between 1×10^6 g mol^{-1} and over 2×10^6 g mol^{-1}, together with low molecular weight compounds between 9000 and 56,000 g mol^{-1} (Halaj et al., 2019). They are mainly negatively charged (hydrophobic) due to the presence of hydrolyzed functional groups. FTIR spectra confirm the presence of carboxyl, hydroxyl, and amino groups (Alam et al., 2014; Halaj et al., 2019).

Polysaccharides are the first component of EPS, which occupies 45–95% of its composition (Naveed et al., 2019). The constitutive sugars found in microalgal EPS are glucose, galactose, mannose, arabinose, rhamnose, and fucose, with glucose and galactose being the major sugars. An exception is the EPS produced by *Dunaliella salina*, which mainly contains fructose (Mishra & Jha, 2009). Monosaccharides can be modified with nonsugar groups such as pyruvate, sulfates, acyl residues, phosphates, O-acetyl, and O-methyl groups (Delattre et al., 2016; Halaj et al., 2019; Lv et al., 2019). Exopolysaccharides from Porphyridium sp. are primarily composed of xylose, glucose, and galactose. Protein is 5.5% of the polymer. The negative charge of EPS is attributed to the presence of negatively charged sulfate groups and glucuronic acids (Pignolet et al., 2013). Extracellular polysaccharides from *Chlorophyta* are distinguished by elevated levels of galactose (Gaignard et al., 2019).

The polysaccharides can have linear or branched configurations with regular or irregular morphologies, which depend on the presence of diverse glyosidic linkages

(Cybulska et al., 2016). Delattre et al. (2016) reported the presence of -(1,3), −(1,3), −(1,4), and (2,1)- connections in EPS. Most EPSs are heteropolymers. However, homopolysaccharides are reported in some cases, for example, the EPS isolated from *Dunaliella tertiocelata*, which are (1,4)- -D-glucans (Geun Goo et al., 2013). Proteins constitute a primary component of microalgal extracellular polymeric substances (EPSs). Functionally, there exist nonenzymatic structural proteins that stabilize the EPS matrix (R. Xiao & Zheng, 2016), as well as enzymatic proteins that degrade exopolysaccharides, serving as a carbon and energy source for cells (Flemming & Wingender, 2010). Xiao and Zheng (2016) reported the concentration of extracellular total protein per mg L^{-1} of supernatant in various microalgae as ranging from 0.5 mg L^{-1} to 11 mg L^{-1}, and when expressed as a percentage of total EPS, it varied from 0.5% to 16.9% (Xiao & Zheng, 2016). Examination of published experimental data related to the number of proteins registered in literature proves large variation for all the studied *Chlorophyta*. Protein in EPS should thus be considered a nonspecies feature rather than a characteristic. EPS composition: An analytical investigation of EPS from *Chlorella vulgaris* found protein concentration equal to 25.6 mg L^{-1} and carbohydrate content amounted to 31.1 mg L^{-1} (Nguyen et al., 2020). Another species in *Chlamydomonadales*, *Chlamydomonas reinhardtii*, produced EPS consisting of 42.1% proteins (Zhu et al., 2012). Liu and Miao (2017) studied the effect of $MgSO_4$ on the production of EPS by *Heynigia riparia* belonging to Chlorophyta. It showed that protein content in EPS varied between 15.27% and 24.19%. The extracellular polymeric substance of Chlorophyta generally consists of glycoproteins (Paulsen et al., 1998).

3 Synthesis of Exopolysaccharides

The production of extracellular polysaccharides by bacteria is made possible by a set of specialized secreted enzymes, such as polymerization and precursor synthesis enzymes, and the process can be either outside the cell or within the cell wall (Roger, 2002). The three major steps in the production of EPS are: (i) precursor substrate synthesis, (ii) polymerization and transfer to the cytoplasmic membrane, and (iii) export via the outer membrane (Fig. 1). The three processes differ based on the carbon source utilized, varying across microorganisms and especially depending on the kinds of polymers.

3.1 Synthesis of Precursor Substrate

The process involves the conversion of intermediate sugar metabolites into the EPS precursor, which includes nucleoside diphosphate sugars, like Guanosine diphosphate (GDP)-sugar, which is equivalent to the absorbed substrate or carbon source.

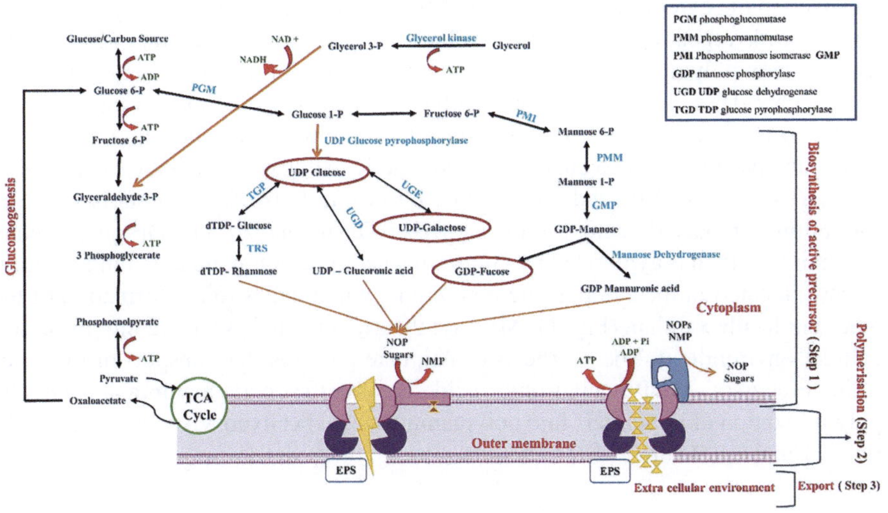

Fig. 1 Steps in the biosynthetic route of bacterial polymers

Sugar nucleosides (nucleoside diphosphate sugars) provide an active form of monosaccharides and allow bacterial cells to interconvert different monosaccharides through epimerization, dehydrogenation, and decarboxylation processes. Specialized enzymes on polymers, the active precursor of biosynthesized polymers, represent an initial committed step that metabolic engineers have targeted as a point for augmentation in producing more polymer as well as providing a route toward the synthesis of customized polysaccharides. In this direction, different types of polymers (dextran, xanthan, and alginate) were produced separately using distinct precursors and different enzymes (Lin & Hassid, 1966). Uridine diphosphate (UDP)-glucose serves as the direct precursor for cellulose synthesis in *Acetobacter xylinum* and pullulan production in *Aureobasidium pullulans*, facilitated by uridine diphosphate glucose (UDPG) pyrophosphorylase and 288 glucosyltransferase activity, respectively (Duan et al., 2008). Each polymer possesses 289 specific precursors and enzymes that differ among organisms.

3.2 *Polymerization and Transfer of Cytoplasmic Membrane*

The process of transfer of the precursor nucleoside diphosphate into the polymer for the polymerization of monomers represents the second stage in EPS biosynthesis. The subsequent successive incorporation of sugars into an isoprenoid lipid, along with the addition of acyl groups simultaneously, follows the activation of monosaccharides by sugar nucleotide complexes. Specific sugar transferase enzymes allow for the transfer of monosaccharides and acyl groups to isoprenoid lipid acceptors localized in the cytoplasmic membrane: bactoprenol, C55-isoprenoid lipid. The oligosaccharide repeating units, which include acetyl, pyruvyl, and other acyl modifications, are polymerized. After the polymerization of the repeating units, the polysaccharide is extruded out through the outer cytoplasmic membrane, possibly by the formation of a multi-protein complex containing cytoplasmic, outer membrane, and periplasmic proteins. The biosynthetic pathway of xanthan is depicted in Fig. 1 (Rosalam & England, 2006). The biosynthesis of Xanthan starts with the formation of repeated pentose units, namely, GDP-mannose and UDP-glucuronate. The units are then polymerized by GumE, the catalytic component of xanthan polymerase, located in the cytoplasmic membrane, which leads to the formation of the macromolecule xanthan (Fig. 1). After synthesis, xanthan is secreted into the extracellular environment. The synthesis of Alginate requires the transport of the cytosolic precursor GDP-mannuronic acid across the cell membrane and the polymerization of monomers into polymannuronate via Glycosyl-transferase (Alg8) (Figs. 1 and 2) (Rehm, 2010).

3.3 Export Through the Outer Membrane

The final step of EPS is the release from the cytoplasmic membrane which involves, crossing the periplasm, and outer membrane, and finally release into the extracellular environment. AlgE is the gene encoding the export enzyme for alginate, while GumJ is the gene encoding the excretion enzyme for xanthan (Fig. 2). During EPS synthesis, lipid transporters bind to the extracellular membrane and facilitate the precise and sequential assembly of the carbohydrate chain as it passes through the

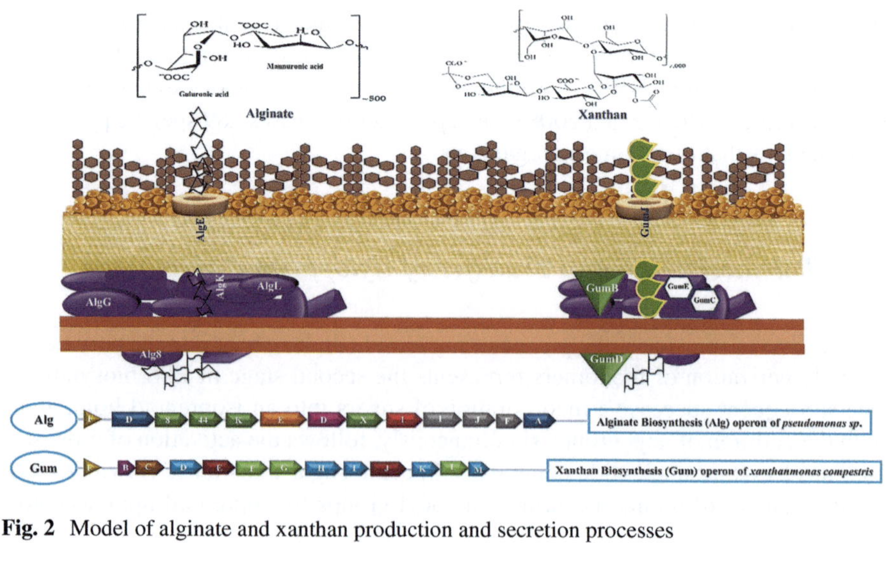

Fig. 2 Model of alginate and xanthan production and secretion processes

cell membrane. Polysaccharides are polymerized on the inner surface of the cytoplasmic membrane and then exported across the membrane in response to a lipid transporter. The transporters consist of long-chain phosphate esters and isoprenoid alcohols, similar to those identified in the biosynthesis of lipopolysaccharides and peptidoglycans (Sutherland, 1972). They are essential for heteropolysaccharide synthesis, which takes place in parallel with the excretion of EPS.

4 Exopolysaccharide Extraction and Purification

The methods of successful separation, recovery, and quantitation of the produced polysaccharide are just as important as the considerations that ensure the successful growth of EPS-producing LAB. Many recommended methods are highly dependent upon the medium used during the fermentation stages and on the end goal of the intended application of the EPS product. During postproduction, it is crucial to separate and clean the EPS from impurities such as medium impurities, microbial biomass, nucleic acids DNA and RNA, and proteins (Sørensen et al., 2022). EPS extraction processes are numerous and can vary with the source of the microorganism, types of EPS, complexity of the production media, and amount of desired purity (Jurášková et al., 2022; Kavitake et al., 2023).

4.1 Isolation and Recovery

EPS extraction from culture is generally divided into three steps: (1) removal of cells and other impurities, (2) polymer precipitation from the cell-free supernatant, and (3) drying of the precipitated polymer (Fig. 3) (Freitas et al., 2011).

4.1.1 Preliminary Treatments

The simplest recovery and isolation methodology is using a predefined medium, with a subsequent centrifugation step to remove cells, and then final basic dialysis using water (Freitas et al., 2011). The recovery from liquid media is highly dependent upon the decision to use either complex or defined media, or some pretreatment steps may be necessary to inactivate certain enzymes or remove certain proteins (Leroy & De Vuyst, 2016). In several instances, especially if a complicated medium is used, a heat treatment at 90–100 °C is used for the elimination of microorganisms and inactivation of any enzyme present (Polak-Berecka et al., 2014). Furthermore, the effect of thermal treatment on the final EPS productivity before isolation has been studied. A heat treatment step was considered necessary for complete EPS recovery to significantly elevated yields (Rimada & Abraham, 2003). This includes the utilization of a vacuum rotary evaporator that concentrates the fermentation

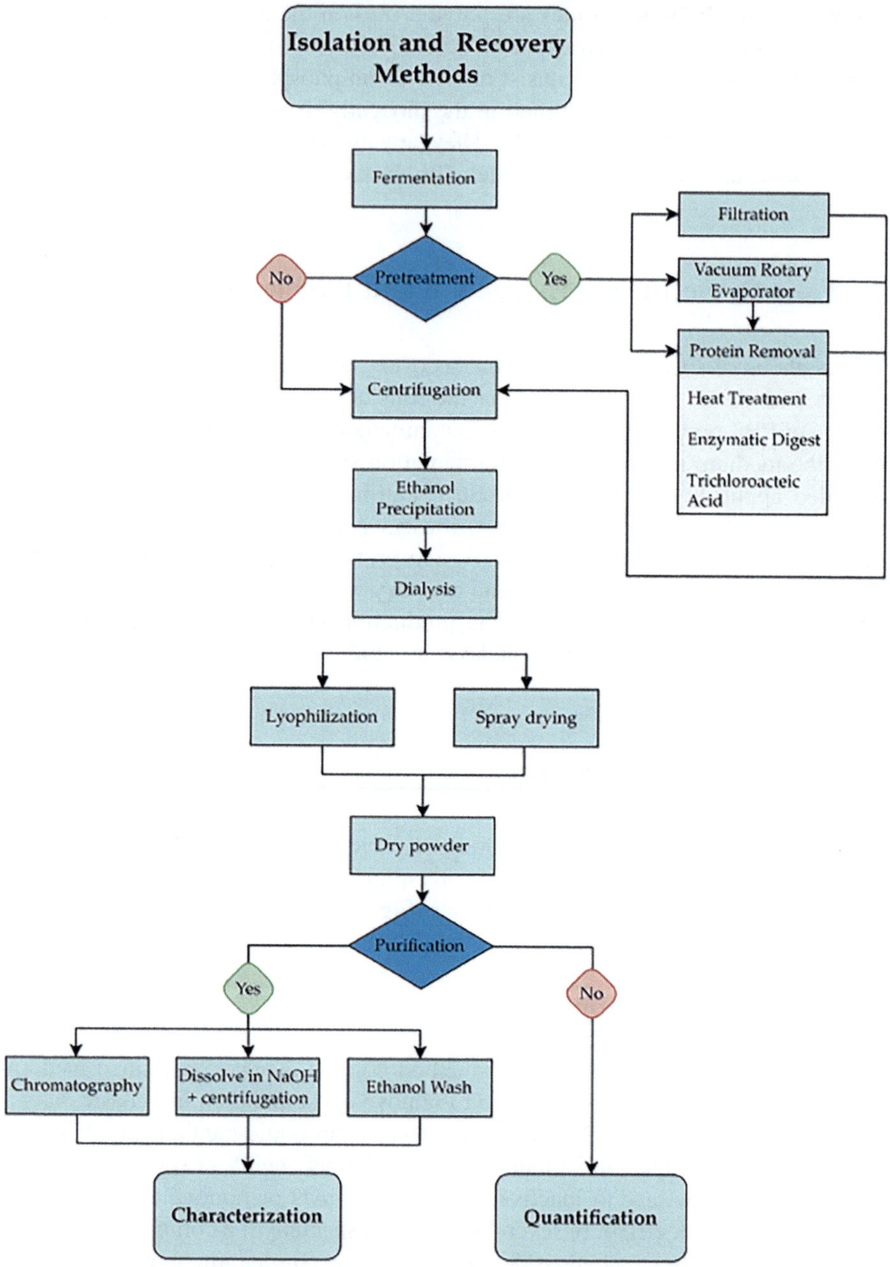

Fig. 3 Isolation and recovery of EPS. (Adopted from Sørensen et al. (2022))

broth before separation. Although vacuum rotary evaporatoration is a very less described technique (Polak-Berecka et al., 2013). Complete removal of the protein fraction usually involves a precipitation step that can be attained either by the addition of trichloroacetic acid (TCA) in mixed conditions with a concentration of 4–14%, or enzymatic digestion through proteases, or by the combination of these two (Leroy & De Vuyst, 2016; Ruas-Madiedo, & de los Reyes-Gavilán, 2005). It can be applied as a preremoval EPS from the culture medium (Li et al., 2016; Polak-Berecka et al., 2015). The necessity of a TCA phase is dependent on the final goal. While it is necessary to identify the exact composition of the EPS, however, for quantitative analysis, it can be neglected (Ruas-Madiedo, & de los Reyes-Gavilán, 2005). Ultrafiltration has also been proposed as a faster and more accurate method for the separation of EPS, instead of TCA and ethanol treatments (Macedo et al., 2002a). The cellular fraction and protein substances can be extracted through centrifugation The liquid phase can be subjected to centrifugation to extract the cellular element and coagulated proteins (Shao et al., 2014), followed by a microfiltration or ultrafiltration stage for additional purification (Li et al., 2014; Shene et al., 2008).

4.1.2 Precipitation and Dialysis

The extraction of EPS using the precipitation method involves stages such as the removal of biomass, degradation of protein, and precipitation (Sørensen et al., 2022). Centrifugation is the primary practice for eradicating cell biomass and attaining the cell-free supernatant (CFS) comprising EPS (Yu et al., 2023), which can be further filtered by 0.45 μm membrane filter (Zaghloul et al., 2023). Some studies exhibited that the culture medium is heated for 10–15 min at 90–100 °C to destroy bacterial cells and denature enzymes, followed by centrifugation for removal of the cells (Choi et al., 2021; Yu et al., 2023).

The second step in the removal of protein is precipitation and degradation of protein to obtain pure EPS (Sørensen et al., 2022). TCA is mainly used for protein precipitation and denaturation, typically at concentrations from 2 to 80% (w/v) at 4 °C for 2 h or overnight (Angelov et al., 2023; Kang et al., 2023). The dehydration of hydration shells around the proteins leads to precipitation. Alternatively, others say that the acidity of TCA changes the protein structure, causing precipitation; the exact mechanism behind the precipitation remains unknown (Hooshdar et al., 2020). Additional precipitating agents and degrading agents for proteins used include Sevag reagent, a mixture of chloroform:n-butanol in the ratio of 4:1, and protease enzyme hydrolysis at a concentration of 50 μg/mL to remove protein impurities from EPSs (Zhan et al., 2023).

The third critical precipitating stage of EPS is precipitation, with the use of one to four volumes of ice-cooled 95% or 100% ethanol (v/v) with overnight incubation at 4 °C (Derdak et al., 2022; Kavitake et al., 2023; Sheng et al., 2022). Ethanol displaces the water molecule hydration shell from the polar surface of the hydrophilic region of the EPS because ethanol is less polar than water; hence, its use reduces EPS solubility (Bramhachari et al., 2007). Further chemicals used as precipitants include propanol, acetone, and isopropanol, in some cases used for the precipitation of EPS (Murugu & Narayanan, 2024; Nachtigall et al., 2021). The participating EPS is separated by centrifugation, dissolved in H_2O or Milli-Q water, and then purified through a series of processes, to be discussed in the following section (Pourjafar et al., 2023). The EPS of *Lpb. plantarum* HDC-01 has been precipitated from the TCA (80% w/v)-treated CFS by three liters of ice-cold 95% ethanol (v/v) at 4 °C overnight (Y.-J. Yu et al., 2018). Al-Nabulsi et al. (2022) used a cooled ethanol solution of 100% (v/v) to precipitate EPS-L from the CFS of *L. bulgaricus*, followed by inactivating the enzymes and proteins with 20% TCA (w/v) (Rahnama Vosough et al., 2021). Precipitated the EPS from the CFS of three different *Enterococcus* strains using three volumes of 96% cold ethanol (v/v) and proteins were eliminated using Sevag reagent.

After this pretreatment next step followed by is usually precipitation which mostly uses ethanol. Rimada and Abraham (2003) tested several methods for the isolation of EPS, including one or two steps of ethanol precipitation, one-step ethanol precipitation followed by dialysis, dialysis using membranes with different cut-offs (6000, 8000, 12,000, and 14,000), and TCA precipitation. The single step using ethanol was not sufficient for isolation, as residual lactose co-precipitated with EPS, resulting in artificially high values. Two-step ethanol precipitation and dialysis using membranes with a molecular weight cut-off below 8000 produced equivalent results and were considered effective procedures for EPS separation. The yield by TCA precipitation was found to be very low, and it was proposed that this may indeed be a false negative in case the treatment by TCA would remove a substantial amount of proteins and impurities, as such treatment would also precipitate some EPSs that could reduce final polymer concentration to an extent of up to 50%. TCA precipitation should be at least washed once. Research shows that ethanol temperature varies, but a two-temperature experiment, −20 °C and 4 °C, showed no difference in the final EPS (Rimada & Abraham, 2003). In some cases, acetone or a combination of ethanol and acetone has been used (De Vuyst et al., 2001; Simova et al., 2004). Dialysis, the final step before drying, removes low molecular mass carbohydrates (Leroy & De Vuyst, 2016). The precipitation approach is preferred for EPS extraction due to its ease of function, effectiveness, least chemical contamination, and high throughput. However, the disadvantages of the process include incomplete precipitation, co-precipitation of proteins, nucleic acids, and other biopolymers.

4.1.3 Ultrafiltration

Ultrafiltration employs size exclusion along with pressure-driven separation (Shoshaa et al., 2023). Studies demonstrate ultrafiltration as a more efficient and precise method than TCA and ethanol for extracting EPS (Costa et al., 2021; Sørensen et al., 2022). Pollutants consisting of salts, organic substances, etc. are separated through size using ultrafiltration (Zhang et al., 2023). Larger-sized molecules are retained by the membrane commonly called the retentate, whereas the smaller molecules along with the solvents are driven across (Bedade & Pawar, 2023). Centrifugation eradicates coagulated proteins and cell fractions from the fluid medium, whereas ultrafiltration gathers EPS (Ziadi et al., 2018). The CFS is ultra-filtrated through a membrane having molecular weight cut-off (MWCOs) in the range of 10–500 kDa according to the EPS molecules' size after filtration using a 0.20 μm filter; TCA denatures the protein after this (Delattre et al., 2016; Shene et al., 2008). Further, concentrated EPS may be subjected to additional processing, including drying and freeze-drying, for further analysis or applications. Ziadi et al. (2018) isolated the EPS from the CFS of *Lactococcus lactis* SLT10, *Leuconostoc mesenteroides* B3. Ultrafiltration, applied to concentrate ninefold the supernatant from *L. crispatus* L1, was successful in reaching 85% of recovery yield. Yu et al. (2018) employed a 100 kDa MWCO membrane to isolate the EPS from *W. cibaria* W27. Macedo et al. (2002a), b) used ultrafiltration through a 30 kDa MWCO membrane for separating the EPS of *L. rhamnosus* RW-9595M. Ultrafiltration is a cost-effective, scalable, and chemical-free method of concentrating and purifying EPSs. However, improvement in the challenges that surround membrane fouling, high upfront costs, and some potential limitations in molecular weight selectivity have to be addressed in future studies to achieve successful implementation.

4.2 Dehydration, Refinement, and Characterization

Generally, lyophilization, spray-drying, or spray-freeze drying is applied to isolate EPS and make a powder (Xiao et al., 2020). EPS lyophilization can be enhanced for particular EPS characterization rather than quantification by either dissolving the powder in sodium hydroxide and then centrifuging to remove contaminants (Notararigo et al., 2013) or washing the powder with ethanol (Tuinier et al., 1999). Ion-exchange chromatography (Ciszek-Lenda et al., 2013; Wang et al., 2015) and size exclusion chromatography (Li et al., 2014; Shao et al., 2014; Zhang et al., 2016) can be employed for characterizing the EPS generated by LAB. Advanced methods including transmission electron microscopy (TEM), nuclear magnetic resonance (NMR), Fourier transform infrared (FT-IR), and scanning electron microscopy (SEM) may be deployed to inspect the structure of EPS (Sharma et al., 2020). The target EPS is extracted and separated from a microbial colony by a multi-step process called purification. The quality and degree of purity needed for the applications may influence the type of purification process (Talbi et al., 2023). The initial purification process includes dialysis. This is a widely adopted procedure for EPS purification using a membrane with MWCO (Kavitake et al., 2023). In the next step of purification, size-exclusion chromatography (SEC) and anion-exchange chromatography (AEC) become absolutely necessary in determining the outstanding level of purity achieved by the EPS (Feng et al., 2023; Pourjafar et al., 2023). The purifications achieve significance when the recovered EPS is utilized for characterization rather than for approximating EPS creation (Leroy & De Vuyst, 2016). Several practices, including hot air drying, microwave drying, vacuum drying, spray drying

(SD), vacuum spray freeze-drying (SFD), and vacuum freeze-drying (FD), have been extensively employed for the dehydration of polysaccharides (Yi et al., 2020). Studies have revealed that polysaccharides, when dried through freeze-drying, possess a higher antioxidant potential (Liu et al., 2020). Pure EPSs are made through freeze-drying for 12 h at −80 °C using a vacuum freeze-drier, and the EPS solution is kept at −20 °C (Sheng et al., 2022; Yu et al., 2023). The pure dried, EPS is stored at -20 °C for characterization and further investigation (Kavitake et al., 2023). Freeze-dried EPS appears white and fluffy (Du et al., 2022; Yu et al., 2023). Figure 3 illustrates a schematic diagram depicting the many stages of purification.

4.2.1 Dialysis

Dialysis is conducted as the last phase before drying and is a crucial technique as it facilitates the removal of low molecular weight carbohydrates (Sørensen et al., 2022). The isolated EPS precipitates are dissolved in H_2O or Milli-Q water before dialysis (Jurášková et al., 2022; Prete et al., 2024). Dialysis is performed using H_2O/Milli-Q water or phosphate-buffered saline (PBS, pH 7.4) to remove small neutral sugars, media contaminants, small proteins, salts, and damaged segments of the EPS via a dialysis membrane with a molecular weight cut-off (MWCO) of 6, 8, 12, or 14 kDa, over a period of 2 days, with three water changes daily (Jurášková et al., 2022; Nguyen et al., 2020). Thereafter, the dialyzed EPS sample is concentrated via freeze-drying, resulting in a white hue and the soft, spongy texture characteristic of pure EPS (Ramos et al., 2023). The current predicted EPS yield is determined by the weight of the acquired EPS. Al-Nabulsi et al. (2022) isolated the EPS-L samples from *L. bulgaricus* using dialysis utilizing a 20 kDa MWCO membrane against H_2O, then freeze-drying and storing them at −20 °C. Tiwari et al. (2024) employed a 12–14 kDa-MWCO dialysis membrane to dialyze the EPS isolated from *E. hirae* OL616073 against H_2O at 4 °C for a duration of 2 days. Dialysis eschews harsh chemicals or extreme conditions, effectively eliminates small contaminants (salts, residual sugars, solvents, and small metabolites), does not necessitate organic solvents, is straightforward, and is cost-effective in small-scale or laboratory settings. However, its protracted process, potential for sample loss, and substantial buffer requirements render it less suitable for large-scale industrial production without additional optimization or supplementary purification methods.

4.2.2 Anion Exchange Chromatography

Anion exchange chromatography is utilized to remove negatively charged molecules, such as proteins and nucleic acids, which bind to positively charged resins, including ion exchange Sepharose, DEAE-cellulose-52, DEAE-Sepharose, and phenyl-Sepharose (Costa et al., 2021; Hou et al., 2021; Kavitake et al., 2023). This method involves dissolving dialyzed and lyophilized crude EPS samples in Tris-HCl (0.02 M, pH 7.4) buffer, which are then loaded onto anion exchange columns.

Then gradient elution is performed using Tris-HCl buffer and NaCl (0–0.3 M) at a flow rate of 1 or 2 mL/min (Jiang et al., 2021; Li et al., 2022). Sheng et al. (2022) purified the EPS extracted from *L. pantheris* TCP102 using a DEAE-Sepharose fast-flow column and AEC. The study identified three different EPSs, which were then purified using size SEC. A dialyzed EPS-F2 sample was purified by researchers using a DEAE-52 column and eluted with an NaCl buffer (0–0.5 M) at a flow rate of 2 mL/min and saw a prominent single peak, indicating the presence of a singular EPS (Jiang et al., 2021). Using the DEAE-52 column with a NaCl solution of 0.1–0.3 M eluted at a flow rate of 1 mL/min, Li et al. (2022) utilized anion exchange chromatography (AEC) for the purification of the dialyzed EPS sample. Three peaks were observed by them corresponding to three different types of EPSs: EPS-1, EPS-2, and EPS-3. AEC effectively removes contaminants while minimizing loss of sample, is a focused and economical method for separating charged EPSs from complex mixtures, and provides high resolution with potential fractionation. Still, it is less feasible for large-scale manufacture without adequate resources and preparation because it requires careful tuning, is not applicable for neutral or weakly charged EPS, may be pricey, and is very sensitive to pH and ionic strength.

4.2.3 Size-Exclusion Chromatography or Gel-Filtration Chromatography

Due to the separation based on the size and molecular weight of EPSs, size-exclusion chromatography, also known as gel-filtration chromatography, is employed for the separation of pure components of EPSs (Costa et al., 2021; Feng et al., 2023). For SEC, Sepharose, Sephadex, Sephacryl, Superdex, Seralose, and Superose are employed (Kaddour et al., 2021). Previous research has employed the use of Sephadex G-75, G-100, Superdex G-200, Sephacryl S-400, and Seralose 6B to purify the EPSs from the LAB strain. Dialyzed and AEC-eluted EPS fractions are passed through the GFC column and eluted with ultrapure water at an optimal flow rate (Feng et al., 2023; Kavitake et al., 2023; Yu et al., 2023). The dialyzed and AEC-eluted fractions of EPS are introduced into the GFC column, and elution is conducted using ultrapure water at an optimum flow rate (Feng et al., 2018; Jiang et al., 2021). Yu et al. (2023) purified dialyzed EPS using gel filtration chromatography with Sephadex G-100, eluting with ddH_2O at 2 mL/min and seeing a single symmetric peak, confirming homogeneity and purity. Sharma et al. (2020) purified dialyzed EPS by Sephadex G-75 and H_2O at a flow rate of 1.5 mL/min using GFC, and they found a single peak corresponding to pure EPS. Jiang et al. (2021) purified EPS-F2 by gel filtration chromatography on a Sephacryl S-400 HR column eluted with ultrapure water at 1 mL/min, and a single symmetrical peak was detected, indicating sample homogeneity. GFC is a size-exclusion technique that eliminates minor impurities and performs desalting in a nondenaturing manner to purify EPSs, requiring no specialized buffer and ensuring no sample loss owing to charge. Nonetheless, it is constrained in resolution for comparably sized pollutants, can be sluggish and resource-demanding, and may not be the most suitable choice for large-scale, high-throughput procedures unless meticulously adjusted.

5 Conclusion

EPSs are a promising and dynamic research area with immense implications in various fields, such as health, biotechnology, agriculture, and environmental science. EPSs produced by LAB have more advantages than other natural agents in industrial applications and health promotion and treatments. The choice of one or more procedures for the EPS will depend on the microbial strain, the culture conditions, and the required degree of accuracy and precision for subsequent research. Ethanol precipitation is a very good method for isolating EPSs due to its simplicity, cost-effectiveness, and scalability. However, it has limitations regarding partial precipitation, potential sample loss, and the need for optimization. Dialysis is the most used method for the purification of EPS, but AEC and gel permeation chromatography (GPC) are applied depending on the level of purity needed. The phenol-sulfuric acid method is widely used in the determination of EPS because of its simplicity and applicability for a wide range of carbohydrate content. The process of EPS extraction by the precipitation technique involves a number of steps, namely: removal of biomass, degradation of proteins, and precipitation. The most direct recovery and isolation technique is to use a preset medium, then a spinning stage to remove cells, and finally basic dialysis with water. Ultrafiltration is generally preferred as a more efficient and sensitive method for the isolation of EPS than TCA and ethanol treatments. Additionally, EPS isolation is usually carried out using lyophilization, spray-drying, or spray-freeze drying to yield a powder. Such a stage is dialysis, which takes place as the last operation before drying and is considered essential for the removal of low molecular weight carbohydrates. The method of elimination of negatively charged molecules, including proteins and nucleic acids, that bind to positively charged resins is anion exchange chromatography. Size-exclusion chromatography or gel-filtration chromatography is a generally applied technique used to separate individual pure EPS constituents since it separates the EPS according to their sizes and molecular weight.

References

Alam, M. A., Wan, C., Guo, S.-L., Zhao, X.-Q., Huang, Z.-Y., Yang, Y.-L., Chang, J.-S., & Bai, F.-W. (2014). Characterization of the flocculating agent from the spontaneously flocculating microalga Chlorella vulgaris JSC-7. *Journal of Bioscience and Bioengineering, 118*(1), 29–33. https://doi.org/10.1016/j.jbiosc.2013.12.021

Al-Nabulsi, A. A., Jaradat, Z. W., Al Qudsi, F. R., Elsalem, L., Osaili, T. M., Olaimat, A. N., Esposito, G., Liu, S.-Q., & Ayyash, M. M. (2022). Characterization and bioactive properties of exopolysaccharides produced by Streptococcus thermophilus and Lactobacillus bulgaricus isolated from labaneh. *LWT, 167*, 113817. https://doi.org/10.1016/j.lwt.2022.113817

Angelin, J., & Kavitha, M. (2020). Exopolysaccharides from probiotic bacteria and their health potential. *International Journal of Biological Macromolecules, 162*, 853–865. https://doi.org/10.1016/j.ijbiomac.2020.06.190

Angelov, A., Georgieva, A., Petkova, M., Bartkiene, E., Rocha, J. M., Ognyanov, M., & Gotcheva, V. (2023). On the molecular selection of exopolysaccharide-producing lactic acid bacteria

from indigenous fermented plant-based foods and further fine chemical characterization. *Food, 12*(18), 3346. https://doi.org/10.3390/foods12183346
Asianezhad, A., Bari, M. R., & Amiri, S. (2023). Bio-producing and characterizing biochemical and physicochemical properties of a novel antioxidant exopolysaccharide by Bacillus coagulans IBRC-M 10807. *Journal of Polymers and the Environment, 31*(10), 4338–4352. https://doi.org/10.1007/s10924-023-02892-0
Bedade, D., & Pawar, S. (2023). Downstream processing of biotechnology products. In *Basic biotechniques for bioprocess and bioentrepreneurship* (pp. 377–390). Elsevier. https://doi.org/10.1016/B978-0-12-816109-8.00026-X
Bramhachari, P. V., Kishor, P. B. K., Ramadevi, R., Kumar, R., Rao, B. R., & Dubey, S. K. (2007). Isolation and characterization of mucous exopolysaccharide (EPS) produced by Vibrio furnissii strain VB0S3. *Journal of Microbiology and Biotechnology, 17*(1), 44–51.
Casillo, A., Lanzetta, R., Parrilli, M., & Corsaro, M. M. (2018). Exopolysaccharides from marine and marine Extremophilic bacteria: Structures, properties, ecological roles and applications. *Marine Drugs, 16*(2), 69. https://doi.org/10.3390/md16020069
Choi, I. S., Ko, S. H., Lee, M. E., Kim, H. M., Yang, J. E., Jeong, S.-G., Lee, K. H., Chang, J. Y., Kim, J.-C., & Park, H. W. (2021). Production, characterization, and antioxidant activities of an exopolysaccharide extracted from spent media wastewater after *Leuconostoc mesenteroides* WiKim32 fermentation. *ACS Omega, 6*(12), 8171–8178. https://doi.org/10.1021/acsomega.0c06095
Cirrincione, S., Breuer, Y., Mangiapane, E., Mazzoli, R., & Pessione, E. (2018). "Ropy" phenotype, exopolysaccharides and metabolism: Study on food isolated potential probiotics LAB. *Microbiological Research, 214*, 137–145. https://doi.org/10.1016/j.micres.2018.07.004
Ciszek-Lenda, M., Nowak, B., Srottek, M., Walczewska, M., Gorska-Fraczek, S., Gamian, A., & Marcinkiewicz, J. (2013). Further studies on immunomodulatory effects of exopolysaccharide isolated from Lactobacillus rhamnosus KL37C. *Central European Journal of Immunology, 3*, 289–298. https://doi.org/10.5114/ceji.2013.37743
Costa, J. A. V., Lucas, B. F., Alvarenga, A. G. P., Moreira, J. B., & de Morais, M. G. (2021). Microalgae polysaccharides: An overview of production, characterization, and potential applications. *Polysaccharides, 2*(4), 759–772. https://doi.org/10.3390/polysaccharides2040046
Cybulska, J., Halaj, M., Cepák, V., Lukavský, J., & Capek, P. (2016). Nanostructure features of microalgae biopolymer. *Starch – Stärke, 68*(7–8), 629–636. https://doi.org/10.1002/star.201500159
De Vuyst, L., De Vin, F., Vaningelgem, F., & Degeest, B. (2001). Recent developments in the biosynthesis and applications of heteropolysaccharides from lactic acid bacteria. *International Dairy Journal, 11*(9), 687–707. https://doi.org/10.1016/S0958-6946(01)00114-5
Delattre, C., Pierre, G., Laroche, C., & Michaud, P. (2016). Production, extraction and characterization of microalgal and cyanobacterial exopolysaccharides. *Biotechnology Advances, 34*(7), 1159–1179. https://doi.org/10.1016/j.biotechadv.2016.08.001
Derdak, R., Sakoui, S., Pop, O. L., Cristian Vodnar, D., Addoum, B., Elmakssoudi, A., Errachidi, F., Suharoschi, R., Soukri, A., & El Khalfi, B. (2022). Screening, optimization and characterization of exopolysaccharides produced by novel strains isolated from Moroccan raw donkey milk. *Food Chemistry: X, 14*, 100305. https://doi.org/10.1016/j.fochx.2022.100305
Du, R., Pei, F., Kang, J., Zhang, W., Wang, S., Ping, W., Ling, H., & Ge, J. (2022). Analysis of the structure and properties of dextran produced by Weissella confusa. *International Journal of Biological Macromolecules, 204*, 677–684. https://doi.org/10.1016/j.ijbiomac.2022.02.038
Duan, X., Chi, Z., Wang, L., & Wang, X. (2008). Influence of different sugars on pullulan production and activities of α-phosphoglucose mutase, UDPG-pyrophosphorylase and glucosyltransferase involved in pullulan synthesis in Aureobasidium pullulans Y68. *Carbohydrate Polymers, 73*(4), 587–593. https://doi.org/10.1016/j.carbpol.2007.12.028
Feng, F., Zhou, Q., Yang, Y., Zhao, F., Du, R., Han, Y., Xiao, H., & Zhou, Z. (2018). Characterization of highly branched dextran produced by Leuconostoc citreum B-2 from pineapple fermented product. *International Journal of Biological Macromolecules, 113*, 45–50. https://doi.org/10.1016/j.ijbiomac.2018.02.119

Feng, L., Qian, T., Yang, G., & Mu, J. (2023). Characteristics of exopolysaccharides produced by isolates from natural bioflocculant of Ruditapes philippinarum conglutination mud. *Frontiers in Microbiology, 13*, 1068922. https://doi.org/10.3389/fmicb.2022.1068922

Flemming, H.-C., & Wingender, J. (2010). The biofilm matrix. *Nature Reviews Microbiology, 8*(9), 623–633. https://doi.org/10.1038/nrmicro2415

Freitas, F., Alves, V. D., & Reis, M. A. M. (2011). Advances in bacterial exopolysaccharides: From production to biotechnological applications. *Trends in Biotechnology, 29*(8), 388–398. https://doi.org/10.1016/j.tibtech.2011.03.008

Gaignard, C., Laroche, C., Pierre, G., Dubessay, P., Delattre, C., Gardarin, C., Gourvil, P., Probert, I., Dubuffet, A., & Michaud, P. (2019). Screening of marine microalgae: Investigation of new exopolysaccharide producers. *Algal Research, 44*, 101711. https://doi.org/10.1016/j.algal.2019.101711

Geun Goo, B., Baek, G., Jin Choi, D., Il Park, Y., Synytsya, A., Bleha, R., Ho Seong, D., Lee, C.-G., & Kweon Park, J. (2013). Characterization of a renewable extracellular polysaccharide from defatted microalgae Dunaliella tertiolecta. *Bioresource Technology, 129*, 343–350. https://doi.org/10.1016/j.biortech.2012.11.077

Halaj, M., Paulovičová, E., Paulovičová, L., Jantová, S., Cepák, V., Lukavský, J., & Capek, P. (2019). Extracellular biopolymers produced by Dictyosphaerium family – chemical and immunomodulative properties. *International Journal of Biological Macromolecules, 121*, 1254–1263. https://doi.org/10.1016/j.ijbiomac.2018.10.116

Hooshdar, P., Kermanshahi, R. K., Ghadam, P., & Khosravi-Darani, K. (2020). A review on production of exopolysaccharide and biofilm in probiotics like lactobacilli and methods of analysis. *Biointerface Research in Applied Chemistry, 10*(5), 6058–6075. https://doi.org/10.33263/BRIAC105.60586075

Hou, C., Yin, M., Lan, P., Wang, H., Nie, H., & Ji, X. (2021). Recent progress in the research of Angelica sinensis (Oliv.) Diels polysaccharides: Extraction, purification, structure and bioactivities. *Chemical and Biological Technologies in Agriculture, 8*(1), 13. https://doi.org/10.1186/s40538-021-00214-x

Jiang, G., Gan, L., Li, X., He, J., Zhang, S., Chen, J., Zhang, R., Xu, Z., & Tian, Y. (2021). Characterization of structural and physicochemical properties of an exopolysaccharide produced by enterococcus sp. F2 from fermented soya beans. *Frontiers in Microbiology, 12*, 744007. https://doi.org/10.3389/fmicb.2021.744007

Jurášková, D., Ribeiro, S. C., & Silva, C. C. G. (2022). Exopolysaccharides produced by lactic acid bacteria: From biosynthesis to health-promoting properties. *Food, 11*(2), 156. https://doi.org/10.3390/foods11020156

Kaddour, H., Tranquille, M., & Okeoma, C. M. (2021). The past, the present, and the future of the size exclusion chromatography in extracellular vesicles separation. *Viruses, 13*(11), 2272. https://doi.org/10.3390/v13112272

Kang, Y. J., Kim, T. J., Kim, M. J., Yoo, J. Y., & Kim, J. H. (2023). Isolation of exopolysaccharide-producing lactic acid bacteria from pa-kimchi and characterization of exopolysaccharides. *Microbiology and Biotechnology Letters, 51*(2), 157–166. https://doi.org/10.48022/mbl.2305.05004

Kavitake, D., Devi, P. B., Delattre, C., Reddy, G. B., & Shetty, P. H. (2023). Exopolysaccharides produced by enterococcus genus — An overview. *International Journal of Biological Macromolecules, 226*, 111–120. https://doi.org/10.1016/j.ijbiomac.2022.12.042

Leroy, F., & De Vuyst, L. (2016). Advances in production and simplified methods for recovery and quantification of exopolysaccharides for applications in food and health. *Journal of Dairy Science, 99*(4), 3229–3238. https://doi.org/10.3168/jds.2015-9936

Li, W., Ji, J., Rui, X., Yu, J., Tang, W., Chen, X., Jiang, M., & Dong, M. (2014). Production of exopolysaccharides by Lactobacillus helveticus MB2-1 and its functional characteristics in vitro. *LWT – Food Science and Technology, 59*(2), 732–739. https://doi.org/10.1016/j.lwt.2014.06.063

Li, D., Li, J., Zhao, F., Wang, G., Qin, Q., & Hao, Y. (2016). The influence of fermentation condition on production and molecular mass of EPS produced by Streptococcus thermophilus 05-34 in milk-based medium. *Food Chemistry, 197*, 367–372. https://doi.org/10.1016/j.foodchem.2015.10.129

Li, Y., Xin, M., Xie, D., Fan, S., Ma, J., Liu, K., & Yu, F. (2021). Variation in extracellular polymeric substances from *Enterobacter* sp. and their Pb $^{2+}$ adsorption behaviors. *ACS Omega, 6*(14), 9617–9628. https://doi.org/10.1021/acsomega.1c00185

Li, M., Li, W., Li, D., Tian, J., Xiao, L., Kwok, L.-Y., Li, W., & Sun, Z. (2022). Structure characterization, antioxidant capacity, rheological characteristics and expression of biosynthetic genes of exopolysaccharides produced by Lactococcus lactis subsp. lactis IMAU11823. *Food Chemistry, 384*, 132566. https://doi.org/10.1016/j.foodchem.2022.132566

Lin, T. Y., & Hassid, W. Z. (1966). Pathway of algnic acid synthesis in the marine brown alga, Fucus gardneri Silva. *The Journal of Biological Chemistry, 241*(22), 5284–5297.

Liu, G., & Miao, X. (2017). Switching cultivation for enhancing biomass and lipid production with extracellular polymeric substance as co-products in Heynigia riparia SX01. *Bioresource Technology, 227*, 214–220. https://doi.org/10.1016/j.biortech.2016.12.039

Liu, Y., Luo, M., Liu, F., Feng, X., Ibrahim, S. A., Cheng, L., & Huang, W. (2020). Effects of freeze drying and hot-air drying on the physicochemical properties and bioactivities of polysaccharides from Lentinula edodes. *International Journal of Biological Macromolecules, 145*, 476–483. https://doi.org/10.1016/j.ijbiomac.2019.12.222

Lv, J., Guo, B., Feng, J., Liu, Q., Nan, F., Liu, X., & Xie, S. (2019). Integration of wastewater treatment and flocculation for harvesting biomass for lipid production by a newly isolated self-flocculating microalga Scenedesmus rubescens SX. *Journal of Cleaner Production, 240*, 118211. https://doi.org/10.1016/j.jclepro.2019.118211

Macedo, M. G., Lacroix, C., & Champagne, C. P. (2002a). Combined effects of temperature and medium composition on exopolysaccharide production by *Lactobacillus rhamnosus* RW-9595M in a whey permeate based medium. *Biotechnology Progress, 18*(2), 167–173. https://doi.org/10.1021/bp0101637

Macedo, M. G., Lacroix, C., Gardner, N. J., & Champagne, C. P. (2002b). Effect of medium supplementation on exopolysaccharide production by Lactobacillus rhamnosus RW-9595M in whey permeate. *International Dairy Journal, 12*(5), 419–426. https://doi.org/10.1016/S0958-6946(01)00173-X

Mishra, A., & Jha, B. (2009). Isolation and characterization of extracellular polymeric substances from micro-algae Dunaliella salina under salt stress. *Bioresource Technology, 100*(13), 3382–3386. https://doi.org/10.1016/j.biortech.2009.02.006

Murugu, J., & Narayanan, R. (2024). Production, purification, and characterization of a novel exopolysaccharide from probiotic Lactobacillus amylovorus: MTCC 8129. *Indian Journal of Microbiology, 64*(3), 1355–1365. https://doi.org/10.1007/s12088-024-01346-y

Nachtigall, C., Rohm, H., & Jaros, D. (2021). Degradation of exopolysaccharides from lactic acid bacteria by thermal, chemical, enzymatic and ultrasound stresses. *Foods, 10*(2), 396. https://doi.org/10.3390/foods10020396

Naveed, S., Li, C., Lu, X., Chen, S., Yin, B., Zhang, C., & Ge, Y. (2019). Microalgal extracellular polymeric substances and their interactions with metal(loid)s: A review. *Critical Reviews in Environmental Science and Technology, 49*(19), 1769–1802. https://doi.org/10.1080/10643389.2019.1583052

Nguyen, T. D. P., Vo, C. T., Nguyen-Sy, T., Tran, T. N. T., Le, T. V. A., Chiu, C.-Y., Sankaran, R., & Show, P. L. (2020). Utilization of microalgae for self-regulation of extracellular polymeric substance production. *Biochemical Engineering Journal, 159*, 107616. https://doi.org/10.1016/j.bej.2020.107616

Nicolaus, B., Kambourova, M., & Oner, E. T. (2010). Exopolysaccharides from extremophiles: From fundamentals to biotechnology. *Environmental Technology, 31*(10), 1145–1158. https://doi.org/10.1080/09593330903552094

Notararigo, S., Nácher-Vázquez, M., Ibarburu, I., Werning, M. L., de Palencia, P. F., Dueñas, M. T., Aznar, R., López, P., & Prieto, A. (2013). Comparative analysis of production and purification of homo- and hetero-polysaccharides produced by lactic acid bacteria. *Carbohydrate Polymers, 93*(1), 57–64. https://doi.org/10.1016/j.carbpol.2012.05.016

Pan, M., Zhu, L., Chen, L., Qiu, Y., & Wang, J. (2016). Detection techniques for extracellular polymeric substances in biofilms: A review. *BioResources, 11*(3), 8092–8115. https://doi.org/10.15376/biores.11.3.8092-8115

Paulsen, B. S., Aslaksen, T., Freire-Nordi, C. S., & Vieira, A. A. H. (1998). Extracellular polysaccharides from *Ankistrodesmus densus* (Chlorophyceae). *Journal of Phycology, 34*(4), 638–641. https://doi.org/10.1046/j.1529-8817.1998.340638.x

Pierre, G., Delattre, C., Dubessay, P., Jubeau, S., Vialleix, C., Cadoret, J.-P., Probert, I., & Michaud, P. (2019). What is in store for EPS microalgae in the next decade? *Molecules, 24*(23), 4296. https://doi.org/10.3390/molecules24234296

Pignolet, O., Jubeau, S., Vaca-Garcia, C., & Michaud, P. (2013). Highly valuable microalgae: Biochemical and topological aspects. *Journal of Industrial Microbiology and Biotechnology, 40*(8), 781–796. https://doi.org/10.1007/s10295-013-1281-7

Polak-Berecka, M., Waśko, A., Szwajgier, D., & Chomaz, A. (2013). Bifidogenic and antioxidant activity of exopolysaccharides produced by Lactobacillus rhamnosus E/N cultivated on different carbon sources. *Polish Journal of Microbiology, 62*(2), 181–188.

Polak-Berecka, M., Waśko, A., & Kubik-Komar, A. (2014). Optimization of culture conditions for exopolysaccharide production by a probiotic strain of Lactobacillus rhamnosus E/N. *Polish Journal of Microbiology, 63*(2), 253–257.

Polak-Berecka, M., Choma, A., Waśko, A., Górska, S., Gamian, A., & Cybulska, J. (2015). Physicochemical characterization of exopolysaccharides produced by Lactobacillus rhamnosus on various carbon sources. *Carbohydrate Polymers, 117*, 501–509. https://doi.org/10.1016/j.carbpol.2014.10.006

Pourjafar, H., Ansari, F., Sadeghi, A., Samakkhah, S. A., & Jafari, S. M. (2023). Functional and health-promoting properties of probiotics' exopolysaccharides; isolation, characterization, and applications in the food industry. *Critical Reviews in Food Science and Nutrition, 63*(26), 8194–8225. https://doi.org/10.1080/10408398.2022.2047883

Prete, R., Dell'Orco, F., Sabatini, G., Montagano, F., Battista, N., & Corsetti, A. (2024). Improving the antioxidant and anti-inflammatory activity of fermented milks with exopolysaccharides-producing Lactiplantibacillus plantarum strains. *Food, 13*(11), 1663. https://doi.org/10.3390/foods13111663

Rahnama Vosough, P., Habibi Najafi, M. B., Edalatian Dovom, M. R., Javadmanesh, A., & Mayo, B. (2021). Evaluation of antioxidant, antibacterial and cytotoxicity activities of exopolysaccharide from enterococcus strains isolated from traditional Iranian Kishk. *Journal of Food Measurement and Characterization, 15*(6), 5221–5230. https://doi.org/10.1007/s11694-021-01092-5

Ramos, I. M., Seseña, S., Poveda, J. M., & Palop, M. L. (2023). Screening of lactic acid bacteria strains to improve the properties of non-fat set yogurt by in situ EPS production. *Food and Bioprocess Technology, 16*(11), 2541–2558. https://doi.org/10.1007/s11947-023-03080-7

Rehm, B. H. A. (2010). Bacterial polymers: Biosynthesis, modifications and applications. *Nature Reviews Microbiology, 8*(8), 578–592. https://doi.org/10.1038/nrmicro2354

Rimada, P. S., & Abraham, A. G. (2003). Comparative study of different methodologies to determine the exopolysaccharide produced by kefir grains in milk and whey. *Le Lait, 83*(1), 79–87. https://doi.org/10.1051/lait:2002051

Roger, O. (2002). *Etude d'oligosaccharides bioactifs issus d'exopolysaccharides bactériens : obtention, caractérisation et relation structure/fonction* (p. 13). Université Paris.

Rosalam, S., & England, R. (2006). Review of xanthan gum production from unmodified starches by Xanthomonas campestris sp. *Enzyme Microbial Technology, 39*(2), 197–207.

Rossi, F., & De Philippis, R. (2016). Exocellular polysaccharides in microalgae and cyanobacteria: Chemical features, role and enzymes and genes involved in their biosynthesis. In *The physiology of microalgae* (pp. 565–590). Springer. https://doi.org/10.1007/978-3-319-24945-2_21

Ruas-Madiedo, P., & de los Reyes-Gavilán, C. G. (2005). Invited review: Methods for the screening, isolation, and characterization of exopolysaccharides produced by lactic acid bacteria. *Journal of Dairy Science, 88*(3), 843–856. https://doi.org/10.3168/jds.S0022-0302(05)72750-8

Shao, L., Wu, Z., Zhang, H., Chen, W., Ai, L., & Guo, B. (2014). Partial characterization and immunostimulatory activity of exopolysaccharides from Lactobacillus rhamnosus KF5. *Carbohydrate Polymers, 107*, 51–56. https://doi.org/10.1016/j.carbpol.2014.02.037

Sharma, K., Sharma, N., Handa, S., & Pathania, S. (2020). Purification and characterization of novel exopolysaccharides produced from Lactobacillus paraplantarum KM1 isolated from human milk and its cytotoxicity. *Journal of Genetic Engineering and Biotechnology, 18*(1), 56. https://doi.org/10.1186/s43141-020-00063-5

Shene, C., Canquil, N., Bravo, S., & Rubilar, M. (2008). Production of the exopolysacchzrides by Streptococcus thermophilus: Effect of growth conditions on fermentation kinetics and intrinsic viscosity. *International Journal of Food Microbiology, 124*(3), 279–284. https://doi.org/10.1016/j.ijfoodmicro.2008.03.013

Sheng, S., Fu, Y., Pan, N., Zhang, H., Xiu, L., Liang, Y., Liu, Y., Liu, B., Ma, C., Du, R., & Wang, X. (2022). Novel exopolysaccharide derived from probiotic Lactobacillus pantheris TCP102 strain with immune-enhancing and anticancer activities. *Frontiers in Microbiology, 13*, 1015270. https://doi.org/10.3389/fmicb.2022.1015270

Shi, Y., Huang, J., Zeng, G., Gu, Y., Chen, Y., Hu, Y., Tang, B., Zhou, J., Yang, Y., & Shi, L. (2017a). Exploiting extracellular polymeric substances (EPS) controlling strategies for performance enhancement of biological wastewater treatments: An overview. *Chemosphere, 180*, 396–411. https://doi.org/10.1016/j.chemosphere.2017.04.042

Shi, N., Li, N., Duan, X., & Niu, H. (2017b). Interaction between the gut microbiome and mucosal immune system. *Military Medical Research, 4*(1), 14. https://doi.org/10.1186/s40779-017-0122-9

Shoshaa, R., Ashfaq, M. Y., & Al-Ghouti, M. A. (2023). Recent developments in ultrafiltration membrane technology for the removal of potentially toxic elements, and enhanced antifouling performance: A review. *Environmental Technology & Innovation, 31*, 103162. https://doi.org/10.1016/j.eti.2023.103162

Shukla, A., Mehta, K., Parmar, J., Pandya, J., & Saraf, M. (2019). Depicting the exemplary knowledge of microbial exopolysaccharides in a nutshell. *European Polymer Journal, 119*, 298–310. https://doi.org/10.1016/j.eurpolymj.2019.07.044

Simova, E. D., Frengova, G. I., & Beshkova, D. M. (2004). Exopolysaccharides produced by mixed culture of yeast Rhodotorula rubra GED10 and yogurt bacteria (Streptococcus thermophilus 13a + Lactobacillus bulgaricus 2-11). *Journal of Applied Microbiology, 97*(3), 512–519. https://doi.org/10.1111/j.1365-2672.2004.02316.x

Sørensen, H. M., Rochfort, K. D., Maye, S., MacLeod, G., Brabazon, D., Loscher, C., & Freeland, B. (2022). Exopolysaccharides of lactic acid bacteria: Production, purification and health benefits towards functional food. *Nutrients, 14*(14), 2938. https://doi.org/10.3390/nu14142938

Sutherland, I. W. (1972). Bacterial exopolysaccharides. *Advances in Microbial Physiology, 8*, 143–213. https://doi.org/10.1016/S0065-2911(08)60190-3

Talbi, C., Elmarrakchy, S., Youssfi, M., Bouzroud, S., Belfquih, M., Sifou, A., Bouhaddou, N., Badaoui, B., Balahbib, A., Bouyahya, A., & Bourais, I. (2023). Bacterial exopolysaccharides: From production to functional features. *Progress In Microbes & Molecular Biology, 6*(1). https://doi.org/10.36877/pmmb.a0000384

Tiwari, S., Kavitake, D., Devi, P. B., Baria, B., Agarwal, K., Ravi, R., Singh, A. K., & Shetty, P. H. (2024). Functional enhancement of yoghurt through incorporation of glucan exopolysaccharide from enterococcus hirae OL616073 of food origin. *Journal of Food Measurement and Characterization, 18*(7), 5462–5476. https://doi.org/10.1007/s11694-024-02580-0

Tuinier, R., Zoon, P., Olieman, C., Stuart, M. A. C., Fleer, G. J., & De Kruif, C. G. (1999). Isolation and physical characterization of an exocellular polysaccharide. *Biopolymers, 49*, 1–9.

Villacorte, L. O., Ekowati, Y., Neu, T. R., Kleijn, J. M., Winters, H., Amy, G., Schippers, J. C., & Kennedy, M. D. (2015). Characterisation of algal organic matter produced by bloom-

forming marine and freshwater algae. *Water Research, 73*, 216–230. https://doi.org/10.1016/j.watres.2015.01.028
Wang, J., Zhao, X., Tian, Z., He, C., Yang, Y., & Yang, Z. (2015). Isolation and characterization of exopolysaccharide-producing Lactobacillus plantarum SKT109 from Tibet Kefir. *Polish Journal Of Food And Nutrition Sciences, 65*(4), 269–279. https://doi.org/10.1515/pjfns-2015-0023
Wang, Q., Pang, W., Mao, Y., Ge, S., Yu, H., Dai, C., & Zhao, M. (2019). Changes of extracellular polymeric substance (EPS) during Microcystis Aeruginosa blooms at different levels of nutrients in a eutrophic microcosmic simulation device. *Polish Journal of Environmental Studies, 29*(1), 349–360. https://doi.org/10.15244/pjoes/102367
Xiao, R., & Zheng, Y. (2016). Overview of microalgal extracellular polymeric substances (EPS) and their applications. *Biotechnology Advances, 34*(7), 1225–1244. https://doi.org/10.1016/j.biotechadv.2016.08.004
Yu, Y.-J., Chen, Z., Chen, P. T., & Ng, I.-S. (2018). Production, characterization and antibacterial activity of exopolysaccharide from a newly isolated Weissella cibaria under sucrose effect. *Journal of Bioscience and Bioengineering, 126*(6), 769–777. https://doi.org/10.1016/j.jbiosc.2018.05.028
Yu, L., Ye, G., Qi, X., Yang, Y., Zhou, B., Zhang, Y., Du, R., Ge, J., & Ping, W. (2023). Purification, characterization and probiotic proliferation effect of exopolysaccharides produced by Lactiplantibacillus plantarum HDC-01 isolated from sauerkraut. *Frontiers in Microbiology, 14*, 1210302. https://doi.org/10.3389/fmicb.2023.1210302
Zaghloul, E. H., Ibrahim, M. I. A., & Zaghloul, H. A. H. (2023). Antibacterial activity of exopolysaccharide produced by bee gut-resident enterococcus sp. BE11 against marine fish pathogens. *BMC Microbiology, 23*(1), 231. https://doi.org/10.1186/s12866-023-02977-9
Zhan, K., Ji, X., & Luo, L. (2023). Recent progress in research on Momordica charantia polysaccharides: Extraction, purification, structural characteristics and bioactivities. *Chemical and Biological Technologies in Agriculture, 10*(1), 58. https://doi.org/10.1186/s40538-023-00433-4
Zhang, Z., Zhou, Z., Li, Y., Zhou, L., Ding, Q., & Xu, L. (2016). Isolated exopolysaccharides from Lactobacillus rhamnosus GG alleviated adipogenesis mediated by TLR2 in mice. *Scientific Reports, 6*(1), 36083. https://doi.org/10.1038/srep36083
Zhang, J., Li, G., Yuan, X., Li, P., Yu, Y., Yang, W., & Zhao, S. (2023). Reduction of ultrafiltration membrane fouling by the pretreatment removal of emerging pollutants: A review. *Membranes, 13*(1), 77. https://doi.org/10.3390/membranes13010077
Zhang, K., Liu, S., Liang, S., Xiang, F., Wang, X., Lian, H., Li, B., & Liu, F. (2024). Exopolysaccharides of lactic acid bacteria: Structure, biological activity, structure-activity relationship, and application in the food industry: A review. *International Journal of Biological Macromolecules, 257*, 128733. https://doi.org/10.1016/j.ijbiomac.2023.128733
Zhu, C., Chen, C., Zhao, L., Zhang, Y., Yang, J., Song, L., & Yang, S. (2012). Bioflocculant produced by Chlamydomonas reinhardtii. *Journal of Applied Phycology, 24*(5), 1245–1251. https://doi.org/10.1007/s10811-011-9769-x
Ziadi, M., Bouzaiene, T., M'Hir, S., Zaafouri, K., Mokhtar, F., Hamdi, M., & Boisset-Helbert, C. (2018). Evaluation of the efficiency of ethanol precipitation and ultrafiltration on the purification and characteristics of exopolysaccharides produced by three lactic acid bacteria. *BioMed Research International, 2018*, 1–11. https://doi.org/10.1155/2018/1896240
Xiao, L., Li, Y., Tian, J., Zhou, J., Xu, Q., Feng, L., ... & Li, W. (2020). Influences of drying methods on the structural, physicochemical and antioxidant properties of exopolysaccharide from Lactobacillus helveticus MB2-1. *International Journal of Biological Macromolecules, 157*, 220–231, https://doi.org/10.1016/j.ijbiomac.2020.04.196
Yi, Y., Xu, W., Wang, H. X., Huang, F., & Wang, L. M. (2020). Natural polysaccharides experience physiochemical and functional changes during preparation: A review. *Carbohydrate Polymers, 234*, 115896. https://doi.org/10.1016/j.carbpol.2020.115896

EPS as Natural Preservatives

Rishav Jana, Sayoni Sutradhar, Satabdi Mallick, and Amit Ghosh

1 Introduction

Exopolysaccharide (EPS) are mostly composed of lipids, proteins, polysaccharides, and nucleic acids. Most significantly, EPS maintain their stable matrix structure and create a three-dimensional network of polymers that facilitate cell interaction and surface adherence. EPS also serve as carbon sinks and sources of extracellular energy. Additionally, they provide a wealth of structurally and compositionally diverse biopolymers with distinct bioactivities for high-value uses, such as antiviral, anticancer, antioxidant, anticoagulant, and anti-inflammatory properties. EPS-producing microorganisms may be isolated from a range of natural sources, such as soil, water, and biofilm, in both aquatic and terrestrial environments. The most widely available EPSs on the market include cellulose, hyaluronic acid, gellen, alginate, and xanthan. EPS is a kind of long-chain biopolymer composed of non-carbohydrate components that are discharged into the environment or attached to the outside of cells, as well as branching, repeating units of sugars (Akpeji et al., 2023). EPS are slimy substances that are classified as either cell-bound or released depending on where they are found (Freitas et al., 2011). Cell-bound EPS are those that are attached to the cell surface. In recent years, EPS has become more widely used in industrial product compositions due to its ability to improve solution viscosity or induce the formation of gels (Ribeiro & Burkert, 2016).

The use of EPS as natural food preservatives to preserve perishable foods has been the subject of research in recent decades due to the increased demand for these

Authors Rishav Jana and Sayoni Sutradhar have been equally contributed to this book

R. Jana · S. Sutradhar · S. Mallick · A. Ghosh (✉)
Department of Biological Sciences, School of Life Science and Biotechnology, Adamas University, Kolkata, India

T. Sarkar et al. (eds.), *Extracellular Polysaccharides*,
https://doi.org/10.1007/978-3-031-94472-7_4

products (Bouarab Chibane et al., 2019). Because consumers, the food business, and food safety authorities are concerned about food-borne illnesses. EPS can inhibit the growth of germs and fungi to improve quality and prolong shelf life. Food safety is primarily concerned with food-borne pathogens, including *Escherichia coli* 0157: H7, *Clostridium botulinum*, *Staphylococcus aureus*, *Campylobacter jejuni*, *Bacillus cereus*, *Listeria monocytogenes*, and *Cryptosporidium hominis* (Bergwerff & Debast, 2021). The desire to live a healthy lifestyle and the requirement for easy, prepared for cooking or consumption foods have led to growth in the consumption of fresh foods, including meat, fish, and horticulture goods. Because fruits, vegetables, seafood, and meat often have a short shelf life, proper preservation methods are needed to increase their shelf lives (Babuskin et al., 2014). Even in affluent nations, outbreaks of foodborne illnesses are increasing as fresh foods like minimally processed fruits and vegetables replace the problems caused by animal meals. Food-borne diseases have risen in tandem with the expansion of the fresh-cut quick salad business; precutting the salad leaves releases nutrients that promote bacteria growth. The package's altered environment increases the pathogenicity of pathogens such as *E. coli* 0157: H7 (Sun et al., 2022) while decreasing spoiling by aerobes. There is ongoing market pressure to use synthetic chemicals as preservatives because of the short shelf life of these fresh products and their association with outbreaks of food-borne illnesses. To stop food from decaying due to lipid oxidation and microorganisms, chemical additions are being called for as preservatives. People are growing more conscious of the limitations of artificial antioxidants and antibacterial agents, particularly in relation to their impacts on human well-being. This is due to a rise in problems including teratogenicity, carcinogenicity, and hepatic, kidney, cardiac, respiratory, or neurological issues, as well as other conditions brought on by artificial ingredients. In this scenario, consumers are calling for partial or whole elimination of chemically produced preservatives, which are used to prolong the shelf life of food (Ahmad et al., 2015). Applying EPS as a food preservative is quite safe.

The existence of distinct functional groups in EPS makes them appropriate for the creation of a range of bionanostructures. Specifically, the use of nanotechnology in the food industry has produced polysaccharide-based nanostructures, which have produced superb outcomes in terms of extending the shelf life and quality of food in addition to improved food protection against the effects of the environment, including dust, gas emissions, heat, light, oxygen, and enzymes. Furthermore, the latest applications for the creation of eco-friendly packaging systems also include the use of hemicelluloses, marine polysaccharides, and bacterial exopolysaccharides (Bilal et al., 2021; Nešić et al., 2019).

To enhance future research and address the issue of food and nutrition insecurity, it is imperative to promote and comprehend the current state of affairs about the possibilities of natural preservation to enhance food safety and shelf life of fresh, minimally processed, and ready-to-eat foods (Aziz & Karboune, 2016). However, detailed studies on the benefits of natural preservation, particularly the antimicrobial and antioxidant substances found in EPS, the application of EPS edible coatings, and the factors affecting the antimicrobial abilities of EPS to improve food safety and extend shelf life are limited (Carocho et al., 2014). Therefore, the objective of this review was to explore the capacity and role of EPS as natural preservatives to enhance food safety and extend shelf life.

2 EPS: Types and Sources

Exopolysaccharides are extracellular polymeric substances with sugar residues and are synthesized and released into the extracellular medium by microorganisms. These biopolymers vary widely in their structure, composition, and function, depending on the producing organism and the environmental conditions (Korcz & Varga, 2021). Exopolysaccharides (EPS) can be broadly classified based on their monomer variety into two categories: Exopolysaccharides (EPS) can be broadly classified into two categories based on the composition of their repeating units: homopolysaccharides (HoPS) and heteropolysaccharides (HePS). HoPS are composed of identical monosaccharides, while HePS consist of varying monosaccharides, typically ranging from 2 to 8 different units (Korcz & Varga, 2021). HoPS and HePS are structurally different in their constituent subunits, as well as in their branching, structure, sugar connections, and chain length. HoPS generally consist of fructose or glucose residues, hence glucans and fructans, respectively. The most common α-glucans are categorized into four subgroups: first dereans, second dereans, mutans, reuterans, and alternans. Fructans are generally categorized into levan-type and inulin-type polysaccharides. Multipotent microorganisms commonly found in complex environments synthesize the EPS. Yeasts of the genera *Lactobacillus, Streptococcus, and Leuconostoc* are known to synthesize EPSs, including species such as dextran and levan. Other known producers of EPS are fungi and algae, with pullulan from *Aureobasidium* pullulans and alginate from brown algae being well-known examples.

2.1 *Bacteria*

Various species of lactic acid bacteria (LAB) synthesize exopolysaccharides (EPS), mainly HePS, with diverse sugar compositions, charges, chain lengths, and branching degrees (Korcz & Varga, 2021). These HePS usually contain repeating units of D-glucose, D-galactose, and L-rhamnose, and sometimes include N-acetylglucosamine, N-acetylgalactosamine, or glucuronic acid (Torino et al., 2015). Additionally, HePS are often modified through acetylation, pyruvylation, and phosphorylation). For instance, EPS produced by *Lactococcus lactis* subsp. cremoris contain rhamnose, galactose, and glucose (Looijesteijn et al., 1999). *Streptococcus thermophilus* strains primarily produce EPS composed of glucose and galactose, with smaller amounts of rhamnose and N-acetylgalactosamine (Pachekrepapol et al., 2017). In a chemically defined medium, *Lactobacillus delbrueckii sub*sp. *Bulgaricus subsp* synthesizes EPS that include galactose and glucose, and *Leuconostoc citreum* strain B-2 synthesizes EPS with specific α-(1 → 6) and α-(1 → 3) linkage, whereas *Lactobacillus reuteri E81* produces a thermally stable α-glucan homopolysaccharide with moisture-absorbing properties, which makes it suitable for applications in the food industry (Lin & Chien, 2007). In addition, *Leuconostoc mesenteroides* NRRL B-512 F synthesizes EPS incorporated with α-(1 → 6), α-(1 → 3), and α-(1 → 4) linkages. Due to this structural diversity, EPS obtained from LAB are useful in many industries (Table 1).

2.2 Algae

The extracellular polymeric substances EPS produced by Chlorophyta species such as *Chlorella sp., Dunaliella salina, Ankistrodesmus angustus*, and *Botryococcus braunii* contain various exopolysaccharides. These exopolysaccharides contain monosaccharides such as hexose, pentose, uronic acid, methyl pentose, methyl

Table 1 Microorganisms such as bacteria (e.g., Lactococcus lactis and Xanthomonas campestris), algae (e.g., Dunaliella salina), diatoms (e.g., Fragilaria), and fungi (e.g., Aureobasidium pullulans) produce a type of sugar called exopolysaccharides (EPS). These sugars are made up of components such as rhamnose, glucose, and alginate. The EPS have useful properties that make them good for food and industry

Microorganisms	Species	EPS details (with reference)
Bacteria	1. *Lactococcus lactis* 2. *Streptococcus thermophilus* 3. *Lactobacillus delbrueckii* subsp. 4. *Xanthomonas campestris* 5. *P. aeruginosa* 6. *Serratia Levanicum*	1. Rhamnose, galactose, and glucose (Looijesteijn et al., 1999) 2. Glucose, galactose, rhamnose, and N-acetylgalactosamine (Pachekrepapol et al., 2017). 3. Glucose, galactose, and rhamnose (Lin & Chien, 2007) 4.Xanthanv (Angelin & Kavitha, 2020) 5. Alginate (Wu et al., 2021b) 6. Fructan (Mensink et al., 2015)
Algae	1. *Dunaliella salina* 2. *Botryococcus braunii* 3. *P. purpureum* 4. *Chlorella sp*	1. Hexose, pentose, and uronic acid (Moore & Tischer, 1964) 2.Glucose, galactose, and arabinose (Moore & Tischer, 1964) 3.Xylose, mannose, and galacturonic acid (Geresh et al., 1992) 4. Arabinose and Fructose (Yalcin et al., 1994)
Diatoms	1. *Fragilaria* 2. *Achnanthes* 3. *Cymbella* 4. *Cylindrotheca fusiformis*	1. Glucose and galactose 2. Mannose 3. Glucose and mannose (Bohórquez et al., 2017) 4. Glucose, Xylose, and Rhamnose (Magaletti et al., 2004)
Fungi	1. *Aureobasidium pullulans* 2. *Sclerotium rolfsii* 3. Mushroom 4. *Sclerotium rolfsii* 5. *Tremella mesenterica*	1. Pullulan 2. Scleroglucan 3. Alpha-glucans (Stoica et al., 2023) 4. Scleroglucan (Kornmann et al., 2003) 5. Pullulan (Sugumaran & Ponnusami, 2017)

hexose, and sulfate, with glucose, galactose, arabinose, fucose, xylose, mannose, glucuronic acid, and galacturonic acid being the most common sugars (Moore & Tischer, 1964). Most of the research related to red microalgae is centered on the *Porphyridium* genus, particularly *Porphyridium purpureum*, *Porphyridium cruentum*, and *Porphyridium aerugineum*, while few researches include *Rhodella* and *Rhodosorus*. These EPS have been shown to have pharmaceutical (e.g., antiviral, antibacterial, antitumor, antioxidant, and anti-inflammatory) and industrial (e.g., drag reducers and biolubricants) applications (Geresh et al., 1992) (Table 1).

2.3 Diatoms

Diatoms are single-cell microalgae, some of which secrete extracellular polymeric substances known as exopolysaccharides during their metabolism. These polysaccharides are secreted into their surrounding environment and play a role in biofilm formation and particle aggregation, as well as carbon flux and motility, colony formation, and desiccate protection. The biofilm provides stabilization and protection against environmental stress. EPS produced by diatoms play a crucial role in nutrient capture, cell adhesion, and ecosystem stability and are of interest for their potential applications in biotechnology and environmental sciences (Table 1).

2.4 Fungi

Various fungi, including pullulan from *Aureobasidium pullulans*, scleroglucan from *Sclerotium rolfsii*, and β-glucans from different mushroom species, create EPS with distinct characteristics. These fungal EPS have many applications in food preservation, as film formers, as antioxidant, and as antibacterial agents. They find application in the synthesis of green preservatives and food packing materials because of their inherent bioactivity and tenability (Stoica et al., 2023) (Table 1).

3 Properties of EPS

EPS demonstrate considerable antibacterial action and provide efficient means of controlling the formation of pathogenic microorganisms. These polysaccharides exert the efficacy to discharge toxicity on bacterial cell walls, inhibit biofilm formation, and prevent microorganisms from settling on surfaces. Diatoms, fungi, bacteria, and algae synthesize EPS. EPS have a broad spectrum in food preservation and medicinal applications and present a viable alternative to synthetic antibacterial agents because of their natural origin and capacity to improve food safety and prolong shelf life (Badel et al., 2011).

3.1 Mode of Action of EPS as Preservative

Exopolysaccharides also make good natural preservatives because they possess a high degree of antecedent antibiotic activity. The chief way most EPS provide antibacterial functions is by their ability to provide excellent chemical and physical barriers to the growth of pathogenic and spoilage microorganisms. On the surface of the food, EPS can develop a chemical layer that acts as a layer of protection and prevents microbes from getting access to nutrients and other vital ingredients in the food. Such layers formed by microbial cells interfere with the normal working of microbial cells. Some EPS can exert direct toxicity of microbial cell membranes, whereby they can cause cell breakage and death, for example. Others seem to arrest enzyme processes in microbial cells, where the effects prevent them from multiplying and extending. Depending on the kind of EPS and the target microorganism, the precise method of antibacterial activity varies, but overall, there is a decrease in microbial contamination, which increases the shelf life of food products (Badel et al., 2011).

Lactic acid bacteria (LAB) that produce exopolysaccharides (EPS) have garnered significant attention, not only because of their various techno-functional applications but also due to their classification as GRAS (generally recognized as safe) (Ruas-Madiedo et al., 2002). Notably, several EPS extracted from LAB have been shown to possess an appreciable degree of antimicrobial activity against many Gram-positive and Gram-negative bacteria. How these EPS demonstrate their antibacterial action is under antagonistic capacity. This antagonism is believed to impair pathogen cell division by disrupting the cell wall and degrading DNA (M.-H. Wu et al., 2010).

As shown in Table 1, a growing number of papers have investigated the antibacterial activity of EPS from LAB throughout the last 10 years. Dextran EPS, a homopolysaccharide produced by *Weissella cibaria*, had an inhibitory effect against Escherichia coli, *Staphylococcus aureus,* and *Bacillus subtilis*. Additionally, two types of EPS from *Limosilactobacillus reuteri and Limosilactobacillus vaginalis* originating from the hen gut cecum had enhanced antibacterial activity against *E. coli and Salmonella typhimurium* (Rajoka et al., 2019; Yu et al., 2018).

EPS derived from LAB species *Bifidobacterium bifidum* and *Lactiplantibacillus plantarum* had antibacterial properties at 300 μg/mL against pathogens such as *Cronobacter sakazakii, E.coli,* and *Listeria monocytogenes* (Li et al., 2014). Similarly, hetero-polysaccharide EPS composed of glucose, mannose, galactose, rhamnose, and fucose, extracted from *Lactobacillus gasseri*, showed good antibacterial activity against various foodborne pathogens, such as *E. coli*, *L. monocytogenes*, *S. aureus*, *and Enterococcus faecalis,* as assessed by agar well diffusion (Rani et al., 2018).

Some of the triggers that determine the bactericidal effectiveness of the EPS include the spectra of the microorganisms from which the EPS is derived, the type of pathogen involved, and properties of the EPS, such as charge and concentration. For instance, research done in vivo proves that these variables influence antibacterial results. Table 2 also presents other information on the antibacterial properties of different LAB EPS against both pathogenic and nonpathogenic microorganisms.

Researchers found that a mixture of *S. thermophilus and L. bulgaricus* (isolated from yogurt) exhibited strong antibacterial properties averse to *Staphylococcus aureus and Escherichia coli.* However, it is more effective when EPS is concentrated because it can increase the overall effectiveness of yogurt. Reduced levels of EPS are unlikely to be bactericidal; instead, they are bacteriostatic, although higher levels are bacteriolytic. However, a higher 1% or 2.5% concentration effectively killed those bacteria, thus a bactericidal effect. The antibacterial capability of EPS is also dependent on the physicochemical properties of EPS and the type of pathogenic microorganisms. EPS with negative charges more effectively killed Gram-positive bacteria than Gram-negative bacteria (Nehal et al., 2019).

3.2 *Microbial Biofilms Disruption by EPS*

The capacity of EPS to break up microbial biofilms is one of its most potent antibacterial qualities. Encased in a self-produced EPS matrix, biofilms are intricate populations of bacteria that offer protection against many environmental stressors, such as antimicrobial chemicals. Biofilms on food items and in food processing facilities can cause deterioration and ongoing contamination. It has been demonstrated that EPS from specific microbes may demolish these biofilms either by generating enzymes that attack the structure of the biofilm or by competing with the native EPS inside the biofilm matrix. In addition to lowering the microbial burden on food goods, EPS also improves the efficiency of other preservation techniques such as pasteurization and refrigeration by upsetting biofilms (Sarikaya et al., 2017).

Table 2 Antibacterial activity of EPS from different bacteria

Bacterial strains	Source	Activity against
Lactobacillus plantarum (EPLB)	Ricotta cheese	*S. aureus* and *S. typhimurium* (Mahdhi et al., 2017)
Lactobacillus kefiranofaciens DN1	Kefir	*Listeria monocytogenes* and *Salmonella Enteritidis* (Jeong et al., 2017)
Streptococcus thermophilus GST-6	Skim milk	*Escherichia coli* and *Salmonella typhimurium* (Zhang et al., 2013)
Bifidobacterium longum BCRC 14634	Infant's intestine	*E. coli BCRC10239* and *Salmonella typhimurium* (M.-H. Wu et al., 2010)
Lactobacillus rhamnosus SHA111	Human breast milk	*E. coli* and *Salmonella typhimurium*(Rajoka et al., 2022)
Streptococcus thermophilus ASCC 1275	Milk	*Listeria monocytogenes CMCC* (Li et al., 2014)
Weissella cibaria 27	Kimchi	*E. coli BL21*, *B. subtilis*, and *S. aureus* (Yu et al., 2018)

Studies have shown that biofilms formed by lactic acid bacteria have antimicrobial effects on pathogens (Guerrieri et al., 2009; Speranza et al., 2009). Additionally, exopolysaccharides (EPS) play several important biological roles. They protect microorganisms from phagocytosis, bacteriophage attacks, antibiotics, dehydration, and toxic substances. For instance, *Lactococcus lactis* cells within EPS are resistant to bacteriophage attacks. Biofilms with antimicrobial properties formed by lactic acid bacteria are not only used as preservatives against pathogens and spoilage microorganisms but also against biofilms formed by these microorganisms. For example, released-exopolysaccharides (r-EPS) can prevent biofilm formation by a wide range of Gram-negative and Gram-positive pathogens (Kim & Kim, 2009). Studies on antipathogenic properties of extracellular polymeric substances have shown that a concentration of 1.0 mg/mL of r-EPS significantly reduced enterohemorrhagic *Escherichia coli* (EHEC) biofilms. r-EPS extracted from *Lactobacillus acidophilus may partially alter the surface characteristics of bacteria, preventing EHEC O157:H7 from adhering and auto-aggregating during biofilm formation.* After that, r-EPS from *L. acidophilus* could be used as food-grade supplements for microbial biofilm management in a variety of food businesses.

Additionally, another study found that EPS from *B. longum* prevented the growth of pathogenic or spoilage microorganisms.

Microbial biofilms are protective bacterial colonies that cling to surfaces, and exopolysaccharides (EPS) are essential in breaking these communities. Antibiotic-resistant biofilms are a common cause of chronic infections. EPS can prevent the production of biofilms in a number of ways. By dissolving the extracellular polymeric materials that keep the biofilm together, they disrupt the structural integrity of the biofilm matrix (Vu et al., 2009). Additionally, EPS stop bacteria from adhering to surfaces in the first place, preventing the formation of biofilms. EPS sometimes can increase the penetration of antimicrobial agents into biofilms, weakening the barrier and increasing the susceptibility of bacteria. EPS are useful in industrial and medical applications, including infection control and sanitation, because of its capacity to destroy biofilms. Different modes of disruptions of biofilm are depicted in Fig. 1. EPS plays an integral role in the formation of biofilms through several mechanisms: electrostatic repulsion, destabilization of the biofilm matrix by interacting with key functional groups, and block biofilm formation by hampering quorum sensing by small amounts of EPS to prevent microbial adhesion. EPS also promote cell detachment, alter bacterial surface properties to prevent aggregation, and mechanically weaken biofilms. Additionally, EPS-associated enzymes can degrade the biofilm matrix, which helps in biofilm disruption.

4 Characteristics of EPS as Natural Preservatives

Natural preservatives are substances derived from natural sources that are used to extend the shelf life of food products by preventing spoilage and microbial growth (Ben Akacha et al., 2023). According to recent studies, using natural antimicrobial

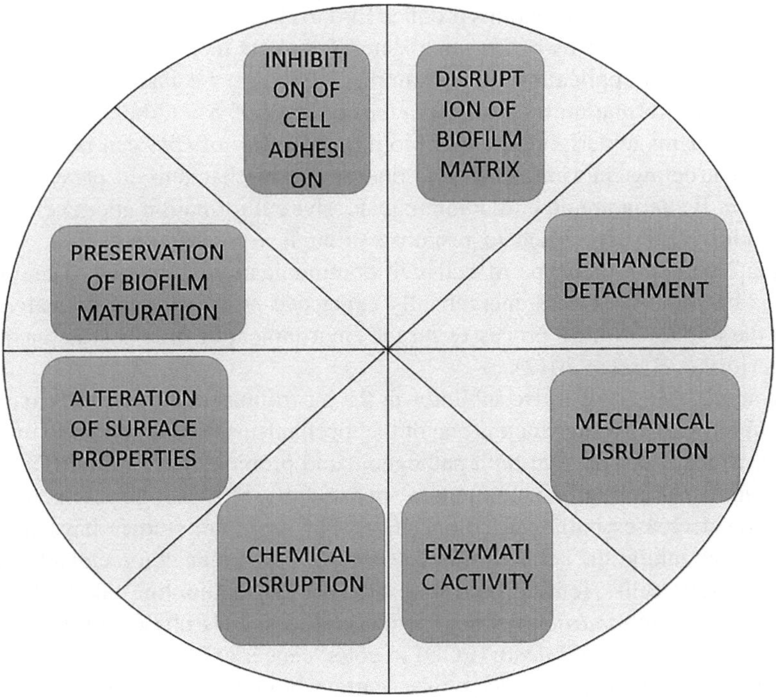

Fig. 1 Various methods for breaking down biofilms, showing mechanical, chemical, enzymatic, and biological techniques. Each method focuses on either the biofilm's structure or the microbes' activity, emphasizing different ways to stop or remove biofilms

agents can improve palatability expand the antibacterial spectrum, and minimize antimicrobial resistance. Among the bioactive substances that are used as biopreservatives, a majority of it is extracted from EPS of bacteria. Microbial exopolysaccharides are utilized as gelling agents, lubricants, thickeners, binders, coagulants, emulsifiers, film formers, and stabilizers based on their rheological properties. EPS exhibit several mechanisms that contribute to their preservative properties, which are as follows.

4.1 Antimicrobial Activity

EPS contain various antibacterial qualities. These features work by either directly preventing the growth of spoilage organisms or by interfering with their biological functions. For example, some EPS may obstruct the development of biofilms, which are bacterial protective sheaths that increase the bacteria's resilience in comparison to traditional preservatives. EPS is a useful tool for reducing microbial corruption in food systems because it inhibits the production of biofilms.

Different *Lactobacillus* strains produce EPS to show significant potential in stopping biofilm formation by pathogenic bacteria, making them useful for use in food safety and health applications. For example, EPS from *Lactobacillus casei* NA-2 inhibit biofilm formation by *Bacillus cereus* by over 95.5% (Abdalla et al., 2021). The mechanisms underlying the anti-biofilm properties of EPS can be of different types. Obstructing quorum sensing is one of the mechanisms to prevent biofilm formation. Bacteria are able to communicate, share information about cell density, and modify gene expression in response through a process known as "Quorum Sensing," which is one type of cell–cell communication technique. This process enables bacteria to express energetically expensive processes as a collective only when the impact of those processes on the environment or on a host is maximized (Rutherford & Bassler, 2012).

By acting as a competitive inhibitor in the gastrointestinal tract, EPS may exert its antimicrobial effects, which is one of the mechanisms. EPS can bind to intestinal mucus, which helps prevent both pathogenic and probiotic bacteria from attaching to the intestinal epithelium. Immune response enhanced through this competitive inhibition decreases pathogen colonization. For example, studies have indicated that EPS can inhibit the adherence of *E. coli* O157 to human colorectal adenocarcinoma (HT-29) cells. Tetrandrine (Tet), a bisbenzylisoquinoline alkaloid isolated from *Stephania tetrandra*, arrests cancerous cells at the G1 phase and subsequently induces programmed cell death (PCD) in colon cancer cells via the caspase family and mitochondrial pathways. Moreover, certain biofilms formed by Lactobacillus strains on abiotic surfaces, such as polystyrene or glass, contribute to their stability in the gastrointestinal tract. The ability of probiotic LAB to form biofilms leads to the inhibition of pathogenic bacteria, allowing them to occupy ecological niches and compete effectively for space and nutrients.

Another main characteristic of EPS is their functionality, which is determined by their structural composition as well. For instance, a branched polysaccharide, kefiran, is known to restrict bacterial toxins from entering the intestinal cell surface receptors. On the other hand, several polysaccharides, such as dextran, do not show this protection. In conclusion, EPS synthesized from Lactobacillus strain systems reduce biofilm-forming bacterial infections in the food industry and improve host immunity. Their unique ability to stop pathogenic biofilm formation, interfere with bacterial communication, and increase colonization resistance makes them valuable food-grade additives in industries.

4.2 Antioxidant Properties

Food deterioration is mostly caused by oxidation, which breaks down proteins, lipids, and other delicate ingredients. By scavenging free radicals, EPS can function as antioxidants, postponing the oxidative degradation of food goods. This characteristic is especially useful for maintaining the nutritional value of fatty foods like oils, nuts, and meat products. Food products are often incorporated with synthetic preservatives to avoid oxidation, which leads to major food spoilage. By preventing oxidation reactions, synthetic antioxidants help preserve the standard of foodstuff. These consist of propyl gallate, butylated hydroxyanisole (BHA), tert-butylhydroquinone (TBHQ), butylated hydroxytoluene (BHT), sodium erythorbate, sodium ascorbate, and curing salts such as nitrites and nitrates. However, current research shows that the excessive addition and misuse of synthetic antioxidants could be associated with DNA damage, apoptosis, and carcinogenicity (Ben Akacha et al., 2023). The immune system and the redox balance, including apoptosis, cell signaling, and ion transport, are significantly impacted by reactive oxygen species (ROS). Excess ROS, however, can harm the body by causing a number of illnesses. Lipid oxidation is seen as a breakdown process in foods that include lipids and is a significant reaction in both biological and dietary systems. Many studies have attempted to prevent lipid oxidation in food products, and adding antioxidants during food processing is one of the most effective strategies (Choi et al., 2021).

Natural sources of antioxidants are increasingly sought after to address dietary deficiencies. Beyond their nutritional benefits, antioxidants function in food preservation by impeding oxidation. Recent studies have highlighted that polysaccharides derived from natural products offer a range of therapeutic effects and health benefits. The antioxidant activity of these polysaccharides can be assessed using various methods, including 2,2-diphenyl-1-picrylhydrazyl (DPPH) assay, ferric reducing antioxidant power (FRAP) test, metal chelation activity, 2,2-azino-bis (3-ethylbenz othiazoline-6-sulfonic acid) (ABTS) assay, hydroxyl radical scavenging activity, and the β-carotene-linoleate bleaching assay. The DPPH radical scavenging activity, reducing power, and metal chelating activity of carboxymethylated polysaccharides demonstrated high antioxidant potential.

Research conducted by scientists has shown that Grifola frondosa SH-05 intracellular zinc polysaccharides have been tested for their antibacterial and antioxidant properties. By scavenging hydroxyl and DPPH radicals and demonstrating Fe2+ chelating activity, the results indicated that IZPPS and IPS have significant antioxidant qualities. Additionally, against foodborne pathogens, IZPS showed possible antibacterial action. Fucoidan, a polysaccharide primarily composed of fucose and extracted from the brown algae *Spatoglossum asperum*, has shown impressive antioxidant properties, particularly in DPPH scavenging assays. When tested at a concentration of 50 mg/mL, it demonstrated a maximum reducing power of 42.63%. These results are consistent with the last studies that address the total antioxidant property of isolated fucoidan being similar to that of fucoidans sourced from *Padina tetrastomatica* and the sulfated polysaccharide derived from *Pterocladia capillacea*.

4.3 *Water Retention and Textural Enhancement*

EPS can alter the water activity of food items, thereby regulating microbial development. EPS bind water; hence, foods remain fresh for longer periods because they reduce the amount of free water available for microbial metabolism, thereby extending shelf life. EPS also enhance food texture by acting as thickening or gelling agents, which improves the mouthfeel and stability of emulsions and foams. Their ability to hold moisture contributes to preservation in addition to enhancing the overall sensory experience of the product, which makes EPS a valuable component in food formulation and safety.

Pullulan is one such example of polysaccharides extracted from EPS. Pullulan, produced by *Aureobasidium pullulans* when developed in cultures of sugar and starch, is a neutral EPS. Pullulan's ability to prevent food decay by dehydrating the food itself is an important feature; it also prevents fungal growth and contamination.. It acts as a low-viscosity filler, enhancing texture without significantly altering flow characteristics in beverages and sauces. Its unique characteristics allow it to stabilize emulsions, making it an attractive choice for balancing the consistency and texture of mayonnaise. Pullulan's viscosity remains stable under various conditions, including heating, pH changes, and exposure to metallic ions like sodium chloride. In addition to its role in food products, pullulan and its derivatives exhibit strong adhesive properties, making them appropriate for applications as stabilizers and binders in food pastes. They can effectively adhere to ingredients, like attaching nuts to cookies, enhancing both texture and presentation. Nonfood sectors also utilize pullulan; for example, it is used as a binding material in tobacco products and is effective for coating seeds and crop fertilizers, promoting adherence and improving agricultural techniques. The multiaspect use of pullulan underscores its value in both food and industrial applications, providing stability, texture, and adhesion.

5 Applications and Benefits of EPS as a Natural Preservative in Food Preservation

Food preservation is an important step that enhances the quality and shelf life of a food product. Food preservation is often done by incorporation of chemical preservatives; however, natural preservatives are coming up to be an excellent alternative to chemical preservatives. Exopolysaccharides are one such source of biopreservatives. EPS has a wide application in food industries due to its versatility for various preservation methods. It is often used in food industries as a firming agent. It is also proven effective against food-borne pathogens and food spoilage bacteria. The flexibility of EPS makes them acceptable for use in a variety of food products. Their natural origin and multifunctional properties allow them to be used not only as preservatives but also as agents that enhance the texture, stability, and overall quality of food. Some important food industries where EPS are widely used as natural preservatives are below.

5.1 Dairy Products

EPS are highly appreciated in the dairy sector for their capacity to enhance the consistency and texture of goods, including cheese, yogurt, and fermented milk. Commonly employed in the fermentation process, EPS-producing lactic acid bacteria add to the product's viscosity and creaminess. For the production of EPS, lactic acid bacteria (LAB) play a pivotal role, which helps in the production of cultured milk products like cheese, yogurt, milk-based desserts, etc.). LAB's production of exopolysaccharides varies widely in chemical composition, charge, quantity, side chains, molecular mass, and molecular rigidity (Suryawanshi et al., 2022). Extension of the shelf life of the product and prevention of the formation of spoiling germs are among the advantages of EPS. Their capacity to inhibit syneresis, which is the separation of liquid from a gel, makes them very advantageous for human use, like whey from yogurt. The firmness and creaminess significantly improved by the use of EPS. By increasing the final product's thickness and mixing water hydration with novel milk components such as proteins and micelles, EPS may act as texturizers and stabilizers by fortifying the casein network's rigidity. EPS can improve product quality and decrease the production of harmful byproducts. Additionally, EPS has been shown to have a major benefit on gut health.

Dextran and xanthan are mostly used as EPS in the dairy industry. Microbial homopolysaccharides with a linear backbone and alpha-linked D-glucopyranosyl repeating units are called dextrans. The malformations of acidified milk gels, as well as their apparent creamy texture, thickness, and reduced syneresis, have been negatively impacted by the water-binding capacity of EPS produced by LAB, particularly dextran. Since dextran may bind and retain water to the fat-free mass and improve hydration, it is a good candidate for lowering the fat content of cheese.

In frozen dairy products, xanthan gum is effective, and it works well with yogurts made via acidification in conjunction with acidification carboxymethylcellulose. The cheese spread's flavor release is enhanced and strengthened by the combination of locust bean gum, xanthan, and guar. Improved heat transfer during processing, long-term stability, appropriate viscosity, enhanced flavor releases, protection from ice crystal control, and heat shock are all benefits of this cost-effective mixing.

5.2 Bakery Products

EPS have been used in bakery products for a long time. EPS, such as dextran, is used as a hydrocolloid and added to sourdough for its thickening properties. Recent studies discovered that EPS is an amazing substance that is used in gluten-free products for maintaining stability in binding and texture. Low water absorption, altered crumb properties, reduced bread volume, and poor stability are features of gluten-free products. By including EPS-producing LAB in gluten-free food, such flaws are removed.

The incorporation of xanthan gum in gluten-free bread formulations can improve the texture and shelf life of the bread by retaining moisture and preventing the bread from becoming dry and crumbly over time. In situ dextran production by *Weissella confusa* was reported to substantially enhance the extensional properties of whole grain pearl millet bread dough, increase the bread-specific volume, and reduce crumb firmness, staling rate, and moisture loss (Wu et al., 2023).

5.3 Meat Products

In raw fermented meat products, LAB, often used for their water-binding properties, help retain moisture, improve texture, and enhance the overall sensory features of the meat. Modified sensory attributes are achieved by adding culture (e.g., improved texture and taste), which also helps control microbial safety. Depending on the type of strain and sugar used, LAB can form EPS, which may lead to structural and textural changes in the products. Use of EPS leads to the possibility of lowering the overall fat content of products while balancing spreadability and taste. The water binding capacity of EPS provides product succulence. Sausages that have been manufactured with an EPS-forming culture appears to be flexible, chewy and softer, which could be attributed to EPS's web-like morphology inside the meat as determined by scanning electron microscopy (SEM). EPS can protect meat products from lipid oxidation, which is a major cause of spoilage and foul flavor. EPS reduce the need for artificial preservatives by aiding overall product stabilization (Wu et al., 2024).

5.4 Beverages

In beverages, EPS have been widely used as emulsifiers and stabilizers as they prevent phase separation. This property becomes significant in fruit pulp–containing beverages. Fermented drinks such as wine and ciders as examples of the products produced via EPS usage. However, only selective LAB species are found in fermented beverages. The development of LAB occurs after the yeasts have driven alcoholic fermentation (AF) in wine and cider. The medium is acidic and includes a high concentration of ethanol and relatively few remaining sugars after the yeasts have been lysed at the end of AF. Resistance to this mechanism has only been observed in a few species, such as LAB of the Lactobacillus and Pediococcus fermentation (MLF) genera. These bacteria proliferate by eating the acids and sugars that the yeasts left behind. They generate either a mixture of lactic acid and acetate or ethanol (heterofermentative bacteria) or simply lactic acid (homofermentative bacteria) based on the metabolic pathways used to break down the carbohydrates. But the most obvious change that lactic acid bacteria in wine and cider cause is turning the L-malic acid into L-lactic acid and CO2. Malolactic fermentation (MLF) is the name given to this process.

This reduces spoilage and reduced acidity of the beverage (Adebayo-Tayo et al., 2024).

5.5 *Plant-Based Foods*

Plant-based diets are becoming more and more popular, and EPS has uses in this field as well. EPS can enhance the mouthfeel and stability of plant-based dairy substitutes such as soy yogurt and almond milk, reducing phase separation and improving the overall customer experience. The interest in various plant-based alternatives to dairy products, such as cheeses and yogurts, due to increasing lactose intolerance, is hindered by difficulties in creating products with acceptable textures and flavors. Plant-based products are low in saturated fats, and one of the reasons for resistance to the embracement of plant-based milk is the presence of dissatisfactory "beany" flavors and undesirable textures. This is due to aldehydes, primarily hexanal, as well as alcohols such as *n*-hexanol and ketones such as ethyl vinyl ketone. Fermentation of plant-based milk by a range of LAB can completely remove hexanal and decrease the concentration of other contributing volatile organic compounds in soy. Antinutritional factors present in plants such as tannins, saponins, phytic acid, α-galactosides, and trypsin inhibitors are also broken down during fermentation by LAB (Angelov et al., 2023). Molecular processes such as proteolysis and EPS formation during fermentation in plant products can be understood by plant systems. Fermented lupin milk experimentally resulted in a yogurt-like product with an exopolysaccharide concentration similar to dairy milk products. Fermented lupin milk shows higher EPS concentration when ultra heated along with increased turbidity and a lower rate of synersis. The final product of oat milk fermented with LAB strains that produce exopolysaccharides was more viscous than that of oat milk fermented with strains that do not produce exopolysaccharides. The highest viscosity is found in oat milk, which contains glucose, which can be used to synthesize exopolysaccharides.

Furthermore, the antimicrobial properties of EPS are valuable in preserving the freshness and safety of plant-based products, which can be more susceptible to spoilage due to their high moisture content (Huang et al., 2024). Table 3 provides a summary of exopolysaccharides (EPS) produced by microorganisms, focusing on their roles as preservatives and their uses.

Table 3 A summary of exopolysaccharides (EPS) produced by microorganisms, focusing on their roles as preservatives and their uses

Microorganism	EPS	Preservative function	Application
Bacteria	Xanthan gum, Gellan gum, Dextran	Antimicrobial, enhances texture, and moisture retention	Dairy products, sauces, baked goods (Abidizadegan et al., 2024)
Fungi	Pullulan, Scleroglucan	Antioxidant and film-forming properties for food coatings	Edible coatings, food packaging, and beverages (Mahapatra & Banerjee, 2013)
Green microalgae	Hexose, pentose, uronic acid, methyl pentose	Selective mobility of particles because of the high porosity of its matrix with regular elongated pores	Various industrial applications related to storage (Wu et al., 2021a).
Red microalgae	Xylans, carrageenan	Antimicrobial, stabilizing, and shelf-life extending properties	Seafood preservation, dairy products, and dressings (Balti et al., 2020)
Diatoms	Diatom-derived EPS containing methylpentose	Antibacterial, biofilm-forming, inhibits microbial growth	Aquatic food systems, bio-based food wraps (Khandeparker & Bhosle, 2001)

6 Conclusion

In the food industry, there is a cumulative demand to avoid artificial chemicals and additives. Therefore, customers' preference for natural and safe products has fostered the rise in the use of natural food additives in recent times. Microbial EPS, particularly those made unswervingly in the food, could be a good substitute for enlightening the physical and chemical properties of food items because of their ability to change the texture and other technical features. EPS can change the way food behaves in different ways, depending on factors such as the size of the molecules, the amount and length of side chains, the type of bonds, and the composition of the basic units. In both laboratory and living organism studies, EPS has demonstrated several beneficial bioactivities, such as antiviral, anticancer, antibiotic, anti-inflammatory, and anticoagulant effects. Despite this impressive set of benefits, further research is needed to determine the toxicity and bioavailability of EPS for use in human health applications.

The dairy, meat, nondairy, and baking industries are just some of the many places where EPS is used, whether made on site or elsewhere. EPS is mainly used in dairy products to make them thicker and creamier. They can also act as stabilizers by holding water, preventing separation, and interacting with other milk components. EPS can improve the texture of meat products, especially those with less fat in the meat industry. To make plant-based products smoother, tastier, and last longer on the shelf, producers are increasingly using LAB strains that make EPS in the making of yogurt-like drinks. EPS helps bind water and form a network with different dough ingredients, which is what gives baked goods their good qualities. This means EPS can improve the texture, structure, and size of bread, making it stay fresh longer and increasing its shelf life. While the LAB can usually make small amounts of EPS, there are several benefits to using them that we should consider. One big challenge is that different bacterial strains can produce biopolymers in different amounts and qualities. Choosing the right strain and optimizing the production process can help solve this problem. The type and number of EPS produced, along with the best conditions for the bacteria to grow, need to be adjusted for specific production methods and food properties. Additionally, we do not know how much EPS is produced and how it changes in different biological and physical conditions. The main problem in current EPS research is that we have not figured out the basic rules that control its production, which makes it hard to predict or control how much EPS is made. To make better use of EPS, we still need a deeper understanding of how their structure affects their function in the food industry.

References

Abdalla, A. K., Ayyash, M. M., Olaimat, A. N., Osaili, T. M., Al-Nabulsi, A. A., Shah, N. P., & Holley, R. (2021). Exopolysaccharides as antimicrobial agents: Mechanism and spectrum of activity. *Frontiers in Microbiology, 12*, 664395. https://www.frontiersin.org/articles/10.3389/fmicb.2021.664395/full

Abidizadegan, M., Peltomaa, E., Ilina, P., Tammela, P., & Blomster, J. (2024). Cryptophytes as potential source of natural antimicrobials for food preservation. *Frontiers in Microbiology, 15*, 1462696. https://www.frontiersin.org/journals/microbiology/articles/10.3389/fmicb.2024.1462696/full

Adebayo-Tayo, B. C., Ogundele, B. R., Ajani, O. A., & Olaniyi, O. A. (2024). Characterization of lactic acid bacterium exopolysaccharide, biological, and nutritional evaluation of probiotic formulated fermented coconut beverage. *International Journal of Food Science, 2024*(1), 8923217. https://doi.org/10.1155/2024/8923217

Ahmad, S. R., Gokulakrishnan, P., Giriprasad, R., & Yatoo, M. A. (2015). Fruit-based natural antioxidants in meat and meat products: A review. *Critical Reviews in Food Science and Nutrition, 55*(11), 1503–1513. https://doi.org/10.1080/10408398.2012.701674

Akpeji, S. C., Adebayo Tayo, B. C., & Ossai, C. O. (2023). Characterisation of some rhizobium EPS and their biopreservative potentials on shelf life extension of Musa species. https://www.preprints.org/manuscript/202303.0263.

Angelin, J., & Kavitha, M. (2020). Exopolysaccharides from probiotic bacteria and their health potential. *International Journal of Biological Macromolecules, 162*, 853–865. https://www.sciencedirect.com/science/article/pii/S014181302033631X

Angelov, A., Georgieva, A., Petkova, M., Bartkiene, E., Rocha, J. M., Ognyanov, M., & Gotcheva, V. (2023). On the molecular selection of exopolysaccharide-producing lactic acid bacteria from indigenous fermented plant-based foods and further fine chemical characterization. *Food, 12*(18), 3346. https://www.mdpi.com/2304-8158/12/18/3346

Aziz, M., & Karboune, S. (2016). Natural antimicrobial/antioxidant agents in meat and poultry products as well as fruits and vegetables: A review. *Critical Reviews in Food Science and Nutrition, 58*, 486–511. https://doi.org/10.1080/10408398.2016.1194256

Babuskin, S., Babu, P. A. S., Sasikala, M., Sabina, K., Archana, G., Sivarajan, M., & Sukumar, M. (2014). Antimicrobial and antioxidant effects of spice extracts on the shelf life extension of raw chicken meat. *International Journal of Food Microbiology, 171*, 32–40. https://www.sciencedirect.com/science/article/pii/S0168160513005321

Badel, S., Bernardi, T., & Michaud, P. (2011). New perspectives for lactobacilli exopolysaccharides. *Biotechnology Advances, 29*(1), 54–66. https://www.sciencedirect.com/science/article/pii/S0734975010001138

Balti, R., Mansour, M. B., Zayoud, N., Le Balc'h, R., Brodu, N., Arhaliass, A., & Masse, A. (2020). Active exopolysaccharides based edible coatings enriched with red seaweed (Gracilaria gracilis) extract to improve shrimp preservation during refrigerated storage. *Food Bioscience, 34*, 100522. https://www.sciencedirect.com/science/article/pii/S2212429218304267

Ben Akacha, B., Michalak, M., Najar, B., Venturi, F., Taglieri, I., Kačániová, M., Ben Saad, R., Mnif, W., Garzoli, S., & Ben Hsouna, A. (2023). Recent advances in the incorporation of polysaccharides with antioxidant and antibacterial functions to preserve the quality and shelf life of meat products. *Food, 12*(8), 1647. https://www.mdpi.com/2304-8158/12/8/1647

Bergwerff, A. A., & Debast, S. B. (2021). Modernization of control of pathogenic micro-organisms in the food-chain requires a durable role for immunoaffinity-based detection methodology—A review. *Food, 10*(4), 832. https://www.mdpi.com/2304-8158/10/4/832

Bilal, M., Gul, I., Basharat, A., & Qamar, S. A. (2021). Polysaccharides-based bio-nanostructures and their potential food applications. *International Journal of Biological Macromolecules, 176*, 540–557. https://www.sciencedirect.com/science/article/pii/S0141813021003871

Bohórquez, J., McGenity, T. J., Papaspyrou, S., García-Robledo, E., Corzo, A., & Underwood, G. J. (2017). Different types of diatom-derived extracellular polymeric substances drive changes in heterotrophic bacterial communities from intertidal sediments. *Frontiers in Microbiology, 8*, 245. https://www.frontiersin.org/articles/10.3389/fmicb.2017.00245/full

Bouarab Chibane, L., Degraeve, P., Ferhout, H., Bouajila, J., & Oulahal, N. (2019). Plant antimicrobial polyphenols as potential natural food preservatives. *Journal of the Science of Food and Agriculture, 99*(4), 1457–1474. https://doi.org/10.1002/jsfa.9357

Carocho, M., Barreiro, M. F., Morales, P., & Ferreira, I. C. F. R. (2014). Adding molecules to food, pros and cons: A review on synthetic and natural food additives. *Comprehensive Reviews in Food Science and Food Safety, 13*(4), 377–399. https://doi.org/10.1111/1541-4337.12065

Choi, I. S., Ko, S. H., Lee, M. E., Kim, H. M., Yang, J. E., Jeong, S.-G., Lee, K. H., Chang, J. Y., Kim, J.-C., & Park, H. W. (2021). Production, characterization, and antioxidant activities of an exopolysaccharide extracted from spent media wastewater after *Leuconostoc mesenteroides* WiKim32 fermentation. *ACS Omega, 6*(12), 8171–8178. https://doi.org/10.1021/acsomega.0c06095

Freitas, F., Alves, V. D., & Reis, M. A. (2011). Advances in bacterial exopolysaccharides: From production to biotechnological applications. *Trends in Biotechnology, 29*(8), 388–398. https://www.cell.com/trends/biotechnology/fulltext/S0167-7799(11)00055-2

Geresh, S., Lupescu, N., & Arad, S. M. (1992). Fractionation and partial characterization of the sulphated polysaccharide of Porphyridium. *Phytochemistry, 31*(12), 4181–4186. https://www.sciencedirect.com/science/article/pii/003194229280439L

Guerrieri, E., de Niederhäusern, S., Messi, P., Sabia, C., Iseppi, R., Anacarso, I., & Bondi, M. (2009). Use of lactic acid bacteria (LAB) biofilms for the control of listeria monocytogenes in a small-scale model. *Food Control, 20*(9), 861–865. https://www.sciencedirect.com/science/article/pii/S0956713508003125

Huang, W., Yang, S., Wätjen, A. P., Gumulya, Y., Fernández-Pacheco, P., Marcellin, E., Prakash, S., Bang-Berthelsen, C. H., & Turner, M. S. (2024). Isolation of an exopolysaccharide-producing Weissella confusa strain from lettuce and exploring its application as a texture modifying adjunct culture in a soy milk alternative. *International Journal of Food Microbiology, 428*, 110992. https://www.sciencedirect.com/science/article/pii/S0168160524004367

Jeong, D., Kim, D.-H., Kang, I.-B., Kim, H., Song, K.-Y., Kim, H.-S., & Seo, K.-H. (2017). Characterization and antibacterial activity of a novel exopolysaccharide produced by lactobacillus kefiranofaciens DN1 isolated from kefir. *Food Control, 78*, 436–442. https://www.sciencedirect.com/science/article/pii/S0956713517300762

Khandeparker, R. D., & Bhosle, N. B. (2001). Extracellular polymeric substances of the marine fouling diatom *amphora rostrata* Wm.Sm. *Biofouling, 17*(2), 117–127. https://doi.org/10.1080/08927010109378471

Kim, Y., & Kim, S. H. (2009). Released exopolysaccharide (r-EPS) produced from probiotic bacteria reduce biofilm formation of enterohemorrhagic Escherichia coli O157: H7. *Biochemical and Biophysical Research Communications, 379*(2), 324–329. https://www.sciencedirect.com/science/article/pii/S0006291X08024340

Korcz, E., & Varga, L. (2021). Exopolysaccharides from lactic acid bacteria: Techno-functional application in the food industry. *Trends in Food Science & Technology, 110*, 375–384. https://www.sciencedirect.com/science/article/pii/S0924224421001151

Kornmann, H., Duboc, P., Marison, I., & Von Stockar, U. (2003). Influence of nutritional factors on the nature, yield, and composition of exopolysaccharides produced by *Gluconacetobacter xylinus* I-2281. *Applied and Environmental Microbiology, 69*(10), 6091–6098. https://doi.org/10.1128/AEM.69.10.6091-6098.2003

Li, W., Ji, J., Rui, X., Yu, J., Tang, W., Chen, X., Jiang, M., & Dong, M. (2014). Production of exopolysaccharides by Lactobacillus helveticus MB2-1 and its functional characteristics in vitro. *LWT-Food Science and Technology, 59*(2), 732–739. https://www.sciencedirect.com/science/article/pii/S0023643814004319

Lin, T. Y., & Chien, M.-F. C. (2007). Exopolysaccharides production as affected by lactic acid bacteria and fermentation time. *Food Chemistry, 100*(4), 1419–1423. https://www.sciencedirect.com/science/article/pii/S0308814605010770

Looijesteijn, P. J., Boels, I. C., Kleerebezem, M., & Hugenholtz, J. (1999). Regulation of exopolysaccharide production by *Lactococcus lactis* subsp. *Cremoris* by the sugar source. *Applied and Environmental Microbiology, 65*(11), 5003–5008. https://doi.org/10.1128/AEM.65.11.5003-5008.1999

Magaletti, E., Urbani, R., Sist, P., Ferrari, C. R., & Cicero, A. M. (2004). Abundance and chemical characterization of extracellular carbohydrates released by the marine diatom *Cylindrotheca fusiformis* under N- and P-limitation. *European Journal of Phycology, 39*(2), 133–142. https://doi.org/10.1080/0967026042000202118

Mahapatra, S., & Banerjee, D. (2013). Fungal exopolysaccharide: Production, composition and applications. *Microbiology Insights, 6*, MBI-S10957. https://journals.sagepub.com/doi/abs/10.4137/MBI.S10957

Mahdhi, A., Leban, N., Chakroun, I., Chaouch, M. A., Hafsa, J., Fdhila, K., Mahdouani, K., & Majdoub, H. (2017). Extracellular polysaccharide derived from potential probiotic strain with antioxidant and antibacterial activities as a prebiotic agent to control pathogenic bacterial biofilm formation. *Microbial Pathogenesis, 109*, 214–220. https://www.sciencedirect.com/science/article/pii/S0882401016305174

Mensink, M. A., Frijlink, H. W., van der Voort Maarschalk, K., & Hinrichs, W. L. (2015). Inulin, a flexible oligosaccharide I: Review of its physicochemical characteristics. *Carbohydrate Polymers, 130*, 405–419. https://www.sciencedirect.com/science/article/pii/S0144861715004257

Moore, B. G., & Tischer, R. G. (1964). Extracellular polysaccharides of algae: Effects on life-support systems. *Science, 145*(3632), 586–587. https://doi.org/10.1126/science.145.3632.586

Nehal, F., Sahnoun, M., Smaoui, S., Jaouadi, B., Bejar, S., & Mohammed, S. (2019). Characterization, high production and antimicrobial activity of exopolysaccharides from Lactococcus lactis F-mou. *Microbial Pathogenesis, 132*, 10–19. https://www.sciencedirect.com/science/article/pii/S0882401018320746

Nešić, A., Cabrera-Barjas, G., Dimitrijević-Branković, S., Davidović, S., Radovanović, N., & Delattre, C. (2019). Prospect of polysaccharide-based materials as advanced food packaging. *Molecules, 25*(1), 135. https://www.mdpi.com/1420-3049/25/1/135

Pachekrepapol, U., Lucey, J. A., Gong, Y., Naran, R., & Azadi, P. (2017). Characterization of the chemical structures and physical properties of exopolysaccharides produced by various Streptococcus thermophilus strains. *Journal of Dairy Science, 100*(5), 3424–3435. https://www.sciencedirect.com/science/article/pii/S0022030217302400

Rajoka, M. S. R., Mehwish, H. M., Hayat, H. F., Hussain, N., Sarwar, S., Aslam, H., Nadeem, A., & Shi, J. (2019). Characterization, the antioxidant and antimicrobial activity of exopolysaccharide isolated from poultry origin lactobacilli. *Probiotics and Antimicrobial Proteins, 11*(4), 1132–1142. https://doi.org/10.1007/s12602-018-9494-8

Rajoka, M. S. R., Mehwish, H. M., Kitazawa, H., Barba, F. J., Berthelot, L., Umair, M., Zhu, Q., He, Z., & Zhao, L. (2022). Techno-functional properties and immunomodulatory potential of exopolysaccharide from Lactiplantibacillus plantarum MM89 isolated from human breast milk. *Food Chemistry, 377*, 131954. https://www.sciencedirect.com/science/article/pii/S0308814621029605

Rani, R. P., Anandharaj, M., & Ravindran, A. D. (2018). Characterization of a novel exopolysaccharide produced by lactobacillus gasseri FR4 and demonstration of its in vitro biological properties. *International Journal of Biological Macromolecules, 109*, 772–783. https://www.sciencedirect.com/science/article/pii/S0141813017335961

Ribeiro, V. A., & Burkert, C. A. V. (2016). Exopolysaccharides produced by rhizobium: Production, composition and rheological properties. *Journal of Polymer and Biopolymer Physics Chemistry, 4*(1), 1–6. https://www.researchgate.net/profile/Maurice-Ekpenyong/post/Hai-DrTereza-I-am-CSARAVANAN-and-interested-to-know-about-the-complex-carbon-sources-and-production-of-exopolysaccharides-in-rhizobium-species/attachment/5c4eb62dcfe4a781a57a82a8/AS%3A719956094238722%401548662317804/download/Exopolysacch+2.pdf

Ruas-Madiedo, P., Hugenholtz, J., & Zoon, P. (2002). An overview of the functionality of exopolysaccharides produced by lactic acid bacteria. *International Dairy Journal, 12*(2–3), 163–171. https://www.sciencedirect.com/science/article/pii/S0958694601001601

Rutherford, S. T., & Bassler, B. L. (2012). Bacterial quorum sensing: Its role in virulence and possibilities for its control. *Cold Spring Harbor Perspectives in Medicine, 2*(11), a012427. https://perspectivesinmedicine.cshlp.org/content/2/11/a012427.short

Sarikaya, H., Aslim, B., & Yuksekdag, Z. (2017). Assessment of anti-biofilm activity and bifidogenic growth stimulator (BGS) effect of lyophilized exopolysaccharides (l-EPSs) from *Lactobacilli* strains. *International Journal of Food Properties, 20*(2), 362–371. https://doi.org/10.1080/10942912.2016.1160923

Speranza, B., Sinigaglia, M., & Corbo, M. R. (2009). Non starter lactic acid bacteria biofilms: A means to control the growth of Listeria monocytogenes in soft cheese. *Food Control, 20*(11), 1063–1067. https://www.sciencedirect.com/science/article/pii/S0956713509000164

Stoica, R. M., Moscovici, M., Lakatos, E. S., & Cioca, L. I. (2023). Exopolysaccharides of fungal origin: Properties and pharmaceutical applications. *PRO, 11*(2), 335. https://www.mdpi.com/2227-9717/11/2/335

Sugumaran, K. R., & Ponnusami, V. (2017). Conventional optimization of aqueous extraction of pullulan in solid-state fermentation of cassava bagasse and Asian palm kernel. *Biocatalysis and Agricultural Biotechnology, 10*, 204–208. https://www.sciencedirect.com/science/article/pii/S1878818116303553

Sun, Y., Zhao, X., Ma, Y., Guan, H., Liang, H., & Wang, D. (2022). Inhibitory effect of modified atmosphere packaging on Escherichia coli O157: H7 in fresh-cut cucumbers (Cucumis sativus L.) and effectively maintain quality during storage. *Food Chemistry, 369*, 130969. https://www.sciencedirect.com/science/article/pii/S0308814621019750

Suryawanshi, N., Naik, S., & Jujjawarapu, S. E. (2022). Exopolysaccharides and their applications in food processing industries. *Food Science And Applied Biotechnology, 5*(1), 22–44. https://www.ijfsab.com/index.php/fsab/article/view/165

Torino, M. I., Font de Valdez, G., & Mozzi, F. (2015). Biopolymers from lactic acid bacteria. Novel applications in foods and beverages. *Frontiers in Microbiology, 6*, 834. https://www.frontiersin.org/articles/10.3389/fmicb.2015.00834/full

Vu, B., Chen, M., Crawford, R. J., & Ivanova, E. P. (2009). Bacterial extracellular polysaccharides involved in biofilm formation. *Molecules, 14*(7), 2535–2554. https://www.mdpi.com/1420-3049/14/7/2535

Wu, M.-H., Pan, T.-M., Wu, Y.-J., Chang, S.-J., Chang, M.-S., & Hu, C.-Y. (2010). Exopolysaccharide activities from probiotic bifidobacterium: Immunomodulatory effects (on J774A. 1 macrophages) and antimicrobial properties. *International Journal of Food Microbiology, 144*(1), 104–110. https://www.sciencedirect.com/science/article/pii/S0168160510005076

Wu, J., Gu, X., Yang, D., Xu, S., Wang, S., Chen, X., & Wang, Z. (2021a). Bioactive substances and potentiality of marine microalgae. *Food Science & Nutrition, 9*(9), 5279–5292. https://doi.org/10.1002/fsn3.2471

Wu, J., Yan, D., Liu, Y., Luo, X., Li, Y., Cao, C., Li, M., Han, Q., Wang, C., & Wu, R. (2021b). Purification, structural characteristics, and biological activities of exopolysaccharide isolated from Leuconostoc mesenteroides SN-8. *Frontiers in Microbiology, 12*, 644226. https://www.frontiersin.org/articles/10.3389/fmicb.2021.644226/full

Wu, J., Han, X., Ye, M., Li, Y., Wang, X., & Zhong, Q. (2023). Exopolysaccharides synthesized by lactic acid bacteria: Biosynthesis pathway, structure-function relationship, structural modification and applicability. *Critical Reviews in Food Science and Nutrition, 63*(24), 7043–7064. https://doi.org/10.1080/10408398.2022.2043822

Wu, Y., Jin, Z., Wang, X., Ji, Q., Bao, D., Jin, G., Shan, B., Mei, L., & Qi, J. (2024). Characterization of the exopolysaccharide produced by Pediococcus acidilactici S1 and its effect on the gel properties of fat substitute meat mince. *International Journal of Biological Macromolecules, 270*, 132262. https://www.sciencedirect.com/science/article/pii/S0141813024030678

Yalcin, I., Hicsasmaz, Z., Boz, B., & Bozoglu, F. (1994). Characterization of the extracellular polysaccharide from freshwater microalgae chlorella sp. *LWT-Food Science and Technology, 27*(2), 158–165. https://www.sciencedirect.com/science/article/pii/S0023643884710322

Yu, Y.-J., Chen, Z., Chen, P. T., & Ng, I.-S. (2018). Production, characterization and antibacterial activity of exopolysaccharide from a newly isolated Weissella cibaria under sucrose effect. *Journal of Bioscience and Bioengineering, 126*(6), 769–777. https://www.sciencedirect.com/science/article/pii/S1389172318302925

Zhang, L., Liu, C., Li, D., Zhao, Y., Zhang, X., Zeng, X., Yang, Z., & Li, S. (2013). Antioxidant activity of an exopolysaccharide isolated from lactobacillus plantarum C88. *International Journal of Biological Macromolecules, 54*, 270–275. https://www.sciencedirect.com/science/article/pii/S0141813012005223

Global Perspectives on EPS Application Showcasing International Case Studies of Successful EPS Application: Analyzing Cultural Influences on the Adoption of EPS in Different Regions

Saikat Mazumder, Navin Kumar, Anjali Lathwal, and Kakali Bandyopadhyay

1 Introduction

Pathogenic and spoilage microorganisms can be present in food products and pose health risks or economic detriment. While a number of chemical additives have been utilized for microbial control, the use of these chemical additives has regulatory implications for safety and health concerns. This explains why nowadays the food processing sector pays more attention to natural compounds as safe alternatives, and thus, it suits consumer demand for environmentally friendly options (Bondi et al., 2017). There have been a number of scholarly studies on the use of biopolymers across these sectors. Polymers obtained or derived from biological sources or degradability, which at the same time may possess biodegradability, are called biopolymers. Biopolymers are classified in terms of their sources into natural, synthetic, and microbial entities. In recent years, there has been a great deal of scholarly interest in biopolymers produced by living microorganisms, including different species of bacteria, yeast, and fungi (Imre & Pukánszky, 2013; Korcz & Varga, 2021; Moradi et al., 2021). Utilization of simple to complex substrates leads to the production of diverse biopolymers with varied chemical properties by bacteria. For cellular location, biopolymers could be either intracellular or extracellular. While the intracellular biopolymers are few and have very limited use, the extracellular biopolymers are many and have wide applications (Nwodo et al., 2012).

S. Mazumder (✉) · N. Kumar · K. Bandyopadhyay
Department of Food Technology, Guru Nanak Institute of Technology, Kolkata, West Bengal, India
e-mail: saikat.mazumder@gnit.ac.in

A. Lathwal
Department of Computer Science and Engineering, Indira Gandhi Delhi Technical University for Women, New Delhi, India

T. Sarkar et al. (eds.), *Extracellular Polysaccharides*,
https://doi.org/10.1007/978-3-031-94472-7_5

Extracellular polysaccharides produced by microorganisms are broadly used in the food industry as well as in the cosmetic and pharmaceutical industries (Korcz et al., 2018). González et al. (2010) state that extracellular biopolymers produced by bacterial and fungal organisms show better chemical quality, bioactivity, relevance, and efficacy than those produced by either plant or algal sources. The EPSs obtained from these extracellular substrates exhibit several distinctive characteristics compared to EPSs of plant and algal origin, most notably for purity, fineness, and environmental sustainability. They enable a facile isolation process at low cost, sustainability, high yields, nontoxicity, and, with enhanced rheological properties, pseudo plasticity. Moreover, exopolysaccharides have been reported to possess a wide variety of biological activities, including antibiotic, antioxidant, antimutagenic, anticoagulant, and immunostimulatory activities (Yang et al., 2015). These polymers exhibit taxonomic relevance and ecological significance due to their relevance in energy storage and utility as biomarkers for environmental strain or alteration (Kambourova et al., 2015). EPSs with favorable thermal stability and improved water retention capacity are promising candidates for biomedical applications (Andrew & Jayaraman, 2022). At present, EPSs may serve as an alternative to synthetic pharmaceuticals and are pharmacologically and medically significant carbohydrates due to the deleterious impact of chemical therapies and increased resistance of existing pharmacological agents (Al-Nabulsi et al., 2022). Also, they are starting to be widely used in drug delivery and scaffold fabrication in tissue engineering (Salimi & Farrokh, 2023). Consequently, they have attracted a great deal of interest as renewable resources in the field of protein engineering and biotechnology, which can be smartly accessed, modified, and no doubt converted into various value-added products that have a positive impact on human welfare (Zong et al., 2012). EPS produced by lactic acid bacteria (LAB) presents a viable and natural alternative to synthetic food additives because they have unique physicochemical properties. Importantly, EPS can be incorporated into the food manufacturing process in which they can influence the rheological properties of the food matrix providing increases in viscosity, suppression of syneresis, and improvements in texture while presenting emulsifying, thickening, and stabilizing properties (Korcz & Varga, 2021). Microbial EPSs have been widely used in recent years to improve textural and rheological attributes of alimentary products, especially dairy products, due to their pseudoplastic rheological behavior and great water-binding capacity. In terms of the application of EPSs in the food sector, it has been claimed that about 20% of the existing literature is related to dairy, as seen in the analysis in 2021 (Abarquero et al., 2022). It may also be employed in the production of food packaging materials. Incorporation of LAB-derived EPS in food products can provide the opportunity to minimize or eliminate synthetic additives in food products, which can be used not only to assist in the optimization of production processes but also fulfill consumer preferences (Korcz & Varga, 2021; Moradi et al., 2021).

Significant production expense from culture media, purification, and packaging plus testing creates a block to the industrial application of microbial exopolysaccharides. Purification challenges are still present, but the inclusion of agricultural byproducts and waste materials may help to reduce costs, albeit for food-grade use

(Ventorino et al., 2019). Specific dependencies of microbial growth during hetero-EPS material formation, especially, are hindering the scale-up to mass production (Seviour et al., 2011). Improving EPS yield is the focus of technological innovations that employ metabolic engineering to modify pathways and genes to increase sugar flux toward EPS synthesis. Microbial EPS-derived nanoparticles has antibacterial, anticancer, and antioxidant functions. EPS bioactivity is further augmented by chemical and enzymatic modifications, that is, acetylation and phosphorylation. In addition, relatively simple physical treatments such as ultrasound and microwave alter the base chemistry of EPS, making it more soluble, more bioactive, and more amenable to extraction. We demonstrate examples such as ultrasound, which promotes EPS production during fermentation, and microwave treatment, which increases microalgal yields. These advancements, coupled with cost-reduction measures, are essential for enhancing the viability of microbial EPSs in industrial contexts (Fukuda & Kono, 2021; Wang et al., 2010).

2 EPS and beyond

Exopolysaccharides (EPSs) are systematically secreted from the cellular membrane into the surrounding environment, making them seemingly unattached, renewable, freely available, and extractable from organic biomasses through specialized fermentation methodology (Orlandelli et al., 2016; Wu et al., 2014). The processes governing the synthesis and translocation of microbial polysaccharides are in a nascent, speculative state. According to Schmid et al. (2015), the chemical condensation of intracellular nucleotide sugars and a variety of precursor metabolites within different metabolic pathways could shed light on how these metabolites are synthesized. In addition, the production of EPSs is naturally subject to modulation by environmental variables and physical conditions (Papinutti, 2010; Polak-Berecka et al., 2015). Studies have shown that the production of EPS is largely determined by culture parameters such as the composition of the culture medium (carbon source, nitrogen source, mineral requirement) and the environmental factors (temperature, pH, dissolved oxygen, salinity, and agitation rate), as well as the bioreactor configuration. These environmental conditions may also affect the production, kind, and size of polysaccharides produced, and the viscosity of the broth culture (García et al., 2022). The process of biosynthesis of microbial exopolysaccharides is also dependent on an extraordinary array of enzymes and regulatory proteins. Sugars are the principal substrates for energy and nitrogen sources are ammonium salts and amino acids for microorganisms that synthesize exopolysaccharides. Fundamentally, the synthesis of exopolysaccharides by microorganisms is highly dependent on the ratio of carbon to nitrogen availability in the external environment. Importantly, under decreased nitrogen concentrations, extracellular production of exopolysaccharides is increased (Rana & Upadhyay, 2020; Shukla et al., 2019).

Exopolysaccharides (EPSs) are categorized into two primary groups: It includes homopolysaccharides and heteropolysaccharides. The first category consists of a single kind of monosaccharide: for example, levan or dextran. Heteropolysaccharides (xanthan, hyaluronic acid), consisting of three or four different monosaccharides, have a backbone structure of a repeating unit having 10 or fewer monomers (Finore et al., 2014). Most homopolysaccharides are neutral, whereas most heteropolysaccharides are polyanionic. α-D-glucans, β-D-glucans, fructans, and polygalactans are the classes into which homopolysaccharides are divided. Heteropolysaccharides, however, consist of a repeating unit of D glucose, D galactose, L-rhamnose, etc. (Prajapat & Patel, 2013). The polymers of homopolysaccharides include pullulan, curdlan, dextran, cellulose, and scleroglucan. Both Curdlan and cellulose are linear homopolysaccharides containing 1,3 β glucan and 1,4 β glycosidic linkages, respectively. Pullulan and scleroglucan are homopolysaccharides having 1,3 α and 1,6 α glycosidic linkages and 1,3 β and 1,6 β glycosidic linkages, respectively. The linkage bonds, chemical composition, synthesis enzymes, and synthesis sites are different among homopolysaccharides and heteropolysaccharides (Rana & Upadhyay, 2020). Exopolysaccharides are classified into the following categories based on functionality: sorptive, surface active, structural, and redox-active. Sorptive exopolysaccharides are charged polymers that impart the capacity to sorb other charged molecules implicated in cell surface interactions. Structural exopolysaccharides function within the matrix to either promote water retention or protect cells. Amphiphilic molecules with diverse chemical structures and surface characteristics may be found as surface-active exopolysaccharides that could contribute to the development of biofilm or possess antibacterial or antifungal effects (Islam & Lam, 2014). However, properties of exopolysaccharides with the characteristics of redox have been described in electroactive bacterial species named *Shewanella, Pseudomonas putida*, and *Escherichia coli.* Electroactive bacterial species from which the exopolysaccharides have been derived are used to facilitate migration and biotransformation of redox-sensitive contaminants (Rana & Upadhyay, 2020).

Microbial EPS producers are eukaryotes (e.g., phytoplankton, fungus, and algae) as well as prokaryotes (e.g., eubacteria and archaebacteria). While not entirely comprehended, microbial extracellular polymeric substances (EPS) have been proposed to play several critical physiological roles, including protection from environmental stresses (e.g., osmotic pressure, temperature fluctuations, pH variation, ultraviolet (UV) radiation, heavy metals, oxidants, and desiccation), cell adherence to surfaces, and as a reserve for carbon of water storage (Freitas et al., 2017). Bacteria and fungi are by far the most common among the microbial EPS producers. Researchers have explored bacterial EPSs extensively. By contrast, the production of EPS from fungi has been well-studied over the last two decades (Mahapatra & Banerjee, 2013).

2.1 Bacteria

Polysaccharide secretion is widely seen in bacteria, including species from the genera *Streptococcus, Xanthomonas, and Acetobacter.* Dextran was the first bacterial exopolysaccharide to be commercialized, a water-soluble glucan made from lactic acid bacteria in the genera *Leuconostoc*, *Streptococcus*, *Weisella*, *Pediococcus*, and *Lactobacillus*, which are generally recognized as safe (GRAS), and frequently employed in the preparation of many food items (Prajapat & Patel, 2013). Several distinct groups of bacteria of the genera *Acetobacter*, *Glucanoacetobacter*, and *Rhizobium* are known to produce cellulose, which is a hydrophobic biopolymer of great mechanical strength, low density, and high water absorption capacity. Although *G. sacchari* is of interest for the production of biological cellulose (BC), the most thoroughly studied strain is *A. xylinum*, which is also the most important in the industrial production of BC (Gomes et al., 2013). *Pseudomonas aeruginosa* and *Azotobacter vinelandii* synthesize alginate, a heteropolysaccharide, the two component residues of which are D-mannosyl and L-glucuronosyl (Manivasagan & Kim, 2014). More EPSs produced from bacterial strains are noted in Table 1.

2.2 *Fungi*

The production of exopolysaccharides (EPS) from fungi is influenced by several interconnected factors, including the type of fungal strains, culture conditions, and nutrient availability (Mahapatra & Banerjee, 2013). EPS production is widely distributed among fungi, including members of the *Genera Aureobasidium*, *Candida*, *Cryptococcus*, among many others. Most of them are homopolysaccharides, but heteropolysaccharides are also very common. Such biopolymers contain D-mannose, either alone or in combination with other sugars (e.g., galactose, xylose), together with uronic acids and nonsugar substituents (Osińska-Jaroszuk et al., 2015). An exopolysaccharide, pullulan, released by the black yeast–like fungus, *Aureobasidium pullulans*, is a linear α-glucan with significant mechanical strength and adhesivity. It is also commercialized as a thickening agent and edible coating by Hayashibara Co., Ltd. (Japan). The plant pathogen *Sclerotium* sp. excretes a water-soluble β 1–3 – 1, 4 – glucan scleroglucan that was first marketed in the 1970s in a few different trademarks (e.g., Clearogel, Polytetran, Polytran FS, Actigum) and used in increased oil recovery, as a food, for cosmetics, and for pharmaceutical purposes (Castillo et al., 2015; Freitas et al., 2017). Figure 1 gives an overview of the production process of exopolysaccharides (EPS) and highlights their applications in the food industry and biomedical fields.

3 EPS and Food Industry

Human society and well-being have always been rooted in food—in its natural and manufactured forms. Microbial EPSs can play a beneficial role in maintaining food quality, production, processing, and preservation (Osemwegie et al., 2020). Exopolysaccharides such as dextran, xanthan, pullulan, gellan, curdlan, and scleroglucan function as natural thickening, stabilizing texturizing, and gelling agents.

Table 1 Application of EPS in food industry

Types of EPS/polymers	Producing microorganism	Type of microorganism	Role/application in the food industry	Reference
Xanthan	*Xanthomonas campestris*	Bacterium	Used as a stabilizer, emulsifier, and suspensor or thickening agent; acts as a cryoprotectant in ice cream	Osemwegie et al. (2020); Suryawanshi et al. (2022)
Dextran	*Leuconostoc mesenteroides*	Bacterium	Improves moisture retention in confectionaries, increases viscosity and sugar crystallization, used as a hydrocolloid with thickening properties in bakeries	Korcz and Varga (2021); Osemwegie et al. (2020)
Pullulan	*Aureobasidium pullulans*	Fungus	Used as a thickener and viscosity stabilizer	Osemwegie et al. (2020)
Alternan	*Leuconostoc mesenteroides, Leuconostoc citreum*	Bacteria	Sweeteners in confectionary manufacturing, bulking agents, stabilizers and emulsifiers, prebiotics	Korcz and Varga (2021)
Reuteran	*Limosilactobacillus reuteri*	Bacterium	Used in bakery products to improve softness and delay staling	
Inulin	*Lactobacillus johnsonii, Limosilactobacillus reuteri, Leuconostoc citreum*	Bacteria	Prebiotics and sugar replacer (in combination with high-intensity sweeteners), fat replacer and texture modifier in low-fat dairy products; improves mouthfeel, enhancing creaminess	
Scleroglucan	*Sclerotium rolfsii*		Employed as a stabilizing and gelling agent or thickener	Basiri (2021); Suryawanshi et al. (2022)
Curdlan	*Alcaligenes faecalis, Rhizobium* spp., *Cellulomonas* spp.	Bacteria	In fish and meat gel-based products, they can act as gel binders and dietary fiber. Unique resilience strength after heating or after freezing and thawing improves chewiness, gumminess, adhesiveness, and viscosity of emulsified meatballs and the quality of tofu, noodles, and surimi	Basiri (2021)

Gellan	*Sphingomonas elodea*	Bacterium	Used as gelling agent in desserts and jams and as stabilizer in dairy, encapsulates probiotics; enhances shelf life in food packaging	
Villian	*Lac. lactis subsp. cremoris*	Bacteria	Used as thickener and stabilizer; improves viscosity, heat-shock resistance, melting characteristics; reduces crystals in ice cream; Reduces syneresis in fermented milk products	
Acetan	*Acetobacter xylinum*	Bacterium	Functions as a viscosifier and gelling agent in confectionaries	
Kefiran	*Lb. kefiranofaciens, Lb. kefirgranum, Lb. parakefir, Lb. kefir and Lb. delbrueckii subsp. Bulgaricus*	Bacteria	Fat replacer decreases syneresis and decreases yeast and mold growth in yogurt. Further, it act as gelling agent, increases viscosity, improve shelf life, act as thickener, and emulsion stabilizer.	
Alginate	*P. aeruginosa, Az. vinelandii*	Bacteria	Used in thickener and stabilizer; increases viscosity and heat–shock resistance; reduces crystal formation; improves melting characteristics in ice creams	
Levan	*P. fluorescens, Serratia Levanicum, Microbacterium laevaniformans, Lb.* spp., *B. stearothermophilus*	Bacteria	Increases *Bifidobacterium* spp. count in prebiotics; assists in the absorption of calcium and magnesium in the gut; Acts as a stabilizer, emulsifier, and flavor enhancer in beverages	
Cellulose	*Komagataeibacter (former Gluconacetobacter), Agrobacterium, Rhizobium, Salmonella, and Sarcina*	Bacteria	Keeps water binding capacity; used as thickener, stabilizer, and fat replacer in meat products; Increases viscosity in yogurt, in icecream act as fat substitute, stabilizers, reduces the melting rate and increase the fiber content. Act as fat replacer and increases the hardness of biscuits."	

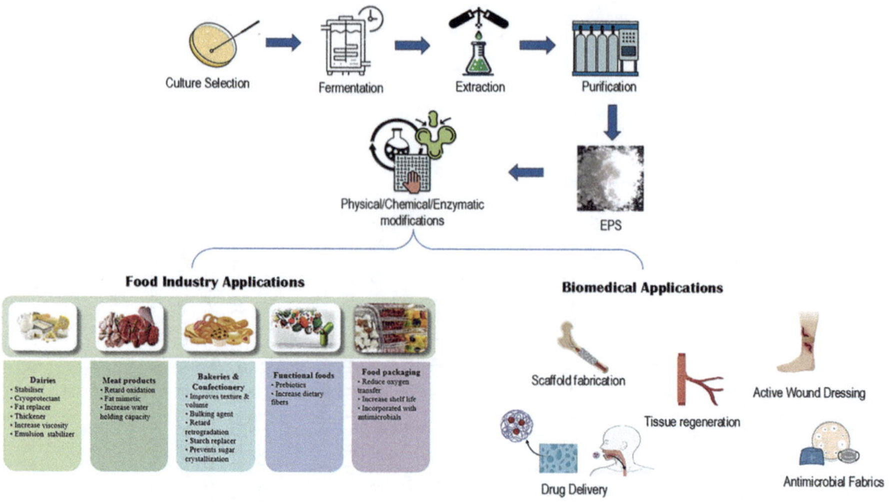

Fig. 1 Production process of EPS and its applications

They also improve the quality of the food by improving texture, rheology, and moisture, which makes them appropriate for use in dairy products, bakery products, beverages, and confectionery. EPSs are also employed in low-calorie and gluten-free applications and have a role to play in extending shelf life. They are biodegradable and flexible materials needed for creating food products that are sustainable and of high quality (Suryawanshi et al., 2022). EPSs can interact with food components to improve rheological and sensory attributes of food processing. They function as texturizers, stabilizers, viscosifiers, biothickeners, or emulsifiers, depending on the temperature, pH, and ionic strength of the medium (Angelin & Kavitha, 2020). EPSs modulate the viscoelastic characteristics of foods according to their monomer composition, bonding types, side chain properties, and molecular weight (Korcz & Varga, 2021). LAB exopolysaccharides have an industrial potential, especially exopolysaccharides produced in situ. They provide a perfect replacement for external hydrocolloids providing effective binding of water and moisture retention in the products (Lynch et al., 2018). A study of exopolysaccharides (EPS) and their effects on sourdough bread quality was done in order to investigate the structural and functional dynamics of reuteran and dextran. Enzymatically synthesized reuteran and dextran were evaluated on bread volume and textural properties. The study used *Lactobacillus reuteri* and *Weissella cibaria* as microbial sources of EPS and performed mutagenesis to produce reuterans with different linkage types and molecular weights. Results showed that dextran significantly increases bread volume and texture, particularly in wheat bread with weaker gluten networks, such as wheat bread containing increasing amounts of rye flour. Among the reuterans, reuteranNS, with a higher proportion of α-(1 $\rightarrow$ 4) linkages and lower molecular weight, showed comparable effects to dextran in improving bread softness and volume. However, the reuterans caused uneven pore distribution in bread crumbs, whereas dextran contributed to a uniform texture (Chen et al., 2016b). In another study in situ fermented ice cream was enhanced by EPS synthesis from *Streptococcus thermophilus* strains during fermentation. The foremost goal was the creation of a functional ice cream with improved texture and stability but lacking the need for conventional stabilizers. Temperature (32–42 °C) and time (2–4 hours) were studied as multiple fermentation parameters, and their effects on EPS yield were assessed along with those on physicochemical, rheological, microstructural, and sensory attributes of ice cream. Results showed that incorporating EPS+ S. thermophilus strains into the ice cream resulted in the increased viscosity and texture of the ice cream. Extended duration over traditional batch fermentation processes and fusion from elevated fermentation temperatures were found to enhance EPS production to produce ice cream with superior consistency and stability. The rheological assessments were found to support the pseudoplastic, non-Newtonian behavior of the ice cream, which should provide favorable flow and mouthfeel behavior. In this work, scanning electron microscopy illustrated a dense, web-like microstructure, which enhanced water retention and stability, while FTIR spectroscopy detected characteristic polysaccharide functional groups within the EPS. Further sensory analyses showed high acceptability of EPS-enriched ice creams with an improved texture, melting resistance, and total acceptance (Dertli et al., 2016). EPSs produced by microorganisms have various applications in the food industry, which is also shown in Table 1.

4 EPS and Biomedical Industry

The biocompatibility and functional properties of EPSs allow them to be put to use in a wide range of biomedical fields, including scaffolds, wound healing, drug delivery systems, coating materials for medical devices, and surgical sealants. EPSs may be used in their natural state, cross-linked, or modified with bioactive substances (Mohd Nadzir et al., 2021). Medical equipment—biosensors and prosthetic implants—are considerably more advanced now than they were earlier. The durability of medical gadgets is improved by production using mechanically stable polymers, or metals, to reduce the frequency of repeated surgeries. Despite this, resilient materials could be so rigid as to elicit an immunological response that could be fatal. Exopolysaccharides, particularly dextran, have been employed as a coating substance for improving the biocompatibility of medical equipment. Bioactive substances including growth factors may be encapsulated within dextran coatings. VEGF was immobilized inside a dextran coating for vascular use by Noel et al. (2016). Thus, the deposition of VEGF into a vascular graft may preferentially recruit endothelial cells to attach to vascular implants and thereby facilitate vascular regeneration (Mohd Nadzir et al., 2021).

The adaptability of exopolysaccharides (EPS) as scaffolding material, transporters of osteogenic agents, and direct biological benefits for bone repair make them a strong candidate in bone regeneration. Porous scaffolds containing exopolysaccharides (EPS) such as hyaluronic acid (HA), bacterial cellulose, gellan, and dextran have been made to support cell colonization and osseous development. These materials are biocompatible and biodegradable and can be modified to enhance mechanical properties and biological interactions. Bacterial cellulose, with its collagen-like structure, increases cell adhesion and directs nutrient diffusion; while HA facilitates osteogenic differentiation of precursor cells, as well as angiogenesis (Bagnol et al., 2022; Lee et al., 2017). Bioactive substances such as hydroxyapatite had already been incorporated with gellan and dextran, which have well-documented gel-forming capability and calcium-binding properties, to enhance mechanical strength and bone mineralization. EPS also has direct effects on bone biology in certain. Specific EPS, such as HE800, produced from *Vibrio diabolicus*, mimics extracellular matrix components, greatly enhancing bone defect repair and promoting vascularization, while high-molecular weight HA enhances bone morphogenetic protein (BMP) activity (Bagnol et al., 2022). Bacterial EPS is a viable medium for the delivery of therapeutic agents, such as growth factors and anticancer pharmaceuticals. Creating EPS for drug delivery is much simpler than designing EPS for biological scaffolds with living cells. Engineering of EPSs can yield additional enhanced regulated drug release, increased drug stability within the organism, and increased therapeutic effectiveness (Mohd Nadzir et al., 2021). Gellan nano hydrogel system was used to deliver poorly water-soluble pharmaceuticals such as an anti-cancer agent, Paclitaxel, and an anti-inflammatory, Prednisolone (Rana & Upadhyay, 2020). Hyaluronic acid (HA) is an anionic linear polysaccharide composed of disaccharide units of 1–4 β D glucuronate and 1–3 β N acetyl D glucosamine residues. Solid lipid nanoparticles coated with hyaluronic acid proved to be a sustained and targeted delivery vehicle for the anticancer medicine Vorinostat toward SCC7 carcinoma cell line, human breast adenocarcinoma (MCF-7) cell line, and human lung epithelial adenocarcinoma (A549) cell line (Tran et al., 2014). The impact of exopolysaccharide (EPS) produced by cold-adapted yeast *R. mucilaginosa* sp. GUMS16 on full-thickness wounds in rats was assessed in a research. The cold ethanol precipitated the EPS of the GUMS16 strain, followed by lyophilization. Nanofibers were processed using single-needle techniques and double-needle techniques from the EPS combined with polycaprolactone (PCL) and gelatin. Hematoxylin and eosin (H&E) and Masson's Trichrome staining were performed to observe wound healing in rats after treating full-thickness wounds with nanofibers. SEM, DLS, FTIR, and TGA results suggested a carbohydrate chemical structure of EPS with an average diameter of 40 nm. Cell viability evaluations showed that the cellular activity for the eggshell powder–loaded sample with 2% EPS is the highest. Additionally, full-thickness wounds from rat models were nanofiber webbed on and implanted in vivo, and healing rates were accelerated with the addition of EPS to the nanofiber web. The findings indicate that the synthesized EPS is applicable for skin tissue engineering purposes (Hivechi et al., 2021). Table 2 illustrates the role of EPSs in the biomedical industry.

Table 2 Application of EPS in biomedical industry

Types of EPS/ polymers	Producing microorganism	Type of microorganism	Role/application in biomedical industry	Reference
Dextran	*Leuconostoc mesenteroides, Streptococcus, Lactobacillus, Gluconobacter*	Bacteria	Delivers drugs to specific action targets or tissue coupled with their role as natural anionic polyelectrolytes; used as a blood plasma expander, blood flow tonic, and anticholecterolics	Osemwegie et al. (2020)
Gellan	*Paenibacillus polymyxa*	Bacteria	Acetylated gellan for antioxidant and anti-tumor properties	Mohd Nadzir et al. (2021)
	Pseudomonas elodea		Gellan film as an implant for insulin delivery	
	Pseudomonas elodea, Sphingomonas paucimobilis		Administration of nasal formulations	
	Sphingomonas paucimobilis		Tissue engineering and bone regeneration	
Schizophyllan	*Schizophyllum commune*	Fungus	Antitumor and immune-stimulating action	Osemwegie et al. (2020)

(continued)

Table 2 (continued)

Types of EPS/ polymers	Producing microorganism	Type of microorganism	Role/application in biomedical industry	Reference
Curdlan	*Paenibacillus polymyxa*	Bacterium	Curdlan sulfate as a vaccine against the infection of hepatitis B virus (HBV) and treatment of severe malaria. Antioxidant and antitumor activities	Mohd Nadzir et al. (2021)
Kefiran	*Lactobacillushilgardii, L.rhamnosus, L. kefir, L. kefiranofascien*	Bacteria	Antitumor properties	
	L.kefiranofascien		Mucosal adjuvant	
Alginate	*Pseudomonas, Azotobacter*	Bacteria	Gel for drug and protein delivery, wound dressing, cell culture, tissue and bone regeneration, islet transplantation for treatment of type1diabetes, chondrocyte transplantation for treatment of urinary incontinence, and vesicoureteral reflux	
Xanthan gum	*Xanthomonas campestris*	Bacterium	Intra-articular injections for the treatment of osteoarthritis, adjuvant in the immune system, hydrogel for antitumor properties, and wound dressing	
Hyaluronic acid	*Streptococcus* sp.	Bacteria	Ophthalmic surgery, wound healing	
	Streptococcusequi, Streptococcus zooepidemicus		Tissue engineering and bone regeneration, drug delivery, cancer diagnosis	
	Cyanobacteria		Wound healing	
Cellulose	*Gluconacetobacter xylinum*	Bacteria	Abacterial cellulose–vaccarin wound dressing	
	Acetobacter xylinum, Pseudomonas sp., *Agrobacterium Rhizobium* sp.		Wound healing and tissue-engineered blood vessels	
	Ochrobactrum pseudintermedium		Surfactant for antibacterial properties	
	Leuconostoc spp.		Nanogels for intracellular drug delivery, drug encapsulating agent	
Levan	*Bacillus subtilis, Streptococcus mutans, Streptococcus salivarius*	Bacteria	Hypocholesterolemic agent; adhesive and antitumor properties	
	Zymomonas mobilis		Nanoparticles for wound healing, encapsulation agent	
	Halomonas smyrnensis		Levan-based scaffold for tissue engineering	

5 Factors Affecting Cost-Benefit Ratio in Industrial Application of Microbial EPSs

5.1 *Production Process*

The major cost in the industrial use of microbial EPSs occurs in the production process. It includes costs for the culture medium, reagents needed for purification procedures, and packaging. In addition, the costs of safety tests and transportation of the products should be considered. In the case that microbes can grow and produce EPS in the required circumstances, such as food matrices, soil, and water, the cost concerning cultivation and purification could be negligible. To reduce the cultivation cost, utilization of agricultural byproducts and waste materials should be a prospective approach (Ventorino et al., 2019). The use of recombinant microbes for EPS production is highly risky in the food industry as yet, especially in such a society that introduction of genetically modified foods is a controversial issue. Thorough purification of EPS is sometimes very difficult and costly, and hence, it could be a big hurdle when it is required for commercialization of EPSs. Furthermore, food-grade stuff should be used when the purified EPS is considered a food additive. Scale-up problem is a common concern to be solved for mass production of EPSs, especially when the culture medium shows high viscosity (Seviour et al., 2011). It should be pointed out that degradation of EPSs during the storage period and/or contamination of endotoxins in the final product could occur. Those should be problematic if they have adverse effects on the final product (Du et al., 2017; Mengi et al., 2020).

5.2 *Technical Breakthrough Expected to Confer Benefits on Industrial Application of Microbial EPSs*

Enormous efforts have been performed and continued to enhance EPS production by optimizing culture conditions, according to the fermentation engineering approach by using central composite design (CCD) integrated with suitable models such as Box-Behnken design (Gomaa & Yousef, 2020), response surface design (Asgher et al., 2020; Yao Ang et al., 2020), Plackett-Burman design (Padmanaban et al., 2015), Taguchi design (Medina-Cabrera et al., 2020), etc. Intriguingly, an economic approach based on Pareto optimality theory was conducted to estimate the optimal productivity, which gave the lowest cost for a culture medium with maintained bioactivity of an EPS (Lin et al., 2019). Discouragingly, no common rule has been figured out yet to optimize the culture conditions of microbes in relation to EPS production. By such conventional approaches, most of the hetero-EPSs achieved several folds greater EPS production than those obtained under preoptimized conditions. These were usually less than required for industrial applications, in contrast to homo-EPSs that were successfully mass-produced at the industrial levels. This is because biosynthesis of hetero-EPS cannot be independent of microbial cell growth, which has a substantial limitation in maximizing the production yields. To overcome this bottleneck, *diverse innovations should be needed. Agrobacterium tumefaciens, Pseudomonas aeruginosa, Vibrio cholerae*

5.2.1 Metabolic Engineering to Enhance EPS Yield

To date, two different pathways, the Wzx/Wzy-dependent pathway and the ABC-transporter-dependent pathway, have been illustrated for hetero-EPS biosynthetic pathways (Schilling et al., 2020; Schmid et al., 2015). Some of the related genes have been demonstrated to modify the sugar flux biased into EPS production. Modulation of Gal metabolism is likely to be a favorable target. For example, a-phosphoglucomutase is an enzyme capable of converting glucose 1-phosphate (Glc-1-P) to glucose 6-phosphate (Glc-6-P) and vice versa. Overexpression of pgm, a gene encoding a-phosphoglucomutase, in combination with galU, encoding UDP glucose pyrophosphorylase, could increase EPS production several folds when lactose (Lac) was used as a sole carbon source in *Streptococcus thermophilus* LY03 (Levander et al., 2002). Gene expression levels of UDP-glucose pyrophosphorylase and UDP-galactose 4-epimerase were highly induced when EPS-producing *Lactobacillus casei* CRL 87 was cultivated at an acidic pH of 5.0 using Gal as a carbon source, giving better production of EPS; therefore, these genes can be candidates for further metabolic engineering in the strain (Mozzi et al., 2003). As is well known, EPS production is induced by abiotic stresses such as bile acid, freezing, peroxides, pH, and salt, and hence, the related metabolism can be a potential target for metabolic engineering. (Jang et al., 2016) showed that increased H_2O_2 levels in alkyl hydroperoxide reductase encoding gene deletion mutant enhanced EPS production in *Acinetobacter oleivorans* DR1. Although the mechanism is still unclear, the elevated expression level of the NADH oxidase encoding gene increased EPS production, probably due to reduced lactate production flux (N. Li et al., 2015). Some nutrient uptake pathways and/or chemosensory systems in microbes are also potential targets for metabolic engineering in terms of enhancing EPS production. A gene, SMc00722, which was putatively involved with the maintenance of intracellular magnesium concentration, was found in a Gram-negative a-proteobaceteria, *Ensifer meliloti* Rm1021 (Hawkins & Oresnik, 2017). Mutation in the gene caused enhanced production of succinoglycan in the strain. A signal transducer protein, DifA, was revealed to regulate EPS production in a Gram-negative -proteobacteria, *Myxococcus xanthus* (Xu et al., 2011). Interestingly, Islam et al. (2020) found that bacterial cellular density could alter the production of different types of EPSs in *Myxococcus xanthus*. Namely, an EPS produced via the Wzx/Wzy-dependent polysaccharide-assembly pathway was preferentially produced by the bacterial cells located at the lower-density swarm periphery, whereas a novel biosurfactant polysaccharide, a type IV pilus-inhibited acidic polymer, was mainly produced in the higher-density swarm interior. This knowledge will provide novel options regarding metabolic engineering for targeting enhanced EPS production levels.

5.2.2 Nanoparticles in EPS Development

A highly attractive technique is the development of EPS covalently conjugated with nanoparticles made of metals such as copper, gold, iron, platinum, selenium, silver, titanium, zinc, and so on, via biosynthesis in microbial cells, so-called green synthesis (Singh et al., 2018). EPS is capable of reducing and stabilizing metal nanoparticles, exhibiting a wide spectrum of beneficial properties, for example, antibacterial (Banerjee et al., 2020; Garza-Cervantes et al., 2019; Li et al., 2020; Navarro Gallón et al., 2019; Rajivgandhi et al., 2020a, b; Riaz Rajoka et al., 2020; Sivasankar et al., 2018; Vazquez-Rodriguez et al., 2020), anticancer (Akturk, 2020; Buttacavoli et al., 2018), antioxidant (Rajivgandhi et al., 2020b; Riaz Rajoka et al., 2020; Singh et al., 2019; Sivasankar et al., 2018; Xiao et al., 2017), bioremediation (Casentini et al., 2019; Wang et al., 2018), ferrous ion supply to truffles (Picceri et al., 2018), food packaging (Shahabi-Ghahfarrokhi & Babaei-Ghazvini, 2019), functionalization in yogurt (Garcés et al., 2020), inhibition of biofilm formation (Abinaya et al., 2018; Bhattacharyya et al., 2018; Cusimano et al., 2020; Garza-Cervantes et al., 2019; Liu et al., 2019; Rajivgandhi et al., 2020b; Vazquez-Rodriguez et al., 2020), mosquito larvicidal (Abinaya et al., 2018), and waste-water treatment (Govarthanan et al., 2020; Saravanan et al., 2017).

A wide structural and functional variety, high biocompatibility, and biodegradability of microbial EPS conjugated with nanoparticles seem to be the most promising technique for future applications in various sorts of industries.

5.2.3 Chemical and Enzymatic Modifications

Chemical and enzymatic modifications are relatively conventional techniques to modulate the physicochemical and bioactive properties of EPSs. For example, EPSs have commonly been acetylated, carboxymethylated, phosphorylated, and sulfated to confer anticoagulant (Brandi, 2011), antimicrobial (Vidiany Aparecida Queiroz et al., 2019), antitumor (Wang et al., 2015), antioxidant (Jia et al., 2017; Vidiany Aparecida Queiroz et al., 2019; Wang et al., 2015; Xiao et al., 2020; Zhang et al., 2016), hypoglycemic, and hypolipidemic effects (Wang et al., 2017). Curiously, N-phthaloyl derivative of polysaccharide derived from Lachnum YM262 could enhance its immunomodulating property in mice treated with cyclophosphamide (Chen et al., 2016a). Zhang et al. (2016) found antitumor activity in a selenized derivative of EPS, which was purified from the liquid culture of an edible mushroom, Grifola frondosa.

Partial degradation of EPS by the action of certain enzymes also potential to promote commercial values of microbial EPSs. A sort of EPS can be hydrolyzed by glucanases and/or depolymerized by polysaccharide lyases via a b-elimination reaction (Michaud et al., 2003) and even modified (Whitfield et al., 2015). Recent findings in phage-encoded depolymerase acting on microbial EPSs can expand the horizons of enzymatic modification of EPS and its application (Hernandez-Morales et al., 2018; Mi et al., 2019; Oliveira et al., 2019, 2020; Pires et al., 2016).

5.2.4 Ultrasound and Microwave Treatments

Ultrasound treatment has been used for increasing aqueous solubility of EPS by means of partial degradation of EPS molecules (Wang et al., 2010). Fortunately, some partially degraded EPSs showed higher bioactivities, including antioxidant (Chen et al., 2014), anti-inflammation (Du et al., 2016), cryo- and bile-protective (Song et al., 2019), and prebiotic (Mao et al., 2020a, b; Song et al., 2018) properties, compared to the intact ones. EPS, were found in ultrasound-treated milk with regard to the biodegradation of milk components by starter bacteria during the fermentation process (Gholamhosseinpour et al., 2020; Potoroko et al., 2018). Lu et al. (2020) indicated direct effects of ultrasound treatment on the promotion of mycelial growth and EPS production in *Agaricus bitorquis* (Quél.) Sacc. Chaidam as a consequence of increased bacterial cell permeability and mass transfer rate. It has also been reported that EPS extraction efficiency could be enhanced by ultrasound treatment (Hasheminya & Dehghannya, 2020; Wang et al., 2019).

Furthermore, Song et al. (2020) showed that ultrasound treatment promoted nano emulsification of EPS with medicinal substances, which could be used for vulvovaginal candidiasis treatment. In a similar manner, microwave treatment was applied for the enhancement of EPS production in a green microalga, *Scenedesmus* sp. (Sivaramakrishnan et al., 2020).

6 Conclusion

Microbial exopolysaccharides (EPSs) are versatile carbohydrates with excellent health and industrial uses largely due to their unique structural and functional properties. They are regarded for their biocompatibility, nontoxicity, thermal stability, and biodegradability, which are primarily derived from microbes. However, EPSs offer viable alternatives to synthetic compounds in medicines as antioxidants and anticancer, antiviral, and antifungal agents. The therapeutic promise of their ability to augment immune responses and attenuate side effects is such that it puts them in this class. Industrially, EPSs have the function of thickeners, stabilizers, and gelling agents for food items; they greatly increase food quality and functioning. Second, they also function as environmentally benign alternatives to environmentally detrimental chemical synthesis methods. However, hurdles remain in their commercialization, and these are driven by greater production costs and the requirement to fully characterize their structure to ensure optimal use. In the future, this work could be extended through genetic and metabolic engineering, optimization of fermentation conditions, and utilizing economical substrates for improved production efficiency. Moreover, the structure–function couplings need to be clarified together by highlighting the potential of EPSs for lucrative pharmaceutical and cosmetic industries where functional attributes are superior to manufacturing costs. The adaptability of microbial EPSs ensures their increasing importance in several sectors, including environmental application of water treatment, food processing, and health care. In the future, as knowledge grows, these biopolymers will become essential in providing sustainable solutions and improving quality standards in many sectors. This demonstrates the importance of these options as environmentally friendly, economical, and with a wide commercial and social impact.

References

Abarquero, D., Renes, E., Fresno, J. M., & Tornadijo, M. E. (2022). Study of exopolysaccharides from lactic acid bacteria and their industrial applications: A review. *International Journal of Food Science & Technology, 57*(1), 16–26. https://doi.org/10.1111/ijfs.15227

Abinaya, M., Vaseeharan, B., Divya, M., Sharmili, A., Govindarajan, M., Alharbi, N. S., Kadaikunnan, S., Khaled, J. M., & Benelli, G. (2018). Bacterial exopolysaccharide (EPS)-coated ZnO nanoparticles showed high antibiofilm activity and larvicidal toxicity against malaria and Zika virus vectors. *Journal of Trace Elements in Medicine and Biology, 45*, 93–103. https://doi.org/10.1016/j.jtemb.2017.10.002

Akturk, O. (2020). Colloidal stability and biological activity evaluation of microbial exopolysaccharide Levan-capped gold nanoparticles. *Colloids and Surfaces B: Biointerfaces, 192*, 111061. https://doi.org/10.1016/j.colsurfb.2020.111061

Al-Nabulsi, A. A., Jaradat, Z. W., Al Qudsi, F. R., Elsalem, L., Osaili, T. M., Olaimat, A. N., Esposito, G., Liu, S.-Q., & Ayyash, M. M. (2022). Characterization and bioactive properties of exopolysaccharides produced by Streptococcus thermophilus and lactobacillus bulgaricus isolated from labaneh. *LWT, 167*, 113817. https://doi.org/10.1016/j.lwt.2022.113817

Andrew, M., & Jayaraman, G. (2022). Molecular characterization and biocompatibility of exopolysaccharide produced by moderately halophilic bacterium Virgibacillus dokdonensis from the Saltern of Kumta coast. *Polymers, 14*(19), 3986. https://doi.org/10.3390/polym14193986

Angelin, J., & Kavitha, M. (2020). Exopolysaccharides from probiotic bacteria and their health potential. *International Journal of Biological Macromolecules, 162*, 853–865. https://doi.org/10.1016/j.ijbiomac.2020.06.190

Asgher, M., Urooj, Y., Qamar, S. A., & Khalid, N. (2020). Improved exopolysaccharide production from bacillus licheniformis MS3: Optimization and structural/functional characterization. *International Journal of Biological Macromolecules, 151*, 984–992. https://doi.org/10.1016/j.ijbiomac.2019.11.094

Bagnol, R., Grijpma, D., Eglin, D., & Moriarty, T. F. (2022). The production and application of bacterial exopolysaccharides as biomaterials for bone regeneration. *Carbohydrate Polymers, 291*, 119550. https://doi.org/10.1016/j.carbpol.2022.119550

Banerjee, A., Das, D., Rudra, S. G., Mazumder, K., Andler, R., & Bandopadhyay, R. (2020). Characterization of exopolysaccharide produced by pseudomonas sp. PFAB4 for synthesis of EPS-coated AgNPs with antimicrobial properties. *Journal of Polymers and the Environment, 28*(1), 242–256. https://doi.org/10.1007/s10924-019-01602-z

Basiri, S. (2021). Applications of microbial exopolysaccharides in the food industry. *Avicenna Journal of Medical Biochemistry, 9*(2), 107–120. https://doi.org/10.34172/ajmb.2021.16

Bhattacharyya, P., Agarwal, B., Goswami, M., Maiti, D., Baruah, S., & Tribedi, P. (2018). Zinc oxide nanoparticle inhibits the biofilm formation of Streptococcus pneumoniae. *Antonie Van Leeuwenhoek, 111*(1), 89–99. https://doi.org/10.1007/s10482-017-0930-7

Bondi, M., Lauková, A., de Niederhausern, S., Messi, P., & Papadopoulou, C. (2017). Natural preservatives to improve food quality and safety. *Journal of Food Quality, 2017*, 1–3. https://doi.org/10.1155/2017/1090932

Brandi, J. (2011). Chemical modification of Botryosphaeran: Structural characterization and anticoagulant activity of a water-soluble sulfonated (1→3)(1→6)-β-D-Glucan. *Journal of Microbiology and Biotechnology, 21*(10), 1036–1042. https://doi.org/10.4014/jmb.1105.05020

Buttacavoli, M., Albanese, N. N., Di Cara, G., Alduina, R., Faleri, C., Gallo, M., Pizzolanti, G., Gallo, G., Feo, S., Baldi, F., & Cancemi, P. (2018). Anticancer activity of biogenerated silver nanoparticles: An integrated proteomic investigation. *Oncotarget, 9*(11), 9685–9705. https://doi.org/10.18632/oncotarget.23859

Casentini, B., Gallo, M., & Baldi, F. (2019). Arsenate and arsenite removal from contaminated water by iron oxides nanoparticles formed inside a bacterial exopolysaccharide. *Journal of Environmental Chemical Engineering, 7*(1), 102908. https://doi.org/10.1016/j.jece.2019.102908

Castillo, N. A., Valdez, A. L., & Fariña, J. I. (2015). Microbial production of scleroglucan and downstream processing. *Frontiers in Microbiology, 6*. https://doi.org/10.3389/fmicb.2015.01106

Chen, X., Siu, K.-C., Cheung, Y.-C., & Wu, J.-Y. (2014). Structure and properties of a (1→3)-β-d-glucan from ultrasound-degraded exopolysaccharides of a medicinal fungus. *Carbohydrate Polymers, 106*, 270–275. https://doi.org/10.1016/j.carbpol.2014.02.040

Chen, T., Wang, Y., Li, J., Su, N., Surhio, M. M., Yang, L., & Ye, M. (2016a). Phthaloyl modification of a polysaccharide from Lachnum YM262 and immunomodulatory activity. *Process Biochemistry, 51*(10), 1599–1609. https://doi.org/10.1016/j.procbio.2016.08.004

Chen, X. Y., Levy, C., & Gänzle, M. G. (2016b). Structure-function relationships of bacterial and enzymatically produced reuterans and dextran in sourdough bread baking application. *International Journal of Food Microbiology, 239*, 95–102. https://doi.org/10.1016/j.ijfoodmicro.2016.06.010

Cusimano, M. G., Ardizzone, F., Nasillo, G., Gallo, M., Sfriso, A., Martino-Chillura, D., Schillaci, D., Baldi, F., & Gallo, G. (2020). Biogenic iron-silver nanoparticles inhibit bacterial biofilm formation due to Ag+ release as determined by a novel phycoerythrin-based assay. *Applied Microbiology and Biotechnology, 104*(14), 6325–6336. https://doi.org/10.1007/s00253-020-10686-w

Dertli, E., Toker, O. S., Durak, M. Z., Yilmaz, M. T., Tatlısu, N. B., Sagdic, O., & Cankurt, H. (2016). Development of a fermented ice-cream as influenced by in situ exopolysaccharide production: Rheological, molecular, microstructural and sensory characterization. *Carbohydrate Polymers, 136*, 427–440. https://doi.org/10.1016/j.carbpol.2015.08.047

Du, B., Zeng, H., Yang, Y., Bian, Z., & Xu, B. (2016). Anti-inflammatory activity of polysaccharide from Schizophyllum commune as affected by ultrasonication. *International Journal of Biological Macromolecules, 91*, 100–105. https://doi.org/10.1016/j.ijbiomac.2016.05.052

Du, A. G.-L., Zykwinska, A., Sinquin, C., Ratiskol, J., Weiss, P., Vinatier, C., Guicheux, J., Delbarre-Ladrat, C., & Colliec-Jouault, S. (2017). Purification of the exopolysaccharide produced by Alteromonas infernus: Identification of endotoxins and effective process to remove them. *Applied Microbiology and Biotechnology, 101*(17), 6597–6606. https://doi.org/10.1007/s00253-017-8364-8

Finore, I., Di Donato, P., Mastascusa, V., Nicolaus, B., & Poli, A. (2014). Fermentation technologies for the optimization of marine microbial exopolysaccharide production. *Marine Drugs, 12*(5), 3005–3024. https://doi.org/10.3390/md12053005

Freitas, F., Torres, C. A. V., & Reis, M. A. M. (2017). Engineering aspects of microbial exopolysaccharide production. *Bioresource Technology, 245*, 1674–1683. https://doi.org/10.1016/j.biortech.2017.05.092

Fukuda, K., & Kono, H. (2021). *Cost-benefit analysis and industrial potential of exopolysaccharides* (pp. 303–339). https://doi.org/10.1007/978-3-030-75289-7_12

Garcés, V., González, A., Sabio, L., Sánchez-Arévalo, C. M., Gálvez, N., & Dominguez-Vera, J. M. (2020). Magnetic and Golden yogurts. Food as a potential nanomedicine carrier. *Materials, 13*(2), 481. https://doi.org/10.3390/ma13020481

García, A., Fernández-Sandoval, M. T., Morales-Guzmán, D., Martínez-Morales, F., & Trejo-Hernández, M. R. (2022). Advances in exopolysaccharide production from marine bacteria. *Journal of Chemical Technology & Biotechnology, 97*(10), 2694–2705. https://doi.org/10.1002/jctb.7156

Garza-Cervantes, J. A., Escárcega-González, C. E., Díaz Barriga Castro, E., Mendiola-Garza, G., Marichal-Cancino, B. A., López-Vázquez, M. A., & Morones-Ramirez, J. R. (2019). Antimicrobial and antibiofilm activity of biopolymer-Ni, Zn nanoparticle biocomposites synthesized using *R. mucilaginosa* UANL-001L exopolysaccharide as a capping agent. *International Journal of Nanomedicine, 14*, 2557–2571. https://doi.org/10.2147/IJN.S196470

Gholamhosseinpour, A., Hashemi, S. M. B., Raoufi Jahromi, L., & Sourki, A. H. (2020). Conventional heating, ultrasound and microwave treatments of milk: Fermentation efficiency and biological activities. *International Dairy Journal, 110*, 104809. https://doi.org/10.1016/j.idairyj.2020.104809

Gomaa, M., & Yousef, N. (2020). Optimization of production and intrinsic viscosity of an exopolysaccharide from a high yielding Virgibacillus salarius BM02: Study of its potential antioxidant, emulsifying properties and application in the mixotrophic cultivation of Spirulina platensis. *International Journal of Biological Macromolecules, 149*, 552–561. https://doi.org/10.1016/j.ijbiomac.2020.01.289

Gomes, F. P., Silva, N. H. C. S., Trovatti, E., Serafim, L. S., Duarte, M. F., Silvestre, A. J. D., Neto, C. P., & Freire, C. S. R. (2013). Production of bacterial cellulose by Gluconacetobacter sacchari using dry olive mill residue. *Biomass and Bioenergy, 55*, 205–211. https://doi.org/10.1016/j.biombioe.2013.02.004

González, A. G., Shirokova, L. S., Pokrovsky, O. S., Emnova, E. E., Martínez, R. E., Santana-Casiano, J. M., González-Dávila, M., & Pokrovski, G. S. (2010). Adsorption of copper on pseudomonas aureofaciens: Protective role of surface exopolysaccharides. *Journal of Colloid and Interface Science, 350*(1), 305–314. https://doi.org/10.1016/j.jcis.2010.06.020

Govarthanan, M., Jeon, C.-H., Jeon, Y.-H., Kwon, J.-H., Bae, H., & Kim, W. (2020). Non-toxic nano approach for wastewater treatment using Chlorella vulgaris exopolysaccharides immobilized in iron-magnetic nanoparticles. *International Journal of Biological Macromolecules, 162*, 1241–1249. https://doi.org/10.1016/j.ijbiomac.2020.06.227

Hasheminya, S.-M., & Dehghannya, J. (2020). Novel ultrasound-assisted extraction of kefiran biomaterial, a prebiotic exopolysaccharide, and investigation of its physicochemical, antioxidant and antimicrobial properties. *Materials Chemistry and Physics, 243*, 122645. https://doi.org/10.1016/j.matchemphys.2020.122645

Hawkins, J. P., & Oresnik, I. J. (2017). Characterisation of a gene encoding a membrane protein that affects exopolysaccharide production and intracellular Mg2+ concentrations in Ensifer meliloti. *FEMS Microbiology Letters, 364*(7). https://doi.org/10.1093/femsle/fnx061

Hernandez-Morales, A. C., Lessor, L. L., Wood, T. L., Migl, D., Mijalis, E. M., Cahill, J., Russell, W. K., Young, R. F., & Gill, J. J. (2018). Genomic and biochemical characterization of Acinetobacter Podophage petty reveals a novel lysis mechanism and tail-associated depolymerase activity. *Journal of Virology, 92*(6). https://doi.org/10.1128/JVI.01064-17

Hivechi, A., Milan, P. B., Modabberi, K., Amoupour, M., Ebrahimzadeh, K., Gholipour, A. R., Sedighi, F., Amini, N., Bahrami, S. H., Rezapour, A., Hamidi, M., & Delattre, C. (2021). Synthesis and characterization of exopolysaccharide encapsulated PCL/gelatin skin substitute for full-thickness wound regeneration. *Polymers, 13*(6), 854. https://doi.org/10.3390/polym13060854

Imre, B., & Pukánszky, B. (2013). Compatibilization in bio-based and biodegradable polymer blends. *European Polymer Journal, 49*(6), 1215–1233. https://doi.org/10.1016/j.eurpolymj.2013.01.019

Islam, S. T., & Lam, J. S. (2014). Synthesis of bacterial polysaccharides via the Wzx/Wzy-dependent pathway. *Canadian Journal of Microbiology, 60*(11), 697–716. https://doi.org/10.1139/cjm-2014-0595

Islam, S. T., Vergara Alvarez, I., Saïdi, F., Guiseppi, A., Vinogradov, E., Sharma, G., Espinosa, L., Morrone, C., Brasseur, G., Guillemot, J.-F., Benarouche, A., Bridot, J.-L., Ravicoularamin, G., Cagna, A., Gauthier, C., Singer, M., Fierobe, H.-P., Mignot, T., & Mauriello, E. M. F. (2020). Modulation of bacterial multicellularity via spatio-specific polysaccharide secretion. *PLoS Biology, 18*(6), e3000728–e3000728. https://doi.org/10.1371/journal.pbio.3000728

Jang, I.-A., Kim, J., & Park, W. (2016). Endogenous hydrogen peroxide increases biofilm formation by inducing exopolysaccharide production in Acinetobacter oleivorans DR1. *Scientific Reports, 6*(1), 21121. https://doi.org/10.1038/srep21121

Jia, X., Wang, C., Bai, Y., Yu, J., & Xu, C. (2017). Sulfation of the extracellular polysaccharide produced by the king oyster culinary-medicinal mushroom, Pleurotus eryngii (Agaricomycetes), and its antioxidant properties in vitro. *International Journal of Medicinal Mushrooms, 19*(4), 355–362. https://doi.org/10.1615/IntJMedMushrooms.v19.i4.60

Kambourova, M., Oner, E. T., & Poli, A. (2015). Exopolysaccharides from prokaryotic microorganisms—Promising sources for white biotechnology processes. In *Industrial biorefineries & white biotechnology* (pp. 523–554). Elsevier. https://doi.org/10.1016/B978-0-444-63453-5.00017-3

Korcz, E., & Varga, L. (2021). Exopolysaccharides from lactic acid bacteria: Techno-functional application in the food industry. *Trends in Food Science & Technology, 110*, 375–384. https://doi.org/10.1016/j.tifs.2021.02.014

Korcz, E., Kerényi, Z., & Varga, L. (2018). Dietary fibers, prebiotics, and exopolysaccharides produced by lactic acid bacteria: Potential health benefits with special regard to cholesterol-lowering effects. *Food & Function, 9*(6), 3057–3068. https://doi.org/10.1039/C8FO00118A

Lee, Y.-J., An, S.-J., Bae, E.-B., Gwon, H.-J., Park, J.-S., Jeong, S., Jeon, Y.-C., Lee, S.-H., Lim, Y.-M., & Huh, J.-B. (2017). The effect of thickness of resorbable bacterial cellulose membrane on guided bone regeneration. *Materials, 10*(3), 320. https://doi.org/10.3390/ma10030320

Levander, F., Svensson, M., & Rådström, P. (2002). Enhanced exopolysaccharide production by metabolic engineering of *Streptococcus thermophilus*. *Applied and Environmental Microbiology, 68*(2), 784–790. https://doi.org/10.1128/AEM.68.2.784-790.2002

Li, N., Wang, Y., Zhu, P., Liu, Z., Guo, B., & Ren, J. (2015). Improvement of exopolysaccharide production in Lactobacillus casei LC2W by overexpression of NADH oxidase gene. *Microbiological Research, 171*, 73–77. https://doi.org/10.1016/j.micres.2014.12.006

Li, S., Xia, H., Xie, A., Wang, Z., Ling, K., Zhang, Q., & Zou, X. (2020). Structure of a fucose-rich polysaccharide derived from EPS produced by Kosakonia sp. CCTCC M2018092 and its application in antibacterial film. *International Journal of Biological Macromolecules, 159*, 295–303. https://doi.org/10.1016/j.ijbiomac.2020.05.029

Lin, T., Chen, C., Chen, B., Shaw, J., & Chen, Y. (2019). Optimal economic productivity of exopolysaccharides from lactic acid bacteria with production possibility curves. *Food Science & Nutrition, 7*(7), 2336–2344. https://doi.org/10.1002/fsn3.1079

Liu, L., Li, J.-H., Zi, S.-F., Liu, F.-R., Deng, C., Ao, X., & Zhang, P. (2019). AgNP combined with quorum sensing inhibitor increased the antibiofilm effect on Pseudomonas aeruginosa. *Applied Microbiology and Biotechnology, 103*(15), 6195–6204. https://doi.org/10.1007/s00253-019-09905-w

Lu, H., Lou, H., Wei, T., Liu, Z., Jiao, Y., & Chen, Q. (2020). Ultrasound enhanced production of mycelia and exopolysaccharide by Agaricus bitorquis (Quél.) Sacc. Chaidam. *Ultrasonics Sonochemistry, 64*, 105040. https://doi.org/10.1016/j.ultsonch.2020.105040

Lynch, K. M., Coffey, A., & Arendt, E. K. (2018). Exopolysaccharide producing lactic acid bacteria: Their techno-functional role and potential application in gluten-free bread products. *Food Research International, 110*, 52–61. https://doi.org/10.1016/j.foodres.2017.03.012

Mahapatra, S., & Banerjee, D. (2013). Fungal exopolysaccharide: Production, composition and applications. *Microbiology Insights, 6*. https://doi.org/10.4137/MBI.S10957

Manivasagan, P., & Kim, S.-K. (2014). *Extracellular polysaccharides produced by marine bacteria* (pp. 79–94). https://doi.org/10.1016/B978-0-12-800269-8.00005-1.

Mao, Y.-H., Song, A.-X., Li, L.-Q., Siu, K.-C., Yao, Z.-P., & Wu, J.-Y. (2020a). Effects of exopolysaccharide fractions with different molecular weights and compositions on fecal microflora during in vitro fermentation. *International Journal of Biological Macromolecules, 144*, 76–84. https://doi.org/10.1016/j.ijbiomac.2019.12.072

Mao, Y.-H., Song, A.-X., Li, L.-Q., Yang, Y., Yao, Z.-P., & Wu, J.-Y. (2020b). A high-molecular weight exopolysaccharide from the Cs-HK1 fungus: Ultrasonic degradation, characterization and in vitro fecal fermentation. *Carbohydrate Polymers, 246*, 116636. https://doi.org/10.1016/j.carbpol.2020.116636

Medina-Cabrera, E. V., Rühmann, B., Schmid, J., & Sieber, V. (2020). Optimization of growth and EPS production in two Porphyridum strains. *Bioresource Technology Reports, 11*, 100486. https://doi.org/10.1016/j.biteb.2020.100486

Mengi, B., Ikeda, S., Murayama, D., Bochimoto, H., Matsumoto, S., Kitazawa, H., Urashima, T., & Fukuda, K. (2020). Factors affecting decreasing viscosity of the culture medium during the stationary growth phase of exopolysaccharide-producing *Lactobacillus fermentum* MTCC 25067. *Bioscience of Microbiota, Food and Health, 39*(3), 160–168. https://doi.org/10.12938/bmfh.2019-051

Mi, L., Liu, Y., Wang, C., He, T., Gao, S., Xing, S., Huang, Y., Fan, H., Zhang, X., Yu, W., Mi, Z., Tong, Y., Bai, C., & Han, F. (2019). Identification of a lytic Pseudomonas aeruginosa phage depolymerase and its anti-biofilm effect and bactericidal contribution to serum. *Virus Genes, 55*(3), 394–405. https://doi.org/10.1007/s11262-019-01660-4

Michaud, P., Da Costa, A., Courtois, B., & Courtois, J. (2003). Polysaccharide Lyases: Recent developments as biotechnological tools. *Critical Reviews in Biotechnology, 23*(4), 233–266. https://doi.org/10.1080/07388550390447043

Mohd Nadzir, M., Nurhayati, R. W., Idris, F. N., & Nguyen, M. H. (2021). Biomedical applications of bacterial exopolysaccharides: A review. *Polymers, 13*(4), 530. https://doi.org/10.3390/polym13040530

Moradi, M., Guimarães, J. T., & Sahin, S. (2021). Current applications of exopolysaccharides from lactic acid bacteria in the development of food active edible packaging. *Current Opinion in Food Science, 40*, 33–39. https://doi.org/10.1016/j.cofs.2020.06.001

Mozzi, F., Savoy de Giori, G., & Font de Valdez, G. (2003). UDP-galactose 4-epimerase: A key enzyme in exopolysaccharide formation by Lactobacillus casei CRL 87 in controlled pH batch cultures. *Journal of Applied Microbiology, 94*(2), 175–183. https://doi.org/10.1046/j.1365-2672.2003.01821.x

Navarro Gallón, S. M., Alpaslan, E., Wang, M., Larese-Casanova, P., Londoño, M. E., Atehortúa, L., Pavón, J. J., & Webster, T. J. (2019). Characterization and study of the antibacterial mechanisms of silver nanoparticles prepared with microalgal exopolysaccharides. *Materials Science and Engineering: C, 99*, 685–695. https://doi.org/10.1016/j.msec.2019.01.134

Noel, S., Fortier, C., Murschel, F., Belzil, A., Gaudet, G., Jolicoeur, M., & De Crescenzo, G. (2016). Co-immobilization of adhesive peptides and VEGF within a dextran-based coating for vascular applications. *Acta Biomaterialia, 37*, 69–82. https://doi.org/10.1016/j.actbio.2016.03.043

Nwodo, U. U., Green, E., & Okoh, A. I. (2012). Bacterial exopolysaccharides: Functionality and prospects. *International Journal of Molecular Sciences, 13*(11), 14002–14015. https://doi.org/10.3390/ijms131114002

Oliveira, H., Costa, A. R., Ferreira, A., Konstantinides, N., Santos, S. B., Boon, M., Noben, J.-P., Lavigne, R., & Azeredo, J. (2019). Functional analysis and Antivirulence properties of a new Depolymerase from a Myovirus that infects Acinetobacter baumannii capsule K45. *Journal of Virology, 93*(4). https://doi.org/10.1128/JVI.01163-18

Oliveira, H., Pinto, G., Mendes, B., Dias, O., Hendrix, H., Akturk, E., Noben, J.-P., Gawor, J., Łobocka, M., Lavigne, R., & Azeredo, J. (2020). A Tailspike with exopolysaccharide Depolymerase activity from a new Providencia stuartii phage makes multidrug-resistant bacteria susceptible to serum-mediated killing. *Applied and Environmental Microbiology, 86*(13). https://doi.org/10.1128/AEM.00073-20

Orlandelli, R. C., Vasconcelos, A. F. D., Azevedo, J. L., Corradi da Silva, M. d. L., & Pamphile, J. A. (2016). Screening of endophytic sources of exopolysaccharides: Preliminary characterization of crude exopolysaccharide produced by submerged culture of Diaporthe sp. JF766998 under different cultivation time. *Biochimie Open, 2*, 33–40. https://doi.org/10.1016/j.biopen.2016.02.003

Osemwegie, O. O., Adetunji, C. O., Ayeni, E. A., Adejobi, O. I., Arise, R. O., Nwonuma, C. O., & Oghenekaro, A. O. (2020). Exopolysaccharides from bacteria and fungi: Current status and perspectives in Africa. *Heliyon, 6*(6), e04205–e04205. https://doi.org/10.1016/j.heliyon.2020.e04205

Osińska-Jaroszuk, M., Jarosz-Wilkołazka, A., Jaroszuk-Ściseł, J., Szałapata, K., Nowak, A., Jaszek, M., Ozimek, E., & Majewska, M. (2015). Extracellular polysaccharides from Ascomycota and Basidiomycota: Production conditions, biochemical characteristics, and biological properties. *World Journal of Microbiology and Biotechnology, 31*(12), 1823–1844. https://doi.org/10.1007/s11274-015-1937-8

Padmanaban, S., Balaji, N., Muthukumaran, C., & Tamilarasan, K. (2015). Statistical optimization of process parameters for exopolysaccharide production by Aureobasidium pullulans using sweet potato based medium. *3 Biotech, 5*(6), 1067–1073. https://doi.org/10.1007/s13205-015-0308-3

Papinutti, L. (2010). Effects of nutrients, pH and water potential on exopolysaccharides production by a fungal strain belonging to Ganoderma lucidum complex. *Bioresource Technology, 101*(6), 1941–1946. https://doi.org/10.1016/j.biortech.2009.09.076

Picceri, G. G., Leonardi, P., Iotti, M., Gallo, M., Baldi, F., Zambonelli, A., Amicucci, A., Vallorani, L., Piccoli, G., Ciccimarra, G., Arshakyan, M., Burattini, S., Falcieri, E., & Chiarantini, L. (2018). Bacteria-produced ferric exopolysaccharide nanoparticles as iron delivery system for truffles (Tuber borchii). *Applied Microbiology and Biotechnology, 102*(3), 1429–1441. https://doi.org/10.1007/s00253-017-8615-8

Pires, D. P., Oliveira, H., Melo, L. D. R., Sillankorva, S., & Azeredo, J. (2016). Bacteriophage-encoded depolymerases: Their diversity and biotechnological applications. *Applied Microbiology and Biotechnology, 100*(5), 2141–2151. https://doi.org/10.1007/s00253-015-7247-0

Polak-Berecka, M., Choma, A., Waśko, A., Górska, S., Gamian, A., & Cybulska, J. (2015). Physicochemical characterization of exopolysaccharides produced by Lactobacillus rhamnosus on various carbon sources. *Carbohydrate Polymers, 117*, 501–509. https://doi.org/10.1016/j.carbpol.2014.10.006

Potoroko, I., Kalinina, I., Botvinnikova, V., Krasulya, O., Fatkullin, R., Bagale, U., & Sonawane, S. H. (2018). Ultrasound effects based on simulation of milk processing properties. *Ultrasonics Sonochemistry, 48*, 463–472. https://doi.org/10.1016/j.ultsonch.2018.06.019

Prajapat, J. B., & Patel, A. (2013). Food and health applications of exopolysaccharides produced by lactic acid bacteria. *Advances in Dairy Research, 1*(2). https://doi.org/10.4172/2329-888X.1000107

Rajivgandhi, G. N., Maruthupandy, M., Li, J.-L., Dong, L., Alharbi, N. S., Kadaikunnan, S., Khaled, J. M., Alanzi, K. F., & Li, W.-J. (2020a). Photocatalytic reduction and anti-bacterial activity of biosynthesized silver nanoparticles against multi drug resistant Staphylococcus saprophyticus BDUMS 5 (MN310601). *Materials Science and Engineering: C, 114*, 111024. https://doi.org/10.1016/j.msec.2020.111024

Rajivgandhi, G. N., Ramachandran, G., Maruthupandy, M., Manoharan, N., Alharbi, N. S., Kadaikunnan, S., Khaled, J. M., Almanaa, T. N., & Li, W.-J. (2020b). Anti-oxidant, anti-bacterial and anti-biofilm activity of biosynthesized silver nanoparticles using Gracilaria corticata against biofilm producing K. pneumoniae. *Colloids and Surfaces A: Physicochemical and Engineering Aspects, 600*, 124830. https://doi.org/10.1016/j.colsurfa.2020.124830

Rana, S., & Upadhyay, L. S. B. (2020). Microbial exopolysaccharides: Synthesis pathways, types and their commercial applications. *International Journal of Biological Macromolecules, 157*, 577–583. https://doi.org/10.1016/j.ijbiomac.2020.04.084

Riaz Rajoka, M. S., Mehwish, H. M., Zhang, H., Ashraf, M., Fang, H., Zeng, X., Wu, Y., Khurshid, M., Zhao, L., & He, Z. (2020). Antibacterial and antioxidant activity of exopolysaccharide mediated silver nanoparticle synthesized by Lactobacillus brevis isolated from Chinese koumiss. *Colloids and Surfaces B: Biointerfaces, 186*, 110734. https://doi.org/10.1016/j.colsurfb.2019.110734

Salimi, F., & Farrokh, P. (2023). Recent advances in the biological activities of microbial exopolysaccharides. *World Journal of Microbiology and Biotechnology, 39*(8), 213. https://doi.org/10.1007/s11274-023-03660-x

Saravanan, C., Rajesh, R., Kaviarasan, T., Muthukumar, K., Kavitake, D., & Shetty, P. H. (2017). Synthesis of silver nanoparticles using bacterial exopolysaccharide and its application for degradation of azo-dyes. *Biotechnology Reports, 15*, 33–40. https://doi.org/10.1016/j.btre.2017.02.006

Schilling, C., Badri, A., Sieber, V., Koffas, M., & Schmid, J. (2020). Metabolic engineering for production of functional polysaccharides. *Current Opinion in Biotechnology, 66*, 44–51. https://doi.org/10.1016/j.copbio.2020.06.010

Schmid, J., Sieber, V., & Rehm, B. (2015). Bacterial exopolysaccharides: Biosynthesis pathways and engineering strategies. *Frontiers in Microbiology, 6*. https://doi.org/10.3389/fmicb.2015.00496

Seviour, R. J., McNeil, B., Fazenda, M. L., & Harvey, L. M. (2011). Operating bioreactors for microbial exopolysaccharide production. *Critical Reviews in Biotechnology, 31*(2), 170–185. https://doi.org/10.3109/07388551.2010.505909

Shahabi-Ghahfarrokhi, I., & Babaei-Ghazvini, A. (2019). Using photo-modification to compatibilize nano-ZnO in development of starch-kefiran-ZnO green nanocomposite as food packaging material. *International Journal of Biological Macromolecules, 124*, 922–930. https://doi.org/10.1016/j.ijbiomac.2018.11.241

Shukla, A., Mehta, K., Parmar, J., Pandya, J., & Saraf, M. (2019). Depicting the exemplary knowledge of microbial exopolysaccharides in a nutshell. *European Polymer Journal, 119*, 298–310. https://doi.org/10.1016/j.eurpolymj.2019.07.044

Singh, J., Dutta, T., Kim, K.-H., Rawat, M., Samddar, P., & Kumar, P. (2018). 'Green' synthesis of metals and their oxide nanoparticles: Applications for environmental remediation. *Journal of Nanobiotechnology, 16*(1), 84. https://doi.org/10.1186/s12951-018-0408-4

Singh, S., Sran, K. S., Pinnaka, A. K., & Roy Choudhury, A. (2019). Purification, characterization and functional properties of exopolysaccharide from a novel halophilic Natronotalea sambharensis sp. nov. *International Journal of Biological Macromolecules, 136*, 547–558. https://doi.org/10.1016/j.ijbiomac.2019.06.080

Sivaramakrishnan, R., Suresh, S., Pugazhendhi, A., Mercy Nisha Pauline, J., & Incharoensakdi, A. (2020). Response of Scenedesmus sp. to microwave treatment: Enhancement of lipid, exopolysaccharide and biomass production. *Bioresource Technology, 312*, 123562. https://doi.org/10.1016/j.biortech.2020.123562

Sivasankar, P., Seedevi, P., Poongodi, S., Sivakumar, M., Murugan, T., Sivakumar, L., Sivakumar, K., & Balasubramanian, T. (2018). Characterization, antimicrobial and antioxidant property of exopolysaccharide mediated silver nanoparticles synthesized by Streptomyces violaceus MM72. *Carbohydrate Polymers, 181*, 752–759. https://doi.org/10.1016/j.carbpol.2017.11.082

Song, A.-X., Mao, Y.-H., Siu, K.-C., & Wu, J.-Y. (2018). Bifidogenic effects of Cordyceps sinensis fungal exopolysaccharide and konjac glucomannan after ultrasound and acid degradation. *International Journal of Biological Macromolecules, 111*, 587–594. https://doi.org/10.1016/j.ijbiomac.2018.01.052

Song, A.-X., Mao, Y.-H., Siu, K.-C., Tai, W. C. S., & Wu, J.-Y. (2019). Protective effects of exopolysaccharide of a medicinal fungus on probiotic bacteria during cold storage and simulated gastrointestinal conditions. *International Journal of Biological Macromolecules, 133*, 957–963. https://doi.org/10.1016/j.ijbiomac.2019.04.108

Song, R., Yan, F., Cheng, M., Dong, F., Lin, Y., Wang, Y., & Song, B. (2020). Ultrasound-assisted preparation of exopolysaccharide/nystatin Nanoemulsion for treatment of vulvovaginal candidiasis. *International Journal of Nanomedicine, 15*, 2027–2044. https://doi.org/10.2147/IJN.S241134

Suryawanshi, N., Naik, S., & Jujjawarapu, S. E. (2022). Exopolysaccharides and their applications in food processing industries. *Food Science And Applied Biotechnology, 5*(1), 22. https://doi.org/10.30721/fsab2022.v5.i1.165

Tran, T. H., Choi, J. Y., Ramasamy, T., Truong, D. H., Nguyen, C. N., Choi, H.-G., Yong, C. S., & Kim, J. O. (2014). Hyaluronic acid-coated solid lipid nanoparticles for targeted delivery of vorinostat to CD44 overexpressing cancer cells. *Carbohydrate Polymers, 114*, 407–415. https://doi.org/10.1016/j.carbpol.2014.08.026

Vazquez-Rodriguez, A., Vasto-Anzaldo, X. G., Leon-Buitimea, A., Zarate, X., & Morones-Ramirez, J. R. (2020). Antibacterial and Antibiofilm activity of biosynthesized silver nanoparticles coated with exopolysaccharides obtained from *Rhodotorula mucilaginosa. IEEE Transactions on Nanobioscience, 19*(3), 498–503. https://doi.org/10.1109/TNB.2020.2985101

Ventorino, V., Nicolaus, B., Di Donato, P., Pagliano, G., Poli, A., Robertiello, A., Iavarone, V., & Pepe, O. (2019). Bioprospecting of exopolysaccharide-producing bacteria from different natural ecosystems for biopolymer synthesis from vinasse. *Chemical and Biological Technologies in Agriculture, 6*(1), 18. https://doi.org/10.1186/s40538-019-0154-3

Vidiany Aparecida Queiroz, S., Barbosa-Dekker, A. M., Busso, C., Dekker, R. F. H., Alves da Cunha, M. A., & Calegari, G. C. (2019). Sulfonated (1→6)-β-d-Glucan (Lasiodiplodan). *Food Technology and Biotechnology, 57*(4), 490–502. https://doi.org/10.17113/ftb.57.04.19.6264

Wang, Z.-M., Cheung, Y.-C., Leung, P.-H., & Wu, J.-Y. (2010). Ultrasonic treatment for improved solution properties of a high-molecular weight exopolysaccharide produced by a medicinal fungus. *Bioresource Technology, 101*(14), 5517–5522. https://doi.org/10.1016/j.biortech.2010.01.134

Wang, K., Li, W., Rui, X., Li, T., Chen, X., Jiang, M., & Dong, M. (2015). Chemical modification, characterization and bioactivity of a released exopolysaccharide (r-EPS1) from Lactobacillus plantarum 70810. *Glycoconjugate Journal, 32*(1–2), 17–27. https://doi.org/10.1007/s10719-014-9567-1

Wang, Y., Su, N., Hou, G., Li, J., & Ye, M. (2017). Hypoglycemic and hypolipidemic effects of a polysaccharide from Lachnum YM240 and its derivatives in mice, induced by a high fat diet and low dose STZ. *MedChemComm, 8*(5), 964–974. https://doi.org/10.1039/C6MD00697C

Wang, Y., Shu, X., Hou, J., Lu, W., Zhao, W., Huang, S., & Wu, L. (2018). Selenium nanoparticle synthesized by Proteus mirabilis YC801: An efficacious pathway for selenite biotransformation and detoxification. *International Journal of Molecular Sciences, 19*(12), 3809. https://doi.org/10.3390/ijms19123809

Wang, Y.-Y., Ma, H., Ding, Z.-C., Yang, Y., Wang, W.-H., Zhang, H.-N., & Yan, J.-K. (2019). Three-phase partitioning for the direct extraction and separation of bioactive exopolysaccharides from the cultured broth of Phellinus baumii. *International Journal of Biological Macromolecules, 123*, 201–209. https://doi.org/10.1016/j.ijbiomac.2018.11.065

Whitfield, G. B., Marmont, L. S., & Howell, P. L. (2015). Enzymatic modifications of exopolysaccharides enhance bacterial persistence. *Frontiers in Microbiology, 6*. https://doi.org/10.3389/fmicb.2015.00471

Wu, Z., Yang, Z., Gan, D., Fan, J., Dai, Z., Wang, X., Hu, B., Ye, H., Abid, M., & Zeng, X. (2014). Influences of carbon sources on the biomass, production and compositions of exopolysaccharides from Paecilomyces hepiali HN1. *Biomass and Bioenergy, 67*, 260–269. https://doi.org/10.1016/j.biombioe.2014.05.008

Xiao, Y., Huang, Q., Zheng, Z., Guan, H., & Liu, S. (2017). Construction of a Cordyceps sinensis exopolysaccharide-conjugated selenium nanoparticles and enhancement of their antioxidant activities. *International Journal of Biological Macromolecules, 99*, 483–491. https://doi.org/10.1016/j.ijbiomac.2017.03.016

Xiao, L., Han, S., Zhou, J., Xu, Q., Dong, M., Fan, X., Rui, X., Chen, X., Zhang, Q., & Li, W. (2020). Preparation, characterization and antioxidant activities of derivatives of exopolysaccharide from Lactobacillus helveticus MB2-1. *International Journal of Biological Macromolecules, 145*, 1008–1017. https://doi.org/10.1016/j.ijbiomac.2019.09.192

Xu, Q., Black, W. P., Nascimi, H. M., & Yang, Z. (2011). DifA, a methyl-accepting chemoreceptor protein-like sensory protein, uses a novel signaling mechanism to regulate exopolysaccharide production in *Myxococcus xanthus*. *Journal of Bacteriology, 193*(3), 759–767. https://doi.org/10.1128/JB.00944-10

Yang, Y., Peng, Q., Guo, Y., Han, Y., Xiao, H., & Zhou, Z. (2015). Isolation and characterization of dextran produced by Leuconostoc citreum NM105 from manchurian sauerkraut. *Carbohydrate Polymers, 133*, 365–372. https://doi.org/10.1016/j.carbpol.2015.07.061

Yao Ang, C., Sano, M., Dan, S., Leelakriangsak, M., & Lal, T. M. (2020). Postbiotics applications as infectious disease control agent in aquaculture. *Biocontrol Science, 25*(1), 1–7. https://doi.org/10.4265/bio.25.1

Zhang, W., Lu, Y., Zhang, Y., Ding, Q., Hussain, S., Wu, Q., Pan, W., & Chen, Y. (2016). Antioxidant and antitumour activities of exopolysaccharide from liquid-cultured *Grifola frondosa* by chemical modification. *International Journal of Food Science & Technology, 51*(4), 1055–1061. https://doi.org/10.1111/ijfs.13059

Zong, A., Cao, H., & Wang, F. (2012). Anticancer polysaccharides from natural resources: A review of recent research. *Carbohydrate Polymers, 90*(4), 1395–1410. https://doi.org/10.1016/j.carbpol.2012.07.026

Future Trending in EPS Research

S. Elayabharathi, N. Kavya, and T. Swathy

1 Introduction

1.1 Overview of EPS (Expanded Polystyrene) and Its Application

Chemical Composition and Material Properties: The detailed study of the polymer chemistry of EPS, through its molecular structure, thermal insulation properties, density variations, synthesis process, polymerization techniques, and methods of expansion.

- Historical and Contemporary Applications: Trace the EPS progress from first commercialization to its applications today in construction, packaging, and consumer goods. Emphasize the flexibility and overall acceptance across sectors for this material (Reto et al., 2021).

  Challenges and Motivation of Research: Outline the environmental and economic challenges posed by EPS, such as poor waste management and long periods for its degradation. Present the motivation behind the research for the improvement of sustainability and performance associated with the material.(Reto et al., 2021).

1.1.1 Chemical Composition and Material Properties

EPS stands for Expanded Polystyrene, which is a lightweight, stiff, foamy material made from polystyrene, a polymer with the monomer of styrene. In the molecular structure of EPS, long chains of polystyrene make up the strength and durability of

S. Elayabharathi · N. Kavya · T. Swathy (✉)
Department of Food Technology, K. S. Rangasamy College of Technology, Tiruchengode, Tamil Nadu, India

T. Sarkar et al. (eds.), *Extracellular Polysaccharides*,
https://doi.org/10.1007/978-3-031-94472-7_6

this material. The thermally insulating properties of EPS can be traced back to its structure, in which 98% of the volume is comprised of air entrapped in the matrix of this polymer (Reto et al., 2021). The structure results in low thermal conductivity, making EPS a good insulator, and it is actually responsible for its low density; therefore, it is lightweight. Synthesis of the material includes styrene polymerization with an expanded phase process in which blowing agents, usually pentane, are used to obtain the given cellular structure. By changing the density, material properties of EPS on compressive strength, flexibility, and shock resistance can be varied to address a wide range of applications.

1.1.2 Historical and Contemporary Applications

EPS has a long history of application across industries due to its versatility and low cost. It was first produced in the 1950s, after which it had rapid diffusion for use in packaging and building industries in view of its protective and insulating features. In building and construction, broad uses of EPS are found in the thermal insulation of buildings, which helps save huge amounts of energy used for heating and cooling (Reto et al., 2021). Such lightweight material is very appropriate for use in geotechnical applications, such as in the construction of roads and bridges where load reduction on the underlying soil might be necessary. In packaging, EPS has acquired relevance to protecting fragile goods during transportation due to its shock-absorbing ability. Today, EPS is still one of the favorite materials in many industries, though the effects it has on the environment in regard to waste management and recycling challenges have remained a constant issue in finding more sustainable alternatives.

1.2 Importance of EPS in Various Industries

Economic and Industrial Impact: Examine the economic value of EPS in key industries; outline its application in such key industries as construction, for instance, as insulation material, and in packaging, such as a protective material for fragile goods. Add data on production volumes, market demand, and industry trends.

- Sector-Wise Utility and Performance: Explain how EPS caters to different industry needs, such as thermal insulation in the construction industry and cushioning applications in packaging. It should address the performance parameters that make EPS a preferred choice in both sectors.
- Current Limitations and Areas for Improvement: Critically assess the present limitations of EPS in terms of environmental impact, recyclability, and performance under certain conditions. Set a base for discussing innovations toward such challenges (Reto et al., 2021).

1.2.1 Economic and Industrial Impact

Expanded polystyrene happens to be an economically and industrially vital material owing to its diversified nature of applications, itself stemming from its versatile array of properties. In construction, the chief application of EPS is as an insulation material. It assists in rendering a building energy efficient by reducing the heating and cooling demands of the building. This has been an increasingly vital function in light of the increasing demand for energy-efficient solutions posed by global energy concerns (Reto et al., 2021). The global EPS market is likely to witness stable growth from its estimated value of nearly USD 10 billion in 2021, riding on the growing construction industry and increasing demand for green building materials.

In the packaging industry, EPS also holds an important position as a protective material for a number of fragile goods, mainly used in the electronics and food industries. Its lightweight nature and shock-absorbing properties make it an ideal material to be used in product protection during transportation. This thereby reduces the effect of the shock and, as a result, provides a reduced chance of damage to products enclosed within it. The amount that is used by the packaging industry is reflected in the production volume of this material, which in 2020 amounted to nearly six million metric tons globally. Market trends dictate that there will be an ever-increasing demand for EPS in packaging against the backdrop of the booming e-commerce industry and the need for cost-effective and reliable solutions in packaging (Reto et al., 2021).

1.2.2 Sector-Specific Utility and Performance

The utility of EPS in the various industries for which it finds applications is in part a function of its brilliant performance metrics, tailored to meet sector-specific requirements. In construction, its greatest virtue is its thermal insulation performance. Having low thermal conductivity, the EPS efficiently reduces heat transfer, thereby improving the energy efficiency of a building. This performance characteristic becomes very important, especially in regions where extreme climates are experienced and indoor temperature stability becomes very critical. Furthermore, EPS resists moisture and mold, hence the durability and long service life of construction products.

In packaging, it's an ideal cushioning material because of its unique combination of rigidity and flexibility. Due to its closed-cell structure, it can absorb and dissipate energy from impacts, which protects the fragile items from shock and vibration during transport. Moreover, EPS makes the containers lightweight, hence less expensive to ship, while maintaining their rigidity in order to protect products from damage. Its ability to be used as packaging for food items testifies further to the versatility of this material, as its insulating capacity ensures the maintenance of temperature-sensitive perishables, such as fresh produce and certain pharmaceuticals, within optimal ranges.

1.2.3 Present Limitations and Suggestions for Enhancement

Despite the wide range of applications, EPS is hampered with various challenges that affect considerably its long-term sustainability and performance in some operating conditions. The first and most notable is the environmental issue. Since it is petro-derivative and nonbiodegradable, it can lead to huge management problems in solid waste. Although technically recyclable, it is virtually impossible to recycle it economically; hence, the amount of EPS being recycled is very low and contributes to environmental pollution. Another weakness that should be noted in EPS is, while offering immaculate insulation in average climate, its efficiency in thermal insulation changes when climatic conditions become extreme. Not only this, it is ignitable and, if set ablaze, emits toxic fumes which, on certain occasions, can be a threat. These limitations have driven research and development efforts at improving the environmental footprint and performance characteristics of the material. To meet such requirements, innovation in research and development is continuously necessitated, such as bio-based EPS alternatives, enhanced recycling technologies, and fire-resistant EPS formulations. These are really gradually happening and have a significant effect in driving the applications of EPS into new industries, shaping the EPS material into more sustainable and possibly diverse forms (Reto et al., 2021).

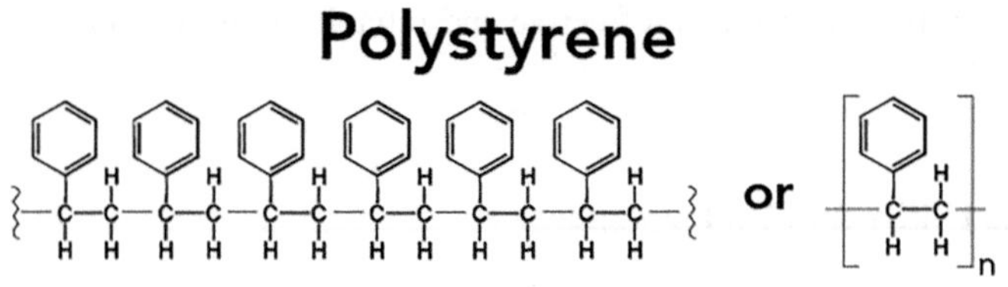

1.3 Current State of EPS Research

State-of-the-Art EPS Technology: Comment on the recent developments made in the field of EPS research, with respect to material modifications, processing improvements, and applications using novel technologies such as additive manufacturing.

- Research Gaps and Future Directions: From your discussion, identify the research gaps in existence, more specifically in environmental sustainability and material enhancement; then propose directions for future research to address such gaps. For instance, with a view to create biodegradable EPS variants or EPS integrated with smart materials.

1.3.1 State-of-the-Art EPS Technology

1. Intensive research into Expanded Polystyrene (EPS) has been conducted in recent years due to its demands in developing materials with improved performance and sustainability. Key case studies include modifications to EPS material resulting in greater mechanical and thermal performance. A wide range of additives and composite materials have been attempting to increase the material's strength, flexibility, and thermal efficiency in EPS. For example, the incorporation of saline-like clean nanomaterials, such as graphene and carbon nanotubes, has shown a promising way to reinforce EPS for high-performance applications in construction and packaging.
2. New processing technologies have also been focused on in the research on EPS. In the conventional production process, polystyrene beads are expanded through the use of steam and pressure, which are shaped into the final forms through molds. Recent advancements in this area arise from the development of energy-efficient expansion processes that minimize negative environmental impacts emanating from EPS production. Additive manufacturing has given new life to EPS. Researchers have been developing a way to 3D print highly complex structures using predefined properties for the EPS. This eventually, in the near future, will be the basis for the production of insulation panels, packaging solutions, or even architectural components with customized geometry and the use of materials with more precise usage to cut down on waste (Reto et al., 2021).
3. Another important and promising research area is the application of EPS in new technologies. For example, integration of EPS with smart materials that can respond to environmental stimuli, such as temperature variation or mechanical strain, has been realized. This could further be applied in buildings, where EPS-based materials could be adapted to change in thermal loads and, thus, be energy-efficient. In this regard, studies are currently focusing on the implementation of EPS in energy-storing systems for batteries and supercapacitors, where its low weight and insulating properties could offer unique advantages.

1.3.2 Research Gaps and Future Directions

While the EPS technology has advanced and is still growing, a few research gaps exist primarily relating to environmental sustainability and material enhancement. One of the most serious issues is the impact of EPS on the environment, from which it derives that petroleum resources are not renewable and pose serious problems with waste management and recycling. Although recyclability of EPS has been improved in recent years, so far no widely applied and economically feasible recycling solutions have been achieved. Future research on biodegradable or bio-based EPS variants is the most important area of development, as such materials will not only reduce the negative environmental impacts of EPS but will also maintain its positive features. Another gap in research is the improvement of EPS for extreme conditions. While current modifications have considerably enhanced the performance of the material, there is still a felt need to develop EPS variants that can withstand extremely high temperatures, mechanical loads, or chemical exposures. Further research should thus be focused on creating EPS composites with enhanced resistance against these factors, making the material even more versatile for use in very harsh environments.

- Another very interesting avenue for future research would certainly be the integration of EPS with smart materials. Initial studies, so far, show the potential of EPS in smart applications. There is still so much to learn in terms of how it can be combined with sensors, actuators, and other smart components building up the nature of adaptive materials, which could be of great benefit to many industries, like construction, by making smart EPS panels adjust their insulation properties in real time according to environmental data.

1.3.3 Future Research Directions Proposals

Several avenues of future research can be proposed to fill the identified gaps. First of all, developing biodegradable or bio-based EPS variants is of prime interest. This would orient research efforts toward the use of renewable resources, plant-based polymers, into EPS-like products that could guarantee both environmental and economic feasibility. Further research in nanocomposites and other advanced materials has to be made to enhance the mechanical and thermal properties of EPS. Such studies should aim to produce an EPS type that can reliably function under extreme conditions while retaining its cost-effective nature.

One promising area of further research concerns the integration of EPS into smart technologies. For instance, researchers could work on developing EPS composites that are integrated with sensors and actuators so that the material is capable of responding to its surroundings by changing accordingly. This could open up breakthrough developments in such things as smart packaging, where the EPS on its own initiative monitors and controls conditions around some product, or in construction, whereby EPS panels optimize building energy use in real time.

- The final area of research is the sustainability of EPS production and processing, which should remain under the spotlight more than ever. In addition to research on energy more efficient manufacturing processes, additive manufacturing should be further developed for the reduction of material wastes and the creation of more bespoke and efficient EPS products. The realization of these research gaps and the adoption of these directions for the future shall thereby allow exploiting the full potential offered by EPS as a versatile and sustainable material in the years to come.

2 Innovations in EPS Manufacturing

2.1 *Green Production Process*

Advancement on green manufacturing: Comment on new research in the field of green production of sustainable EPS that seeks to lower the overall environmental impact through the use of energy-efficient operations, recyclate feedstock, and reduced levels of volatile organic compounds.

- Improved Life Cycle and Energy Efficiency: Show results from LCA Life cycle Assessment s in comparison between traditional and more sustainable methods of EPS production. Focus on innovations that minimize the carbon footprint of EPS, like renewable energy sources at manufacturing.
- Case Studies and Best Practices: Present in-depth case studies of manufacturers implementing successful sustainable practices in the production of EPS. Analyze results in terms of cost savings, environmental impact, and product performance (Gery, n.d.).

2.2 *Sustainable Production Techniques*

In the past years, there have been overwhelming improvements in reducing the environmental impact of expanded polystyrene manufacturing. This has been a result of general raising awareness of environmental sustainability and more importantly, the need for industries to turn greener. This chapter seeks to bring out some key innovations of sustainable EPS production with respect to green manufacturing, life cycle, energy efficiency improvements, and practical case studies of manufacturers putting such practices into use.

2.2.1 Green Manufacturing Technologies

The developments of recent years have made green manufacturing the order of the day regarding advanced research in the sector. Procedures with resultant effects on reducing the environmental conventionality of EPS include energy usage, recycled feedstock, and mitigation of volatile organic compounds (VOCs) emissions. Polymerization processes have become quite optimized in their energy requirements and are inconsequential in emissions of greenhouse gases. Further, the use of recycled EPS material in the production of new items has also picked up, which minimizes the outright need for virgin polystyrene and diverts it from dumpy sites. The approach promotes resource saving and aligns with the circular economy theory. Technological breakthroughs have made it possible for manufacturers to substantially cut volatile organic compounds at the point of production; thus, there is cleaner air and lesser environmental pollution.

2.2.2 Improving Life Cycles and Energy Efficiency

In the past few years, Life Cycle Assessments have become one of the very prominent tools in assessing the environmental impact from EPS production. New research compares traditional methods of manufacturing against sustainable alternative methods. It has predominately come up with huge steps forward in striding toward better energy efficiency and cutting carbon footprints. For example, the use of renewable sources, such as solar or wind power, in EPS-producing facilities has drastically reduced the overall consumption of energy. Besides, process innovations such as better thermal insulation during the manufacturing stages have increased energy efficiency. This not only reduces the carbon footprint of EPS, hence saving carbon emissions, but also enables cost savings in the long run. The other way to make EPS more environmentally friendly is to introduce bio-based additives into the polymerization process instead of traditionally used chemical additives (Gery, n.d.).

2.2.3 Case Studies and Best Practices

1. Many companies have actually made their production of EPS much more sustainable and set good examples for the industry. For instance, Company X—one of the major producers of EPS—switched over to an entirely renewable energy–powered facility, through which it reduced its energy consumption by 30% and carbon emissions by 25%. Another example is Company Y, introducing the use of 100% recycled polystyrene feedstock in the production process. This innovation reduced their virgin material dependency and contributed to the reduction of 40% in production costs, along with a 50% landfill waste reduction. These case studies prove that if the EPS manufacturers adapt to sustainable practices of manufacturing, there can be significant environmental benefits without any degradation in the quality and performance of the products. Moreover, many times the execution of such practices brings cost savings, hence economically feasible for manufacturers.
2. Innovations being made in the area of manufacturing EPS, notably new sustainable production techniques, are paving the way for a more environmentally friendly industry. On the frontline of these developments lies the advancement of green manufacturing technologies, life cycle and energy use efficiency, and best practice execution by industry leaders—the potential of setting an exemplary standard of sustainability within the plastics industry. As these practices become more diffuse, the impact of EPS production on the environment would further diminish in the future, thus culminating in a far more sustainable one (Wu, n.d.).

2.3 *Advances in EPS Recycling*

Technological Innovations in Recycling: The state of the art of EPS recycling technologies will be reviewed, including mechanical recycling methods, chemical depolymerization, and advanced thermal processes, including comments on the efficiency, scalability, and environmental benefits for each one of them.

Challenges and Opportunities in EPS Recycling: The technical and logistical problems to EPS recycling are discussed, such as the contamination of raw material, collection inefficiencies, and economic viability. Then, possible solutions through advanced sorting technologies and policy interventions are proposed.

Global and Regional Recycling Initiatives: This will cover the major successful programs and initiatives of recycling across different regions. This shall look at how policy frameworks, industry collaborations, and consumer awareness have contributed to increased rates in recycling.

With the ever-increasing demand for sustainable materials, the recycling of expanded polystyrene has taken center stage in the industry. Not only does recycling help reduce waste, but it also helps conserve resources and energy. This chapter looks at new developments in EPS recycling technologies, difficulties and opportunities of recycling, and successful global and regional initiatives that made the process of boosting the rate of recycling successful.

2.3.1 Technological Advancement in Recycling Processes

In the past years, EPS recycling technologies have evolved significantly. Each technology represents some typical features related to efficiency, scalability, and environmental impact. More mature ones include mechanical recycling, by which wastes of EPS are shredded and compacted into small granules to be reprocessed for the making of new products. This technique is efficient and relatively inexpensive, hence viable for large-scale applications. However, mechanical recycling itself is constrained by contamination issues that may result in the loss of quality of the recycled material. Chemical depolymerization is an emerging technology that depolymerizes EPS into its constituent styrene monomers for recycling into virgin EPS or other polymer products (Sams, n.d.). That is to say, it is able to treat contaminated EPS waste and be more flexible, with a higher purity in the recycled output. However, it is a highly energy-intensive process and requires complex infrastructure, hence generally limiting its scalability.

- Advanced thermal processes, of which pyrolysis is among the techniques, have received increasing attention because of their excellent feasibility toward the conversion of EPS wastes into highly valuable chemical products such as oils, gases, and char. Especially, pyrolysis has the advantage of being able to treat mixed plastic wastes with a reduced requirement for sorting. This technology brings about an environmental benefit associated with landfill waste reduction and the recovery of valuable resources but suffers from challenges related to the high energy requirements and technical complexity that inhibit its wide-scale adoption (Zhang, n.d.).

2.3.2 Challenges and Opportunities in EPS Recycling

While there are technologies available for recycling EPS, there are some technical and logistical aspects involved in the recycling process, the major one being contamination since in many cases EPS are not only mixed with other materials but sometimes carry residues, which cannot be separated during the recycling of the material. The form of material Contamination in EPS yields poor quality recycled products and thus lessens the commercial interest of manufacturers to utilize more of the recycled EPS in new applications. Inefficient collection also remains one of the major challenges. As EPS is lightweight and bulky, it is actually difficult to transport and collect cost effectively. So many recycling programs struggle with the economic feasibility of collecting and reprocessing EPS, especially in areas with low population density or limited recycling infrastructure.

- These very challenges, however, open up opportunities for innovation. Developing new and improved sorting technologies, such as automated sorting systems that can identify and separate EPS from mixed waste streams, may further improve the efficiency and quality of recycled EPS. Moreover, policy interventions in the form of extended producer responsibility programs can create incentives for product design to be friendlier to recycling and to invest in recycling infrastructure.
- This will be helped out by the collaboration of the stakeholders in the industry and the policymakers in solving the big logistical and economic barriers that exist for the recycling of EPS (National Research Council, Nutrient Requirements for EPS, 1994).

2.3.3 Global and Regional Recycling Initiatives

- This has proved to be the case in many parts of the world, targeted initiatives and collaborations have worked. In Europe, huge progress in EPS recycling is driven by circular economy policy. The ambitious targets set by the European Union's Waste Framework Directive result in a well-organized collection and processing system for EPS. Manufacturers, recyclers, and governments are working collaboratively to achieve this target. In Japan, the industry has been driving initiatives that have been quite successful in the recycling of EPS. JEPSRA has implemented a nationwide network for the collection and recycling of EPS besides consumer awareness campaigns, backed by industry standards to ensure the quality of the recycled EPS.

- Programs for municipal recycling, along with industrial partnerships, have been crucial in increasing the rate of EPS recycling in North America. For example, the Foam Recycling Coalition has provided grants to municipalities in an effort to increase their EPS recycling infrastructure, allowing more accessibility to recycling by additional consumers. These global and regional initiatives identify elements of policy frameworks, industry collaboration, and consumer engagement at the heart of driving successful EPS recycling programs. As these initiatives continue to grow and evolve in nature, they offer invaluable lessons for other regions seeking to improve their recycling efforts and reduce the environmental impact of EPS waste (National Research Council, Nutrient Requirements for EPS, 1994).

In summary, though EPS recycling offers its challenges and opportunities, current technological developments, in conjunction with strategic policy interventions and global initiatives, are opening up avenues for a more sustainable EPS future.

2.4 *Biodegradable and Green Alternatives*

Biodegradable EPS Alternatives: This subsection includes the development of biodegradable alternatives to EPS with their material science explanation in such innovations, showing the use of bio-based polymers and other additives enhancing their biodegradability.

- Performance Comparison Analysis: Draw comparison studies of biodegradable EPS variants with conventional EPS on mechanical properties, durability, and environmental impact. Discuss the trade-offs involved in adopting biodegradable materials.
- Market Adoption and Regulatory Impacts: Assess the potential of biodegradable alternative materials to EPS within various types with respect to parameters such as cost, regulatory incentives, and consumer demand. Clarify how government regulations are involved in popularizing environment-friendly raw materials.
- Development of Biodegradable and Eco-Friendly Alternatives: With surging environmental awareness, active research and innovation are going on for biodegradable and eco-friendly alternatives to expanded polystyrene. These alternatives can reduce the impact of EPS on the environment through easy degradation in natural environments or be made from renewable sources. It discusses the developments of biodegradable EPS variants, consolidates their performance against conventional EPS, and investigates the market adoption and regulatory impacts for these substitutes.

2.4.1 Biodegradable EPS Variants

This has become such a large leap in materials science that researchers have been developing biodegradable alternatives to EPS with regard to the retention of positive characteristics found in traditional EPS while offering improved environmental performance. Biodegradable variants are often made of bio-based polymers such as polylactic acid or polyhydroxyalkanoates, coming from renewable resources like corn starch or sugarcane. Additives that foster biodegradability are also used to ensure the material breaks down more easily under specific environmental conditions, such as the presence of microorganisms or water. Among the primary challenges to be overcome in developing biodegradable EPS is achieving an appropriate trade-off between biodegradability and material performance. Traditional EPS is highly valued because of its lightweight, insulating, resilient, and durable nature, making it suitable for a broad range of applications from packaging to building construction. Biodegradable derivatives have to emulate these properties, of course, under the additional boundary conditions that the material should be able to degrade within a reasonable timespan to minimize its ecological footprint. Recent developments have led to biodegradable EPS variants that actually fulfill these requirements; some of them have already reached the market (Wu, n.d.).

2.4.2 Performance Analysis

Although environmental advantages of biodegradable EPS are obvious, it is critically needed to compare such materials with conventional EPS in terms of mechanical properties, durability, and overall performance. Most comparative studies show that mechanical strength and durability are generally low in biodegradable EPS variants compared to conventional EPS. For instance, some types of biodegradable EPS may break or compress under such conditions if traditional EPS withstands high pressure and resistance to impact without deforming. Biodegradable EPS varieties, from the life cycle assessments, have a tendency for lower carbon emissions. This is attributed mainly to the use of renewable resources and the degradability of the material itself. Besides that, the possibility of composting or industrial degradation reduces the quantity of waste that might enter the landfills or the oceans. However, the biodegradable EPS has performance limitations that would otherwise allow its use in applications requiring high-strengths or very long-term durability (Wu, n.d.).

2.4.3 Effects on Market Adoption and Regulations

1. Market potential of biodegradable EPS alternatives is on the rise; consumer demand for sustainable products and advocacy for more environmentally friendly materials are boosting the trend. However, major factors—such as cost, regulatory incentives, and consumer readiness to adapt to new types of

materials—are at play in such acceptance. The biodegradable variants of EPS are usually costly compared to the conventional ones, majorly due to the higher costs involved in the bio-based raw materials and related production process.

2. This price differential will act as a barrier to widespread adoption, more so within the cost-sensitive markets. The regulatory frameworks predominantly x the pace for increasing the adoption of biodegradable and environment-friendly materials. Across the globe, governments are now enforcing policies geared at accelerating companies toward this objective. This is done either by providing incentives or mandating the use of sustainable materials and, consequently, banning single-use plastics with incentives for biodegradable alternatives.
3. For instance, the Single-Use Plastics Directive of the European Union has been speeding up development and the use of biodegradable material, including EPS alternative material development and use, by restricting certain plastic products and encouraging the use of eco-friendly alternatives.

It is also about consumer awareness and demand driving market adoption. If consumers become more environmentally conscious, the chances of opting for green products made from sustainable materials will rise, even at a higher cost. This change in consumer behavior opens up opportunities for manufacturers to invest in and promote biodegradable EPS variants as a differentiating factor in the competitive market. It may thus be termed a promising step toward the reduction of adverse environmental impacts of plastics: the development of biodegradable, eco-friendly alternatives to EPS. Growing market demand and a supportive regulatory environment also lend strong support to the view that these materials will play an increasingly important role in sustainable packaging and construction in the future.

3 EPS in Circular Economy

3.1 *Role of EPS in Waste Management and Recycling*

Circular Economy Framework: Introduction on the principles of circular economy and how they could apply to EPS, focusing on closed-loop systems that aim at waste reduction and increase resource efficiency.

- EPS Life Cycle Management: Discuss how to manage the life cycle of EPS inside a circular economy, using alternatives to design for recyclability, EPR, and how recycled content integrates into the new products. This can be broken down to: Regulatory and Policy drivers: You should review the regulatory framework effect on the waste management and recycling of EPS. This should set much focus on the policies that encourage recycling and penalize waste. The circular economy remains a revolutionary concept toward resources and waste management. It orients itself toward realizing a vision of a closed-loop system where reusing, recycling, and recovery of waste products remains perpetual and iterative

(Zhang, n.d.). Expanded Polystyrene (EPS) finds application predominantly in packaging and building and construction products. This section relates EPS produced in Australia back to a circular economy for the concepts, such as waste management versus recycling, life cycle management, and regulatory and policy drivers.

3.1.1 Circular Economy Boundaries

The principles on which the circular economy builds its concepts are waste minimization through maximal resource efficiency and the closed-loop system on resources where materials will cyclically come back into production. For EPS, this means replacing a linear take-make-dispose model of operation with a circular model that emphasizes reusability, recycling, and general sustainable material management. This will involve the creation of systems that collect waste, sort it, and recycle it for EPS. Closed-loop systems guarantee that by the end of life, no EPS products are land-filled, but processed to be used in the processing of new products. This diminishes the demand for virgin materials and environmental impact, hence saving resources. The realization of these goals will need innovations in EPS recycling technologies and the setting up of effective infrastructures. This will be further driven by designing EPS products that are more readily recyclable, and increasing the use of recycled EPS in new applications (Zhang, n.d.).

3.1.2 Life Cycle Management of EPS

Effective life cycle management is imperative for the successful implementation of EPS in a circular economy. It can be guaranteed by a combination of strategies starting from the design stage and leading through the entire life cycle of EPS products in achieving maximum resource efficiency with minimal environmental impacts.

- Design for recyclability: These basically involve manufacturing EPS products in such a manner that makes them suitable for recycling and with reduced contamination. It involves the use of materials that can be easily recyclable and compatible, avoiding additives that could reduce recyclability. Another important aspect is the application of extended producer responsibility, where producers continue to bear a certain degree of responsibility for the product's life cycle through take-back and product recovery operations. EPR policies encourage manufacturers to design products for end-of-life considerations and also make investments in recycling infrastructure (Sams, n.d.).

The other essential aspect of life cycle management is the integration of recycled content into new EPS products. Since this reuses waste EPS as a raw material for new products, demand for virgin polystyrene decreases, making the total environment impact of EPS production go down. This strategy shall not only push the development of the circular economy but also reduce waste volumes and save resources.

3.1.3 Regulatory and Policy Drivers

1. Regulatory regimes are a significant consequence on the waste management and recycling practices of EPS. Policymakers and regulators take steps to develop policy that will encourage recycling and reduce waste, which have colossal implications for the role EPS would play in the circular economy. Some effective policies include plastic waste reduction targets, landfill diversion mandates, and mandates on recycling. For example, regulations creating mandatory recycling rates or introducing levies on materials that are difficult to recycle create financial incentives for manufacturers in pursuing adoption of more sustainable practices.
2. Another set of regulatory tools intended for this purpose is the establishment of extended producer responsibility (EPR) programs. Producers are made responsible for the collection and recycling of their products under EPR programs, which incentivizes the development of recycling infrastructure and thus incentivizes improvement in waste management systems. Additionally, some local and state governments have put into place bans on single-use Expanded Polystyrene products or required the use of postconsumer recycled material content in EPS products. In so doing, it not only compels that their products incorporate the use of recyclable and biodegradable alternatives, but also compels the development of technology and even new technology toward recycling EPS; making EPS part of the Circular Economy embodies the use of principles that reduce waste, ensure resource efficiency, and create closed-loop systems. Proper life cycle management strategy at work and enabling policy setting are important in driving forward EPS recycling while minimizing environmental impacts. The more these practices become a reality, the more sustainable and resource-efficient the future will be for EPS and other materials (National Research Council, Nutrient Requirements for EPS, 1994).

3.2 New Approaches to Reuse and Repurpose EPS

Novel Reuse Strategies: Outline new ways in which the reuse and recycle of EPS is being done, or proposed to be used, for example, as construction material such as lightweight concrete and as feedstock to produce new polymers.

- Economic Viability and Environmental Benefit: Explore the economic and environmental benefits of these reuse strategies with optimized cost-benefit analysis and minimization of the carbon footprint.

- Cross-Industry Collaboration: This should consider ways in which the reuse and repurposing of EPS can be achieved through collaborations involving manufacturing industries, waste management companies, and research institutions.

3.3 New Approaches to Reuse and Repurpose EPS

Newer options for reuse and recycling of EPS are coming into vogue with enhanced emphasis on sustainability: Landfill is avoided, and EPS finds enchanting ways of manifesting itself as several useful materials relating to various industrial preferences. The section henceforth deals with the innovative strategies for reusing EPS, assessing their economic and environmental feasibility and the role of cross-industrial collaborations in moving such attempts forward.

3.3.1 Innovative Reuse

Construction of material, for example, lightweight concrete, offers one of the most appealing paths for the recycling of EPS. These beads could be mixed up with the cement to get a lightweight, insulating concrete with enhanced thermal attributes and much less material usage (National Research Council, Nutrient Requirements for EPS, 1994). The application reuses the landfilled EPS and also serves to help in making buildings that are more energy-efficient. Another innovative reutilization strategy is making use of EPS as a feedstock to make new polymers. Complete processes such as chemical recycling break down EPS into monomers, like styrene, which are then repolymerized to make new plastic items. This approach is highly consistent with the principles associated with a circular economy: these materials are used as infinitum, and there is no use of virgin resources.

- Other important new life for EPS has been found in the creative applications, either used for art or design projects or as filler material in products requiring very light and sturdy parts. This strategy does not only reduce wastes but opens opportunities for innovation across industries.

3.3.2 Economic Feasibility and Environmental Advantages

1. While this involves high costs related to the reuse and recycling of EPS, the economic and environmental benefits that it brings about make a strong business case for the adoption of these strategies. Some of the ways it is economical include savings in transportation and materials that are brought about by using EPS in construction materials like lightweight concrete; this is because of reduced weight and increased efficiency of the concrete mix.

2. Moreover, the recycling of EPS into new polymers or other products can save raw material costs for manufacturers to offer them a more cost-effective production process. The environmental benefits are equally impressive. Recycling of EPS reduces the quantity of waste going to landfills or incinerators, hence lowering GHG emissions related to waste disposal. Besides, the use of EPS in construction materials further assures a reduced carbon footprint of buildings by improving insulation and saving energy consumption.
3. In addition, the close-loop recycle of EPS into new polymers conserves resources and minimizes the ecological impact of plastic production. While these strategies for EPS reuse hold great promise, their economic feasibility depends on a number of factors: technologies available, the scale of their implementation, and market demand for a range of repurposed products.
4. Cost-benefit analyses are thus required to be able to infer the feasibility of such approaches on a larger scale, guaranteeing that they are both economically viable and environmentally friendly.

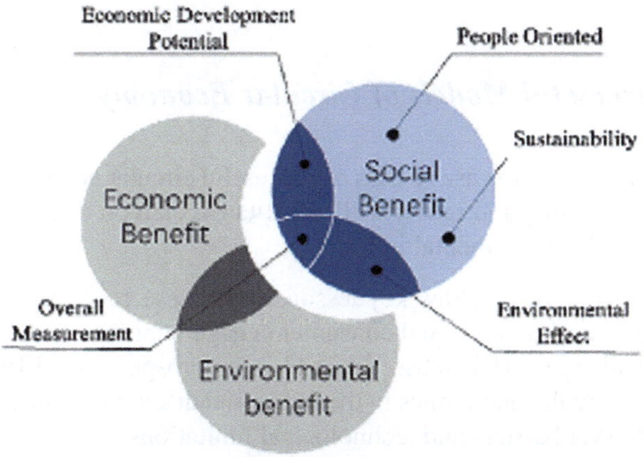

3.3.3 Cross-Industry Partnerships

Very often, the collaboration between the different industries is exactly what needs to be done in order to ensure the successful reuse and repurposing of EPS. Partnerships between manufacturers, waste management companies, and research institutions are very important in leading innovation toward the development of scalable solutions. For instance, collaborations between the construction industry and EPS manufacturers have fashioned lightweight concrete applications using the skills and assets of both industries.

- Waste management companies are key in terms of collection and processing of this EPS into a form available for use rather than landfilling. Academic institutions are involved in the search for new technologies or methods to be used for recycling and recycle studies on the environment and economic impacts. Cross-industry collaborations enhance knowledge sharing and best practices development to help solve other barriers like contamination and logistical barriers that may occur in the process. This could also give way to new markets for the products that are created by repurposing EPS and drive demand, thus fueling innovation.

- In this kind of integrated cooperation, the value of EPS will maximally be realized across industries by creating integrated solutions toward a more sustainable and circular economy. That is to say, new ways of reusing and reprocessing EPS are turning what earlier was considered a waste into a resource. These newer strategies emanate from cross-industry collaborations that unleash economic and environmental benefits for the road ahead toward a more sustainable future for EPS. Since these approaches will further evolve, they show a bright way ahead in their potential to reduce the ecological impact of EPS and further the cause of a circular economy (National Research Council, Nutrient Requirements for EPS, 1994).

3.4 Case Studies: Successful Models of Circular Economy

Detailed Case Studies: Provide in-depth case studies of successful circular economy models that integrated EPS recycling and reuse, with quantitative data on resource savings, economic benefits, and environmental impacts.

- Lessons Learned and Best Practices: Draw key lessons from these case studies on best practices for high recycling rates and consumer engagement.
- Future Prospects and Challenges: This subsection addresses prospects for EPS integration into models of circular economies in the future and discusses some of the challenges, such as market barriers and technological limitations.

Already, several areas and industries have presented circular economy models in which expanded polystyrene is incorporated in the most promising way. This chapter discusses some real case studies of successful implementation and indicates some major lessons and best practices learned from them, as well as future prospects and challenges for this EPS circular economy framework.

3.4.1 Detailed Case Studies

These experiences and successes of models in a circular economy are very valuable in terms of the lessons learned and best practices that could well apply to other regions and industries. One clear lesson learned from the study is building a good collection and sorting infrastructure. Stybenex and NexKemia have been able to demonstrate that efficient collection systems are very necessary in attaining high recycling rates and low contamination of EPS waste. Additionally, partnerships between local authorities and waste management companies have been instrumental in the development and maintenance of the various infrastructure.

3.4.2 Lessons Learned and Best Practices

1. A second key factor is consumer participation. Public awareness campaigns were conducted in both case studies that called for consumers to take part in recycling EPS. Clearly communicating ways of proper disposal of EPS, and provision of access to recycling, are among the strategies that work effectively to improve participation rates.
2. Increasing the use of recycled content in the manufacturing of new products is a best practice that will help increase the sustainability of the EPS manufacturing process. By using recycled EPS to create new products through new production processes, firms will be able to reduce their virgin raw material usage and production costs while at the same time minimizing their environmental impact. In this direction, this approach also supports regulatory frameworks encouraging the use of recycled materials in manufacturing.
3. Models that integrate EPS recycling and reuse are already truly successful for a circular economy, thus showing high potential in this material in regard to resource conservation and environmental sustainability. One famous Dutch company is Stybenex, which pioneered the closed-loop system of EPS recycling. In close cooperation with local municipalities and waste management companies, Stybenex has managed to build a nationwide collection network that ensures high recovery rates of EPS waste. After being collected, the EPS is reprocessed to make new products, including insulation panels and packaging materials. According to quantitative data from this model, the consumption of raw materials was reduced by 30%, while CO_2 emissions were reduced by 50% compared to landfills and traditional incineration routes.

3.4.3 Future Prospects and Challenges

While the prospects of the future look bright for EPS in circular economy models, there are a number of challenges that must be overcome to engender its full potential. One of the major problems is market acceptance. While EPS opens up so many environmental perspectives, resistance is often found in using recycled EPS under pretexts concerning quality or poor performance. Continued research and development will be needed to overcome this barrier by making sure that measurements of recycled EPS measure up as much as virgin raw materials do.

One challenge is the technological limitation, especially in processing and recycling contaminated EPS wastes. Another one is innovations of sorting and cleaning technologies to enhance the quality of the recycled EPS and, hence, enlarge its potential applications. More efficient and scalable recycling process developments are also called for to make the economics of EPS recycling more viable at larger scales (Wu, n.d.).

Policy interventions will, thus, be key to determining the future of EPS in a circular economy. Governments can use regulations that encourage recycling, such as extended producer responsibility schemes and mandates for products with recycled content, to integrate EPS recycling. Policies can thus be adopted to help create a market for recycled EPS and drive investment in recycling infrastructure.

- In summary, successful case studies of recycling and reusing EPS prove that there is great potential for it to be integrated into models of circular economy. These models bring lessons learned that could be applied to provide a roadmap for other regions and industries. To secure a future for EPS in a circular economy, it would have to overcome market barriers, develop recycling technologies, and ensure the implantation of supportive policies,

4 Technological Advancements in EPS

4.1 Smart EPS: Integration with IoT and Smart Systems

The emerging applications of smart EPS include when its potentials are harnessed by new applications such as smart packaging, where embedded sensors can monitor the conditions of the product in real time, or serve as building materials with dynamic thermal properties to enhance energy efficiency.

- Integration of Technologies in IoT with EPS: This section should explain the integration of IoT technologies in EPS, including integration-related challenges like sensor embedding, data transmission, and energy management.
- Case Studies and Experimental Results: Present a few case studies and experimental results on the successful implication of smart EPS in real life. Evaluate performance metrics and the possibility of scaling up.

4.1.1 Emerging Applications of Smart EPS

The development of smart EPS is set to revolutionize the application of such a material, more so in domains such as packaging and construction. One of these promising applications for smart packaging is to integrate EPS with sensors to monitor the state of products in real time. For example, temperature, humidity, and shock level sensors can be integrated into EPS packaging during transportation. This kind of real-time monitoring ensures the integrity and quality of sensitive products like pharmaceuticals, foods that will readily perish, and electronics. Smart EPS packaging continuously provides information about the preventative actions taken to prevent spoilage or damage. In construction, the dynamic thermal properties of smart EPS are employed in building materials. By embedding sensors and actuators into the EPS insulation, one can realize walls and roofs that would regulate their thermal conductivity according to whatever heat gains are arriving from the outside. That is, achieve maximum energy efficiency: exactly as much heating or cooling in any one place at any one time, which consumes energy gently, and therefore, costs less. Besides that, Smart EPS can be applied to prefabricated construction elements, making buildings adjustable to changing environmental conditions (National Research Council, Nutrient Requirements for EPS, 1994).

4.1.2 Integrating the Technologies and Associated Challenges

The integration of IoT technologies with EPS raises a number of technical challenges if the full potential of smart EPS is to be realized. Among the primary difficulties is sensor embedding. As mentioned, EPS is a rather lightweight and relatively fragile material; hence, it might be a bit tricky to embed sensors in it without compromising its structural integrity. Their placing and durability should be well managed in order for the sensors to work without affecting the performance of the EPS.

Another challenge is in data transmission, particularly in applications where constant real-time observation is necessary. Ensuring that data from the embedded sensors is reliably and effectively transmitted to external systems is key to the effectiveness of smart EPS. There is therefore a need to develop low-power, wireless communication technologies that can operationalize within the material without requiring a great deal of energy input. Energy management is also a question of prime consideration. The sensors and other IoT components that must be embedded in EPS need a power supply; the integration of batteries or energy-harvesting technologies into the material is complex. In this regard, development along lines of self-sustaining power systems—for instance, harvesting from environmental sources such as solar or kinetic energy—comes as key to realizing smart EPS for long-term applications.

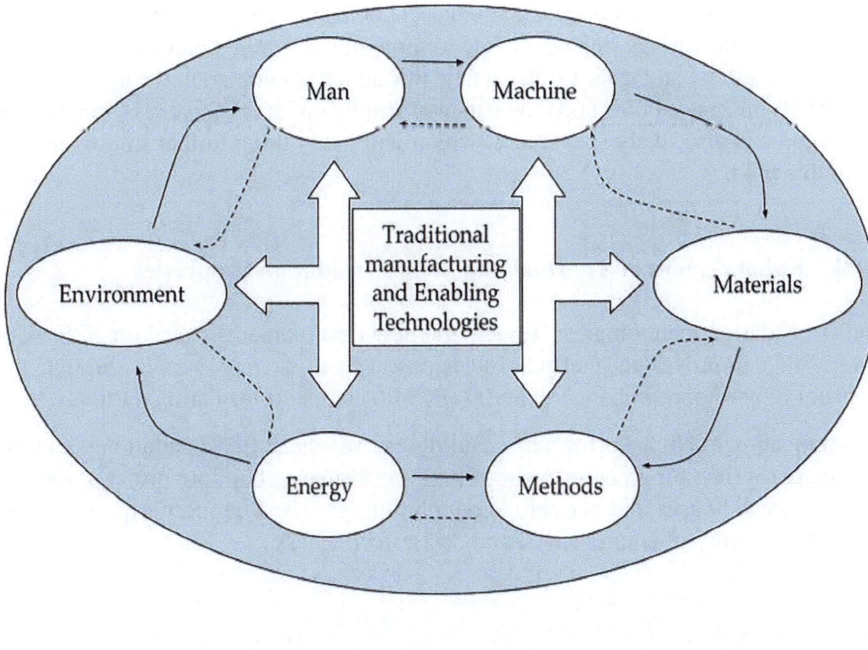

4.1.3 Case Studies and Experimental Findings

1. Application of smart EPS in real-life cases and within experimental projects have seen tremendous potentials in the EPS sector. For instance, smart EPS in the transportation of pharmaceutical products as part of their cold chain logistics. A European logistic company has done a sample test by implanting EPS containers with temperature sensors for transporting vaccines. While such sensors provided temperature continuously, these sensors allow for the storage of vaccines within the desired temperature range over time. Results from the trials showed a 30% reduction in temperature excursions than those from previous or traditional packaging methods, proving the effectiveness of smart EPS in maintaining product integrity (Kim et al., n.d.).
2. Another application case study involves smart-treated expanded polystyrene for use in building insulation. In this particular project, smart EPS panels with embedded thermal sensors and actuators were developed for the retrofitting of a commercial building in Japan. The panels had the ability to change their thermal conductivity, relative to changes in the temperature, so as to realize optimization in energy consumption by the building.
3. From the experimental results, 15% of heating and cooling costs were saved, thus proving the possibility of smart EPS contributing to energy-efficient building design. These case studies are just the epitome of real-life benefits accruable from Smart EPS and offer great lessons on its performance and scalability. The results are very positive, meaning that with further development and fine-tuning, Smart EPS may become standard material for any industry, be it logistics or construction.
4. In this sense, smart EPS consists of this recent technological development entering large-scale applications. Challenges, such as sensor embedding, data transmission, and energy management, among others, hold potential benefits in product safety and energy efficiency through real-time monitoring by smart EPS. With case studies and experimental results on its effectiveness, smart EPS adoption is also likely to increase, which will likely drive further improvements in this field.

4.1.4 Enhancement of Thermal and Sound Insulation Properties

Latest insulation technologies: Review the new developments based on EPS technology that improves its thermal and acoustic insulation properties through the incorporation of phase-change material (PCMs) and nano-insulating particles.

- Application to High-Performance Buildings: Advanced EPS insulation materials are appropriate for an assortment of high-performance building projects, including passive houses and net-zero energy buildings. These properties provide data on the quantity of energy saved and cost-effectiveness.

- Future Research Directions: Show how to do better in the future in the field of EPS insulation. This includes development of hybrid materials and finding new manufacturing techniques. The insulation properties of expanded polystyrene have been long known to different industries, especially building and construction. With the new developments in EPS technology, enhancement in thermal and acoustic insulation became more brilliant, thus highly regarded as one of the materials for high-performance buildings. The section will discuss innovations in insulation technologies, advanced EPS materials in modern construction, and future research directions.

4.1.5 Application to High-Performance Buildings

Development in EPS insulation materials has found significant applications in high performance building projects, be it in passive houses or net-zero energy buildings. This aspect of buildings tries to reduce energy consumption and lower the ecological footprint; therefore, advanced materials for insulation become critical in this context.

For example, passive house construction has utilized EPS insulation with PCMs to great effect in maintaining interior temperatures without active heating or cooling. A number of passive house projects in Europe returned data showing that this EPS insulation can reduce energy consumption for heating and cooling by up to 90% compared to traditional building methods. Not only does this provide huge savings in costs over the building's lifetime, but also helps reduce carbon footprint.

Advanced thermal insulation properties of EPS are also put into practice in net-zero energy buildings designed to produce as much energy as they consume. Since EPS insulation minimizes the loss of heat and maximizes energy gain, it makes the building achieve energy balance. Acoustic insulation with EPS would further add to occupant comfort because of the reduction of noise pollution, especially in an urban setting.

5 Environmental Impact and Mitigation Strategies

5.1 Life Cycle Assessment of EPS

The overall environmental impact of expanded polystyrene needs to be understood in terms of its whole life cycle—from raw material extraction through to disposal at the end of life. This section sets out a detailed life cycle assessment for EPS, identifying environmental hotspots and comparing them with alternative materials to present the relative environmental impacts.

5.1.1 Full Life Cycle Analysis

In this regard, an EPS life cycle assessment will be immensely useful for qualifying the environmental impacts associated with each life cycle stage of the material, from raw material extraction and processing to manufacturing, transportation, use phase, and end-of-life disposal. As such, everything begins at the initial stage of extraction of petroleum-based raw materials, followed by the processing of those into styrene monomers and their subsequent polymerization into beads. This is highly energy-intensive and results in huge greenhouse gas emissions associated with the application of fossil fuels. In the manufacture phase, EPS beads are expanded and molded into products such as insulation panels and packaging materials. This process also requires steam and other energy inputs and thus further increases the carbon footprint of the material. Transportation of the EPS products to the end users adds additional emissions, although the light weight of EPS makes this more benign compared to heavier materials (Kim et al., n.d.).

Generally, low environmental impact is related to the EPS use phase because of the inertness and stability of this material under most conditions. However, the end-of-life stage will be very challenging. Generally, EPS is nonbiodegradable and can persist in the environment for centuries if not properly managed. Options for its disposal range from landfilling and incineration to recycling; each has a different environmental impact. Landfills can lead to long-term environmental contamination, while incineration is releasing pollutants into the environment, even though energy recovery can be attained. Recycling, although more preferable, is limited by inefficiencies in collection and contamination issues.

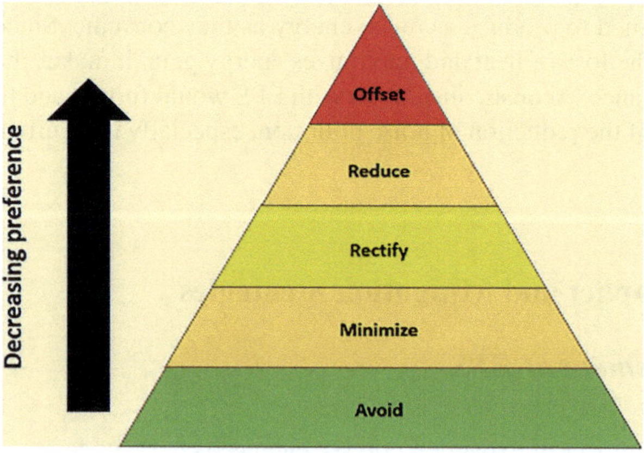

5.1.2 Environmental Hotspots and Mitigation Strategies

Some of the environmental hotspots identified in the LCA of EPS entail the most significant impacts of the material. Some of these impacts are at the beginning of the life cycle, resulting from raw material extraction and production, associated with high energy use and GHG emissions. Another critical environmental hotspot is the end-of-life disposal stage, where improper management can result in environmental pollution. The following methods can help to reduce these impacts. Process optimization at the production stage, such as enhancement of energy efficiency and switching to cleaner energy sources, is able to lower emissions by a large margin. Replacing some of the petroleum-based raw materials with bio-based ones or adding recycled EPS into primary production can also cut the environmental impact (Crowley, n.d.).

At the end-of-life stage of the product, improved collection systems and consumer education raise recycling rates. This can be complemented by the development of more advanced recycling technologies, like chemical recycling, in order to recover and reutilize EPS, hence diminishing virgin material use and waste generation at their very source. Secondly, better design can enable the easy recycling of EPS products, while environmental impacts can be reduced by taking the whole life cycle of the products into account at the design stage.

5.1.3 Comparative LCAs

A comparison of the LCA of EPS with that related to alternative materials indicates a relative environmental performance. Important alternative materials to EPS are those such as extruded polystyrene (XPS), polyurethane foam, and biodegradable plastics. In fact, comparative LCAs often find that EPS, despite numerous associated environmental disadvantages, mainly during its end-of-life phase, are also associated with advantages in terms of the energy efficiency and resource use during production and transportation.

- For example, compared to XPS and polyurethane foam, EPS generally requires a lower amount of energy in production, while in many uses, its lightweight nature could reduce emissions from transportation. Biodegradable plastics possibly have advantages for end-of-life disposal, then, since they are designed to break down in the environment. These very alternatives could again have trade-offs of their own kinds, for instance, increased energy requirements for production or poor applicability in some uses.

This involves the disclosure of environmental impacts and identification of the improvement areas of EPS. It is essential because a reduction in environmental impact can be achieved through the identification and reduction of impacts in hotspots. Additionally, the comparative LCAs put the performance of EPS into perspective by comparing it with alternative materials, thus offering very valuable insights for decision-makers in choosing the most sustainable options for specific applications (Kim, n.d.).

5.2 Reductions of the Carbon Footprint in EPS Production

Reduction of the carbon footprint in the manufacture of expanded polystyrene is very necessary for a wide material usage class with less heavy environmental impact. This section will cover reduction strategies for the carbon footprint, success case studies of carbon management in the EPS industry, and future directions for research and policy (Charalabidis et al., n.d.).

5.2.1 Strategies for Carbon Footprint Reduction

Several strategies could be implemented to reduce the associated carbon emissions from EPS production. One successful approach would be through the application of low-carbon raw materials. Conventionally, EPS is produced from feedstocks based on petroleum; these might be key sources of GHG emissions. A shift toward bio-based or recycled feedstocks significantly reduces the carbon footprint of EPS. For instance, if the manufacturing process were to incorporate bio-based styrene or include recycled polystyrene, this would help cut down on total carbon emissions.

Energy-efficient manufacturing processes are a second crucial area of carbon reduction. Several options also exist for reducing energy use in EPS production, including the use of more efficient steam generators, waste heat recovery, and advanced process controls that help minimize energy consumption. Other carbon reduction opportunities still rest in switching to renewable sources of energy such as solar or wind with regard to satisfying the electricity needs at EPS facilities. Finally, reductions in total usage, by lightweighting and efficiency material gains, reduce the energy used in production (Zhang, n.d.).

5.2.2 Carbon Management Case Studies

Several industries already have evidence of implemented strategies, which are working successfully to reduce the carbon footprint of the EPS that is being manufactured, including but not limited to: Recently, a European manufacturer of EPS had shifted their all type of production to 100% renewable energy. This company deployed solar panels and wind turbines in renewable energy generation installations, and the emissions originating from carbon power were thus reduced to a very dominant extent. The same manufacturer also had optimized production in optimized energy use, and hence at the same time twofold carbon reduction was achieved.

This makes another interesting example about an Asia-based EPS producer that integrates recycled polystyrene during manufacture. To this end, the company uses its recycled material in order to minimize the use of virgin petroleum-based feedstocks and the carbon emanating from extraction and processing activities. This effort was combined with the application of upgraded energy-efficient equipment, which went one step in minimizing production-related carbon.

5.3 *Innovations in EPS Waste Management*

Effective waste management can control the environmental impacts related to expanded polystyrene. Several recent innovations in advanced techniques give more sustainable solutions for the management of EPS wastes. There are several technologies presently in use for waste management—mechanical recycling, chemical depolymerization, and waste-to-energy routes—which in this section have been reviewed with regard to their environmental and economic impacts (Eodanable & Lauchlan, n.d.).

5.3.1 Mechanical Recycling of EPS

One of the most established techniques of EPS-waste management is mechanical recycling. In this process, collected EPS products are cleaned, ground down into smaller bits, and then reprocessed to make new EPS. Mechanical recycling is more effective in managing post-industrial EPS-waste, which normally is clean and uncontaminated.

There are some immense benefits associated with the mechanical recycling of environmental sustainability. By reprocessing the EPS waste to new products, the process reduces demand for virgin raw materials supply, hence conserving resources and reducing the carbon footprint for EPS production. Moreover, mechanical recycling diverts EPS waste from landfills and reduces environmental pollution, probably the event of long-lasting contaminations.

However, the economic feasibility of mechanical recycling is often difficult to attain, particularly for post-consumer EPS wastes. Lightweight and bulky nature of EPS, coupled with high collection and transportation costs, contamination from food or other materials, may make it difficult or impossible to recycle. Although it seems difficult, advancements in sorting technologies and regional recycling programs continue to better the efficiency and applicability of mechanical recycling.

5.3.2 Chemical Depolymerization of EPS

Chemical depolymerization is an entirely new route for the management of EPS wastes by undergoing polymer chain breakdown into its respective monomers, for instance, styrene. It may be done in solvents or catalysts to depolymerize EPS, from which high-purity monomers may be recovered and reused to produce virgin EPS or other polystyrene products.

Moreover, chemical depolymerization reduces the environmental effect, as it gives a possibility to completely recycle the used EPS without the loss of material properties that is taking place in mechanical recycling (Etkin, n.d.).

- This leads to a higher cost of chemical depolymerization compared to mechanical recycling. These methods turn out to be complicated in nature and require special equipment; hence, they increase the cost of treatment. At the same time, the possible recovery of high added-value monomers and the potential inclusion of this operation into large-scale chemical production forms makes this step reasonably affordable. Current research deals with making chemical depolymerization more effective and scalable. In fact, with time, it seems that it can prove to be an attractive solution of waste management.

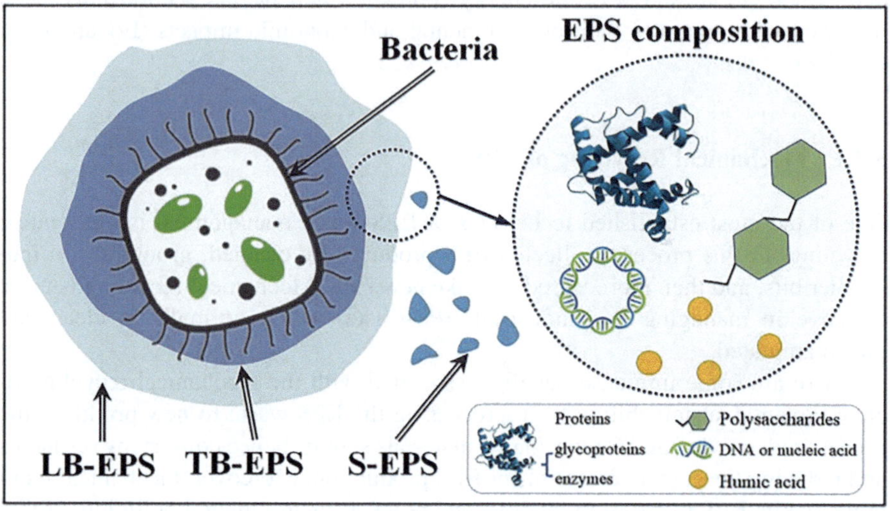

5.3.3 Waste-to-Energy Technologies for EPS

Waste-to-energy (WTE) technologies provide an alternative process in countries with minimal infrastructural capacities to recycle EPS wastes. The technologies require the incineration of EPS waste, resulting in energy in the form of heat or electricity. Although WTE is not a recycling process, it is the recovery of energy in cases where waste to be incinerated would be considered as a candidate for landfilling.

WTE technologies have both positive and negative environmental aspects. On one hand, WTE reduces land use and the danger of environmental contamination resulting from reduced volumes of EPS wastes buried. Further, the energy recovering from an incineration process could replace the use of fossil fuel in producing energy through more sustainable means. On the other hand, incineration of EPS may result in the atmospheric release of some harmful pollutants, including dioxins and furans. Modern WTE facilities have state-of-the-art systems to control emissions, consequently reducing the impacts; however, this has not been able to put to rest the fact that the air quality and long-term sustainability of incineration is still in question.

Economically, WTE may turn out to be a very viable solution for the handling of EPS wastes, particularly in countries with large costs involved in waste disposal and high energy demands. The ability to derive revenues from energy production creates a financial incentive for adopting WTE technologies. Despite that, the initial investment to set up WTE infrastructure can be huge, while the long-run economic viability will depend on the prices of energy, the availability of the waste, and the regulatory settings. While not exhaustive, some of the most interesting innovations that exist today for waste management in EPS include mechanical recycling, chemical depolymerization, and waste-to-energy technologies. These methods have different environmental impacts and costs. Of course, the selection of method would fully depend on such factors as the composition of the waste, the regional facilities, and the policies; there is a place for each of these methods as technology continues to develop in respect of the sustainable management of EPS waste.

6 EPS in Construction and Infrastructure

6.1 Future Applications of EPS in Green Building

In the near future, expanded polystyrene will feature prominently in green building processes where sustainability and energy efficiency are critical concerns. This section covers state-of-the-art uses of EPS in green construction related to insulation systems, modular prefabricated components, and energy-efficient building envelopes.

6.1.1 Advanced Insulation Systems

One major advantage of EPS in green building lies in its exemplary functions of thermal insulation. Among these is a high R-value, which characteristically measures the resistance to heat flow of a material. This kind of material can be suitably used in wall, roofing, and floor insulation, as it controls the loss of heat during winter and influx during summer. As energy efficiency in building design becomes paramount, it is possible that future applications of EPS will be implicated by more sophisticated insulation systems wherein EPS is built as a central component.

One example is that EPS can be sandwiched into insulated panels used to build high-performance envelopes. Such panels can be prefabricated off-site with assured quality and reduced on-site construction wastes. These can also be further coupled with phase change materials or reflective coatings or other systems that help not only in energy efficiency but also regulate the indoor temperature and prevent artificial heating or cooling, hence reducing the need for them. The demand for EPS will grow in advanced insulation systems as building codes continue to emphasize energy conservation (Eodanable & Lauchlan, n.d.).

6.1.2 Modular and Prefabricate

Another factor making EPS a perfect material for modular and prefabricated construction is the fact that it is lightweight but highly robust. The construction industry thus moves toward more sustainable practices, and together with the lessening of material waste, energy consumption and construction timeline because of prefabrication, it also means that EPS is a good material for environments that considers the aspect of sustainability. The same technique applies to EPS, as it is easily shaped into numerous different forms and sizes, hence the designing of limitless custom-made, pre-fabricated elements to meet specific design considerations.

In green building, EPS utilizes modular components based on wall panels and building sections. These components are made in controlled environments, so precision is guaranteed and the environmental impact of construction activities is minimized. The use of EPS in prefabrication also enhances the insulation performance of a building, whereby the material can be used in structural elements that aid in boosting the overall energy efficiency in the building. Besides, the modularity in these members allows for a very rapid, smooth fit-together and disassembly, since the reuse and recycling of the building parts are actually promoted in this way, thereby furthering the cause of sustainable construction.

6.1.3 Energy-Efficient Building Envelopes

In the building design for energy efficiency, there comes the concept of a building envelope—the barrier between a building's inside and outside environment. EPS is being increasingly used in the creation of high-performance building envelopes to reduce energy consumption through increased thermal insulation, airtightness, and moisture resistance. Probably most uses of EPS in the future will relate to the optimization of the energy efficiency of building envelopes by innovative design and integration of materials (Etkin, n.d.).

This can be done by using EPS in continuous insulation applications, whereby it is installed across a building's exterior to create a continuous thermal barrier that eliminates any thermal bridges responsible for energy loss. Such a move enhances the energy efficiency of a building through improved comfort of occupancy based on the maintenance of stable temperatures indoors. Further, EPS can be combined with other materials, such as SIPs or ICFs, to achieve building envelopes that not only meet but in most cases far exceed even the strictest energy codes. As sustainability remains top of mind in the building industry, this role of EPS in developing energy-efficient building envelopes will only continue to grow. Future research and development efforts are expected to focus on further improving the thermal performance of EPS, enhancing its fire performance, and more innovative integration of the insulation type into building systems that help provide overall sustainability in the built environment.

Finally, future applications of EPS in green building can be considered diverse and full of opportunities. From advanced insulation systems to modular prefabricated components and energy-efficient building envelopes, EPS is realized with a great number of benefits aligned to the goals for sustainable construction. The clear path forward is that EPS will undoubtedly play a very vital role in shaping the future with regard to green building practices as this industry continues to evolve (Schaeffer, n.d.).

6.2 EPS in Sustainable Urban Development

The role that EPS can play in contributing toward the goals of sustainable urban development is increasingly recognized, notably by its applications in the infrastructural and urban planning sectors. The next section examines more in detail how EPS may contribute to urban sustainability: its use as lightweight fill material in construction, improving resilience in cities, and contribution toward circular economy policies.

6.2.1 Lightweight Fill Materials for Infrastructure

EPS has evolved into an exceedingly useful material for work in urban infrastructure, particularly due to its lightweight properties, along with structural stability. One of the most innovative uses of EPS in sustainable urban development is its application as a lightweight fill material in road construction, embankments, and drainage systems. Traditional fill materials, like soil and gravel, are massive and require large amounts of energy and resources for their transportation and laying. On the other hand, the EPS blocks are lightweight and easy to handle, and can be made according to the various needs of the building [09].

By using EPS in these applications, cities can lessen the environmental impact from the development of infrastructures. For example, with a lesser weight, EPS will put less load onto underlying soils, reducing ground preparation and, therefore, the total carbon footprint from construction activities. At the same time, this material is perfect for huge long-lasting projects of infrastructure due to its strength and resistance to moisture, which a city development must have to be sustainable and resilient.

6.2.2 Enhancing Resilience in Cities

Enhancing resilience in cities is also essential in the development of more and more resilient urban environments against environmental shocks, such as floods, earthquakes, and temperature changes. Resilience, in relation to sustainable urban development, can be explained as the capacity for the urban system to absorb disturbance and renew itself. Shock absorption and thermal insulation are very useful properties among the variety of EPS features in engineering resilient city infrastructure.

It can also be used to construct flood barriers and retaining walls that help in managing the flow of water during heavy rainfall, thus reducing the risk of urban flooding. Its lightweight nature further allows it to be suitable for earthquake-resistant structures by reducing the seismic load on buildings. Indoor temperatures can also be controlled through EPS insulation systems to ensure energy efficiency and comfort in occupants of urban buildings. By integrating EPS within city design, achievement of the sustainability goals could enhance resilience to environmental challenges (National Research Council, Nutrient Requirements for EPS, 1994).

6.2.3 Supporting Circular Economy Initiatives

The recyclability of EPS supports circular economy initiatives; it aims for zero waste and to create closed loops of material. In a circular economy, materials such as EPS are meant to be designed for their subsequent reuse, for recycling, or for repurposing in such a manner that virgin resource use is reduced and, to the greatest extent, environmental impacts are minimized. EPS has excellent potential to be recycled into new quality products.

In sustainable urban development, EPS recycling programs can be part of the waste management systems to ensure that such material is gathered, treated, and reinjected into the production cycle once again. By doing so, EPS waste reaching the landfills may not only decrease but also give a constant supply of recycled material for new uses associated with building and packaging. Public works can contribute even more to the demand from cities by including reprocessed EPS in infrastructure works. This enhances the local industry that reprocesses it and, in return, helps to increase the rate of used EPS collection. It is only through the promotion of circulatory economy initiatives that cities can reduce their carbon footprint, save on natural resources, and move further in the direction of sustainable growth. The integration of EPS into such initiatives of urban development is mainly based on its recyclability to enhance creation of a more resourceful and ecological urban ecosystem.

6.3 Case Studies: EPS in Innovative Building Projects

Expanded polystyrene has become a game-changing material in modern construction, solving some of the most burning issues and challenges of building design and urban planning. In this section, detailed case studies will be included in the book, revealing the versatility and efficiency of EPS in various building innovative projects related to the sphere of constructing high-performance green roofs and earthquake-resistant buildings.

6.3.1 EPS in High-Performance Green Roofs

Green roofs, which add layers of vegetation and soil over a waterproofing membrane, are increasingly put into practice in urban areas as a way to enhance building sustainability. Among the important challenges of green roof construction are those of adequate thermal insulation and stormwater management. EPS has been shown to be an ideal solution for both, as demonstrated through several high-performance green roof projects (Zhang, n.d.).

In these projects, EPS is used as a thin, lightweight insulation layer under top soil and vegetation. With its high thermal ability to maintain an internal temperature, it reduces energy demand for heating and cooling. For instance, in cities, most buildings remain under the attack of extreme temperature changes; this thermal stability becomes extremely important. Creating green roofs with EPS saves a lot of energy for the buildings, which will help attain overall sustainability.

Moreover, EPS has a critical role in the stormwater management function of green roofs. Lightweight, it does not put any undue load on the building structure; similarly, its permeability will let water seep through in a controlled manner. This reduces the possible risks of leakage in a roof and assures that excess rainwater is managed properly, lowering urban runoff, and thus helping to reduce the problem of flooding in thickly built environments. One outstanding example of EPS in green roof construction is the Biosolar Roof project, which combined EPS insulation with solar panels. By putting EPS insulation together with renewable energy technology, not only was the energy efficiency of the building enhanced but a model of sustainable urban development was also provided. This example shows how, in an urban environment, EPS is used innovatively to answer a set of diverse construction challenges related to the environment.

6.3.2 EPS in Earthquake-Resistant Structures

The other area in which EPS has exhibited its potential for transformation is earthquake-resistant building construction. Earthquakes pose significant risks to buildings, notably where seismic occurrences are more frequent. The challenge was to devise entities that can withstand seismic forces yet remain cost-effective and easy to build. In this regard, EPS, with its lightweight and shock-absorbing qualities, has borne out its effectiveness.

A good number of earthquake-resistant building projects have utilized EPS as a core component in their design. One prominent example is in the construction of EPS-Core Shear Walls, wherein EPS panels are used as a lightweight core material sandwiched between reinforced concrete layers. The design aim of these EPS-core shear walls is that, in case of an earthquake, seismic energy will be absorbed and dissipated out of the structure, putting less stress on the structure itself. Lightweight EPS also reduces the mass of the building, which is the vital factor in controlling seismic forces the structure may be subjected to (National Research Council, Nutrient Requirements for EPS, 1994). This has become quite attractive in building projects in earthquake-prone areas where time is of essence. The use of EPS in these projects not only makes the building safer and resilient but also simply tries to express the capability of this material in application in solving such complex engineering problems. Another such example is the Chile Modular Housing Initiative, using EPS in earthquake-resistant structures to build low-cost, earthquake-resistant homes in seismic zones. Seismic activity of high magnitude can be withstood by these houses, which have delivered safe and rugged housing solutions to communities in need. What this really demonstrates is that EPS can lend itself to disaster-resistant architecture, providing a route to safer, more resilient urban development.

These case studies on high-performance green roofs and earthquake-resistant structures show the versatility and effectiveness of EPS as a modern construction material. With the special properties of EPS, modern building solutions to some of the most critical environmental and structural challenges can be developed by architects and engineers. As demand continues surge for sustainable and resilient construction, so does EPS take the front line in shaping a different future for urban development.

7 Regulatory and Policy Trends

7.1 Emerging Regulations Impacts on EPS Research

The regulatory landscape on EPS has undergone considerable evolution driven by the current and growing concerns about the environment. Governments and international bodies bring new regulations that will minimize the environmental impacts of EPS at all stages of its use, mainly in terms of disposal and recyclability. This section shall investigate the changing regulatory scenario and its implications for the research on EPS in relation to waste management, recycling mandates, and the drive for sustainable alternatives.

7.1.1 Waste Management Regulations

One of the main focuses of regulation in this space is the waste management of EPS. Because it is light and bulky, EPS has, traditionally, been a huge problem for waste management and quite often becomes landfilled or littered into the environment. Hence, a number of jurisdictions are now developing more stringent regulations on waste management with a view to limiting the impact of EPS on the environment. These regulations often require manufacturers to be responsible for the entire life cycle of EPS products, this including their disposal and recycling. For instance, EPR programs are being rolled out across the world, placing a legal obligation on businesses to develop and finance collection and recycling infrastructure. When such policies take effect, academics are motivated to study new technical developments in the recycling of EPS waste, such as chemical depolymerization and new mechanical methods of recycling, in an attempt to make its waste management more effective and sustainable. Ultimately, all this is about closing loops—processes by which the polymer can be recycled numerous times while causing minimal damage to the environment (Eodanable & Lauchlan, n.d.).

7.1.2 Recycling Mandates and Standards

It is now emerging that one of the critical focus areas in the regulatory framework for EPS relates to its recycling. With increased awareness about the environment, particularly related to plastic wastes, most regulators are increasingly mandating an increase in the recycling rate of EPS products. This mandate usually comes with developing standards defining the recyclability of EPS materials, making sure they are conformed to a particular criterion on being reprocessed into useful materials again.

As an example, the Circular Economy Action Plan elaborates ambitious goals in the plastic recycling sector, including EPS, within the European Union. Member states, under this plan, are obligated to establish mechanisms that facilitate collection, sorting, and recycling of wastes from EPS products. This led to the introduction of new standards for EPS products, focused on easier recycling and reduced environmental impacts.

It is these trends in regulations that are forcing researchers into innovating new formulation and additive developments to improve EPS recyclability. The aim is the production of EPS materials that are not only compliant with regulatory legislation but also offer better performance in terms of sustainability and functionality (National Research Council, Nutrient Requirements for EPS, 1994).

7.1.3 Push for Sustainable Alternatives

Besides the regulations in the domain of waste management and recycling, there is a regulatory drive to develop and adopt sustainable alternative materials to traditional EPS. This is part of a wider global agenda aimed at the reduction in the consumption of nonrenewable resources and the reduction in carbon footprint. The regulators are, therefore, invoking active promotion of research into alternative materials that can replace or complement traditional EPS in various end-use applications.

As an example, bioplastics based on renewable feedstock are emerging to provide an answer and probably as a substitute for the future—in place of EPS. Regulatory support is leading to R&D investments in this direction for replacing traditional EPS to eco-friendly substitutes. Additionally, several standards are being developed for assessment regarding sustainability in terms of biodegradability, life cycle impacts, and overall performance on the environment.

This run by regulators for sustainability drives the direction of EPS research and challenges scientists and engineers simultaneously to investigate new raw materials along with innovative production processes, which may be developed in compliance with forthcoming standards. In this regard, researchers on sustainability issues do not only assure that they satisfy the regulations to contribute to the reduction of the environmental impact of the plastics industry but also help to assure sustainability in the scientific and economic sense.

7.2 Global Policy Initiatives on EPS Sustainability

1. The Single-Use Plastics Directive of the European Union

   The European Union has been leading in the global drive toward the reduction of plastic wastes and, consequently, of expanded polystyrene. Adopted in 2019, the Single-Use Plastics Directive especially focuses on items identified to contribute a large proportion of marine pollution. Among the main targets of this directive are food containers and packaging, widely made of EPS. It limits EPS waste by restricting its use in some applications, replacing it with alternatives, and enhancing the practice of recycling. This directive has accelerated innovation in biodegradable EPS variants and other sustainable materials to replace them, shifting the economy toward a circular model within the EU (Kim et al., n.d.).

2. National Policies for Promoting Recycled EPS

   Beyond the EU, many countries have established policies to encourage the use of recycled EPS. The Green Purchasing Law in Japan, for example, incentivizes the government and businesses to preferentially procure products containing recycled materials, including EPS. This has led to the establishment of a proper recycling infrastructure and the development of a market for reprocessed EPS products in the country. In South Korea, for instance, the Extended Producer Responsibility system creates obligations on manufacturers to be responsible for end-of-life management of their products, including packaging that includes EPS. These national policies not only reduce environmental impacts but also create economic opportunities within the recycling sector.

3. Global Collaboration and Future Directions

   Global collaboration is very important in terms of dealing with the environmental challenges posed by EPS. International initiatives, like the United Nations Environment Programme, are leading an effort to harmonize regulations as well as promote best practices for management of EPS between borders. All these efforts are oriented toward a coherently framed support for sustainable production, use, and disposal of EPS. While countries forge ahead with such initiatives and policies through refinement, embedding these initiatives within the global supply chains is increasingly paramount for bringing sustainability benefits to the global sphere. To what extent would cooperation among governments, industry players, and researchers be key to developing future innovative solutions for the sustainability of EPS.

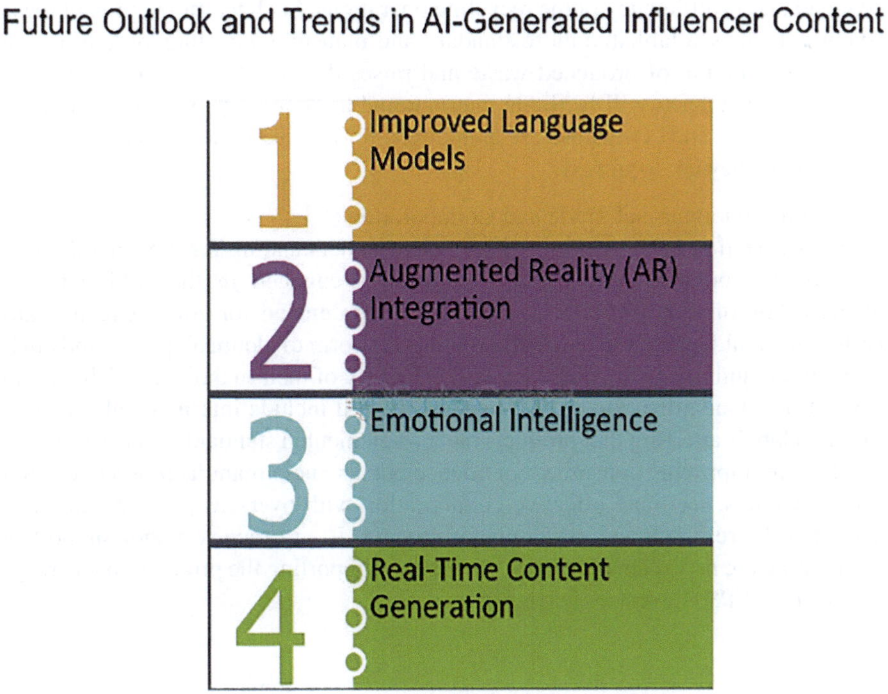

7.3 Future Directions for EPS Standardization

1. Develop Standardized Test Procedures

   The EPS industry is changing; in that, there is an increasing need to define standardized test methods. Such test methods are relevant, indeed, in terms of safety, quality, and sustainability of a large variety of applications involving EPS. Today, no uniformity of test protocols and, more specifically, the methods to study environmental impact during all stages of the life of EPS, exists. Future efforts will focus on also developing full-range testing methods, measuring not just the physical properties, such as thermal insulation and compressive strength of EPS, but also the environmental performance related to the carbon footprint and recyclability. These standard test methods will, therefore, enable manufacturers to deliver EPS all over the world under strict standards for both complete safety and sustainability (Crowley, n.d.).

2. Development of Global Recycling Specifications

   One of the major challenges in handling the wastes of EPSs is variation among different regions in their method and the structure of recycling. With this in mind, there is a growing push toward establishing global benchmarks for EPS operations. This will help in setting minimum rates of recycling, purification of the recycled

EPS, and energy efficiency of the recycling operations. Such harmonization of these benchmarks may establish a more standard and plausible recycling system, which reduces the amount of produced waste and raises the circular economy. Standard recycling benchmarks will facilitate international trade in recycling the EPS material, as countries and companies are guaranteed for the quality and sustainability of the products they are to import.

3. Enabling International Trade and Collaboration

Standardization in this respect gives a clear understanding for the smooth business among countries and different companies engaged in the EPS industry. Globalization of supply chains leads to increased demand for homogeneous standards that should guarantee that EPS products conform to identical quality and environmental requirements, irrespective of the country of their make or use. The future ways of the standardization of EPS very likely will include international certifications and labels attesting that products meet implemented standards. This would not only help in improving consumer confidence but also help manufacturers to expand in new markets, facilitate effective collaboration with overseas partners, and also smoothen the research and development process. It will receive major support in accelerating the new technologies or innovation supporting the production and recycling of new EPS (Lavery & Davies, n.d.).

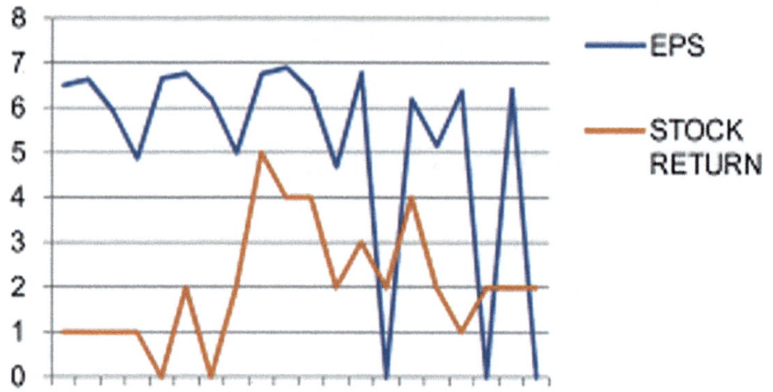

8 Market Trends and Economic Impacts

8.1 EPS Future Market Projections

1. Continuous Market Growth Across Industries

The EPS market is expected to grow continuously with construction, packaging, and consumer goods placed at the forefront among other key growing industries. Very huge demand for EPS is projected from the construction industry because of its marvelous insulation abilities, very low density, and reduced expense costs. As global urbanization continues, so does the demand for effective construction materials still; in exchange, it shall remain to be the best material for use in residential and commercial development projects. In the area of packaging, the padding property and the versatility of design, with EPS, will not go away in the face of its array of variations and applications, as from food to food appliances and gadgets. In addition, the consumer goods industry, which uses EPS in a range of products from toys to appliances, is also expect to contribute to market growth, supported by the continued expansion of e-commerce and retail sectors (Etkin, n.d.).

2. Influence of Technological Advancements

Major drivers for market expansion are developed technologies of EPS production and applications, including the introduction of nanotechnology to improve the mechanical and fire retardancy results of the material, expected to open new routes for EPS in high-performance applications. The development of environmentally friendly EPS grades, including recycled content or biodegradable additive, will feed into the growing demand for sustainable materials. In addition, these advances give not only prominent improvements to the functional properties of EPS but also resonate with the environmental consciousness of consumers and businesses. Adoption of rigid standards in the manufacturing process for energy efficiency and smart EPS with embedded sensor applications in the future probable way indicates pushing the market further with more sophisticated and value-added applications (Schaeffer, n.d.).

3. Investment and Market Expansion Opportunities

Attractive investment is likely to be resulted in by the growing demand for innovative and sustainable EPS products. There would most likely be investments focused on companies that pioneer the advancement of new EPS technologies, more specifically those with a focus on sustainability and environmental protection. Regions like Europe and North America are likely to have stronger support and investment activity for green building practices, driven by regulatory support. Growth opportunities for EPS manufacturers are also voluminous in emerging markets—Asia-Pacific, Latin America, and Africa—where urbanization and industrialization are accelerating. In these regions, increasing adoption in construction and packaging is likely to happen as they seek cheaper solutions to meet the needs of fast-growing economies. This will eventually drive the global EPS market growth with technology innovation, regulatory support, and growing demand from various applications.

8.2 Impact of Global Economic Trends on Demand for EPS

1. Raw Material Price Changes and Supply Chain Ruptures

The demand for EPS is strongly driven by raw material costs and availability, mainly due to the styrene monomer coming from petroleum. Fluctuations in the price of oil, influenced by geopolitical events, disruptions in the supply chain, and changes in global demand, might considerably change the cost of production for

EPS. Higher prices for raw materials lead to higher production costs for EPS manufacturers, which will now be transferred to consumers, thus decreasing demand. Supply chain issues, such as transportation bottlenecks or a lack of some critical input, may additionally complicate the situation by bringing about delays and increasing prices. Companies reliant on EPS for products or packaging may try to reduce its consumption by seeking alternative materials; this could impact overall market demand.

2. Trade Policy and International Relations Influences

   The global trade policies and international relations have very strong control over the shaping of the EPS market. Tariffs, trade restrictions, and changes to import-export regulations may alter the flow of EPS and its raw materials across borders. For instance, a trade war between two giant economies like the USA and China can disrupt the global EPS supply chain, resulting in price fluctuation and uncertainty. Countries that slap tariffs on EPS or its raw materials may inadvertently affect the market competitiveness of this polymer, thereby rendering the products expensive and therefore unattractive to the global market. In contrast, regions with agreed favorable trade pacts are bound to attract steady supply and prices, thereby spurring demand for EPS within that region (Charalabidis et al., n.d.).

3. Economic Uncertainties and Their Bearing on Manufacturing

   These uncertainties can hit hard on the manufacturing sector and, thus, the production and consumption of EPS, in cases of recession or just economic slowdown. When the economy is experiencing a downturn, construction and consumer spending are reduced, which in turn reduces demand for EPS building materials and packaging, and consumer goods. This could also lead to tighter margins for manufacturers, forcing them to cut down their production or seek cost-saving measures that could include cutting back the use of EPS. Nevertheless, the role that EPS plays in various industries is essential; while demand may be changed, it is unlikely actually to diminish. Those companies that will adopt flexible manufacturing strategies and continue to find new, innovative applications for EPS will be better positioned to weather economic challenges and maintain demand (National Research Council, Nutrient Requirements for EPS, 1994).

4. Sustaining Demand Through Sustainability and Green Building Practices

   Despite economic hardships, the market demand for EPS is likely to remain driven by the growing global concern toward sustainability and green building practices, especially in countries where environmental regulations are very strict. Due to the fact that EPS is one of the insulation materials able to save energy, it will have a big role in sustainable construction projects from passive houses up to net-zero energy buildings. With carbon reduction and care for the environment topping the agenda of governments and industries across the world, EPS will likely remain in demand due to its propitious properties and potential recyclability. Demand for high-performance, ecofriendly EPS products is expected to rise in countries where green building certifications and standards become more and more of a requirement, counterbalancing some of the adverse effects of economic uncertainties. This trend underlines how innovation and sustainability will determine future demand for EPS independent of the broader economic conditions (Etkin, n.d.).

8.3 Investment Opportunities in EPS Innovation

1. Advancements in EPS Recycling Technologies

   The demand for effective EPS recycling solutions has increased amid tightening environmental regulations and growing consumer awareness of plastic waste. By making investments in such recycling technologies, huge opportunities for innovation and profitability are opened within the EPS industry. Mechanical recycling and chemical recycling are those two areas that are primed and ready to explode in terms of technological innovation, turning EPS back into its raw form and breaking it down into basic chemicals, respectively. In case of development of more effective and cheaper recycling processes, the ecological footprint of EPS could be reduced, potentially confirming this material as more sustainable for most industries. Investors who focus on companies and technologies that develop the recycling of EPS have the chance to take advantage of the growing market for materials received in the process of recycling due to regulatory requirements and customer demand for "green" products.

2. Development of Biodegradable EPS Alternatives

   The other very promising area for investment is in the development of biodegradable EPS alternatives. Traditional EPS is not biodegradable, and the environmental pollution caused by it is long-lasting, especially within the marine environment. However, recent research has managed to come up with biodegradable and compostable EPS-like materials that are made from renewable resources such as plant-based polymers. Those alternatives, however, hold on to the good points of EPS, like lightness, thermal insulation, and shock absorbing properties, adding to it the advantage of environmental degradation following its use. The investors in this field have the potential to tap into a market that is much aligned with the global goals on sustainability and the increasing preference of consumers for biodegradable products. Demand is therefore bound to increase due to the urge by governments and corporations to slash plastic waste. These create lucrative opportunities for early investors in biodegradable alternatives to EPS. Sustainability will be a key driver of future demand for EPS, irrespective of the broader economic conditions (Kim et al., n.d.).

3. Technology Development for Materials in Building Based on EPS

   Given this, EPS has been widely practiced in construction because of its superiority in insulation properties, durability, and light weight. Yet, innovations in EPS-based construction materials introduce new investment opportunities. Such advancements include the introduction of EPS products that actually enhance energy efficiency, minimize greenhouse gas emissions, and generally enhance the sustainability of the building projects. For example, EPS can be manufactured as much more fire-resistant, structurally sound, and energy-efficient than strict recent requisites of building codes and green building skills. Furthermore, the use of EPS in modular construction and prefabrication activities can also expedite construction by a large margin, reduce the amount of wastes generated in a project, and cut down on costs. Increasing investors are those with supporting companies that move the constructions' limits of EPS, where benefits come through increasing demand for sustainable building materials, either due to environmental regulations or market preferences for green construction solutions.

9 Challenges and Opportunities in EPS Research

9.1 Addressing Environmental Concerns

These days, expanded polystyrene proves to be a very versatile material in wide application, but it poses a serious environmental problem. Taking into consideration the persistence and low recycling rate of EPS, it is feared that it is contributing to environmental pollution, particularly within the marine ecosystems. In that respect, research increasingly focuses on more sustainable solutions for these challenges. The section covers environmental concerns with EPS, developments of recycling methods, and creation of biodegradable EPS alternatives.

1. Environmental Impact of EPS

   The fact that what is often prized about EPS is that it is very hard to degrade has the darker side of the coin that it persists in the environment for a longer time. There can be fragmentation of the discarded EPS into smaller and microsized particles, thus adding to the microplastic pollution load mainly within the oceans. It has low density and is lightweight; therefore, the risks associated with it are most likely to be blown away or swept by water to contaminate the environment. Also, the production processes of EPS are related to nonrenewable resources, with an involvement of greenhouse gases, which makes the situation even worse. Considering such impacts is very important for decreasing the ecologic burden of EPS and adjusting the use toward sustainability goals (Schaeffer, n.d.).

2. Advances in EPS Recycling Techniques

   There are huge difficulties in recycling EPS because it is so low in density, making it uneconomical to transport and process by conventional methods of recycling. However, recent developments help make the recycling of EPS more practical and efficient. Innovations in mechanical recycling, whereby EPS is compacted and remolded into new products, are creating great hopes for reducing the environmental impact of this material. Furthermore, chemical recycling techniques that finally close the loop with respect to EPS by breaking it down into its monomers for producing new items are also in the works. Developments like these help in better handling of EPS wastes and reduce the demand for virgin raw materials, lowering the environmental impact from production of the material. It is going to be a matter of continued research and investment in these technologies to make EPS recycling more common and economically feasible (Zhang, n.d.).

3. Development of Biodegradable EPS Alternatives

Since traditional EPS is associated with difficulties during its recycling process, there is a rising interest in developing biodegradable alternatives that can deliver similar benefits without the long-term environmental impact. In particular, researchers are looking into the uses of biopolymers derived from renewable resources, like starch or polylactic acid, for making EPS-like materials that can degrade more easily in natural environments. Any alternative biodegradable materials should be lightweight, have insulating/cushioning properties related to those of EPS, and be much more environmentally friendly at the end of life. In terms of matching the performance and cost of traditional EPS for these solutions, it is still a difficult task. Further research is needed to fine-tune these materials and ensure they can be produced and used successfully at scale. Development of a suitable alternative with biodegradable EPS can do much to reduce the impact of packaging and other applications where it is typically used (Zhang, n.d.).

9.2 *Technical Barriers to the Development of EPS*

A few technical barriers beleaguer the development of expanded polystyrene. The development of EPS has reached a bottleneck in property improvement and application development. Broadly applied, EPS still suffers from many limitations in material performance and processing technologies in order to meet the changing industry demand. This section goes on to discuss such technical barriers and puts forward some current research efforts into overcoming them.

1. Physical Property Improvement

One of the major technical challenges facing the development of EPS is to improve its physical properties, for example, in terms of its thermal insulation and mechanical strength. On the other hand, despite EPS's high performance in thermal insulation, it has continuously undergone research to enhance its performance further. For example, the incorporation of phase change materials will make EPS more suitable for applications with larger thermal management demands. Moreover, high mechanical strength with respect to compressive strength and impact resistance is very necessary for construction and packaging. Research in this regard is focused on modifying the formulation and structure of EPS through additives or creating composite materials that would improve these properties without significantly affecting cost or manufacturing processes.

2. Processing Efficiency
   Another prominent barrier in the development of EPS is processing efficiency. Traditional EPS manufacturing involves energy-intensive processes related to high production costs and environmental impact. Hence, innovations in processing technologies are of prime importance for the production of EPS in a more sustainable and cost-effective manner. For instance, improvements in extrusion and molding techniques can increase the efficiency of EPS production and reduce waste generation. In this line, researchers are also developing alternative processing routes that not only use renewable sources of energy but also optimize process parameters toward reducing environmental impact and increasing efficiency of EPS production. Besides, more effective recycling processes of this type of foam can contribute to waste handling and recycling of such materials in new products (Zhang, n.d.).

3. Scalability and Cost Challenges
   Of major importance in the development of EPS are challenges associated with scaling up new technologies and reduction of costs. Results that look very promising in research may be extremely difficult to convert into commercially viable entities. Production volume, material costs, and market demand are some of the most influencing factors that determine the feasibility of any new EPS technology. In these respects, countering these challenges will require collaborative effort on the part of the researchers, the industry players, and the policymakers in translating research into commerce. Strategies such as pilot projects, technology demonstrations, and financial incentives could be used to help further develop new EPS technologies and ensure their successful integration into the market.

9.3 Opportunities for Collaboration and Innovation

Collaboration between researchers and industrial players and obligations of policy makers are very important to develop EPS so as to ostracize significant challenges in its production. Through collaboration, the stakeholders can introduce innovations and develop new technologies. This part focuses on how collaboration in development of EPS can facilitate development of the same and emphasize key areas in which the same shall be significantly achieved.

1. Encouraging Research and Development Partnerships
   It requires R&D collaboration among academia, industry champions, and government institutions. This may unlock advances in material science—new EPS formulations in much-needed properties or more advanced additives. For instance, universities bring in the latest research from novel materials or recycling methods, whereas the industry brings applicability insights in terms of scaling and market needs. Government funding programs and policy support will further foster such efforts. Such partnerships pool resources and expertise in order to fast-track the process of innovation to bring new technologies related to EPS efficiently into the marketplace.

2. Improving Recycling and Waste Management
   Such collaborative efforts shall also significantly improve EPS recycling and waste management practices. In this regard, industry players and waste management companies may collaborate on new and more effective technologies and systems for recycling. For instance, it could be possible to have a closed-loop system between the producers of EPS and recyclers to develop more sustainably producible and effectively recyclable EPS material. Besides, research institutions could engage in the investigation of such new recycling process technologies as chemical depolymerization and advanced sorting, but only if industries work on the adequate practical implementation and integration of these processes into existent scheme frameworks on waste management. Such partnerships offer the opportunity to manage and lower environmental impacts by paying attention to the entire life cycle of EPS products and embracing the principles of the circular economy.

3. To Develop a Driving Policy and Standards
   Policymakers, researchers, and industry representatives can collaborate in defining regulations and standards that set a level playing field for EPS sustainable use. Common efforts might prompt the development of standards to ensure safety, quality, and environmental performance with regard to EPS products. For instance, setting standardized test methodologies for establishing recyclability and sustainability of EPS would drive uniformity in its practice across regions or industries. This could also mean that collaborative advocacy will continue to drive changes in policies for incentivizing greater use of recycled EPS and will support innovation in biodegradable alternatives. The stakeholders' collaboration in influencing policy and standards can, in this vein, help ensure regulatory frameworks stay aligned with new technological developments and sustainability objectives.

10 Conclusion

There is an increasing focus on sustainability, technological innovation, and regulatory compliance that can be tracked in the shifting landscape of expanded polystyrene research. These are the principal trends of far-reaching changes instructed in developing, using, and managing EPS across different industries. This section summarizes these trends and gives their meaning for the prospects of EPS.

1. Focus on Sustainability
   The growing emphasis on sustainability in EPS research changes the material's role within the industry. It is driven by environmental considerations and regulatory pressures for the need for more sustainable EPS solutions. This trend is noticed in several areas: Sustainable Production Practices: Green manufacturing techniques developed in this regard make the production of EPS more eco-friendly. It includes enhancement in energy efficiency, reduction in carbon emission, and usage of recycled or biodegradable materials. Improver Recycling and Waste Management: Much focus is given to the development of its waste management system in

developing efficient recycling methods for EPS. Depolymerization by means of chemicals, mechanical recycling, and other such innovative methods are some of the ways by which scientists are trying to eliminate the environmental impact of EPS. Regulatory Compliance: New regulations are mandating increased recyclability and lesser environmental footprints for EPS products. Industry players will, therefore, have to be compliant in order to avert any penal provisions and, at the same time, move in tandem with global sustainability objectives.

2. Technological Innovation

- Technological innovation is going to alter the abilities and uses for EPS. The new technologies being integrated into EPS create an increment in performance and functionality in the following ways: Smart EPS Technologies: Sensing IoT and other smart systems have given EPS the capability to monitor performance in real-time and dynamically change performance as required. This type of innovation has more value in packaging and building applications, especially where performance needs to be measured. Advanced Insulation Materials: Innovations, such as the use of phase change materials and nanotechnology, further improve the thermal and acoustic insulation properties of EPS. Developments targeting more energy-efficient, high-performance buildings are underway.
- New Manufacturing Techniques: New manufacturing processes allow for the production of EPS with better properties and less harm to the environment. Techniques like 3D printing and advanced polymerization open new vistas of applications for EPS. These technological innovations in EPS increase the functional properties, market opportunities, and industry growth. Tracking the development is, therefore, very crucial to remain competitive and at par with the benefits that can be harnessed in their application.

3. Trends in Regulatory Compliance and Policy

The EPS industry is driven by regulatory compliance and evolving policy frameworks. Some of the key developments in this area include: Upcoming Regulations: Emphasis of new regulations is on minimizing the EPS ecological footprint, particularly with regard to disposal and recycling. Compliance with these regulations would be prerequisite to lawful operation and access to the market.

- Global Policy Initiatives: Initiatives such as the European Union's Single-Use Plastics Directive set global momentum for changes to improve the sustainability of EPS. Policies emphasize the use of recycled raw materials and foster alternatives. Future Standardization: Standardization in testing methods and global benchmarks related to sustainability for EPS would make international trade easier and help products achieve higher quality standards. This will also help solve some cross-border regulatory problems.

The regulators are very nice, so industry players shall closely adapt themselves to such regulatory and policy trends in order to ensure compliance and seize arising opportunities. Contribution to policy developments and active standardization underpin a more sustainable and innovative EPS industry.

4. Adapting to the Industry and Evolving the Market

The ability of the EPS industry to adapt itself and change in the light of market demands and emerging expectations about sustainability will define its future trajectory. The key dimensions of industry adaptation include: Sustainable Practices: The industry is increasing its adoption of sustainable practices; this means reduction in the carbon footprint of production processes and promotion of the use of renewable resources. This shift is encouraged by consumer demand and pressure from regulatory environments in a bid to produce more eco-friendly EPS products. Collaboration and Industry Initiatives: Industry players, researchers, and policy framers must collaborate in innovating and solving challenges. Industry initiatives toward attaining sustainability, such as collaborative recycling programs and research partnerships, shall contribute much to developing more sustainable EPS solutions.

Market Trends and Consumer Preferences: The changing consumer preference toward environmentally responsible products is influencing the market trend. As consumers increasingly put priority into sustainability, the demand for eco-friendly EPS products is likely to increase, thus driving the industry to innovate and adapt to changing consumer preference.

References

Charalabidis, Y., Lampathaki, F., Koussouris, S., & Mouzakitis, S. (n.d.). Envisioning future internet enterprise systems: Visionary scenarios and new research areas.

Crowley, R. P. (n.d.). Method of preparing expandable polystyrene (Patent).

Eodanable, M., & Lauchlan, F. (n.d.). The advance of research and evaluation skills by EPs: Implications for training and professional development.

Etkin, B. (n.d.). Research on an aerodynamic particle separator (the eps). https://doi.org/10.1080/02726358808906505

Gery, G. (n.d.). The future of EPSS.

Kim, Y. Y. (n.d.). Expanded polystyrene manufacturing device (Patent).

Kim, D. H., Kim, I. J., Kim, S. H., Cho, S. E., & Kim, J. H. (n.d.). Expandable polystyrene having good thermal insulation and workability, method for preparing the same and foam thereof (Patent).

Lavery, C., & Davies, G. (n.d.). Extracts from the notebook of Xavier Valery. https://doi.org/10.1080/13528160902875671

National Research Council, Nutrient Requirements for EPS. (1994).

Reto, A., Sanabria, R., Rodriguez, J., & Hinostroza, A. (2021). Lightweight concrete precast panels for the improvement of thermal insulation of housing with expanded polystyrenebeads.1033.

Sams, M. W. (n.d.). EEG technique of the future.

Schaeffer, P. (n.d.). Method of molding expandable polystyrene (Patent).

Wu, Y. (n.d.). The future of EPSCoR.

Zhang, L. (n.d.). A method for completing data exchange between EPS and VCU based on data communication of different CAN baud rate by gateway (Patent).

EPS in Confectionery and Snacks: Investigating the Role of EPS in Confectionery and Snack Production and Examining Texture Enhancement and Shelf-Life Extension

Saikat Mazumder, Debasmita Bhattacharya, Dibyajit Lahiri, and Moupriya Nag

1 Introduction

Bacteria inhibit several ecological sites, including niches with adverse environmental circumstances, possessing a cell envelope that functions as a tolerance to ecological stress and physical barriers and enhances their survival while also facilitating adaptability and communication within their environment. Most bacteria possess an exterior surface covered with exopolysaccharides (EPS), a polymeric network made up of carbohydrate chains. Despite the discovery of dextran, the first bacterial exopolysaccharide, by Louis Pasteur in 1861 in wine, and the identification of dextran-producing *Leuconostoc mesenteroides* by Van Tieghem in 1878, the stress-resistant characteristics of biofilms were not acknowledged until 1973 (Singh et al., 2021). EPS facilitates several roles for bacteria, including aiding in cell attachment, providing protection against adverse environmental circumstances, preventing desiccation, and exhibiting antibacterial properties. EPS promotes bacterial survival, adaptability, persistent colonization, and endurance to biotic and abiotic stress by

D. Lahiri (✉) · M. Nag (✉)
Department of Biotechnology, Institute of Engineering and Management, Kolkata, University of Engineering and Management, Kolkata, West Bengal, India
e-mail: dibyajit.lahiri@uem.edu.in; moupriya.nag@uem.edu.in

S. Mazumder
Department of Biotechnology, Institute of Engineering and Management, Kolkata, University of Engineering and Management, Kolkata, West Bengal, India

Department of Food Technology, Guru Nanak Institute of Technology, Kolkata, West Bengal, India

D. Bhattacharya
Department of Basic Science and Humanities, Institute of Engineering and Management, Kolkata, University of Engineering and Management, Kolkata, West Bengal, India

T. Sarkar et al. (eds.), *Extracellular Polysaccharides*,
https://doi.org/10.1007/978-3-031-94472-7_7

facilitating biofilm formation (Dogsa et al., 2005). When incorporated into the cell wall, EPS can confer structural integrity and defensive attributes to the bacterial cell. The capsular EPS is the primary element that enables bacterial cell adhesion to surfaces, resulting in biofilm development in most instances.

Microbial extracellular polysaccharides have extensive application in food, especially in bakery and confectionery industries (Korcz & Varga, 2021). Biopolymers known as EPS, generated by lactic acid bacteria (LAB), have garnered significant interest for their relevance in the food sector. EPS may create a loosely adhered layer or be released into the extracellular environment. Numerous LAB, particularly *Fructilactobacillus* spp., *Lactococcus* spp., *Lacticaseibacillus* spp., *Lactiplantibacillus* spp., *Lentilactobacillus* spp., *Leuconostoc* spp., *Lactobacillus* spp., *Latilactobacillus* spp., *Limosilactobacillus* spp., *Pediococcus* spp., *Streptococcus* spp., and *Weissella* spp., can produce diverse exopolysaccharides (Angelin & Kavitha, 2020; Zheng et al., 2020). Nevertheless, no *Weissella* spp. has yet been deemed "generally considered safe" (GRAS) by the US Food and Drug Administration (Fessard & Remize, 2017) or has received Qualified Presumption of Safety (QPS) status from the European Food Safety Authority (Koutsoumanis et al., 2020), despite their prevalence in spontaneously fermented foods, in which they are believed to enhance the qualities of the final product. *Weissella* spp. exhibit a substantial array of technical and functional attributes that can augment food safety and boost sensory qualities. *Weissella cibaria* and *Weissella confusa* are noted for their significant production of EPS (Fessard & Remize, 2017).

EPS produced by LAB provides a natural substitute for synthetic food additives because of its physicochemical properties. Employing EPS in food processing has enormous benefits, including the enhancement of rheological properties such as enhanced viscosity, decreased syneresis, and improved texture, alongside its emulsification, thickening, and stabilizing capabilities. Furthermore, EPS offers favorable body texture, hardness, creaminess, and mouthfeel and may also be utilized in food packaging production (Angelin & Kavitha, 2020; Moradi et al., 2021). The utilization of LAB-derived EPS in food products may diminish or eliminate the necessity for additives, hence optimizing production and fulfilling consumer demands. Commercially accessible hydrocolloids in food packaging contradict customer expectations for fewer additives in food products, so they must be labeled. Further, EPS use, generated in situ by LAB, is devoid of labeling (Lynch et al., 2018a). It is crucial to comprehend microbial EPS production methodology to appreciate the diverse class of polymers and enable them to be used efficiently. Further, EPS comes under the category of alternative bio-thickeners, with functional attributes, extensively employed in the food industry (Lynch et al., 2018a).

2 Bacterial Exopolysaccharides: A Deep Insight

The terms "exo," implying "external," and "polysaccharide," implying "carbohydrate," consisting of multiple sugar monomers, are conjugated to form "exopolysaccharide." EPS, a class of biopolymers, are naturally occurring substances primarily found in microbes, plants, and animals (Chen & Huang, 2018). Biopolymers produced by bacteria can be divided into two groups based on their presence within or outside of cells: intracellular and extracellular, respectively (Cerning, 1995), and have a varied array of characteristics, utilizing substrates from basic to complicated (Nwodo et al., 2012). Extracellular biopolymers, also referred to as "extracellular polymeric substances," are categorized into four types: polyester-based materials, inorganic polyanhydrides (such as polyphosphates), polyamides, and polysaccharides (Rehm, 2010) which are more prevalent among other subgroups (Dogsa et al., 2005). EPS are polysaccharides composed of monosaccharide units connected by glycosidic linkages (Singh et al., 2019). Polysaccharides can be further categorized based on their location in cells. Gram-positive bacteria include structural polysaccharides in their cell wall, such as lipoteichoic acids and peptidoglycan. The outermost layer of gram-negative bacteria contains lipopolysaccharides. "Glycocalyx," a combination of a few glycoproteins and polysaccharide layers secreted by some bacteria (Ruas-Madiedo & de los Reyes-Gavilán, 2005). EPS are large molecular weight polymers with a range of 10–1000 kDa (Nwodo et al., 2012). Polysaccharides such as "glycogens" are cytoplasmic polymers deposited in the microbial envelope, and structural polymers like "glycans" may be synthesized by bacteria (Castro-Bravo et al., 2018). Polysaccharides released by bacteria and their cell wall–anchored enzymes outside of their cells are known as bacterial exopolysaccharides (Singh et al., 2021).

Microbial polysaccharides are attached to the cell surface as a cohesive layer covalently, forming a capsule outside the cell wall (Ruas-Madiedo et al., 2002) and producing an exopolysaccharide layer that is weakly adhered to the cell surface or expelled into the extracellular environment (Donot et al., 2012). Presence of sugar influences EPS production due to the presence of carbon and source of energy (Flemming & Wingender, 2010). However, bacterial cells exhibit peak metabolic activity during the exponential and stationary stages of bacterial growth, and EPS is produced (Angelin & Kavitha, 2020). EPS production depends on several factors, such as nutrient content, pH, temperature, and carbon/nitrogen ratios including bacterial strain (Patel et al., 2012).

Further, survival of microbial cells affects biofilm development (Burdman, 2000), EPS production (Decho, 2000), cell accumulation and attachment (Frank & Belfort, 2003; Tsuneda et al., 2003), and protection of cells from adverse environments (Nicolaus et al., 2010). The polysaccharides protect bacterial cells from desiccation, phagocytosis, osmotic stress, and bacteriophage attacks (Ruas-Madiedo & de los Reyes-Gavilán, 2005).

3 Classification of Exopolysaccharides: A Perspective on Component Monomers

Polysaccharides may consist of many ratios of sugars and sugar derivatives, including arabinose, galactose, glucose, galacturonic acid, xylose, mannose, rhamnose, ribose, fructose, glucosamine, fucose, glucuronic acid, and galactosamine (Hu

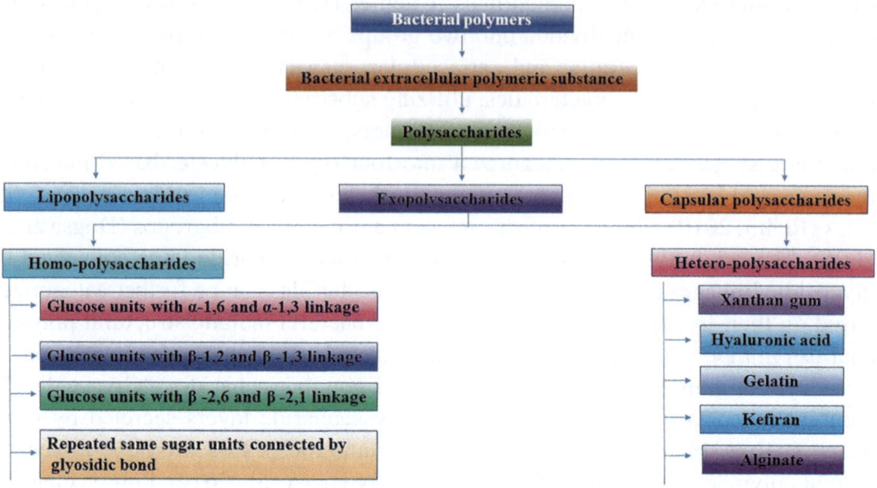

Fig. 1 Classification of exopolysaccharides

et al., 2020; Min et al., 2019). Heteropolysaccharides (HePS) contain 2–8 distinct units of monosaccharides, and homopolysaccharides (HoPS), which contain similar monosaccharides, are the primary types of EPS (Baruah & Das, 2016). Moreover, HoPS and HePS may vary not only in their component subunits but also in their framework, chain length, degree of branching, and glycosidic connections. HoPS typically consists of fructose or glucose units, which frequently aggregate to produce glucans or fructans, respectively (Fig. 1). The predominant types are α-glucans, which may be categorized into four subgroups: dextrans, mutans, reuterans, and alternans. Levan and inulin are the primary subgroups of fructans (Ryan et al., 2015). The repeating monomer units include L-rhamnose, D-glucose, D-galactose, and seasonal presence of N-acetylgalactosamine, N-acetylglucosamine, or glucuronic acid (Torino et al., 2015). The LAB-spawned EPS are HePS that have the ability to produce EPS with different sugar compositions, charges, chain lengths, and degrees of branching (Sánchez et al., 2006) and demonstrate changes including acetylations, pyruvylations, and phosphorylations (Torino et al., 2015). EPS produced by *Lactococcus lactis subsp. cremoris* comprises rhamnose, galactose, and glucose (Looijesteijn & Hugenholtz, 1999). The *Streptococcus thermophilus*–produced EPS consists of glucose and galactose, and to some extent rhamnose and N-acetylga-lactosamine (Pachekrepapol et al., 2017). The *Lactobacillus delbrueckii subsp. Bulgaricus*–generated EPS contains galactose, glucose, and rhamnose (Grobben et al., 1997). Levans and dextrans are often produced by *Streptococcus mutans* and *Leuconostoc mesenteroides*, respectively (Cerning, 1990).

EPS-generated polysaccharide properties may vary in response to environmental changes (Looijesteijn & Hugenholtz, 1999). Furthermore, bacterial strains often exhibit flexibility in their EPS properties, and many of them can produce polysaccharides. Marshall et al. (1995) discovered that *Lactococcus lactis subsp. Cremoris*

LC330 simultaneously produced two distinct types of EPS with varying sugar compositions and molecular weights: a high-molecular-mass (>106 Da) neutral EPS and a smaller (104 Da), phosphorus-rich, negatively charged EPS. Sánchez et al. (2006) indicated that *Lactiplantibacillus pentosus LPS26* (previously *Lactobacillus pentosus*) (Zheng et al., 2020) synthesized two types of EPS contingent upon environmental factors such as medium composition or pH: a high-molecular-weight EPS (>106 Da) comprising glucose and rhamnose and a low-molecular-weight EPS (3.3 × 104 Da) consisting of both glucose and mannose.

Several processes are used to identify and evaluate EPS generated by LAB. Qualitative techniques for EPS detection include electron microscopy and confocal laser scanning microscopy (Loeffler et al., 2020). However, for quantification of the composition and structure of EPS, various colorimetric methods are employed, including the phenol–sulfuric acid method, high-performance liquid chromatography, gas chromatography, size-exclusion chromatography, ion-exclusion chromatography, Fourier transform infrared spectroscopy, and nuclear magnetic resonance spectroscopy (Loeffler et al., 2020).

4 Synthesis and Biochemical Attributes of Exopolysaccharides

The enzymes responsible for EPS synthesis by LAB may be categorized into two primary classes. The initial category comprises proteins essential (not exclusive to EPS) for the synthesis of fundamental sugar nucleotides used by various cellular processes (Zeidan et al., 2017). The other category of EPS-specific enzymes contains glycosyl- and acetyl-transferases, which are accountable for synthesizing monomeric glycosyl donor molecules, as well as molecules involved in polymerization and export of the product, thus controlling the complete process. However, the function of some EPS-specific proteins still needs more research (Ryan et al., 2015).

Certain bacterial genes control enzymes that are unique to EPS (Badel et al., 2011). LAB genes responsible for EPS creation are usually clustered in nature and typically located on the chromosome (genophore) in thermophilic strains, along with plasmids in mesophilic LAB. In the absence of pressure control, the bacteria's ability to synthesize EPSs is likely to become unstable, which leads to failure in EPS production (Kumar et al., 2020). The EPS genes are normally located in the same region of the genome within an operon arrangement with the genes positioned in the concurrent direction and translated into singular mRNA. The process is governed by the promoter region before the first eps gene. The EPS genes execute four specific functions: (1) control of EPS synthesis, (2) determination of chain length, (3) biosynthesis of repetitive units, and (4) polymerization and export (Jolly & Stingele, 2001). The synthesis of HoPS is a process where energy is only used in the production of the required extracellular enzymes since no active transit steps are involved in the synthetic route. These enzymes, which utilize fructose and glucose, respectively, as glycosyl donors in the synthesis process are referred to as glycosyltransferases, encompassing glucosyltransferases and fructosyltransferases (Ryan et al., 2015). In the first step of the synthesis, sugar is internalized; followed by the second step, in which glucose-1-phosphate and fructose-6-phosphate are synthesized into sugar nucleotide precursors, which supply the energy for the polymerization reaction; followed by the third step, in which various glycosyl-transferases at the cytoplasmic membrane level work together to assemble the monosaccharide repeating unit and link it to the glycan subunit carrier; and fourth, the extracellular polymer is exported (Pessione, 2012; Zhou et al., 2019). The process by which carrier-bound repeating units transit from the cytoplasm to the extracellular environment remains incompletely elucidated. Flippase enzymes may be implicated in the process (Pessione, 2012; Ryan et al., 2015). In nonoptimized growth circumstances, the EPS production of various LAB strains ranges from 10 mg/L to 400 mg/L; in perfect conditions, this quantity can be multiplied (Cirrincione et al., 2018; De Vuyst et al., 2001).

The EPS production for *Lactobacillus* strains was measured by Abdellah et al. (2014) to be 160–740 mg/L, *Streptococcus* strains to be 126–319 mg/L, and *Pediococcus* strains to be 132–134 mg/L. Mıdık et al. (2020) demonstrated that a strain of *Lactiplantibacillus plantarum* (previously *Lactobacillus plantarum*). Zheng et al. (2020) generated 515 mg/L of EPS under optimal circumstances. The first EPS producer identified was a strain of *Lactica-seibacillus rhamnosus* (formerly *Lactobacillus rhamnosus*), designated RW-9595M, which demonstrated the ability to synthesize 2.8 g/L of EPS (Macedo et al., 2002). *Lactobacillus kefiranofaciens* WT-2BT was the second-best producer, yielding 2.5 g/L of EPS (Maeda et al., 2004).

Despite the exceptional results noted across LAB, the EPS yield must reach a minimum of 10–15 g/L to render extraction, purification, and use as a food additive commercially viable. *Xanthomonas* cam-pestris, a species of bacterium in the phylum *Proteobacteria*, is now thought to be the one with the greatest amount of EPS production (30–50 g/L). The extracellular HePS generated by *Xanthomonas campestris*, xanthan gum, is mostly utilized in the food sector as a rheology modifier (Caggianiello et al., 2016). Regarding LAB-derived EPS, only dextran generated by *Leuconostoc mesenteroides* is permitted as a food additive (Torino et al., 2015). The EPS produced in situ by LAB could be a good substitute to get the required texture, despite its poor yields (Cirrincione et al., 2018). Remarkably, though, the degree of in situ polysaccharide formation is mostly determined by the culture conditions (temperature, pH, length of incubation, medium composition, etc.), each of which may be adjusted to maximize output (Sánchez et al., 2006).

Alongside quantity, the overall quality of EPS is crucial, particularly when their physicochemical features are to be utilized by the food business. Nguyen et al. (2019) determined that the avoidance of textural defects, such as graininess, in yogurt was not influenced by final EPS concentrations or acidification kinetics, but rather by the shape of bacterial chains and the structural characteristics of EPS. EPS exhibits varying effects on the sensory, rheological, and textural properties of products, contingent upon their structural diversity and interactions with other food elements (Hundschell & Wagemans, 2019; Mende et al., 2016; Nguyen et al., 2019; Oleksy & Klewicka, 2016). In conclusion, meticulously selected LAB-based EPS can significantly enhance food processing or product attributes. Nonetheless, other facets require more investigation to fully realize the advantages of their application. Consequently, academic research has to be integrated with insights from food producers to pinpoint the particular issues that may be mitigated by the application of EPS (Zannini et al., 2016).

5 Exopolysaccharides from LAB

5.1 Homopolysaccharides

Homopolysaccharides (HoPSs) represent the most structurally elementary category of exopolysaccharides (EPSs) since they consist of a singular variety of monosaccharide repeating unit of D-fructose in fructans or D-glucose in glucans joined to create either linear or branched configurations (Abid et al., 2018; Nguyen et al., 2020). HoPSs are generated in substantial quantities, with concentrations of up to 10 g L^{-1} (Lynch et al., 2018b) and differentiated through molecular weights, exceeding 10^6 Da. LAB HoPS biosynthetic route employs the extracellular synthesis route to produce HoPSs. Zhou et al. (2019) through employing extracellular glycosyl transferases and fructosyl transferases (Angelin & Kavitha, 2020). Biosynthesis of HoPSs includes two processes: First, the substrate (like sucrose) is hydrolyzed, and the consequential monosaccharide residues are bound to a glycan acceptor chain; second, polymerized chains are discharged directly into the extracellular environment (Zhou et al., 2019). β-glucan production is performed under the catalysis of β-glycosyltransferase (Laura et al., 2012), which uses nucleotide sugars as substrates via the synthase-dependent route (Zeidan et al., 2017) and varies from other homopolysaccharides. Nucleotide sugars are included in the polymerizing monomers in the cytosol and then carried out of the cell by membrane transport (Laura et al., 2012).

5.2 Types of HoPS

Glucans are a type HoPS that are high in molecular mass and can be divided into α-D glucans and β-D glucans based on the connection in the primary chain. α-D glucans can be further classified as dextrans, mutans, reuterans, and alternans (Angelin & Kavitha, 2020). Reuteran is a α-glucan including α-(1 → 4) and α-(1 → 6) glycosidic linkages with α-(1 → 4)/α-(1 → 6) branching sites (Wangpaiboon et al., 2018; Zannini et al., 2016). Mutan is a water-insoluble glucan consisting of α-D-glucopyranose, predominantly featuring a-(1 → 3) glycosidic linkages. Dextran is a polymer of α-D-glucopyranose mostly connected by α-(1 → 6) bonds in the main chain, with a variable number of branching linkages including α-(1 → 2), α-(1 → 3), and α-(1 → 4) links. Alternan consists of D-glucopyranosyl residues interconnected by alternating α-(1 → 3) and α-(1 → 6) linkages. Conversely, β-D-glucans consist of units of glucose that are interconnected by β-(1 → 3) linkages and β-(1 → 2) branching (Lynch et al., 2018b). Levans is a type of fructans (Angelin & Kavitha, 2020) that are synthesized by levansucrase, which catalyzes the transfer of D-fructosyl residues from fructose, resulting in β-(2 → 6) glycosidic linkages. Inulins are another type of fructans, made up of β-D-fructose monosaccharides connected by b-(2 → 1) glycosidic linkages. (Angelin & Kavitha, 2020).

5.3 *Heteropolysaccharides*

A type of polysaccharide consisting of several repeating monosaccharides is known as heteropolysaccharide (HePSs) and exhibits branching at sites C2, C3, C4, or C6, or may remain unbranched (Zannini et al., 2016). The HePSs consist of 3–8 repeated units of of monosaccharides such as N-acetyl-D-glucosamine, N-acetyl-D-galactosamine, and glucuronic acid D-glucose, D-galactose, L-rhamnose, and non-carbohydrate substituents (Mende et al., 2016; Torino et al., 2015). Pyranose or furanose are monosaccharides that can exist as either as α- or β-anomer (Laura et al., 2012), rather than carbohydrate molecules are made from intracellular sugar nucleotide precursors, not from HoPSs (Bajpai et al., 2015). The production and molecular weight of HePSs generally occurs within the range of 25–600 mg L^{-1} (Angelin & Kavitha, 2020) and from 10^4 to 10^6 Da, respectively (Leroy & De Vuyst, 2016; Torino et al., 2015).

5.4 *Biosynthesis Pathway of HePSs*

The biosynthesis of HePSs in LAB transpires via the Wzx/Wzy-dependent route (Schmid et al., 2015; Zhou et al., 2019) in a process that requires significant energy. This process occurs in five stages, commencing within the cell and concluding outside. (i) Sugar is imported into the cytoplasm until Glucose-1-P is produced; (ii) sugar nucleotides are generated; (iii) repeating units are formed through the sequential addition of triggered sugar residues by specific glycosyltransferases (GTF), which catalyze the glycosidic bond with attachment to the glycosyl carrier lipid; (iv) repeating components are translocated to the extracellular surface by the Wzx-protein, exhibiting flippase activity; and (v) extracellular polymerization of HePSs occurs via the Wzy-protein, which cleaves the bond linking the HePSs to the lipid carrier, releasing them into the medium or anchoring them to the wall of the bacteria (Silva et al., 2019; Zhou et al., 2019).

5.5 *Types of HePSs*

Kefiran is an extensively researched heteropolysaccharide synthesized by bacteria found in kefir grains. It is a water-soluble branching glucogalactan consisting of a branching hexa- or heptasaccharide repeating unit that contains roughly equal proportions of D-glucose and D-galactose residues (Patel et al., 2012). Figure 2 shows the classification of exopolysaccharides from lactic acid bacteria.

6 Application of Exopolysaccharides in Confectionery and Snacks: Perspective on Texture Enhancement and Shelf-Life Extension

Bacterial EPS are widely utilized as gelling agents (gellan gum, agar), thickeners (xanthan, pectin) (Ngouémazong et al., 2015), and emulsifiers (pectin) (Banerjee & Bhattacharya, 2012; Radziej et al., 2021). Improving texture, sensory qualities, mouthfeel, odor, and physical appearance may also be employed to retain moisture, modify rheology, induce hardness, and change other biomechanical food characteristics. The demand for LAB EPS has risen as a result of its antibacterial and antioxidant properties, which can improve and enhance food industry applications (Korcz & Varga, 2021; Ngouémazong et al., 2015). An anionic branching heteropolymer, xanthan is made up of five repeating units of sugar with differing concentrations of residues of glycolysis and acetate (de Mello Luvielmo et al., 2016).

Xanthan was first employed for large-scale oil recovery, but it subsequently became significant as a food additive in America in 1969, following the acquisition of necessary authorization from the FAO (Moscovici, 2015). Alginate, often extracted from seaweeds, may also be produced by Pseudomonas and Azotobacter.

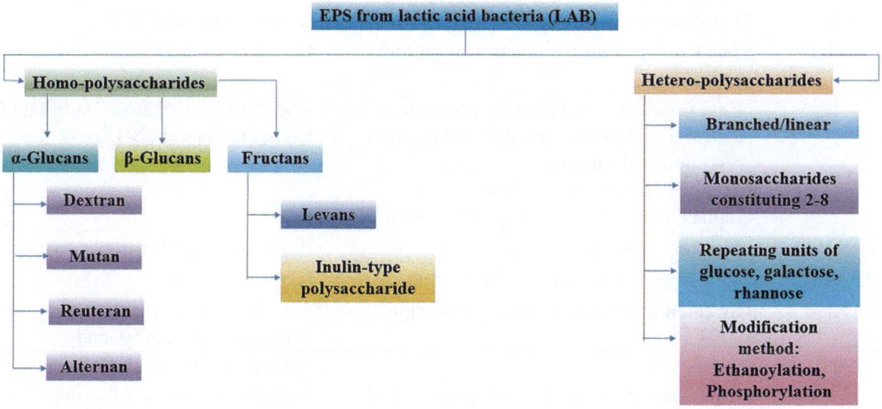

Fig. 2 Classification of exopolysaccharides from lactic acid bacteria

Alginate serves several functions in the food business, such as a thickening and gelling agent, stabilizer, and emulsifier. Another significant application of alginate is the coating of different vegetables and fruits to enhance their antimicrobial protection and prolong shelf life. Alginate-based films serve as packaging materials, particularly for tea, coffee, and milk-based goods (Gheorghita Puscaselu et al., 2020). Carrageenan is a notable edible biopolymer commonly utilized as an emulsifier and gelling agent in the food business, particularly for fat stability, and serves a crucial role in food processing (Saha et al., 2020). Bacterial EPS is widely utilized for packing dry or liquid food products with diverse shelf life, including flour, coffee beans, baked goods, cheese, yogurt, ice cream, and dairy products (Table 1). Commonly utilized EPS in daily life encompasses semi-transparent films for fruits and vegetables and protective films for meat, fish, and cheese, in addition to containers for

Table 1 Application of EPS in bakery and confectionery products

Type of EPS	Applications in bakery/confectionary	References
Dextran	Enhances the softness, crumb structure, and loaf volume	Angelin and Kavitha (2020), Wang et al. (2019), and Zheng et al. (2020)
	Inhibits sugar crystallization and utilizes gelling agents in gelatinous confections	Maina et al. (2011) and Santos et al. (2000)
Pullulan	Starch substitute reduces retrogradation	Karakaş-Budak (2019)
Curdlan	Minimizes oil absorption, thickening agents	Basiri (2021) and Burey et al. (2009)
Reuteran	Enhances the quality of bread made with gluten-free sorghum flour. Dietary fiber improves the nutritional value of bread	Sandra et al. (2012)
	Utilized in baked goods; enhancing tenderness, postponing staling	Badel et al. (2011) and Galle et al. (2012)
Scleroglucan	Thickening and gelling agent or stabilizer	Meftahi et al. (2010)
Cellulose	Fat substitute in biscuits, enhances firmness	Tarancón et al. (2015)
Inulin	Substitute for sugar in chocolate	Le Bastard et al. (2020)
Xanthan	Thickening, stabilizing, and suspending agent	Mohammadi et al. (2014)
	Cakes, muffins, and biscuits have a regular distribution of moisture, enhancing water retention and aeration Stability in batter Chocolate: cocoa replacement, elevates the melting point	Abdullah et al. (2014), Habibi and Khosravi-Darani (2017)
Gellan	Thickening agents	Burey et al. (2009)
Acetan	Viscosity enhancer and gelling agent	Giavasis (2013)
Levan	Sweetener in confectionery production	Angelin and Kavitha (2020), Bounaix et al. (2009), and Wangpaiboon et al. (2018)
	Enhancing volume, delaying staling, and augmenting texture, softness, crumb structure, and flavor of baked goods	Badel et al. (2011), Mensink et al. (2015), Ryan (2011), and Zheng et al. (2020)

confectionery products, among others. The composite films for food packaging produced using EPS curdlan have elevated tensile strength, elongation, improved viscosity, and effective moisture barrier properties (Kaur & Dey, 2023).

Certain EPS can substitute or diminish the necessity for costly hydrocolloidal polysaccharides in the baking industry (Das et al., 2014), which are crucial for enhancing dough machinability, texture, volume, rheology, and the shelf life of bread. Therefore, the in situ synthesis of EPS by LAB during fermentation is advocated to enhance the textural qualities of bread, facilitating cost reduction and eliminating the need for extra additions. Sourdough is used in baking and contributes to improved nutrition, texture, fragrance, and shelf life (Gänzle, 2014). The main roles of LAB in sourdough include fragrance generation, acidity, and unique metabolic processes such as EPS synthesis (Minervini et al., 2014). LAB creates homopolysaccarides by the activity of glycosyl hydrolases, in the presence of sucrose, during sourdough fermentation (Broadbent et al., 2001). To create glucans (dextrans), fructans (levans), or alternans, the resulting glucose and/or fructose are used (Zarour et al., 2017). Additionally, LAB uses leftover fructose to produce acetate, which has antifungal properties. However, excessive acetate concentrations can negatively affect crumb hardness, flavor, dough rheology, and bread volume (Broadbent et al., 2001). The selection of LAB that can effectively compete and prevail over contaminants is a crucial aspect in determining the ultimate quality of bread, and it should be kept in mind while utilizing certain sourdough fermentation methods (Karimi et al., 2015). *L. sanfranciscensis*, *L. pontis*, *L. reuteri*, *L. panis*, and *L. frumenti* are important LAB varieties that are employed because of their capacity to produce homopolysaccarides in the sourdough (Zarour et al., 2017).

Customers' demands for healthy eating include requesting high-fiber foods. Bran used as a source of fiber in bread baking lowers the hydration of the gluten and, as a result, interferes with the creation of the gluten network, altering the structure of the wheat dough adversely. Because *Weisella confusa* can create dextran in wheat bran, a hydrocolloid that enhances the quality of high-fiber wheat bread, it is employed to solve this issue (Costa et al., 2010). People with celiac disease who are unable to eat gluten present another difficulty for the baking industry: creating gluten-free goods. The bulk of those patients are forced to use the gluten-free items that are already on the market, and they are of poor quality with dry crumbs, poor texture, and bland flavor (Costa et al., 2010). Utilizing EPS, such as reuteran (from *L. reuteri*), levan (from *L. sanfranciscensis*), and dextran (from *Leuconostoc mesenteroides* and *Weisella cibaria*), improves bread texture and dough rheology. It can also reduce or replace the use of hydrocolloids, which are used as bread improvers in wheat and gluten-free baking (Rühmkorf et al., 2012b; Zarour et al., 2017). Moreover, EPS can enhance the rheology of dough in fermented meals made from cereals and legumes. The rheological qualities, texture, mouthfeel, taste perception, and durability of fermented foods like idli batter are enhanced by the polymers released by *W. confusa*, *W. cibaria*, and *Pediococcus parvulus* (Kavitake et al., 2016).

Dextran is a type of HoPS produced from *Leuconostoc*, *Streptococcus*, *Weissella*, *Pediococcus*, and *Lactobacillus* (Baruah et al., 2017; Heperkan et al., 2020; Kothari et al., 2015) and comprises D-glucose units (Wolter et al., 2014). The LABs associated with EPSs are dextran, levan, kefiran, and hyaluronic acid (Moradi et al., 2021). There is growing interest in employing dextran to enhance the rheological properties and quality of bread (Galle et al., 2012; Wolter et al., 2014). EPSs from bacteria that produce lactic acid during sourdough fermentation enhance the rheology, texture, and mouthfeel of bread products (Bejar et al., 2013; Du et al., 2018). Dextran uses in the food industry which include texturizing, emulsifying, prebiotics, cloud-forming, and edible coating purposes (Davidović et al., 2018; Eckel et al., 2019; Tang et al., 2018) including baking industry (Baruah et al., 2017) to establish the prebiotic role of dextran derived from *Weissella cibaria RBA12* isolated from pummelo, along with its possible use in sourdough fermentation (Rühmkorf et al., 2012a) to evaluate as a substitute for commercially available hydrocolloids in gluten free bread. The quality was enhanced by dextran generated by *Weissella confusa*. Dextran synthesized in situ *Weissella confusa* was helpful for the improvement of wheat-fab which is a bread comprises of bean and wheat, and wholegrain sorghum bread and extended the shelf-life of Chinese steam bread with suitable sensory attributes (Tang et al., 2018). Further, hydrocolloids help to improve the textural and rheological qualities of dough and bread-making processes, including the shelf-life of bakery products (Tang et al., 2018).

Weissella sp. and *Leuconostoc mesenteroides* produced dextran in situ, which improved the sourdough wheat bread's loaf capacity, mouthfeel, consistency, and shelf life, including smoothness (Galle et al., 2012). Further, dextran has little branching bonds and an elevated molecular mass used in sourdough (Lacaze et al., 2007). Dextran may also be used in the baking industry to create gluten-free food items for individuals with celiac illness (Galle et al., 2010; Rao & Goyal, 2013).

Xanthan Gum is mostly used in the bakery sector to increase the amount of water retained in baked goods after baking, which extends the shelf life of chilled dough and bakery goods. In particular in bread processing, eggs can be replaced with xanthan gum while not affecting the product's flavor or texture at the same time. Xanthan gum makes for better bulk, texture, calorie-conscious baked items, and gluten-free bread. Xanthan gum hot or cold-processed baked goods and fruit pie fillings increase the release of flavor and texture. For softness, air implementation, and retention, xanthan gum is often used to give mixtures for muffins, bread mixes, biscuits, and cake method persistence. Fruit and cream fillings provide the added advantages of longer shelf life, frozen-thaw stability, and syneresis management (Suryawanshi et al., 2022).

Hydrocolloids, including xanthan gum, are frequently employed in the formulation of bakery goods, an important aspect of the quality of cake, pasta, and bread is their texture. Milde et al. (2020) developed a gluten-free pasta utilizing cassava starch, corn flour, and xanthan gum. The use of 0.6% xanthan gum in the pasta formulation enhanced dough handling, reduced cooking loss, and increased physical and textural characteristics. Mohammadi et al. (2014) improved formulations for gluten-free flatbread, using rice flour, corn starch, soy flour, xanthan gum, and xanthan gum-carboxymethyl cellulose. The optimal dough production, bread yield, and minimal bread weight loss were achieved with formulations including xanthan gum and a xanthan gum-carboxymethyl cellulose combination. In a different research, adding 1% xanthan gum to a recipe for sponge cakes made with wheat flour and cornstarch increased the product's firmness but dropped the sensory evaluation's overall approval ratings (Noorlaila et al., 2020). Other culinary applications of xanthan gum include potential use as a sodium salt substitute in low-sodium meat product formulations and the development by Santos et al. (2020) of food-grade Pickering emulsions utilizing zein-xanthan gum and sunflower oil. To keep fresh-cut pears from browning and to extend their shelf life, Sharma and Rao (2015) created edible coatings with xanthan gum and cinnamic acid incorporated in them. The three main factors that determine the EPS function in bread improvement are sucrose concentration, presence of carbon, and incubation period (Yu et al., 2018). Wheat flour or quinoa and amaranth in a 1:1 ratio with 3% sucrose, liquid sourdoughs (LSs) with a dough yield (DY) of 500 were tested against the *L. plantarum ITM21B* strain, which does not produce EPS in mMRS (Kang et al., 2020). The study further revealed that 10% sucrose was added to a liquid medium containing the *Weisella cibaria C43-11*, which showed significant EPS production. While the *L. plantarum ITM21B* strain causes protein breakdown at a dough yield of 250, *W. cibaria* hydrolyzes protein fermentations.

EPSs influence technical and nutritional characteristics; likewise, phytases enhance mineral absorption by reducing phytate content in bread (Kaur et al., 2014). The integration of EPS and pseudo cereals significantly improves the quality indices of food items, including shelf-life, dough rheology, nutritional content, taste, and scent of the final goods (Khanal et al., 2018). Consequently, the incorporation of pseudo cereals augmented the synthesis of EPS, which is subsequently utilized in baked items as texture enhancers. The bakery sector worldwide has the significant challenge of reducing levels of salt in yeasted bread (Kiani et al., 2010). The industries have often attempted to address this issue; nevertheless, they have encountered adverse consequences on organoleptic qualities, including diminished shelf-life, altered crumb structure, and insufficient saltiness.

To lessen yeasted bread quality issues, these sourdough combinations contain functional EPS of two species: *Weisella cibaria MG1* and *Lactobacillus amylovorus*. Bread's volume, shelf life, and crumb texture all increased when the salt content was lowered to 0.3%, enhancing the bread's sensory profile (Kim et al., 2019). Consequently, functional sourdoughs enriched with EPS are the optimal means to offset the reduced salt content in yeasted bread. Furthermore, we can inherently preserve a superior quality of bread without the incorporation of synthetic ingredients. Second, to determine the functionality of the bread, the sourdoughs were fermented with EPS-producing lactic acid bacteria species (*Weissella cibaria*, *Leuconostoc mesenteroides*, *Fructilactobacillus sanfranciscensis*, and *Latilactobacillus sakei LS81*). EPS generation by these *Lactobacilli* spp. improves texture and specific volume while decreasing stalling rate and crumb hardness in bread (Arserim Ucar et al., 2021). In a similar vein, in situ EPS manufacturing offers a natural hydrocolloid substitute, improving the physicochemical characteristics of both gluten and gluten-free bread goods (Konuk Takma & Korel, 2017).

7 Conclusion

Microbial polysaccharides provide novel options in the food sector that are eco-friendly, sustainable, and multifunctional. Microbial EPS is utilized in the food sector owing to its distinctive properties, including emulsification, stabilization, thickening, texturization, encapsulation, and transport of bioactive substances, as well as coating and film-forming capabilities. The rheological and techno-functional qualities of microbial EPS, particularly those generated in situ, can serve as a significant alternative for enhancing physicochemical characteristics. The beneficial impact of EPS on baked products is attributed to their capacity to retain water and establish a network with various dough constituents. Consequently, EPS may enhance the rheological properties, structure, and volume of bread, leading to reduced staling rates and a prolonged shelf life.

References

Abdellah, M., Ahcne, H., Benalia, Y., Saad, B., & Abdelmalek, B. (2014). Screening for exopolysaccharide-producing strains of thermophilic lactic acid bacteria isolated from Algerian raw camel milk. *African Journal of Microbiology Research, 8*(22), 2208–2214. https://doi.org/10.5897/AJMR2014.6759

Abdullah, M. S., Amir, I. Z., & Sharonc, W. X. (2014). Mixture experiment on rheological properties of dark chocolate as influenced by cocoa butter substitution with xanthan gum/corn starch/glycerin blends. *International Food Research Journal, 21*(5), 1887–1892.

Abid, Y., Casillo, A., Gharsallah, H., Joulak, I., Lanzetta, R., Corsaro, M. M., Attia, H., & Azabou, S. (2018). Production and structural characterization of exopolysaccharides from newly isolated probiotic lactic acid bacteria. *International Journal of Biological Macromolecules, 108*, 719–728. https://doi.org/10.1016/j.ijbiomac.2017.10.155

Angelin, J., & Kavitha, M. (2020). Exopolysaccharides from probiotic bacteria and their health potential. *International Journal of Biological Macromolecules, 162*, 853–865. https://doi.org/10.1016/j.ijbiomac.2020.06.190

Arserim Ucar, D. K., Konuk Takma, D., & Korel, F. (2021). *Exopolysaccharides in food processing industrials* (pp. 201–234). https://doi.org/10.1007/978-3-030-75289-7_8

Badel, S., Bernardi, T., & Michaud, P. (2011). New perspectives for Lactobacilli exopolysaccharides. *Biotechnology Advances, 29*(1), 54–66. https://doi.org/10.1016/j.biotechadv.2010.08.011

Bajpai, V. K., Rather, I. A., Majumder, R., Shukla, S., Aeron, A., Kim, K., Kang, S. C., Dubey, R. C., Maheshwari, D. K., Lim, J., & Park, Y.-H. (2015). Exopolysaccharide and lactic acid bacteria: Perception, functionality and prospects. *Bangladesh Journal of Pharmacology, 11*(1), 1. https://doi.org/10.3329/bjp.v11i1.23819

Banerjee, S., & Bhattacharya, S. (2012). Food gels: Gelling process and new applications. *Critical Reviews in Food Science and Nutrition, 52*(4), 334–346. https://doi.org/10.1080/10408398.2010.500234

Baruah, R., & Das, D. (2016). Heteropolysaccharides from lactic acid bacteria: Current trends and applications. *Journal of Probiotics & Health, 04*(02). https://doi.org/10.4172/2329-8901.1000141

Baruah, R., Maina, N. H., Katina, K., Juvonen, R., & Goyal, A. (2017). Functional food applications of dextran from Weissella cibaria RBA12 from pummelo (Citrus maxima). *International Journal of Food Microbiology, 242*, 124–131. https://doi.org/10.1016/j.ijfoodmicro.2016.11.012

Basiri, S. (2021). Applications of microbial exopolysaccharides in the food industry. *Avicenna Journal of Medical Biochemistry, 9*(2), 107–120. https://doi.org/10.34172/ajmb.2021.16

Bejar, W., Gabriel, V., Amari, M., Morel, S., Mezghani, M., Maguin, E., Fontagné-Faucher, C., Bejar, S., & Chouayekh, H. (2013). Characterization of glucansucrase and dextran from Weissella sp. TN610 with potential as safe food additives. *International Journal of Biological Macromolecules, 52*, 125–132. https://doi.org/10.1016/j.ijbiomac.2012.09.014

Bounaix, M.-S., Gabriel, V., Morel, S., Robert, H., Rabier, P., Remaud-Siméon, M., Gabriel, B., & Fontagné-Faucher, C. (2009). Biodiversity of exopolysaccharides produced from sucrose by sourdough lactic acid bacteria. *Journal of Agricultural and Food Chemistry, 57*(22), 10889–10897. https://doi.org/10.1021/jf902068t

Broadbent, J. R., McMahon, D. J., Oberg, C. J., & Welker, D. L. (2001). Use of exopolysaccharide-producing cultures to improve the functionality of low fat cheese. *International Dairy Journal, 11*(4–7), 433–439. https://doi.org/10.1016/S0958-6946(01)00084-X

Burdman, S. (2000). Extracellular polysaccharide composition of Azospirillum brasilense and its relation with cell aggregation. *FEMS Microbiology Letters, 189*(2), 259–264. https://doi.org/10.1016/S0378-1097(00)00294-9

Burey, P., Bhandari, B. R., Rutgers, R. P. G., Halley, P. J., & Torley, P. J. (2009). Confectionery gels: A review on formulation, rheological and structural aspects. *International Journal of Food Properties, 12*(1), 176–210. https://doi.org/10.1080/10942910802223404

Caggianiello, G., Kleerebezem, M., & Spano, G. (2016). Exopolysaccharides produced by lactic acid bacteria: From health-promoting benefits to stress tolerance mechanisms. *Applied Microbiology and Biotechnology, 100*(9), 3877–3886. https://doi.org/10.1007/s00253-016-7471-2

Castro-Bravo, N., Wells, J. M., Margolles, A., & Ruas-Madiedo, P. (2018). Interactions of surface exopolysaccharides from bifidobacterium and lactobacillus within the intestinal environment. *Frontiers in Microbiology, 9*. https://doi.org/10.3389/fmicb.2018.02426

Cerning, J. (1990). Exocellular polysaccharides produced by lactic acid bacteria. *FEMS Microbiology Letters, 87*(1–2), 113–130. https://doi.org/10.1111/j.1574-6968.1990.tb04883.x

Cerning, J. (1995). Production of exopolysaccharides by lactic acid bacteria and dairy propionibacteria. *Le Lait, 75*(4–5), 463–472. https://doi.org/10.1051/lait:19954-536

Chen, F., & Huang, G. (2018). Preparation and immunological activity of polysaccharides and their derivatives. *International Journal of Biological Macromolecules, 112*, 211–216. https://doi.org/10.1016/j.ijbiomac.2018.01.169

Cirrincione, S., Breuer, Y., Mangiapane, E., Mazzoli, R., & Pessione, E. (2018). "Ropy" phenotype, exopolysaccharides and metabolism: Study on food isolated potential probiotics LAB. *Microbiological Research, 214*, 137–145. https://doi.org/10.1016/j.micres.2018.07.004

Costa, N. E., Hannon, J. A., Guinee, T. P., Auty, M. A. E., McSweeney, P. L. H., & Beresford, T. P. (2010). Effect of exopolysaccharide produced by isogenic strains of Lactococcus lactis on half-fat Cheddar cheese. *Journal of Dairy Science, 93*(8), 3469–3486.

Das, L., Raychaudhuri, U., & Chakraborty, R. (2014). Effects of hydrocolloids as texture improver in coriander bread. *Journal of Food Science and Technology*. https://doi.org/10.1007/s13197-014-1296-8

Davidović, S., Miljković, M., Tomić, M., Gordić, M., Nešić, A., & Dimitrijević, S. (2018). Response surface methodology for optimisation of edible coatings based on dextran from Leuconostoc mesenteroides T3. *Carbohydrate Polymers, 184*, 207–213. https://doi.org/10.1016/j.carbpol.2017.12.061

de Mello Luvielmo, M., Borges, C. D., de Oliveira Toyama, D., Vendruscolo, C. T., & Scamparini, A. R. P. (2016). Structure of xanthan gum and cell ultrastructure at different times of alkali stress. *Brazilian Journal of Microbiology, 47*(1), 102–109. https://doi.org/10.1016/j.bjm.2015.11.006

De Vuyst, L., De Vin, F., Vaningelgem, F., & Degeest, B. (2001). Recent developments in the biosynthesis and applications of heteropolysaccharides from lactic acid bacteria. *International Dairy Journal, 11*(9), 687–707. https://doi.org/10.1016/S0958-6946(01)00114-5

Decho, A. W. (2000). Microbial biofilms in intertidal systems: An overview. *Continental Shelf Research, 20*(10–11), 1257–1273. https://doi.org/10.1016/S0278-4343(00)00022-4

Dogsa, I., Kriechbaum, M., Stopar, D., & Laggner, P. (2005). Structure of bacterial extracellular polymeric substances at different pH values as determined by SAXS. *Biophysical Journal, 89*(4), 2711–2720. https://doi.org/10.1529/biophysj.105.061648

Donot, F., Fontana, A., Baccou, J. C., & Schorr-Galindo, S. (2012). Microbial exopolysaccharides: Main examples of synthesis, excretion, genetics and extraction. *Carbohydrate Polymers, 87*(2), 951–962. https://doi.org/10.1016/j.carbpol.2011.08.083

Du, R., Qiao, X., Zhao, F., Song, Q., Zhou, Q., Wang, Y., Pan, L., Han, Y., & Zhou, Z. (2018). Purification, characterization and antioxidant activity of dextran produced by Leuconostoc pseudomesenteroides from homemade wine. *Carbohydrate Polymers, 198*, 529–536. https://doi.org/10.1016/j.carbpol.2018.06.116

Eckel, V. P. L., Vogel, R. F., & Jakob, F. (2019). In situ production and characterization of cloud forming dextrans in fruit-juices. *International Journal of Food Microbiology, 306*, 108261. https://doi.org/10.1016/j.ijfoodmicro.2019.108261

Fessard, A., & Remize, F. (2017). Why are Weissella spp. not used as commercial starter cultures for food fermentation? *Fermentation, 3*(3), 38. https://doi.org/10.3390/fermentation3030038

Flemming, H.-C., & Wingender, J. (2010). The biofilm matrix. *Nature Reviews Microbiology, 8*(9), 623–633. https://doi.org/10.1038/nrmicro2415

Frank, B. P., & Belfort, G. (2003). Polysaccharides and sticky membrane surfaces: Critical ionic effects. *Journal of Membrane Science, 212*(1–2), 205–212. https://doi.org/10.1016/S0376-7388(02)00502-1

Galle, S., Schwab, C., Arendt, E., & Gänzle, M. (2010). Exopolysaccharide-forming *Weissella* strains as starter cultures for sorghum and wheat sourdoughs. *Journal of Agricultural and Food Chemistry, 58*(9), 5834–5841. https://doi.org/10.1021/jf1002683

Galle, S., Schwab, C., Dal Bello, F., Coffey, A., Gänzle, M. G., & Arendt, E. K. (2012). Influence of in-situ synthesized exopolysaccharides on the quality of gluten-free sorghum sourdough bread. *International Journal of Food Microbiology, 155*(3), 105–112. https://doi.org/10.1016/j.ijfoodmicro.2012.01.009

Gänzle, M. G. (2014). Enzymatic and bacterial conversions during sourdough fermentation. *Food Microbiology, 37*, 2–10. https://doi.org/10.1016/j.fm.2013.04.007

Gheorghita Puscaselu, R., Lobiuc, A., Dimian, M., & Covasa, M. (2020). Alginate: From food industry to biomedical applications and management of metabolic disorders. *Polymers, 12*(10), 2417. https://doi.org/10.3390/polym12102417

Giavasis, I. (2013). Production of microbial polysaccharides for use in food. In *Microbial production of food ingredients, enzymes and nutraceuticals* (pp. 413–468). Elsevier. https://doi.org/10.1533/9780857093547.2.413

Grobben, G. J., van Casteren, W. H. M., Schols, H. A., Oosterveld, A., Sala, G., Smith, M. R., Sikkema, J., & de Bont, J. A. M. (1997). Analysis of the exopolysaccharides produced by Lactobacillus delbrueckii subsp. bulgaricus NCFB 2772 grown in continuous culture on glucose and fructose. *Applied Microbiology and Biotechnology, 48*(4), 516–521. https://doi.org/10.1007/s002530051089

Habibi, H., & Khosravi-Darani, K. (2017). Effective variables on production and structure of xanthan gum and its food applications: A review. *Biocatalysis and Agricultural Biotechnology, 10*, 130–140. https://doi.org/10.1016/j.bcab.2017.02.013

Heperkan, Z. D., Bolluk, M., & Bülbül, S. (2020). Structural analysis and properties of dextran produced by Weissella confusa and the effect of different cereals on its rheological characteristics. *International Journal of Biological Macromolecules, 143*, 305–313. https://doi.org/10.1016/j.ijbiomac.2019.12.036

Hu, T., Cui, Y., Zhang, Y., Qu, X., & Zhao, C. (2020). Genome analysis and physiological characterization of four streptococcus thermophilus strains isolated from Chinese traditional fermented milk. *Frontiers in Microbiology, 11*. https://doi.org/10.3389/fmicb.2020.00184

Hundschell, C. S., & Wagemans, A. M. (2019). Rheology of common uncharged exopolysaccharides for food applications. *Current Opinion in Food Science, 27*, 1–7. https://doi.org/10.1016/j.cofs.2019.02.011

Jolly, L., & Stingele, F. (2001). Molecular organization and functionality of exopolysaccharide gene clusters in lactic acid bacteria. *International Dairy Journal, 11*(9), 733–745. https://doi.org/10.1016/S0958-6946(01)00117-0

Kang, Z.-L., Wang, T., Li, Y., Li, K., & Ma, H. (2020). Effect of sodium alginate on physical-chemical, protein conformation and sensory of low-fat frankfurters. *Meat Science, 162*, 108043. https://doi.org/10.1016/j.meatsci.2019.108043

Karakaş-Budak, B. (2019). Effect of starch substitution with pullulan on confectionery starch gel texture of lokum. *Mediterranean Agricultural Sciences, 32*(3), 323–327. https://doi.org/10.29136/mediterranean.609017

Karimi, R., Azizi, M. H., Ghasemlou, M., & Vaziri, M. (2015). Application of inulin in cheese as prebiotic, fat replacer and texturizer: A review. *Carbohydrate Polymers, 119*, 85–100. https://doi.org/10.1016/j.carbpol.2014.11.029

Kaur, N., & Dey, P. (2023). Bacterial exopolysaccharides as emerging bioactive macromolecules: From fundamentals to applications. *Research in Microbiology, 174*(4), 104024. https://doi.org/10.1016/j.resmic.2022.104024

Kaur, V., Bera, M. B., Panesar, P. S., Kumar, H., & Kennedy, J. F. (2014). Welan gum: Microbial production, characterization, and applications. *International Journal of Biological Macromolecules, 65*, 454–461. https://doi.org/10.1016/j.ijbiomac.2014.01.061

Kavitake, D., Devi, P. B., Singh, S. P., & Shetty, P. H. (2016). Characterization of a novel galactan produced by Weissella confusa KR780676 from an acidic fermented food. *International Journal of Biological Macromolecules, 86*, 681–689. https://doi.org/10.1016/j.ijbiomac.2016.01.099

Khanal, B. K. S., Bhandari, B., Prakash, S., Liu, D., Zhou, P., & Bansal, N. (2018). Modifying textural and microstructural properties of low fat Cheddar cheese using sodium alginate. *Food Hydrocolloids, 83*, 97–108. https://doi.org/10.1016/j.foodhyd.2018.03.015

Kiani, H., Mousavi, M. E., Razavi, H., & Morris, E. R. (2010). Effect of gellan, alone and in combination with high-methoxy pectin, on the structure and stability of doogh, a yogurt-based Iranian drink. *Food Hydrocolloids, 24*(8), 744–754. https://doi.org/10.1016/j.foodhyd.2010.03.016

Kim, W.-S., Han, G. G., Hong, L., Kang, S.-K., Shokouhimehr, M., Choi, Y. J., & Cho, C.-S. (2019). Novel production of natural bacteriocin via internalization of dextran nanoparticles into probiotics. *Biomaterials, 218*, 119360. https://doi.org/10.1016/j.biomaterials.2019.119360

Konuk Takma, D., & Korel, F. (2017). Impact of preharvest and postharvest alginate treatments enriched with vanillin on postharvest decay, biochemical properties, quality and sensory attributes of table grapes. *Food Chemistry, 221*, 187–195. https://doi.org/10.1016/j.foodchem.2016.09.195

Korcz, E., & Varga, L. (2021). Exopolysaccharides from lactic acid bacteria: Techno-functional application in the food industry. *Trends in Food Science & Technology, 110*, 375–384. https://doi.org/10.1016/j.tifs.2021.02.014

Kothari, D., Tingirikari, J. M. R., & Goyal, A. (2015). In vitro analysis of dextran from Leuconostoc mesenteroides NRRL B-1426 for functional food application. *Bioactive Carbohydrates and Dietary Fibre, 6*(2), 55–61. https://doi.org/10.1016/j.bcdf.2015.08.001

Koutsoumanis, K., Allende, A., Alvarez-Ordóñez, A., Bolton, D., Bover-Cid, S., Chemaly, M., Davies, R., De Cesare, A., Hilbert, F., Lindqvist, R., Nauta, M., Peixe, L., Ru, G., Simmons, M., Skandamis, P., Suffredini, E., Cocconcelli, P. S., Fernández Escámez, P. S., Maradona, M. P., et al. (2020). Update of the list of QPS-recommended biological agents intentionally added to food or feed as notified to EFSA 12: Suitability of taxonomic units notified to EFSA until March 2020. *EFSA Journal, 18*(7). https://doi.org/10.2903/j.efsa.2020.6174

Kumar, R., Bansal, P., Singh, J., & Dhanda, S. (2020). Purification, partial structural characterization and health benefits of exopolysaccharides from potential probiotic Pediococcus acidilactici NCDC 252. *Process Biochemistry, 99*, 79–86. https://doi.org/10.1016/j.procbio.2020.08.028

Lacaze, G., Wick, M., & Cappelle, S. (2007). Emerging fermentation technologies: Development of novel sourdoughs. *Food Microbiology, 24*(2), 155–160. https://doi.org/10.1016/j.fm.2006.07.015

Laura, M., Notararigo, S., Ncher, M., de Palencia, P. F., Aznar, R., & Lpez, P. (2012). Biosynthesis, purification and biotechnological use of exopolysaccharides produced by lactic acid bacteria. In *Food additive*. InTech. https://doi.org/10.5772/32412

Le Bastard, Q., Chapelet, G., Javaudin, F., Lepelletier, D., Batard, E., & Montassier, E. (2020). The effects of inulin on gut microbial composition: A systematic review of evidence from human studies. *European Journal of Clinical Microbiology & Infectious Diseases, 39*(3), 403–413. https://doi.org/10.1007/s10096-019-03721-w

Leroy, F., & De Vuyst, L. (2016). Advances in production and simplified methods for recovery and quantification of exopolysaccharides for applications in food and health. *Journal of Dairy Science, 99*(4), 3229–3238. https://doi.org/10.3168/jds.2015-9936

Loeffler, M., Hilbig, J., Velasco, L., & Weiss, J. (2020). Usage of *in situ* exopolysaccharide-forming lactic acid bacteria in food production: Meat products—A new field of application? *Comprehensive Reviews in Food Science and Food Safety, 19*(6), 2932–2954. https://doi.org/10.1111/1541-4337.12615

Looijesteijn, P. J., & Hugenholtz, J. (1999). Uncoupling of growth and exopolysaccharide production by Lactococcus lactis subsp. cremoris NIZO B40 and optimization of its synthesis. *Journal of Bioscience and Bioengineering, 88*(2), 178–182. https://doi.org/10.1016/S1389-1723(99)80198-4

Lynch, K. M., Coffey, A., & Arendt, E. K. (2018a). Exopolysaccharide producing lactic acid bacteria: Their techno-functional role and potential application in gluten-free bread products. *Food Research International, 110*, 52–61. https://doi.org/10.1016/j.foodres.2017.03.012

Lynch, K. M., Zannini, E., Coffey, A., & Arendt, E. K. (2018b). Lactic acid bacteria exopolysaccharides in foods and beverages: Isolation, properties, characterization, and health benefits. *Annual Review of Food Science and Technology, 9*(1), 155–176. https://doi.org/10.1146/annurev-food-030117-012537

Macedo, M. G., Lacroix, C., Gardner, N. J., & Champagne, C. P. (2002). Effect of medium supplementation on exopolysaccharide production by Lactobacillus rhamnosus RW-9595M in whey permeate. *International Dairy Journal, 12*(5), 419–426. https://doi.org/10.1016/S0958-6946(01)00173-X

Maeda, H., Zhu, X., Suzuki, S., Suzuki, K., & Kitamura, S. (2004). Structural characterization and biological activities of an exopolysaccharide kefiran produced by *Lactobacillus kefiranofaciens* WT-2B ^{T}. *Journal of Agricultural and Food Chemistry, 52*(17), 5533–5538. https://doi.org/10.1021/jf049617g

Maina, N. H., Virkki, L., Pynnönen, H., Maaheimo, H., & Tenkanen, M. (2011). Structural analysis of enzyme-resistant isomaltooligosaccharides reveals the elongation of α-(1→3)-linked branches in weissella confusa dextran. *Biomacromolecules, 12*(2), 409–418. https://doi.org/10.1021/bm1011536

Marshall, V. M., Cowie, E. N., & Moreton, R. S. (1995). Analysis and production of two exopolysaccharides from *Lactococcus lactis* subsp. *cremoris* LC330. *Journal of Dairy Research, 62*(4), 621–628. https://doi.org/10.1017/S0022029900031356

Meftahi, A., Khajavi, R., Rashidi, A., Sattari, M., Yazdanshenas, M. E., & Torabi, M. (2010). The effects of cotton gauze coating with microbial cellulose. *Cellulose, 17*(1), 199–204. https://doi.org/10.1007/s10570-009-9377-y

Mende, S., Rohm, H., & Jaros, D. (2016). Influence of exopolysaccharides on the structure, texture, stability and sensory properties of yoghurt and related products. *International Dairy Journal, 52*, 57–71. https://doi.org/10.1016/j.idairyj.2015.08.002

Mensink, M. A., Frijlink, H. W., van der Voort Maarschalk, K., & Hinrichs, W. L. J. (2015). Inulin, a flexible oligosaccharide I: Review of its physicochemical characteristics. *Carbohydrate Polymers, 130*, 405–419. https://doi.org/10.1016/j.carbpol.2015.05.026

Mıdık, F., Tokatlı, M., Bağder Elmacı, S., & Özçelik, F. (2020). Influence of different culture conditions on exopolysaccharide production by indigenous lactic acid bacteria isolated from pickles. *Archives of Microbiology, 202*(4), 875–885. https://doi.org/10.1007/s00203-019-01799-6

Milde, L. B., Chigal, P. S., Olivera, J. E., & González, K. G. (2020). Incorporation of xanthan gum to gluten-free pasta with cassava starch. Physical, textural and sensory attributes. *LWT, 131*, 109674. https://doi.org/10.1016/j.lwt.2020.109674

Min, W.-H., Fang, X.-B., Wu, T., Fang, L., Liu, C.-L., & Wang, J. (2019). Characterization and antioxidant activity of an acidic exopolysaccharide from Lactobacillus plantarum JLAU103. *Journal of Bioscience and Bioengineering, 127*(6), 758–766. https://doi.org/10.1016/j.jbiosc.2018.12.004

Minervini, F., De Angelis, M., Di Cagno, R., & Gobbetti, M. (2014). Ecological parameters influencing microbial diversity and stability of traditional sourdough. *International Journal of Food Microbiology, 171*, 136–146. https://doi.org/10.1016/j.ijfoodmicro.2013.11.021

Mohammadi, M., Sadeghnia, N., Azizi, M.-H., Neyestani, T.-R., & Mortazavian, A. M. (2014). Development of gluten-free flat bread using hydrocolloids: Xanthan and CMC. *Journal of Industrial and Engineering Chemistry, 20*(4), 1812–1818. https://doi.org/10.1016/j.jiec.2013.08.035

Moradi, M., Guimarães, J. T., & Sahin, S. (2021). Current applications of exopolysaccharides from lactic acid bacteria in the development of food active edible packaging. *Current Opinion in Food Science, 40*, 33–39. https://doi.org/10.1016/j.cofs.2020.06.001

Moscovici, M. (2015). Present and future medical applications of microbial exopolysaccharides. *Frontiers in Microbiology, 6*. https://doi.org/10.3389/fmicb.2015.01012

Ngouémazong, E. D., Christiaens, S., Shpigelman, A., Van Loey, A., & Hendrickx, M. (2015). The emulsifying and emulsion-stabilizing properties of pectin: A review. *Comprehensive Reviews in Food Science and Food Safety, 14*(6), 705–718. https://doi.org/10.1111/1541-4337.12160

Nguyen, A. T.-B., Picart-Palmade, L., Nigen, M., Jimenez, L., Ait-Abderahim, H., & Marchesseau, S. (2019). Effect of mono or co-culture of EPS-producing Streptococcus thermophilus strains on the formation of acid milk gel and the appearance of texture defects. *International Dairy Journal, 98*, 17–24. https://doi.org/10.1016/j.idairyj.2019.06.007

Nguyen, P.-T., Nguyen, T.-T., Bui, D.-C., Hong, P.-T., Hoang, Q.-K., & Nguyen, H.-T. (2020). Exopolysaccharide production by lactic acid bacteria: The manipulation of environmental stresses for industrial applications. *AIMS Microbiology, 6*(4), 451–469. https://doi.org/10.3934/microbiol.2020027

Nicolaus, B., Kambourova, M., & Oner, E. T. (2010). Exopolysaccharides from extremophiles: From fundamentals to biotechnology. *Environmental Technology, 31*(10), 1145–1158. https://doi.org/10.1080/09593330903552094

Noorlaila, A., Hasanah, H. N., Asmeda, R., & Yusoff, A. (2020). The effects of xanthan gum and hydroxypropylmethylcellulose on physical properties of sponge cakes. *Journal of the Saudi Society of Agricultural Sciences, 19*(2), 128–135. https://doi.org/10.1016/j.jssas.2018.08.001

Nwodo, U. U., Green, E., & Okoh, A. I. (2012). Bacterial exopolysaccharides: Functionality and prospects. *International Journal of Molecular Sciences, 13*(11), 14002–14015. https://doi.org/10.3390/ijms131114002

Oleksy, M., & Klewicka, E. (2016). Exopolysaccharides produced by *Lactobacillus* sp.: Biosynthesis and applications. *Critical Reviews in Food Science and Nutrition*, 1–13. https://doi.org/10.1080/10408398.2016.1187112

Pachekrepapol, U., Lucey, J. A., Gong, Y., Naran, R., & Azadi, P. (2017). Characterization of the chemical structures and physical properties of exopolysaccharides produced by various Streptococcus thermophilus strains. *Journal of Dairy Science, 100*(5), 3424–3435. https://doi.org/10.3168/jds.2016-12125

Pessione, E. (2012). Lactic acid bacteria contribution to gut microbiota complexity: Lights and shadows. *Frontiers in Cellular and Infection Microbiology, 2*. https://doi.org/10.3389/fcimb.2012.00086

Radziej, S., Scherb-Forster, J., Schlicht, C., & Eisenreich, W. (2021). Fast identification of food thickeners by nontargeted NMR-spectroscopy. *Journal of Agricultural and Food Chemistry, 69*(12), 3761–3775. https://doi.org/10.1021/acs.jafc.0c07760

Rao, T. J. M., & Goyal, A. (2013). A novel high dextran yielding *Weissella cibaria* JAG8 for cereal food application. *International Journal of Food Sciences and Nutrition, 64*(3), 346–354. https://doi.org/10.3109/09637486.2012.734289

Rehm, B. H. A. (2010). Bacterial polymers: Biosynthesis, modifications and applications. *Nature Reviews Microbiology, 8*(8), 578–592. https://doi.org/10.1038/nrmicro2354

Ruas-Madiedo, P., & de los Reyes-Gavilán, C. G. (2005). Invited review: Methods for the screening, isolation, and characterization of exopolysaccharides produced by lactic acid bacteria. *Journal of Dairy Science, 88*(3), 843–856. https://doi.org/10.3168/jds.S0022-0302(05)72750-8

Ruas-Madiedo, P., Hugenholtz, J., & Zoon, P. (2002). An overview of the functionality of exopolysaccharides produced by lactic acid bacteria. *International Dairy Journal, 12*(2–3), 163–171. https://doi.org/10.1016/S0958-6946(01)00160-1

Rühmkorf, C., Jungkunz, S., Wagner, M., & Vogel, R. F. (2012a). Optimization of homoexopolysaccharide formation by lactobacilli in gluten-free sourdoughs. *Food Microbiology, 32*(2), 286–294. https://doi.org/10.1016/j.fm.2012.07.002

Rühmkorf, C., Rübsam, H., Becker, T., Bork, C., Voiges, K., Mischnick, P., Brandt, M. J., & Vogel, R. F. (2012b). Effect of structurally different microbial homoexopolysaccharides on the quality of gluten-free bread. *European Food Research and Technology, 235*(1), 139–146. https://doi.org/10.1007/s00217-012-1746-3

Ryan, E. P. (2011). Bioactive food components and health properties of rice bran. *Journal of the American Veterinary Medical Association, 238*(5), 593–600. https://doi.org/10.2460/javma.238.5.593

Ryan, P. M., Ross, R. P., Fitzgerald, G. F., Caplice, N. M., & Stanton, C. (2015). Sugar-coated: Exopolysaccharide producing lactic acid bacteria for food and human health applications. *Food & Function, 6*(3), 679–693. https://doi.org/10.1039/C4FO00529E

Saha, T., Hoque, M. E., & Mahbub, T. (2020). Biopolymers for sustainable packaging in food, cosmetics, and pharmaceuticals. In *Advanced processing, properties, and applications of starch and other bio-based polymers* (pp. 197–214). Elsevier. https://doi.org/10.1016/B978-0-12-819661-8.00013-5

Sánchez, J.-I., Martínez, B., Guillén, R., Jiménez-Díaz, R., & Rodríguez, A. (2006). Culture conditions determine the balance between two different exopolysaccharides produced by *Lactobacillus pentosus* LPS26. *Applied and Environmental Microbiology, 72*(12), 7495–7502. https://doi.org/10.1128/AEM.01078-06

Sandra, G., Schwab, C., Bello, F. D., Coffey, A., Gänzle, M., & Arendt, E. (2012). Comparison of the impact of dextran and reuteran on the quality of wheat sourdough bread. *Journal of Cereal Science, 56*(3), 531–537. https://doi.org/10.1016/j.jcs.2012.07.001

Santos, M., Teixeira, J., & Rodrigues, A. (2000). Production of dextransucrase, dextran and fructose from sucrose using Leuconostoc mesenteroides NRRL B512(f). *Biochemical Engineering Journal, 4*(3), 177–188. https://doi.org/10.1016/S1369-703X(99)00047-9

Santos, J., Alcaide-González, M. A., Trujillo-Cayado, L. A., Carrillo, F., & Alfaro-Rodríguez, M. C. (2020). Development of food-grade Pickering emulsions stabilized by a biological macromolecule (xanthan gum) and zein. *International Journal of Biological Macromolecules, 153*, 747–754. https://doi.org/10.1016/j.ijbiomac.2020.03.078

Schmid, J., Sieber, V., & Rehm, B. (2015). Bacterial exopolysaccharides: Biosynthesis pathways and engineering strategies. *Frontiers in Microbiology, 6*. https://doi.org/10.3389/fmicb.2015.00496

Sharma, S., & Rao, T. V. R. (2015). Xanthan gum based edible coating enriched with cinnamic acid prevents browning and extends the shelf-life of fresh-cut pears. *LWT – Food Science and Technology, 62*(1), 791–800. https://doi.org/10.1016/j.lwt.2014.11.050

Silva, L. A., Lopes Neto, J. H. P., & Cardarelli, H. R. (2019). Exopolysaccharides produced by Lactobacillus plantarum: Technological properties, biological activity, and potential application in the food industry. *Annals of Microbiology, 69*(4), 321–328. https://doi.org/10.1007/s13213-019-01456-9

Singh, S., Sran, K. S., Pinnaka, A. K., & Roy Choudhury, A. (2019). Purification, characterization and functional properties of exopolysaccharide from a novel halophilic Natronotalea sambharensis sp. nov. *International Journal of Biological Macromolecules, 136*, 547–558. https://doi.org/10.1016/j.ijbiomac.2019.06.080

Singh, S., Datta, S., Narayanan, K. B., & Rajnish, K. N. (2021). Bacterial exo-polysaccharides in biofilms: Role in antimicrobial resistance and treatments. *Journal of Genetic Engineering and Biotechnology, 19*(1), 140. https://doi.org/10.1186/s43141-021-00242-y

Suryawanshi, N., Naik, S., & Jujjawarapu, S. E. (2022). Exopolysaccharides and their applications in food processing industries. *Food Science and Applied Biotechnology, 5*(1), 22–44. 10.30721/fsab2022.v5.i1.165

Tang, X., Liu, R., Huang, W., Zhang, B., Wu, Y., Zhuang, J., Omedi, J. O., Wang, F., & Zheng, J. (2018). Impact of in situ formed exopolysaccharides on dough performance and quality of Chinese steamed bread. *LWT, 96*, 519–525. https://doi.org/10.1016/j.lwt.2018.04.039

Tarancón, P., Hernández, M. J., Salvador, A., & Sanz, T. (2015). Relevance of creep and oscillatory tests for understanding how cellulose emulsions function as fat replacers in biscuits. *LWT – Food Science and Technology, 62*(1), 640–646. https://doi.org/10.1016/j.lwt.2014.06.029

Torino, M. I., Font de Valdez, G., & Mozzi, F. (2015). Biopolymers from lactic acid bacteria. Novel applications in foods and beverages. *Frontiers in Microbiology, 6*. https://doi.org/10.3389/fmicb.2015.00834

Tsuneda, S., Aikawa, H., Hayashi, H., Yuasa, A., & Hirata, A. (2003). Extracellular polymeric substances responsible for bacterial adhesion onto solid surface. *FEMS Microbiology Letters, 223*(2), 287–292. https://doi.org/10.1016/S0378-1097(03)00399-9

Wang, Y., Compaoré-Sérémé, D., Sawadogo-Lingani, H., Coda, R., Katina, K., & Maina, N. H. (2019). Influence of dextran synthesized in situ on the rheological, technological and nutritional properties of whole grain pearl millet bread. *Food Chemistry, 285*, 221–230. https://doi.org/10.1016/j.foodchem.2019.01.126

Wangpaiboon, K., Padungros, P., Nakapong, S., Charoenwongpaiboon, T., Rejzek, M., Field, R. A., & Pichyangkura, R. (2018). An α-1,6-and α-1,3-linked glucan produced by Leuconostoc citreum ABK-1 alternansucrase with nanoparticle and film-forming properties. *Scientific Reports, 8*(1), 8340. https://doi.org/10.1038/s41598-018-26721-w

Wolter, A., Hager, A.-S., Zannini, E., Czerny, M., & Arendt, E. K. (2014). Influence of dextran-producing Weissella cibaria on baking properties and sensory profile of gluten-free and wheat breads. *International Journal of Food Microbiology, 172*, 83–91. https://doi.org/10.1016/j.ijfoodmicro.2013.11.015

Yu, Y.-J., Chen, Z., Chen, P. T., & Ng, I.-S. (2018). Production, characterization and antibacterial activity of exopolysaccharide from a newly isolated Weissella cibaria under sucrose effect. *Journal of Bioscience and Bioengineering, 126*(6), 769–777. https://doi.org/10.1016/j.jbiosc.2018.05.028

Zannini, E., Waters, D. M., Coffey, A., & Arendt, E. K. (2016). Production, properties, and industrial food application of lactic acid bacteria-derived exopolysaccharides. *Applied Microbiology and Biotechnology, 100*(3), 1121–1135. https://doi.org/10.1007/s00253-015-7172-2

Zarour, K., Vieco, N., Pérez-Ramos, A., Nácher-Vázquez, M., Mohedano, M. L., & López, P. (2017). Food ingredients synthesized by lactic acid bacteria. In *Microbial production of food ingredients and additives* (pp. 89–124). Elsevier. https://doi.org/10.1016/B978-0-12-811520-6.00004-0

Zeidan, A. A., Poulsen, V. K., Janzen, T., Buldo, P., Derkx, P. M. F., Øregaard, G., & Neves, A. R. (2017). Polysaccharide production by lactic acid bacteria: From genes to industrial applications. *FEMS Microbiology Reviews, 41*(Supp_1), S168–S200. https://doi.org/10.1093/femsre/fux017

Zhang, Y., Li, D., Yang, N., Jin, Z., & Xu, X. (2018). Comparison of dextran molecular weight on wheat bread quality and their performance in dough rheology and starch retrogradation. *LWT, 98*, 39–45. https://doi.org/10.1016/j.lwt.2018.08.021

Zheng, J., Wittouck, S., Salvetti, E., Franz, C. M. A. P., Harris, H. M. B., Mattarelli, P., O'Toole, P. W., Pot, B., Vandamme, P., Walter, J., Watanabe, K., Wuyts, S., Felis, G. E., Gänzle, M. G., & Lebeer, S. (2020). A taxonomic note on the genus Lactobacillus: Description of 23 novel genera, emended description of the genus Lactobacillus Beijerinck 1901, and union of Lactobacillaceae and Leuconostocaceae. *International Journal of Systematic and Evolutionary Microbiology, 70*(4), 2782–2858. https://doi.org/10.1099/ijsem.0.004107

Zhou, Y., Cui, Y., & Qu, X. (2019). Exopolysaccharides of lactic acid bacteria: Structure, bioactivity and associations: A review. *Carbohydrate Polymers, 207*, 317–332. https://doi.org/10.1016/j.carbpol.2018.11.093

Extracellular Polysaccharides (EPS) and Fermented Foods

V. Vijayalakshmi, S. Sibikumar, and K. Prabha

1 Introduction

1.1 Extracellular Polysaccharides (EPS) in Fermented Foods

EPS are long chained structures/sugar units as produced by some categories of microorganisms, mainly lactic acid bacteria, and are also referred to as polysaccharides produced from the cell. The newly formed structures or polymers are dispersed within the space outside the cells and contribute immensely toward the extension of the structure as well as the functionality and nutritional value of the fermented products. Structured or functional existence is due to the presence of specific sugars most notably glucose, galactose, and fructose as well as several other simple sugars in a straight chain or branched chain configuration. For that reason, EPS production by fermentation is of the interest of research work as it can be used to enhance the food quality and safety without synthetic additives. Of course, this activity is also important in the fermented products such as yogurt, kefir, sauerkraut, and sourdough, where it affects the EPS content, thus the texture, mouthfeel, stability, and nutritional aspects of the final product.

1.1.1 EPS in the Fermentation Process

- Microorganisms also excrete EPS during fermentation as part of their metabolism. The fermentation is the interplay of the microorganisms and the substrate under specific environmental conditions such as temperature and pH. EPS can be

V. Vijayalakshmi · S. Sibikumar · K. Prabha (✉)
Department of Food Technology, K. S. Rangasamy College of Technology, Tiruchengode, India
e-mail: prabha@ksrct.ac.in

T. Sarkar et al. (eds.), *Extracellular Polysaccharides*,
https://doi.org/10.1007/978-3-031-94472-7_8

synthesized and secreted in one of the two ways, namely, intra-cell or "out of the cell," which is released into the food medium.

- Synthesis Mechanism: EPS production process involves combinations of two major processes, which are the processes of heteropolysaccharides that involve incorporation of various sugar monomers and homopolysaccharide including polymerization of single sugar types. The concentration of EPS is also affected by the composition of the medium especially death-association of sugar sources, nitrogen, and other micronutrients required for growth.

1.2 EPS in Fermentation Process

Role in Fermentation: In the course of the fermentation, EPS carry out some important functions, such as:

1. Stabilization: EPS have the ability to emulsify, prevent phase separation, and increase the viscosity of the final product.
2. Structural Development: EPS form some of the structures inside the gel present in dairy products like yogurt, thus helping it attain a smooth thick texture.
3. Microbial Interactions: EPS may assist microorganisms to form biofilms as well that improve their persistence and activity in high-stress environments such as high acidity in the case of fermentation.
4. Strain-Specific Production: Certain strains of LAB including, for instance, *Lactobacillus delbrueckii ssp. bulgaricus* and *Streptococcus thermophilus*.

1.3 Importance of EPS in Food Fermentation

1. Texture and Mouthfeel: One of the most apparent benefits of EPS in food fermentation is that they provide texture improvement. For instance, yogurt is a dairy product and when EPS are present in the product, there is assurance of forming a thicker and creamier layer enhancing the mouthfeel. For bread products, EPS produced by sourdough bacteria help to hold moisture in the dough and the structure of the crumb as well.
2. Nutritional Benefits: There are EPS that are prebiotic, which means that they feed beneficial bacteria in the gut. Therefore, foods that are fermented and high in EPS such as kefir and sauerkraut are not just probiotics but prebiotics as well, thus boosting the health of the gut. Some EPS are also known to have cholesterol-lowering properties and are also believed to modulate the immune system.
3. Natural Food Additive: EPS are also natural thickeners, stabilizers, and gelling agents thus offering a clean label option to artificial additives such as xanthan girls and pectin gum. This is in response to the growing consumer trend and the desire for less processed food that has fewer artificial components.

4. Safety and Shelf Life: EPS can even improve the quality of fermented foods by increasing their viscosity and structure, and thus protect the product from microbial spoilage, thus prolonging the shelf life. EPS also reduce the sedimentation of whey in dairy products during storage, which helps increase the stability of the products.

2 Molecular and Biochemical Characterization of EPS

EPS stand for "extracellular polysaccharides," a polysaccharide that is synthesized by the majority of the microorganisms particularly lactic acid bacteria (LAB). It is hard to restrict EPS to a single monomeric subunit because of variations such as the contents and binding of various monomers, which are polysaccharide sugar polymers. It is important to study the molecular and biochemical properties and composition of bacterially produced EPS to assess how they can be derived and to use them effectively in food systems.

2.1 Classification and Functional Properties of EPS

Such EPS formed by LAB can be structured into classes according to their molecular structure, charge, and type of monomers. Such classification is necessary for being in a position to understand their different functional properties in systems of fermented foods (Fig. 1).

2.2 Structural Diversity and Composition of EPS

EPS is noted for its considerably polysaccharide indistinguishabilities, which likewise has a direct impact in the functional properties of the substrates in food and other fermentation processes. The nomenclature of the EPS can be recognized in two definite forms, i.e., HoPS and HePS (Fig. 2).

2.2.1 Homopolysaccharides (HoPS)

- HoPS is made up of one class of structural monosaccharides such as glucose, fructose, and galactose synthesized into either branched or linear chain polymers. The following types of homologous polysaccharides participate as illustrations of HoPS.
- Dextran: This is a polysaccharide of glucose and is formed by organisms like *Leuconostoc mesenteroides*. It is highly branched and has gum-like properties. It

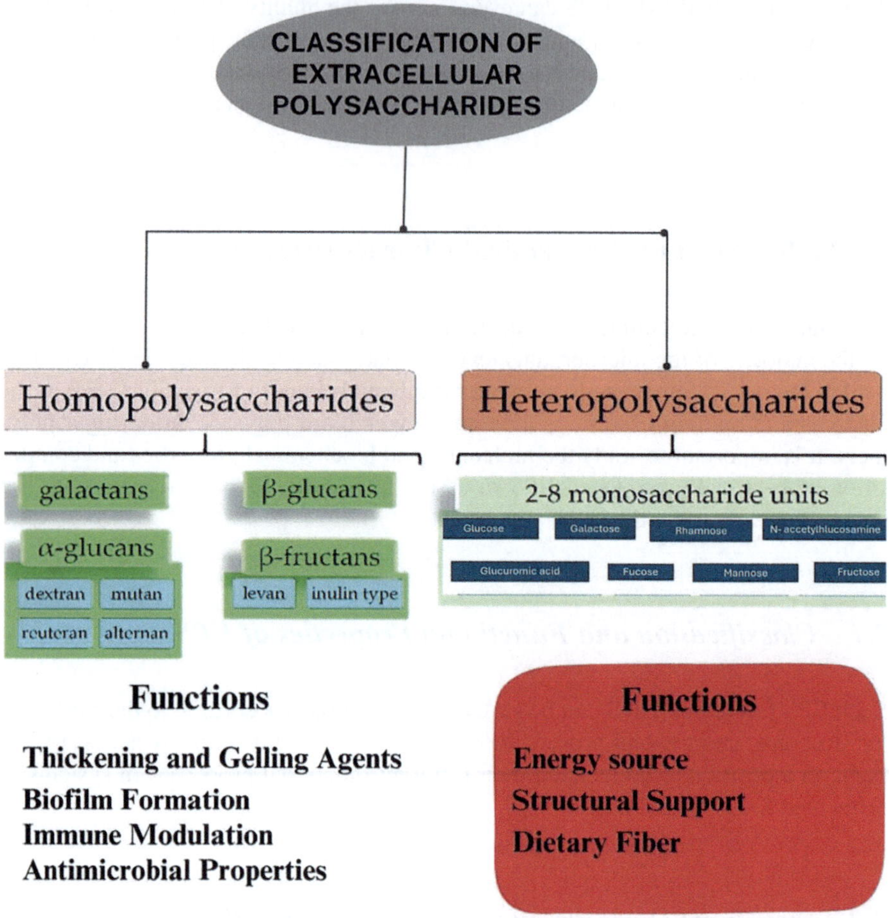

Fig. 1 Classification of extracellular polysaccharides and its function

is also used in fermented food products like dairy items and bakery products for viscosity enhancement and tactility.

- Levan: It is a polysaccharide composed of fructose end blocks and synthesized by lactic acid bacteria like *L. reuteri*. In vitro evaluation showed that levan exerts some prebiotic effects and helps in improving the textural attributes of fermented products such as beverages.
- Mutan: Mutan is produced by the Streptococcus species and is structurally similar to dextran; however, the structure has a variation in the degree of branching. This property makes it one of the important constituents of biofilms in microbial communities.

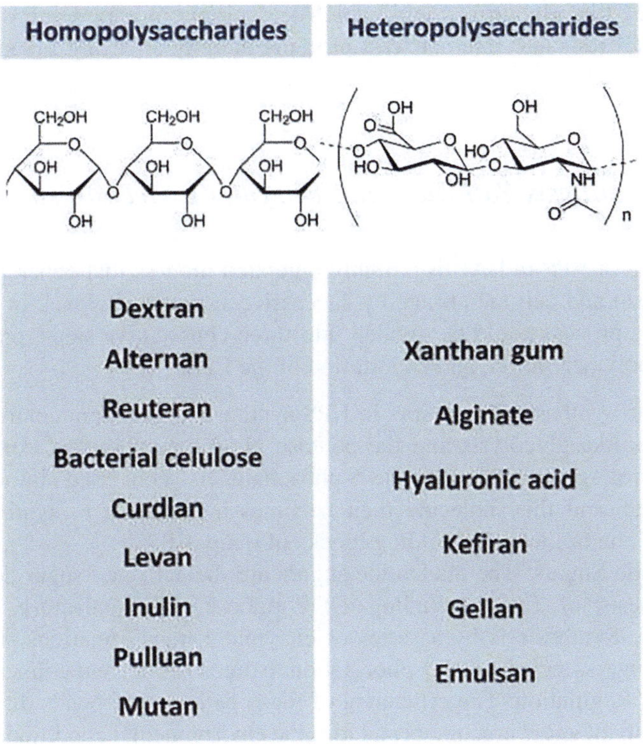

Fig. 2 Difference between HoPS and HePS

2.2.2 Heteropolysaccharides (HePS)

- HePS consist of two or more than two types of monosaccharides (glucose, galactose, mannose, rhamnose), and each type may be present in many different forms. Although there are structures with a linear form, those with branched types are more common. General properties of the HePS include:
- Linear vs. Branched Structures: HePS can either be linear, which means that a constant period of sugars is followed by other periods or high complex branching can occur. This branching affects their solubility, gelling abilities, and their interactions with other food components.
- Monomer Variability: As researchers have suggested, HePS comprises a combination of neutral and acidic sugars. For instance, *S. thermophilus* is known to synthesize HePS that contains glucose, galactose, and rhamnose sugars, the latter two making certain dairy products such as yogurts and creams more viscous.

The charge of EPS is notable within their structural context. Charged molecules tend to endow EPS with functional properties, which neutral polysaccharides such as dextran do not, as these contain no charged groups, which makes them useful exclusively as thickening and gelling agents. Acidic EPS also contain uronic acids, like glucuronic acid, which are anionic in nature and can therefore chelate metal ions and interact with proteins thereby affecting the structure and stability of food.

2.3 Biosynthesis Pathways and Enzymatic Mechanisms in LAB

Biosynthesis of EPS in LAB is a highly regulated process that generally occurs in the cytoplasm and cell membrane by successive steps of enzymes. In general, the biosynthesis process could be divided into three consecutive steps: precursor biosynthesis, polymerization, and exportation of the EPS.

- Precursor Synthesis: Precursors in EPS synthesis come from central metabolic pathways like glycolysis and the pentose phosphate pathway. For example, a product of glycolysis, glucose-6-phosphate, is converted into glucose-1-phosphate, and this molecule then becomes a precursor to synthesize UDP-glucose, one of the main building blocks of many EPS.
- Nucleotide Sugars: The nucleotide sugars are the activated sugar donors in the biosynthesis of EPS, including UDP-glucose, UDP-galactose, and GDP-mannose. Synthesis is by a series of enzymatic transformations starting from simple sugars, including phosphorylation, isomerization, and epimerization.
- Pathway Regulation: The efficiency of these pathways depends directly on the availability of sugar precursors and also the environmental conditions such as pH and temperature. While high sugar concentrations may serve to induce EPS production, an imbalance in precursor supply may result in incomplete or defective synthesis of EPS.

2.3.1 Polymerization

The polymerization of sugar monomers into polysaccharide chains is catalyzed by membrane-bound enzymes called glycosyltransferases that add activated sugars, nucleotide sugars, to the growing polysaccharide chain. This enzyme is specific to the sugar type that has to be added, as well as to the linkage type, α or β, which defines the general structure and function of the produced EPS. Different LAB strains will produce a different set of glycosyltransferases, which explains the big diversity in EPS produced across species. Chain length is controlled by specific enzymes, which determine the number of repeating units to be incorporated into the polymer. Long-chain EPS produced by some bacteria may have different properties than short-chain ones produced by other bacteria.

2.3.2 Export and Assembly

After synthesis, this produced EPS has to be transferred across the cell membrane into the extracellular environment. This is assisted through the respective transport system in which Wzy-dependent pathway mechanisms are involved in Gram-positive bacteria like LAB. Wzy-dependent pathway: It includes the action of flippases that flip the growing chain of polysaccharide across the membrane and polymerases that complete the assembly of the EPS in the extracellular matrix. It plays a critical role in the process of biofilm formation, allowing bacteria to attach to surfaces and acquire resistance against environmental stress. In fermented foods, this can give protection to the microbial community during long fermentations, enhancing product stability.

3 Regulation of EPS Production

Production of EPS by LAB is a very complex process, as its generation will be highly regulated, depending on environmental conditions, genetic regulation, and properties of the microbial strain. The optimization of EPS production requires detailed knowledge about the interaction of these factors and their impact on microbial metabolism with respect to yield, composition, and functional properties of the EPS produced. The next section considers how environmental and genetic regulation parameters may modulate EPS production and describes methodologies to optimize processes.

3.1 Environmental Parameters of EPS Synthesis

The environment where LAB are grown is a very relevant factor for EPS production. Changes in temperature, pH, and substrate availability, among other factors, directly influence the metabolic routes to biosynthesis. Such environmental parameters can be altered in order to enhance or reduce the yield of EPS and change their molecular structure.

- Temperature
  - Temperature is one of the most relevant parameters influencing microbial growth and metabolic activity. In most cases, LAB able to produce EPS show optimal temperature comprised between 30 and 37 °C, even though some thermophilic.
- pH
  - pH of the fermentation medium affects enzymatic activities responsible for production of EPS. Most LAB tend to grow well at slightly acidic pH. Thus,

pH values in the range of 5.5–6.5 were identified as generally optimal for an effective production of EPS. pH-Enzyme Activity: The enzymes responsible for polymerization by glycosyltransferases are pH-sensitive. A considerable deviation from the optimal pH may lead to a lower efficiency of the enzymes, causing either lower amounts of produced EPS or an altered structure of the produced polymers. Molecular Weight Impact: Under acidic conditions, low-molecular-weight EPSs are produced, while high-molecular-weight polymers may form at near-neutral pH-values, which are of greater interest for some food applications.

- Substrate Availability
  - Availability of the carbon source, especially sugars, is one of the critical regulatory elements in EPS synthesis. LABs can produce different types of EPS from various sugars such as glucose, fructose, and lactose. Type of sugar: Various sugars serve as substrates for various types of EPS. For instance, Leuconostoc species utilize sucrose to produce dextran, while Lactobacillus species have the capability to synthesize levan or fructan from fructose. When the concentration of these specific sugars is high in the medium, a higher production of EPS directly relates to it. Carbon to Nitrogen Ratio: Carbon and nitrogen sources are balanced; a high carbon content supports the synthesis of more EPS as it spreads energy resources to the synthesis of polymers, while inadequate nitrogen could result in a limitation to microbial growth, hence low overall yield of EPS.
- Oxygen Availability
  - Although generally microaerophilic or anaerobic, LAB may also be sensitive to oxygen availability for the production of EPS in some species. Conditions of Aerobes vs. Anaerobes: A few LAB like Leuconostoc spp. can produce higher yields of EPS under microaerophilic conditions, while strict anaerobic conditions might suppress the production of EPS.
- Other Nutrients and Mineral Ions
  - Phosphate and Magnesium Ions: These minerals are crucial to the enzymatic processes in EPS biosynthesis. The mineral phosphate acts as an energy carrier.

3.2 Genetic Regulation and Strain Variability in LAB

The genetic makeup of LAB basically determines the ability of a particular strain to produce EPS. Indeed, owing to genomic content, especially genes encoding the biosynthetic, polymerizing, and exporting enzymes, great variability exists between different strains with respect to the yields, composition, and functionality of the produced EPS.

1. Genetic Regulation of EPS Biosynthesis
    - Its biosynthesis is based on highly specific gene clusters, which encode the enzymes responsible for the synthesis of precursors, for glycosyltransferase activity, and for the export of polymers. Often these genes are encoded in operons to ensure a coordinated expression. EPS gene clusters: The LAB eps gene cluster usually encodes glycosyltransferases, regulatory proteins, and enzymes responsible for the polymerization and transport processes of EPS. Variability within the eps gene clusters results in differences in types of sugars incorporated and their linkages, thus influencing properties of the produced EPS. Transcriptional Regulators: The majority of the major regulators of eps operon transcription are proteins, which include two-component regulatory systems and sigma factors. These regulators sense environmental changes, such as elevated or low temperature or pH, and switch on or off the production of EPS.
2. Variability among Strains
    - LAB strains vary in the type and quantity of EPS elaborated due to their unique eps gene clusters. For example, *L. rhamnosus* and *L. plantarum* have HePS in high-molecular-weight and possess better viscosity and gelation properties. *Streptococcus thermophilus* is the prime producer of EPS among all in dairy fermentations, in which the produced polysaccharides enhance the texturing and stability of yoghurt.
3. Genetic Mutations and EPS Production
    - These genes may naturally or artificially undergo spontaneous or induced mutations. Such genetic modifications change the structure and function of EPS. For example, mutation in genes encoding for glycosyltransferases could result in truncation of polysaccharide chains or a shift in the branching patterns. Such mutations could affect the viscosity of the produced biopolymer or its ability to form gels.
4. Horizontal Gene Transfer (HGT)
    - HGT has been and remains crucial for LAB strain evolution to improve the performance of their EPS production. Those receiving eps gene clusters make new types of EPS, thus becoming more competitive in specific fermentative environments. It has indeed been more clearly realized in mixed cultures where exchange between strains of genetic material is an almost common event.

3.3 Process Optimization for Enhanced EPS Production

Optimization of fermentation processes will be quite important for the maximum production of EPS for their industrial and food applications. Process optimization involves the variation of environmental parameters, genetic manipulation, and the use of particular microbial strain types for improving yields of EPS and functional properties for intended applications.

- Optimization of Fermentation Parameters

  For the improvement in the production of EPS, it is necessary to control environmental variables such as temperature, pH, and substrate concentration. Generally, the following techniques will be employed: Fed-Batch Fermentation: In this type of fed-batch fermentation process, the slow addition of sugars will help the growth of the microorganism and give a constant rate of production of EPS. This evades nutrient limitation and also avoids any metabolic shift that could result in low yield of EPS. pH Control: The optimum pH lies in the range of approximately pH 6.0. Maintenance of a constant pH during fermentation would facilitate maximum activity of the enzymes responsible for the production of EPS. Most of the industrial fermentations possess automatic pH control systems that avoid possible fluctuations, which could inhibit the synthesis of EPS. Temperature Optimization: Maintenance of fermentation temperature in the optimal range-30–37 °C in most LAB-can ensure the optimum activity of the enzymes concerned with the biosynthesis of EPS. Thermophilic LAB need higher temperatures, and thus thermal control is fundamental for maximum EPS yield.

4 Functional Role of EPS in Fermentation Processes

Extracellular polysaccharides synthesized by lactic acid bacteria have a great impact on the texture and viscosity of fermented foods. They influence the physical properties of food matrices, modulate fermentation kinetics, and contribute to the stability and quality not only of dairy but also nondairy fermented products. In fact, the functional role of EPS has to be understood in order to further optimize food fermentation processes for obtaining desired product characteristics.

4.1 Impact of EPS on Dairy Fermentation: Texture and Viscosity

One of the most important roles of EPS in fermented dairy products pertains to its function as a texturizer and viscosifier. Indeed, the molecular structure and concentration of EPS is directly related to the rheological properties of these products, hence their creamy mouthfeel and smooth and stable nature.

1. Texture Improvement: Improved creaminess and mouthfeel: EPS enhance the smooth and creamy texture of dairy products due to the creation of a gel network that entraps water and prevents syneresis. The gel matrix, hence formed in

response to the presence of EPS, interacts with proteins and droplets of fat to give a homogeneous, creamy consistency. Reduction of Syneresis: One of the main functional roles played by EPS in dairy products is related to their ability to reduce syneresis, a term describing the separation of whey from yogurt or cheese curds. High-molecular-weight, branched EPS retain water within the protein matrix, improving overall product stability during storage.

2. Viscosity Enhancement: Higher viscosity: EPS, high molecular weight polysaccharides, greatly increases the viscosity of fermented milk products. The ability to form a dense and coherent network in the water phase gives rise to thickening. In yogurt, for example, EPS produced by the actions of *S. thermophilus* and *L. bulgaricus* lead to a thicker and more viscous consistency. Type of EPSS-influence: The type of EPS-homopolysaccharide or heteropolysaccharide influences the level of viscosity. For example, dextran is a type of homopolysaccharide synthesized by *L. mesenteroides* and increases viscosity highly due to its high molecular weight and large extent of branching. However, heteropolysaccharides from *L. rhamnosus* form more complex networks, which also contribute to viscosity.

4.2 *EPS in Nondairy Fermented Foods: Functional Contributions*

Besides dairy products, EPS also have major contributions in terms of texture, stability, and other functional properties in nondairy fermented foods. Examples are fermented vegetables, beverages, cereals, and soybean-derived products. Fermented vegetables, for example, Kimchi and Sauerkraut. In fermented vegetables, EPS produced by LAB enhances the general structure and texture of the product through improvement of water retention and the microbial community's stability in an unchanged status during fermentation.

- Texture Improvement: The presence of EPS helps to retain water within the vegetable matrix. The product will not become very soft or mushy in nature. EPS helps give a crispy texture to many products, such as kimchi and sauerkraut, which is highly desired by consumers. It acts as a means of protection for the biofilms formed, consisting of the beneficial LAB in fermented vegetables, thus ensuring consistent and prolonged fermentation. This, therefore, stabilizes the fermentation process and improves the product's shelf life.
- Beverages (e.g., Kefir, Kombucha): In beverages like kefir and kombucha, EPS contributes to viscosity and mouthfeel, enhancing the sensory of the drink.
- Viscosity and Mouthfeel: EPS produced by LAB in kefir increase the viscosity of the beverage, making it thicker and more velvety. In kombucha, where LAB usually play an apparently less dominating role, the produced EPS can also take part in maintaining the stability of the microbial biofilm, commonly called

SCOBY, symbiotic culture of bacteria and yeast, a structure crucial for the fermentation process.
- Fermented Cereals and Soy-Based Food: It has also been reported to perform application functions in fermented cereal-based foods like sourdough and soy-based foods such as miso and tempeh.
- Sourdough Bread: In sourdough, EPS produced by LAB like *L. sanfranciscensis* contribute viscoelastic properties to the dough, enhancing loaf volume, crumb texture, and moisture retention. These EPS also serve as natural emulsifiers enhancing dough stability and improving shelf life of the final product.
- Fermented Soy-Based Foods: In fermented soy-based foods, EPS improves the product, such as tempeh and miso, in terms of consistency and texture. Besides that, EPS interacts with soy proteins to form a stable matrix that inhibits the separation of liquid during storage.

4.3 Modulation of Fermentation Kinetics by EPS

Besides affecting the physical properties, EPS play a significant role in the modulation of fermentation kinetics. Generally, modulation in rates and efficiency of fermentation are achieved by the action of EPS on microbial ecology and nutrient control (Fig. 3).

- Biofilm Formation and Microbial Stability: EPS establish biofilms in microbial communities and are an important mechanism to stabilize the fermentation environment. The biofilm protects the LAB and other beneficial microorganisms from the environmental stressors, such as pH fluctuations and temperature variation, throughout the course of fermentation.
- Increased Microbial Interactions: Biofilms mediated by EPS allow for good exchange of nutrition and metabolic by-products between different microbial species for a better and more effective fermentation. In mixed fermentations, for example, kefir-multiple bacterial and yeast species are present, while EPS will ensure that microbial interactions are optimized in view of consistent and high-quality products.
- Protection Against Adverse Conditions: In long fermentations, such as sourdough and miso, EPS protects the microbial community from adverse conditions so that fermentation proceeds without disruptions due to fluctuations in either temperature or acidity. Rate of Fermentation: The presence of EPS can be considered by changing the availability of sugars to the microorganisms involved.
- Slow Fermentation: Sometimes, the formed EPS creates a viscous matrix around the cells, reducing the rate of sugar diffusion into the cells and therefore slowing down the rate of fermentation. This aspect is useful in some applications, for example, in sour-dough bread, where slower fermentation rates give rise to more complex flavors.

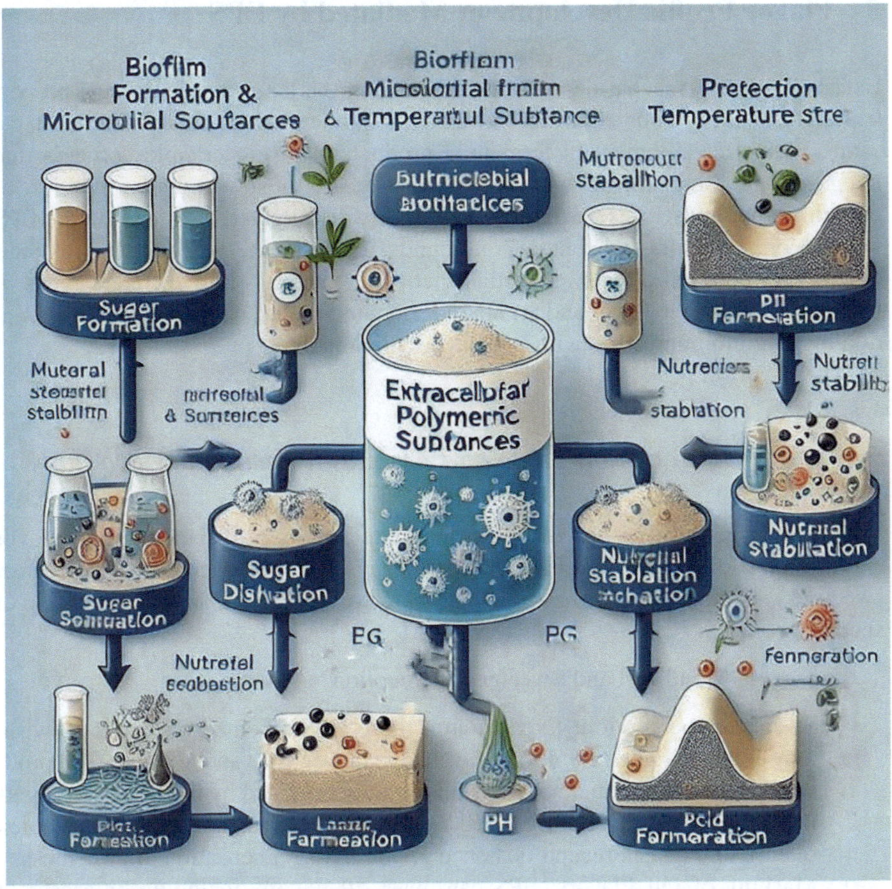

Fig. 3 Fermentation kinetics by EPS

- Stabilization of Nutrient Availability: Another way in which EPS exerts its influence is by trapping nutrients within the food matrix, hence guaranteeing a more constant supply of substrates to the microbial population in the course of fermentation. It inhibits fast depletion of sugars and prolongs the fermentation, hence giving the possibility of better control over product characteristics.
- Influence on pH and Acid Production: LAB producers of EPS very often present slower acidification during fermentation due to the competition between metabolic pathways for EPS synthesis and lactic acid production.
- Modulated Acidification: Due to the competition between synthesis of EPS and lactic acid production, there is retarded pH drop during fermentation. This may be advantageous in those products where a gradual acidification may be required, such as in particular yogurt and cheese varieties. Indeed, this slower pH drop also enhances stability of the microbial association and hinders over-acidification that jeopardizes texture and flavor.

5 Flavor Profile Development Mediated by EPS

Extracellular polysaccharides besides influencing the texture and physical properties of fermented foods contribute to the development of the flavor profile of the same. Flavor development in fermented products is a very complex process that integrates microbial metabolism, transformation of substrates, and formation of volatile compounds. These processes may be directly affected by interaction of EPS with flavor precursors and volatile compounds, and indirectly affected through their effects on microbial dynamics. In this section, we discuss how different types of EPS contribute to the flavor and aroma of fermented foods and their sensory perception in regard to interaction with key flavor compounds.

5.1 EPS Influence on Taste and Aroma in Fermented Products

The EPS formed during processing in fermented foods contributes to both flavor and aroma, adding to overall flavor in many ways. The polysaccharides will affect sweetness, acidity, and bitterness perception, as well as umami and the formation of specific aromas.

1. Improving Mouthfeel and Sweetness Perception

   - Sweetness perception: Even though these EPSs themselves are not sweet, they enhance sweetness due to the change in viscosity and texture of the product. The increased thickness and smoothness provided by the EPS in products like yogurt, kefir, and fermented beverages protects the sugars from quick release to the taste receptors, hence giving a longer sensation of sweetness.
   - Masking of Off-Flavors: EPS can mask off-flavors, including bitterness or excessive acidity, by providing a more even and rounded flavor profile. This is important in low-sugar or low-fat products where defects in flavor are more pronounced.

2. Modulation of Acidity

   - Perceived Sourness: In fermented products, such as yogurt, cheese, or pickles, EPS can reduce perceived sourness by changing the interaction of lactic acid with its taste receptors. Rich in EPS, products have a thick, gel-like consistency that slows the diffusion of acids to taste buds, making the sourness better balanced and smooth.
   - Buffering Capacity: EPS can serve as buffering agents that, while fermenting, there is stabilization of pH changes. This could provide a better controlled and slower process of acid production, leading to milder acidity and flavors that are in balance in the case of long-fermentation products, such as sourdough and fermented soy products.

3. Aroma Development

   - Interaction with Aromatic Compounds: EPS can bind to or encapsulate volatile aroma compounds, hence affecting their release upon consumption. For instance, in milk fermentations, bacterial species like *S. thermophilus* and *L. bulgaricus* were found to produce EPS that controlled the release of diacetyl, acetoin, and acetic acid, responsible for buttery and creamy and fruity-type flavor compounds of yogurt and cheese.
   - Aroma Retention: The presence of EPS can retain the aroma of volatile compounds and cause retarded release, hence improving the overall aroma profile of fermented foods. In this respect, in products such as fermented beverages (e.g., kefir or kombucha) where aroma compounds significantly contribute to the consumer's sensory experience, this may be an important benefit.

5.2 Sensory Evaluation of EPS-Enriched Foods

Sensory evaluation can provide a systematic procedure for assessing the effect of EPS on flavor, texture, and acceptability of foods with these polymers. Generally, consumers like products enriched with EPS; this is caused by a creamier mouthfeel, which makes them more acceptable and/or by a good balance in the flavor profile. The sensory characteristics of EPS-containing foods can be established by testing some key attributes including texture, taste, aroma, and aftertaste.

1. Texture and Mouthfeel

   - Creaminess and Thickness: Sensory analysis typically demonstrates that products containing high levels of EPS are considered creamier and smoother. As an example, yogurt and kefir to which EPS-producing LAB have been added score higher in mouthfeel, since the polysaccharides develop a velvety texture more pleasing to the consumer.
   - Reduction of Grittiness: In fermented plant-based products, like soy-based yogurts or fermented almond milk, EPS can reduce sensations of grittiness or graininess, thus improving the general texture and consumer acceptance of such products.

2. Flavor Balance and Intensity

   - Flavor Balance: In general, fermented products containing EPS show more balanced flavor profiles with reduced sharpness, while there is also harmony between sweetness, acidity, and umami. For example, in the case of yogurt, this can soften the acidity of the product, making it more acceptable to those who cannot tolerate highly acid flavors.
   - Flavor Complexity: Some works have indeed considered the role of EPS in contributing to a more complex flavor profile in fermented foods, such as in sourdough bread and kombucha, where the interaction of microbial activity with EPS production enhances the depth of flavor.

3. Aroma Perception

   - Aroma Intensity and Persistence: By sensory evaluation, fermented products containing EPS usually present a more intense and long-lasting aroma, because the polysaccharides can trap and slowly release volatile compounds. This is highly desirable in products such as fermented dairy and vegetable-based foods, since aroma is an important contributor to overall flavor perception.

5.3 *Interaction of EPS with Volatile Compounds and Flavor Precursors*

This is one of the critical aspects of the role that EPS plays in the development of a fermented product's final flavor profile, through the interaction with flavor compounds. EPS can interact not just with volatile compounds and flavor precursors but also with other molecules that are involved in the process of fermentation, in particular ways, to affect both intensity and release of flavor (Fig. 4).

1. Binding and Encapsulation of Volatile Compounds

   - EPS have an extremely high capacity to bind volatile compounds, virtually encapsulating them within their polysaccharide matrix. The latter can impact the release rates of volatile compounds upon consumption and subsequently its sensory perception with respect to both aroma and taste.
   - Encapsulation of Flavor Molecules: The volatile molecules including esters, alcohols, and acids may be entrapped by EPS within the food matrix. This is an important aspect in the context of dairy fermentations, where some of the aroma compounds like diacetyl, acetaldehyde, and ethyl butyrate are responsible for the buttery, fruity, or floral flavor notes characteristic of yogurt and cheese.
   - Controlled Release of Aromas: EPS encapsulates some volatile compounds, which guarantees a controlled release of the aromas along with time and thus a more durable sensorial experience. This slow release is extremely valued in fermented beverages and long-fermented products, like cheese, where the retention of aroma is a major factor affecting consumer acceptance.

2. Interaction with Flavor Precursors

   - The interaction of EPS with flavor precursors will modulate the formation of key compounds in fermentation. Such an interaction may result in new flavors, or an intensified existence of a current flavor profile.
   - Modulation of Proteolysis and Lipolysis: In fermented dairy products, EPS could affect enzymatic reactions such as proteolysis and lipolysis, both of

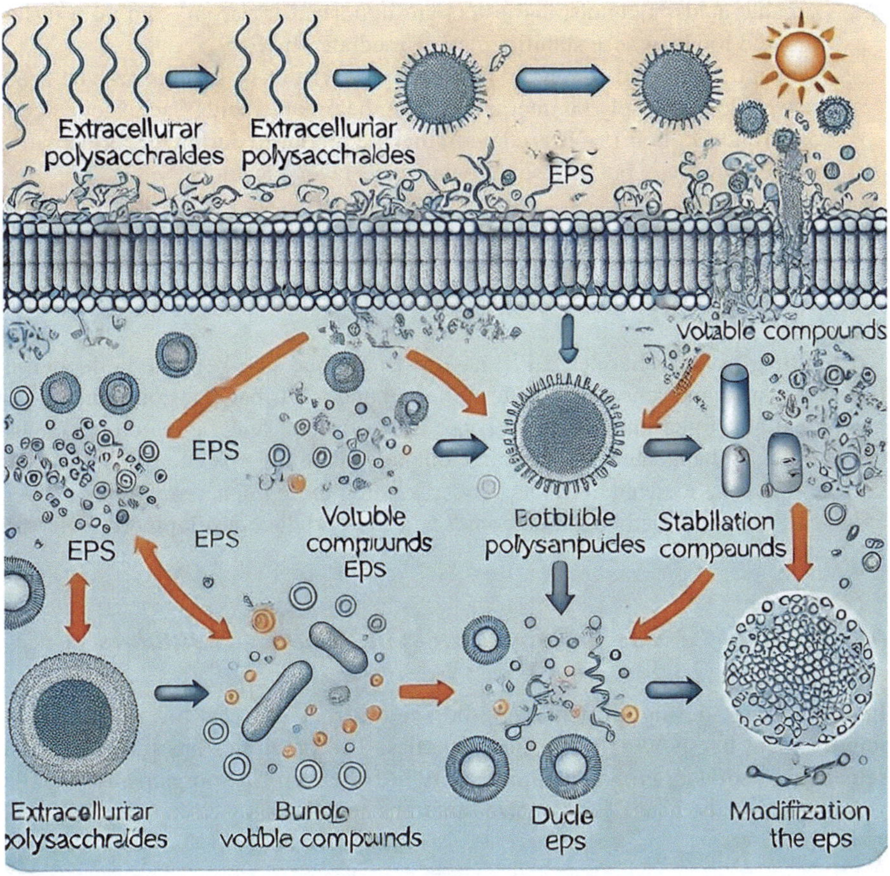

Fig. 4 Interaction of EPS with volatile compounds

which are events that are fundamental for the development of flavor. For example, in cheese manufacturing, EPS could affect the hydrolysis of milk proteins into peptides and amino acids, which are precursors of flavor compounds like sulfurous, cheesy, and nutty aromas.

- Maillard Reactions: LAB-produced EPS may interact with reducing sugars and amino acids during bread fermentation, thus affecting the course of the Maillard reactions during baking. This interaction would further develop a complex flavor, especially in sourdough bread, having caramel or roasted notes.

3. Synergistic Effects with Other Microbial Metabolites

- Presence of EPS may enhance the production of other microbial metabolites contributing to flavor development. For example, metabolic by-products of organic acids, alcohols, and esters are important for aroma and taste in fermented foods, whose stability may be mediated by EPS.

- Organic Acids and Alcohols: For example, in kefir or kombucha-type products, EPS will likely be interacting with the organic acids (lactic acid, acetic acid) and alcohols (such as ethanol) that are produced during the fermentation process. A tangy but refreshing balance of flavors develops.

6 Nutritional and Functional Impacts of EPS

Extracellular polysaccharides (EPS) have not only played a key role in the definition of physical and sensory attributes in fermented foods but also confer important nutritional and functional benefits. These range from serving as prebiotics, nutritional value enhancement to reaching out and affecting the bioavailability and digestibility of associated nutrients. Understanding these impacts can help in leveraging EPS for improved health outcomes and food product development.

6.1 Role of EPS as Prebiotics and Gut Health Modulators

In many cases, during fermentation, the synthesized EPS acts toward gut health improvement by showing prebiotic properties. The prebiotics are substances that selectively stimulate growth and/or activity of beneficial microorganisms from the gut, enhancing the health of digestive functions and general well-being.

1. Prebiotic Effects

   - Selective Fermentation by Beneficial Bacteria: It can also be used as a substrate for selective fermentation of gut beneficial bacteria, including Bifidobacteria and Lactobacilli. The fermented products of EPS are the SCFAs like acetate, propionate, and butyrate, which are believed to be gut-friendly. For example, EPS produced from Lactobacillus and Bifidobacterium species has been found to selectively enhance the gut growth of these friendly bacteria while inhibiting the growth of pathogenic microorganisms.
   - Increase SCFAs Production: Fermentation by gut microbiota increases the production of SCFAs important in health functions such as gut barrier integrity, low inflammatory level, and acting as an energy source to colonocytes. Furthermore, SCFA participate in appetite and metabolism regulation, thus being very important in general health.

2. Modulation of Gut Microbiota

   - Diverse Microbial Populations: EPS can enhance the diversity of the gut microbiota by providing nutrition to a wide range of microbial species. Diversity is an important element to maintain an appropriate gut microbiome

and avoid dysbiosis—a condition associated with several gastrointestinal and systemic diseases.
- Anti-pathogenic Effects: Considering the enhancement in the growth of beneficial bacteria and production of antimicrobial compounds during fermentation, EPS could inhibit the growth of pathogenic bacteria. This protective effect develops a healthier gut environment, further providing overall immune support.

6.2 Nutritional Improvement by EPS Incorporation

The incorporation of EPS into various food product formulations could create significant nutritional improvements in addition to their functional roles related to texture and flavor.

1. High Fiber Content
   - Source of Dietary Fiber: EPS is one form of dietary fiber that may be associated with overall intake of dietary fiber. Dietary fiber has, therefore, been associated with a number of beneficial effects to health, such as the promotion of bowel regularity, the reduction of risks associated with cardiovascular diseases, and the improvement of glycemic control. For instance, the EPS from Lactobacillus sp. will add value to fiber in the fermentation process of dairy and plant-based products.
   - Functional Fibers: EPS can also act as functional fibers, providing extra physiological benefits not conferred by traditional dietary fiber. For exam ple, some EPS has been found to enhance satiety and lower appetite, which generally serves to facilitate weight management and metabolic health.
2. Improving Nutrient Density
   - Nutritional Fortification: Addition of EPS to a food formulation may have an added nutritional value. EPS could serve as a potential vehicle for food fortification with further nutrients, like vitamins and minerals, by either acting as a carrier or interacting with other nutritional components to enhance their stability and bioavailability.
   - Synergistic Nutrient Interaction: EPS can interact with other nutrients within the food matrix to enhance their bioavailability and activity. As an example, the binding of minerals like calcium, magnesium, and zinc with EPS may enhance their absorption and utilization by the body.
3. Functional Food Development
   - Health Claims: Foods fortified with EPS can carry health claims associated with their prebiotic effects, as related to fiber and other related aspects connected with gut health. This might attract a number of health-oriented con-

sumers who look for functional foods with added value from health perspectives.

- Product Innovation: EPS application in product development can enable the formulation of various innovative functional foods, such as high-fiber snacks, low-calorie alternatives to dairy products, and probiotic-enhanced beverages, which satisfy the growing demand for health-promoting foods.

6.3 Nutritional Bioavailability and Digestibility of Food Products Involving EPS

The consumption of fermented foods that have been produced with EPS may affect the bioavailability and digestibility of associated nutrients due to changes in the accessibility of these nutrients to digestive enzymes and/or transport mechanisms across the intestine.

- Nutrient Bioavailability: Impact Exopolysaccharides Increase Mineral Absorption: EPS can enhance mineral bioavailability by forming soluble complexes and/or inhibiting precipitation in the gastrointestinal tract. In this respect, exopolysaccharides from fermented dairy products could enhance the solubility of calcium, magnesium, and other relevant minerals, improving their absorption and utilization.
- Improved Vitamin Stability: EPS can act protectively for those sensitive vitamins, such as vitamins A, C, and B-complex, during food processing and storage. This helps maintain the nutritional quality of the product and ensures full value from these vitamins is realized by the consumer.
- Digestibility of EPS.
- SCFAs Production and Fermentation: Fermentation of EPS by gut microbiota produces SCFAs, which generally contributes to gut health but also enhances the overall digestibility of the food matrix. SCFAs may have a positive effect on gut motility, improving nutrient absorption. This supports the overall digestive function.
- Potential Prebiotic Effects on Digestive Enzymes: Some studies show a trend in the ability of EPS to change digestive enzymatic activities, hence affecting digestion and the absorption of other nutrients. This may be considered one of the pathways through which improved digestion and nutritional assimilation take place.
- Impact on Nutritional Balance: Interaction with Other Dietary Components: EPS can also interact with other nutritional food components and modify the overall food nutritional balance. For instance, EPS can be used to bind and then release nutrients in a controlled manner, therefore giving a more steady supply of essential vitamins and minerals throughout the digestion process.

- Improvement of gut health and nutritional status, leading to a reduction in chronic diseases, may be achieved by adding EPS into foods due to the summative added effects of nutrient bioavailability, increased fiber intake, and positive modulation of gut microbiota.

7 Industrial and Technological Applications of EPS

Extracellular polysaccharides represent versatile biopolymers with huge applications in various industries, especially in the production of food. Indeed, their unique properties allow them to obtain value in the enhancement of quality and functionality in both fermented dairy and nondairy products. This chapter covers the industrial and technological applications of EPS according to their roles in the dairy industry, fermented foods out of dairy, and challenges plus advances in the production of EPS.

7.1 EPS in the Dairy Industry: Applications and Innovations

EPSs have found applications in the dairy industry to improve product quality, texture, and functionality; their role ranges from yogurt to cheese and milk beverages.

1. Dairy Product Applications

   - Yogurt Production: EPS enhance the texture, creaminess, and viscosity of yogurt. In addition, EPS can provide a smoother mouthfeel and prevent whey separation. Indeed, these effects have been achieved in yoghurt with EPS produced by *S. thermophilus* and *L. bulgaricus*.
   - Cheese Production: EPS has varied roles to play in the improvement of texture and defects that are unwanted in cheese. For example, EPS produced by *Lactococcus lactis* can contribute to the smoothness of cheese curd and reduce the occurrence of graininess.
   - Dairy Beverages: In beverages, kefir or flavored milk, for example, EPS can enhance product stability and mouthfeel. This is important because it provides a homogeneous distribution of flavorings, while preventing sedimentation.

2. Innovations in Dairy Applications

   - Reduced-Fat and Low-Sugar Products: EPS increasingly find their application in reduced-fat and low-sugar dairy products, mainly for replacing the losses in texture and creaminess associated with these reductions. EPS also can deliver similar sensory properties like fat and sugar; therefore, such products are gaining popularity among customers.

- Probiotic Dairy Products: The integration of EPS with probiotic strains enhances the stability and viability of probiotics in dairy products. In this way, it is able to protect the probiotic bacteria during processing and storage so that they may remain active and keep on exerting beneficial effects on gut health.

3. Processing and Formulation Techniques

   - Fermentation Optimization: Several fermentation technological innovations are under discussion for the optimization of EPS production and functionality in dairy products, including controlled fermentation conditions and specific EPS-producing strains. Among recent developments is the improvement in encapsulation technologies to enhance the stability and release of EPS in these products. This involves microencapsulation to protect the EPS during processing and enhance their functional properties in the final product.

7.2 Application of EPS in Nondairy Fermented Food Products

EPS have further been applied to various nondairy fermented food products in enhancing texture, stability, and quality of the product.

1. Application of EPS in Nondairy Products

   - Fermented Vegetables: Exopolysaccharides have been used to enhance the texturization and stability of fermented vegetables such as kimchi, sauerkraut, and pickles. They help in maintaining crispiness against loss of texture in storage.
   - Beverages with Fermentation: EPS contributes to the viscosity and mouthfeel in beverages like kombucha and fermented plant-based beverages. They also stabilize the microbial cultures during fermentation, thereby helping achieve consistency in quality for the end product.
   - Fermented Plant-Based Foods: EPSs are increasingly used in a variety of plant-based fermented foods such as soy yogurt, tempeh, and miso. Improvement of texture, stability, and overall sensory properties makes these products more desirable to consumers.

2. Nondairy Applications: Recent Developments

   - Improved Functional Properties: There is continuous research in the identification and utilization of novel EPS with unique functional properties in nondairy applications. This includes EPS with a high water-holding capacity, improved texturization capacity, and high stability in plant-based matrices.
   - Improved Flavor Profiles: For this reason, EPS are being researched for application in the modification of flavor profiles in nondairy products. The interaction between the polysaccharides and flavor precursors and volatile

compounds can thus result in the creation of more complex and interesting flavors.

3. Formulation and Processing Techniques

   - The Integration with Other Ingredients: Innovations in formulation focus on integrating EPS with other functional ingredients, including plant proteins and fibers, to enhance the overall quality of nondairy fermented foods.
   - Advanced Processing Methods: High-pressure processing and enzyme treatments, among other advanced techniques, are being used to optimize the functionality of EPS and enhance the quality of the nondairy fermented products.

7.3 *Technological Challenges and Advances in the Production of EPS*

For the production of EPS, several are the issues related to technologies, but recent developments in the field of biotechnology and various processing techniques are overcoming these challenge. EPS Production Yield and Cost: High yields of the production of EPS at reasonable costs is still a challenge. In many cases, this requires complicated fermentation conditions, and substrate costs, which make the production of EPS on an industrial scale both difficult and economically nonviable.

1. Purification and Extraction: Generally, the processes for purification and extraction of EPS from fermentation broths are labor-intensive and costly. Consequently, there is a dire need for efficient separation and purification techniques of the produced EPS in order to reduce the production costs and improve the quality of the product.
2. Consistency and Quality Control: In addition to being in reliable quality and functionality, sometimes EPS from different batches might not be easy to main tain. Microbial variability, for example, together with shifts in the conditions of fermentation and raw materials, may influence the nature of the final product, that is, EPS.
3. Improvement Within the Production of EPS

   - Biotechnology Advancement: New biotechnology-like genetic engineering and synthetic biology are being applied to optimize production. Development of new microbial strains with increased functionality in the biosynthesis of EPS and with improved fermentation properties will lead to more efficient and sustainable industrial processes.
   - Process Optimization: The research is targeted on enhancing the fermentation process to provide better yields at reduced production costs by increasing alternative substrates, improvement in fermentation technology, and automation of processes.

- Improved Purification Techniques: New techniques in the purification and extraction of EPS are being discovered and explored. Advanced filtration, techniques of precipitation, and chromatographic techniques are studied for an improved outcome. The benefits will consist of cost reduction with enhancement of purity and quality of the EPS.
- Application-specific formulations include the development of different forms of EPS targeting specific applications, such as specifically formulated EPS for dairy, nondairy products, and functional foods. It involves using blended EPS and modified polysaccharides to meet the specific demand of a particular product.

References

Ahmadul Kabir, U., Shamsuddeen, A. M., & Magashi. (2022). Screening for potential exopolysaccharide producers from Lactobacillus spp isolated from locally fermented milk (nono). *UMYU Journal of Microbiology Research, 7*(1), 84–92.

Bazaka, K., Crawford, R. J., Nazarenko, E. L., & Ivanova, E. P. (2011). Bacterial extracellular polysaccharides. *Adv. Exp. Med. Biol., 715*, 213–226.

Behare, P., & Kumar, S. (2019). Exopolysaccharides producing lactic acid cultures for indigenous fermented milk products. *Journal of AgriSearch, 6*, 142–145.

Berthold-Pluta, A., St. Pluta, A., Garbowska, M., & Stasiak-Różańska, L. (2019). Exopolysaccharide-producing lactic acid bacteria – Health-promoting properties and application in the dairy industry. *Advancements of Microbiology, 58*, 191–204.

Bibi, A., Xiong, Y., Rajoka, M. S. R., Mehwish, H. M., Radicetti, E., Umair, M., Shoukat, M., Khan, M. K. I., & Aadil, R. M. (2021). Recent advances in the production of exopolysaccharide (EPS) from Lactobacillus spp. and its application in the food industry: A review. *Sustainability, 13*(22), 12429.

Cázares-Vásquez, M. L., Rodríguez-Herrera, R., Aguilar-González, C. N., Sáenz-Galindo, A., Solanilla-Duque, J. F., Contreras-Esquivel, J. C., & Flores-Gallegos, A. C. (2021). Microbial exopolysaccharides in traditional Mexican fermented beverages. *Fermentation, 7*(4), 249.

Chen, L., & Shu, G. (2023). Purification and identification of EPS produced by five lactic acid bacteria and evaluation of their effects on the texture of fermented goat milk. *Fermentation, 9*(6), 527.

Choi, I. S., Ko, S. H., Lee, M. E., Kim, H. M., Yang, J. E., Jeong, S.-G., Lee, K. H., Chang, J. Y., Kim, J.-C., & Park, H. W. (2021). Production, characterization, and antioxidant activities of an exopolysaccharide extracted from spent media wastewater after Leuconostoc mesenteroides WiKim32 fermentation. *ACS Omega, 6*, 8171–8178.

da Silva, L. A., Neto, J. H. P. L., & Cardarelli, H. R. (2019). Exopolysaccharides produced by Lactobacillus plantarum: Technological properties, biological activity, and potential application in the food industry. *Annals of Microbiology, 69*, 321–328.

Dertli, E. (2013). *Biochemistry and functional analysis of exopolysaccharide production in Lactobacillus johnsonii* (Ph.D. thesis).

Dimopoulou, M., & Dols-Lafargue, M. (2021). Exopolysaccharides producing lactic acid bacteria in wine and other fermented beverages: For better or for worse? *Foods, 10*(9), 2204.

Giavasis, I. (2013). Production of microbial polysaccharides for use in food. In *Microbial production of food ingredients, enzymes and nutraceuticals*. Woodhead Publishing Limited.

Grosu-Tudor, S. S., & Zamfir, M. (2014). Exopolysaccharide production by selected lactic acid bacteria isolated from fermented vegetables. *Scientific Bulletin. Series F. Biotechnologies, 18*, 107–114.

Jaiswal, P., Sharma, R., Sanodiya, B. S., & Bisen, P. S. (2014). Microbial exopolysaccharides: Natural modulators of dairy products. *Journal of Applied Pharmaceutical Science, 4*(10), 105–109.
Jurášková, D., Ribeiro, S. C., & Silva, C. C. G. (2022). Exopolysaccharides produced by lactic acid bacteria: From biosynthesis to health-promoting properties. *Foods, 11*(2), 156.
Korcz, E., & Varga, L. (2021). Exopolysaccharides from lactic acid bacteria: Techno-functional application in the food industry. *Trends in Food Science and Technology, 110*, 375–384.
Li, Y., Xiao, L., Tian, J., Wang, X., Fang, Y.-z., & Li, W. (2022). Structural characterization, rheological properties and protection of oxidative damage of an exopolysaccharide from Leuconostoc citreum 1.2461 fermented in soybean whey. *Foods, 11*(15), 2283.
Loeffler, M., Hilbig, J., Velasco, L., & Weiss, J. (2020). Usage of in situ exopolysaccharide-forming lactic acid bacteria in food production: Meat products—A new field of application? *Comprehensive Reviews in Food Science and Food Safety, 19*(6), 2932–2954.
Miao, W., Li, N., & Jian-Lin, W. (2022). Food polysaccharides utilization via in vitro fermentation: Microbiota, structure, and function. *Current Opinion in Food Science, 48*, 100911.
Molina-Santiago, C., de Vicente, A., & Romero, D. (2021). Bacterial extracellular matrix as a natural source of biotechnologically multivalent materials. *Computational and Structural Biotechnology Journal, 19*, 2796–2805.
Netrusov, A. I., Liyaskina, E. V., Kurgaeva, I. V., Yang, G., & Revin, V. V. (2023). Exopolysaccharides producing bacteria: A review. *Microorganisms, 11*(6), 1541.
Nikitina, E. D., Petrova, T., Sungatullina, A., Bondar, O. V., Kharina, M., Mikshina, P., Gavrilova, E., & Kayumov, A. R. (2023). The profile of exopolysaccharides produced by various Lactobacillus species from silage during not-fat milk fermentation. *Fermentation, 9*(2), 197.
Ozogul, Y., El Abed, N., Montanari, C., & Ozogul, F. (2022). Contribution of polysaccharides from crustacean in fermented food products. Advances in food and nutrition research. *Advances in Food and Nutrition Research, 102*, 47–92.
Ricciardi, A., & Clementi, F. (2000). Exopolysaccharides from lactic acid bacteria: Structure, production and technological applications. *Italian Journal of Food Science, 12*(1), 23–45.
Roca, C. F. A., Alves, V. D., Freitas, F., & Reis, M. A. M. (2015). Exopolysaccharides enriched in rare sugars: Bacterial sources, production, and applications. *Frontiers in Microbiology, 6*, 288.
Ryan, P. M., Ross, R. P., Fitzgerald, G. F., Caplice, N. M., & Stanton, C. (2015). Sugar-coated: Exopolysaccharide producing lactic acid bacteria for food and human health applications. *Food & Function, 6*(3), 679–693.
Salazar, N., Gueimonde, M., Hernández-Barranco, A. M., Ruas-Madiedo, P., & de los Reyes-Gavilán, C. G. (2008). Exopolysaccharides produced by intestinal Bifidobacterium strains act as fermentable substrates for human intestinal bacteria. *Applied and Environmental Microbiology, 74*, 4737–4745.
Salazar, N., Gueimonde, M., de los Reyes-Gavilán, C. G., & Ruas-Madiedo, P. (2016). Exopolysaccharides produced by lactic acid bacteria and Bifidobacteria as fermentable substrates by the intestinal microbiota. *Critical Reviews in Food Science and Nutrition, 56*(9), 1440–1453.
Shah, N. P. (2003). The exopolysaccharide production by starter cultures and their influence on textural characteristics of fermented milk. *International Dairy Federation Special Issue, 1*, 101–115.
Sørensen, H. M., Rochfort, K. D., Maye, S., MacLeod, G. G., Brabazon, D., Loscher, C. E., & Freeland, B. (2022). Exopolysaccharides of lactic acid bacteria: Production, purification and health benefits towards functional food. *Nutrients, 14*(14), 2938.
Stauffer, K. R., & Leeder, J. G. (1978). Extracellular microbial polysaccharide production by fermentation on whey or hydrolyzed whey. *Journal of Food Science, 43*, 756–758.
Sugiyama, M. (2019). *Extracellular polysaccharide of lactic acid bacteria and use thereof.* Paent No. AU 2018281512 B2.
Suryawanshi, N., Naik, S., & Satya Eswari, J. (2022). Exopolysaccharides and their applications in food processing industries. *Food Science and Applied Biotechnology, 5*, 22–44.

Tiwari, O. N., Sasmal, S., Kataria, A. K., & Devi, I. (2020). Application of microbial extracellular carbohydrate polymeric substances in food and allied industries. *3 Biotech, 10*(5), 221.

Vandamme, E., De Baets, S., & Steinbüchel, A. (2002). *Polysaccharides II: Polysaccharides from Eukaryotes*. Wiley-Blackwell.

Wa, Y., Chanyi, R. M., Nguyen, H. T. H., Gu, R., Day, L., & Altermann, E. (2022). Extracellular polysaccharide extraction from Streptococcus thermophilus in fermented milk. *Microbiology Spectrum, 10*, e02280-21.

Wang, G., Xie, L., Huang, Z., & Xie, J. (2023). Recent advances in polysaccharide biomodification by microbial fermentation: Production, properties, bioactivities, and mechanisms. *Critical Reviews in Food Science and Nutrition, 64*(33), 12999–13023.

Zhao, X., Chen, G.-Q., Wang, F., Zhao, H., Wei, Y., Liu, L., & Zhang, H. (2022). Extraction, characterization, antioxidant activity and rheological behavior of a polysaccharide produced by the extremely salt tolerant Bacillus subtilis LR-1. *LWT, 162*, 113413.

Zheng, T., Ji, W., Zhe, Z., & ZhenNai, Y. (2013). Research advances on structure and function of exopolysaccharides produced by lactic acid bacteria. *Journal of Food Safety and Quality, 4*, 783–790.

Harnessing the Power of Exopolysaccharides for Improved Gut Health

Saikat Mazumder, Debasmita Bhattacharya, Dibyajit Lahiri, and Moupriya Nag

1 Introduction

Lactic acid bacteria (LAB) are known to produce a wide variety of polysaccharides. High-molecular-weight molecules composed of monosaccharide units linked through a glycosidic bond show a wide range of architectures, functional characteristics, and biological activities (Welman & Maddox, 2003). Polysaccharides are one of the primary ingredients in the creation of the extracellular biofilm matrix. They play important roles in microbial cells' adherence to solid surfaces as well as the protection of bacteria from detrimental environmental conditions (Harapanahalli et al., 2015; Oleksy & Klewicka, 2018). Polysaccharides are structural components of the cell surface that have been reported as either lipopolysaccharides or capsular polysaccharides in bacterial structures (Juráŝková et al., 2022). They establish direct contact with the environment and make up the outermost layer of the bacterial cell, thereby providing protection. Although some capsular polysaccharides may be released from the cell, they are less likely to be purified or separated from the cell biomass because of their adhesion to the cell surface. Exopolysaccharides are

S. Mazumder
Department of Biotechnology, Institute of Engineering and Management, Kolkata, University of Engineering and Management, Kolkata, West Bengal, India

Department of Food Technology, Guru Nanak Institute of Technology, Kolkata, West Bengal, India

D. Bhattacharya
Department of Basic Science and Humanities, Institute of Engineering and Management, Kolkata, University of Engineering and Management, Kolkata, West Bengal, India

D. Lahiri (✉) · M. Nag (✉)
Department of Biotechnology, Institute of Engineering and Management, Kolkata, University of Engineering and Management, Kolkata, West Bengal, India
e-mail: dibyajit.lahiri@uem.edu.in; moupriya.nag@uem.edu.in

T. Sarkar et al. (eds.), *Extracellular Polysaccharides*,
https://doi.org/10.1007/978-3-031-94472-7_9

polysaccharides that are secreted into the extracellular media or weakly bound to the cell surface (Daba et al., 2021). They can be produced intracellularly and secreted from the cells, or they can be produced extracellularly by the secreted enzymes of the bacterium (Lynch et al., 2018). Very recently, microbial envelope glycans were identified as key effector chemicals in intestinal health (Verma et al., 2018; Zhou et al., 2017). There are two categories of glycans: capsular polysaccharides, which are attached firmly to the cell surface, and exopolysaccharides (EPS), which are loosely attached or dispersed in the environment. Because of their relevance to humans, there has been increasing interest in EPS (Poole et al., 2018). Investigations of EPS composition and health effects have made investigations contemporary and relevant as the health advantages of probiotics may be connected to specific EPS structures they create (Rossi et al., 2015). The majority of the bacteria create EPS, which changes microbiota compositions and interacts with the intestinal epithelial barrier, and the intestinal immune system (Castro-Bravo et al., 2018).

A wide variety of symbiotic microbes colonize the human GIT (Proctor et al., 2019). Marchesi and Ravel (2015) estimate that the gut microbiome contains around 1000 species of bacteria, fungi, archaea, protozoa, and viruses. Most of these species are bacterial and have favorable effects on the host (Sánchez et al., 2017). The health of the host depends on a balanced microbiota because dysbiosis or altered bacterial communities have been linked to more than 100 diseases or disorders (Rojo et al., 2017). *Lactobacillus* and *Bifidobacterium* in the gut microbiome improve host health. *Lactobacillus* is the most abundant in the small intestine, while *Bifidobacterium* predominates in the colon (Milani et al., 2017). For example, the gut microbiome can be modified by transferring microbiota from a healthy person or by the inclusion of beneficial bacteria in a patient's diet in order to alleviate symptoms among "Irritable Bowel Syndrome (IBS)" and "Inflammatory Bowel Disease (IBD)" patients (Staudacher et al., 2017). Changes in the abundance of *Lactobacillus*, *Bifidobacterium*, and other bacteria are linked to health concerns. This does not imply those bacterial species and genera are the only health-beneficial microorganisms. Gut microbiota inhabit a specific community and ecology and produce essential molecules, such as SCFAs (Morrison & Preston, 2016), and indole derivatives (Agus et al., 2018). Supplementation of good strains of bacteria in the diet also reduces illness by replenishing the gut microbiota. In IBS, diminutive levels of bifidobacteria and lactobacilli worsen symptoms (Liu et al., 2017). Among the species, a reduced abundance of *Bifidobacterium bifidum* is a hallmark of ulcerative colitis (UC) (Duranti et al., 2016). Dysbiosis, restored with probiotics such as bifidobacteria and lactobacilli, reduces symptoms of UC through alteration of gut microbial ecology (Staudacher et al., 2017). Providing prebiotic fibers also improves health through the repair of microbial ecosystems. Most prebiotic fibers are carbohydrate polymers that resist digestion in the upper digestive tract and arrive in the colon intact, where they can feed the gut microbiota and support species growth (Holscher, 2017). Intervention in gut microbiota has been associated with health effects including enhancing specific species, preventing dysbiosis, and maintaining species diversity (Salazar et al., 2008, 2009). This tends to correspond with increased health, although understanding the bacteria's effector chemicals is still missing, as shown above. Some research suggests components of the cell wall, such as polysaccharide-A and lipoteichoic acid, might increase health.

2 Beneficial Effects of EPS in the Gastrointestinal Tract

This recent research has explored many positive effects that bacterial extracellular polysaccharides (EPS) impose on the host's gastrointestinal tract (Fig. 1). Even minor molecular structures of EPS may change or completely exclude their biological effects. This indicates that the effects rely significantly on those molecular structures (Salazar et al., 2008).

2.1 Intestinal Microbes Ferment EPS to Produce Short-Chain Fatty Acids (SCFAs)

According to Gibson et al. (2017), EPS should be "substrates that are selectively utilized by host microorganisms conferring a health benefit" as prebiotic dietary supplements. To do this, EPS must perform two critical roles. They should not be

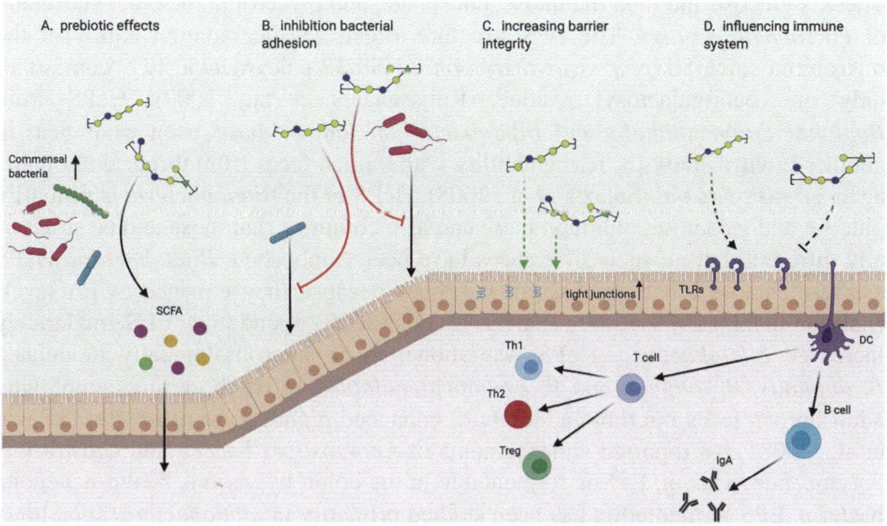

Fig. 1 An outline of the advantages that EPS provides to the digestive system. By upregulating tight junctions, EPS can (**a**) promote the development of good bacteria, (**b**) prevent bacteria from adhering to the intestinal epithelium, and (**c**) strengthen the intestinal barrier. By interacting with Toll-like receptors both directly and indirectly, EPS can also (**d**) affect the immune system. (Adopted from Oerlemans et al. (2021))

digested or absorbed in the upper gastrointestinal tract and enter the colon. Because HoPS is not broken down by human enzymes attacking fructans and β-glucans, it can enter the colon in an undigested form. EPS from the *Lactobacillus sanfranciscensis*, which is a β-2,6 linked fructan, was tested for indigestibility: stable at low pH and after heating treatments (Korakli et al., 2002). Because of its complexity, HePS may resist digestion better than HoPS, although this has not been demonstrated in vitro or in vivo (Castro-Bravo et al., 2018). To be classified as prebiotic dietary supplements, EPS should be utilized by specific gut microbiota once it reaches the colon. For some HoPS, microbiota-derived enzymes break down EPS structures into host-beneficial compounds (Korakli et al., 2002). For instance, Korakli et al. (2002) examined high molecular weight ferment ability. Studies examined HoPS generated by *Lactobacillus sanfranciscensis* by *bifidobacteria* and *lactobacilli*. This study employed 99.3% fructose and 0.7% glucose HoPS. *Bifidobacteria* such *B. adolescentis*, *B. bifidum*, *B. breve*, and *B. infantis* fermented *Lactobacillus sanfranciscensis*' EPS and formed acetate, lactate, and formate, whereas lactobacilli did not. HePS fermentability was examined by Ruijssenaars et al. (2000) using *streptococci*, *lactobacilli*, and *lactococci* (Fig. 1). HePS from *Streptococcus thermophilus* SFi12—rhamnose, galactose, and glucose—was easily destroyed by adult human microbiota. Due to their composition, EPS from *Lactococcus lactis ssp.* cremoris (rhamnose, galactose, and glucose) and *Lactobacillus sakei* (rhamnose, glucose, phosphate, and glycerol) were less accessible to degrading enzymes (Ruijssenaars et al., 2000). *Lactococcus lactis* ssp. Cremoris had two side chains; these were rhamnose and galactose, with phosphate. There is glucose and then rhamnose, phosphate, and glycerol in the other side chain of *Lactobacillus sakei*. The EPS structure affects the degradation ability of the microbiota since *Streptococcus thermophilus* SFi12's degradable EPS consists of only one beta-galactosyl residue (Ruijssenaars et al., 2000). HePS from *Bifidobacterium animalis* and *Bifidobacterium longum* have been examined, in another in vitro study for fermentability with human feces from three adult donors aged 31–40 years old (Salazar et al., 2008). HePS of the *Bifidobacteria* is primarily glucose and galactose, although these can also comprise rhamnose and be structurally different, but no such differences have been established. *Bifidobacteria* HePS are fermentable substrates for gut microbes because they enhance SCFA levels (Salazar et al., 2008, 2009). Salazar et al. (2008) found that EPS moderately increased *bifidobacteria*. HePS was shown to be microbiologically modulated *B. animalis*, *B. longum*, and *B. pseudocatenulatum*. In batch cultures employing adult human feces microbiota, all HePS enhanced *bifidobacterial species*. Salazar et al. (2008) also reported enhancements of *Anaerostipes caccae* and *Clostridium polysaccharolyticum*. EPS is fermentable in the colon by various health-beneficial bacteria. EPS fermentation has been studied primarily in *bifidobacteria* spp., however other species can break down EPS structures. Rios-Covian et al. (2016) report that Bacteroides fragilis DSMZ 2151 can grow on pure EPS from *B. animalis ssp. lactis* and contains glucose, galactose, rhamnose, and phosphate while *B. longum*

E44 contains only glucose and galactose in a minimal medium. The fermentation increased populations of bacteria and discharged SCFA as well as organic acids. EPS fermentation by specific bacteria can produce SCFAs, which feed colonocytes, maintain the intestinal barrier, and modulate host metabolism and immune responses (HAMER et al., 2008). Recently, the fermentation of *B. breve* EPS in an infant colon model with feces from healthy, full-term breast-fed infants increased the formation of SCFAs acetate, propionate, and butyrate, which may promote infant health (Püngel et al., 2020). EPS's complex role in the composition of the microbiota and bacterial metabolite production is as depicted in Fig. 2.

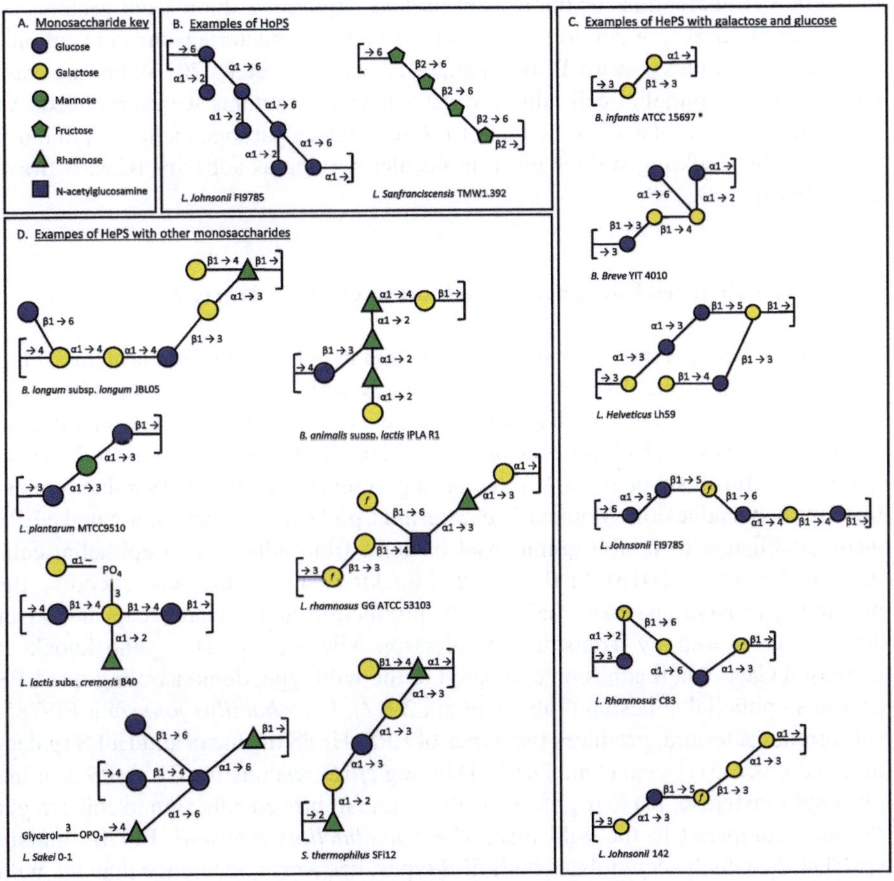

Fig. 2 HoPS and HePS structural examples. (**a**) HoPS and HePS are made up of monosaccharides. (**b**) HoPS structure examples. (**c**) Examples of HePS made of glucose and galactose. (**d**) Examples of HePS made of different monosaccharides. (Adopted from Oerlemans et al. (2021)). *Backbone is partially substituted with β-D-Glucose in the carbon C-6 of both monosaccharides

2.2 Interactions Between EPS and the Intestinal Epithelial Barrier

EPS may influence intestinal microbiota compositions and intestinal epithelial barrier functions, including pathogenic and helpful bacterial strains. Studies have explored how EPS directly impacts epithelial cell barrier integrity (Zhou et al., 2018).

2.2.1 Effect of EPS on Intestinal Epithelial Bacterial Adhesion

Quantifying the adhesion of EPS-producing bacteria in vitro can be done by labeling with 3 H nucleotides or fluorescent probes, expressing fluorescent proteins on the bacterial surface, or counting the number of adhered bacteria using conventional culturing techniques (Castro-Bravo et al., 2017; Živković et al., 2016). For the comparison of gut epithelial cell adhesion, EPS-negative mutants were produced. All these methods have demonstrated that EPS may reduce pathogen adhesion to intestinal cells by shielding surface macromolecules serving as adhesins (Castro-Bravo et al., 2018).

2.2.2 EPS Reduces Intestinal Epithelium Bacterial Adherence

Because EPS covers the cell wall of bacteria and may hide adhesion proteins, it has been hypothesized that EPS interferes negatively with intestinal epithelial bacterial adherence. Even though shielding surface proteins does not contribute to an increase in bacterial adhesion, EPS is necessary to avoid the host immune system and survive in the gut (Murofushi et al., 2015). According to several studies, EPS reduces intestinal epithelial adhesion. Mutated *L. plantarum* Lp90 bacteria that cannot make EPS were established to inhibit commensal bacteria from adhering to epithelial cells in vitro (Lee et al., 2016). In *B. longum*, knocking out cps, the gene encoding the priming glycosyltransferase, halted EPS production and lowered capsular layer density, as viewed by Transmission Electron Microscopy. This gene knockout increased Caco-2 cell adhesion compared to the wild type, demonstrating that EPS prevents epithelial adhesion (Tahoun et al., 2017). *Lactobacillus johnsonii* FI9785, a different bacterium, produces two types of EPS: HoPS α-glucan and HePS (galactose and glucose) (Dertli et al., 2013). Deleting *epsE*, responsible for HePS priming glycosyltransferase, led to α-glucan synthesis and improved adhesion to chicken gut explants compared to the wild type. The *Lactobacillus johnsonii* FI9785 mutant epsCD88N, which accumulates both EPS types, has worse adherence than the wild type (Dertli et al., 2015). Finally, in vitro experiments by Lebeer et al. (2011) indicated that an EPS-deficient mutant of *L. rhamnosus* GG adhered better to Caco-2 cells. EPS decreases the adherence of bacteria to epithelial cells. In vivo experiments demonstrated that EPS prevents bacterial adhesion to the intestinal epithelium. *B. breve* UCC2003's EPS, whose chemical structure is not known, failed to enhance mouse colonization compared with the non-EPS-producing mutant strain (Fanning et al., 2012). Other in vivo studies support the adhesion-inhibiting action of EPS to gut epithelium (Dertli et al., 2015) and depict its complex role in host bacteria survival. As will be discussed in later sections, gut bacteria require EPS to survive and avoid the host immune system.

2.2.3 EPS Modulates Intestinal Epithelial Pathogen Adhesion

The epithelial adhesion of commensal bacteria renders EPS useful for avoiding pathogenic bacteria adhesion. For example, Živković et al. (2016) reported that *Lactobacillus paracasei subsp. Paracasei* BGSJ2 EPS completely prevents *Escherichia coli* adherence to Caco-2 cells. This strain produced two HePS of unknown compositions: one was a low-molecular-weight molecule with 13,600 Da and the other had a high molecular weight of 41,200 Da. It also reduced the adhesion of *E. coli* to Caco-2 cells. A mutant *Lactobacillus paracasei subsp. Paracasei* EPS7, which lacked the low molecular weight HePS, was unable to inhibit *E.coli* adherence, indicating that just one of the EPS of BGSJ2 was accountable for the observed anti-adhesion effects Živković et al. (2016). In vivo studies confirm EPS diminishes pathogen adhesion. Supplementation with *Clostridium perfringens* colonization and persistence was reduced in 20-day-old chicks by a poultry-derived isolate, *L. johnsonii* FI9785, which produced HoPS α-glucan and HePS from galactose and glucose (Dertli et al., 2013). Fanning et al. (2012) reported that mice were infected with *C. rodentium* after being treated with *B. breve* UCC2003 and the non-EPS-producing isogenic mutant UCC2003-EPSdel. *B. breve* UCC2003 protected mice against infection, and *C. rodentium* was absent from the colon and cecum at day 14. *B. breve* UCC2003-EPSdel failed to block *C. rodentium*. Studies indicate that EPS can inhibit pathogenic bacteria from adhering to the host intestinal epithelium.

2.2.4 EPS Decoy Effect

EPS can act as a decoy receptor by binding adhesion molecules to gut-pathogenic bacteria (Lynch et al., 2018). EPS decoy capacity has been studied in various in vitro models. Early research explored the capacity of probiotic bacteria to bind to pathogens (García-Almendárez et al., 2008). In the study, 126 strains of *Lactobacillus plantarum* from raw milk cheeses were isolated and their binding capabilities to *Staphylococcus aureus*, *Listeria monocytogenes*, and *E. coli O157:H7* were investigated. Different *Lactobacillus plantarum* strains have different pathogen selectivity and binding levels. IFPL156 bound *L. monocytogenes* better than IFPL162, which bound *S. aureus* better. Various *Lactobacillus plantarum* strains create various EPS structures, which may explain these effects. Another study found that *Lactobacillus acidophilus* A4, which contains unknown EPS, reduced *E. coli* O157:H7 adherence to HT-29 cells (Kim et al., 2008). Since pretreatment of cells with supernatant did not decrease adhesion, inhibition was only observed after treatment. This indicates how pathogenic bacteria are impaired, most likely due to a decoy effect that prevents them from adhering to epithelial cell layers (Kim et al., 2008).

2.2.5 EPS Effects on Epithelial Barrier Integrity

Bacterial EPS can interfere with the host epithelial cell layer integrity. Junctions between epithelial cells compromise the intestinal barrier. Complex lipoprotein structures, like claudins and occludin act as main tight junction transmembrane proteins, adjoining adjacent proteins at the same site. Claudins are a large family (at least 24 members) of intercellular adhesion molecules where claudins 1,4,5,7,8 and 14 reduce, while claudins 2 and 16 enhance the permeability of the intestinal barrier Clauids 1, 4, 5, 7, 8, and 14 reduce, while clauids 2 and 16 enhance the permeability of the intestinal barrier (Van Itallie & Anderson, 2006). The researchers probed HePS from *Lactobacillus rhamnosus* GG, *Bifidobacterium animalis*, and *Bifidobacterium longum* for enhancing integrity. All HePS contained glucose, galactose, and rhamnose, while ratios of monosaccharides varied. The HePS of *B. animalis* contained phosphate but not *B. longum* nor *L. rhamnosus* (Ruas-Madiedo et al., 2010). In Caco-2 cells, HePS from the three strains studied decreased the cytotoxic effect of *Bacillus cereus* toxin on barrier integrity (Ruas-Madiedo et al., 2010). In another study, *Streptococcus thermophilus* MN-BM-A01 HePS, with its composition of rhamnose, glucose, galactose, and mannose, repaired the integrity of the Caco-2 cell barrier following LPS stimulation (Chen et al., 2019). The LPS-treated cells exhibited reduced transepithelial electrical resistance and increased permeability by higher dextran-FITC flux across the cell layer. Pretreatment of the cells with HePS decreased both indicators of barrier integrity, thus indicating that EPS protects against disruption. Several tight junction genes were downregulated by LPS, but *S. thermophilus* HePS restored barrier integrity in vivo (Chen et al., 2019). EPS from *Lactobacillus plantarum* NCU116 increased the TEER and tight junction gene expression of Caco-2 cells. HePS includes mannose, glucose, glucuronic acid, glucosamine, and galactosamine (Zhou et al., 2017). This HePS enhanced the expression of occludin and ZO-1 and increased the TEER of Caco-2 cells when compared to the control condition (Zhou et al., 2018). Another research study showed that the conditioned media with *Bifidobacterium infantis* EPS enhances the TEER of human intestinal cell line T84 through upregulation of claudin-4, ZO-1, and occludin expression and downregulation of claudin. IFNγ and TNFα activation caused intracellular redistribution of claudin-1 and occludin, which in turn decreased TEER. Pretreatment of T84 cells with *B. infantis* conditioned medium suppressed intracellular redistribution of claudin-1 and occludin and epithelial resistance loss (Ewaschuk et al., 2008). EPS has also been reported to maintain the intestinal barrier in vivo. Chen et al. (2019) investigated *Streptococcus thermophilus* MN-BM-A01 HePS in a barrier-disrupting mice-DSS model. EPS comprises rhamnose, glucose, galactose, and rhamnose at a molar ratio of 12.9:26.0:60.9:0.25, with a molecular weight of 4.23 × 105 Da. It was investigated in DSS-reduced gut barrier models. DSS decreases colonic epithelial tight junction expression of claudin-1, occludin, and E-cadherin. HePS of *Streptococcus thermophilus* MN-BM-A01 prevented colitis development in mice (Chen et al., 2019). In another DSS mouse model investigation, three *Bifidobacteria longum subsp. longum* strains with varying EPS status were examined. In this study, DSS lowered tight junction gene expression and MUC2, a mucus-producing gene (Yan et al., 2019). The expression was not restored by HAN4-25, a *Bifidobacteria longum subsp. longum* strain that does not generate EPS. The uncharacterized EPS-producing strain C11A10B may enhance claudin-1 and MUC2. This did not restore the mucosal barrier. Treatment with the spontaneous mutant of C11A10B, the YS108R strain, which produces a ropy EPS with a higher molecular weight, maintained or increased claudin-1, ZO-1, and MUC-2 expression and reduced DSS-induced colonic damage. The specific composition of various strains' EPS was not characterized, but this work demonstrates the unique functions of different EPS. In a DSS mouse model, *Lactobacillus plantarum* NCU116 HePS was studied. This HePS, rich in mannose, glucose, glucuronic acid, glucosamine, and galactosamine, prevented intestinal barrier disruption by reducing serum dextran-FITC levels. Claudin-1, occludin, and ZO-1 were upregulated, but claudin-2 was downregulated (Zhou et al., 2018). *Bifidobacterium infantis* was also tested in IL-10-deficient mice, which spontaneously developed colitis at 6–8 weeks. *B. infantis* conditioned medium decreased barrier permeability disruption and alleviated colitis symptoms (Ewaschuk et al., 2008). *Streptococcus thermophilus* ST10, which produces a galactose-rhamnose HePS, was given to healthy volunteers in a double-blind, placebo-controlled study. Supplementation with HePS-producing bacteria significantly improved intestinal barrier function. The lactitol/mannitol ratio decreased, showing an improvement in small intestinal barrier function. HePS supplementation reduced sucralose excretion, an indicator of intestinal permeability (Del Piano et al., 2014). All these studies suggest that certain EPS enhances the integrity of the barrier.

3 Application of EPS-Producing LAB in Food Products Prebiotics

3.1 Yoghurt

LAB-produced EPS thickens and stabilizes fermented dairy products. Syneresis can arise from casein micelle destabilization during yogurt fermentation by lowering pH. EPS generation in situ improves rheological qualities compared to adding it as a component (Berthold-Pluta et al., 2019; Korcz & Varga, 2021). *Lactobacillus delbrueckii subsp. bulgaricus* and *Streptococcus thermophilus*, the starting cultures for yogurt, produce exopolysaccharides 60–150 mg/L and 30–890 mg/L, respectively (Daba et al., 2021; Zarour et al., 2017). Novel EPS producing starting or adjunct cultures for yogurt production have been recently investigated (Daba et al., 2021; Tiwari et al., 2021). Han et al. (2016) evaluated 19 high-EPS *S. thermophilus* strains from traditional Chinese fermented milk products used to make yogurt. A starter with high EPS production (SH-1) had lower syneresis, improved texture, and improved sensory evaluation compared to commercial yogurt starting cultures (Han et al., 2016). Low-fat fermented dairy products can be improved by using starters with high EPS production since milk fat impacts flavor, texture, and rheology (Patel & Prajapati, 2013). For example, EPS-producing Limosi *Lactobacillus mucosae* (formerly *Lactobacillus mucosae*) DPC 6426 as a co-culture in low-fat yogurt manufacture decreased syneresis and enhanced functional properties (London et al., 2015).

3.2 Kefir

Eastern European kefir is a self-carbonated, weakly alcoholic fermented milk beverage. Kefir grains are proteins and polysaccharides with a symbiotic association of homofermentative and heterofermentative lactic acid bacteria, acetic acid bacteria, and yeasts (Patel & Prajapati, 2013). Kefiran, exopolysaccharides produced during fermentation, controls viscosity. *Lactobacillus kefiranofaciens* produces kefiran, a branched water-soluble gluco-galactan with equal parts glucose and galactose (Daba et al., 2021; Patel & Prajapati, 2013). *Lactiplantibacillus plantarum*, *Lacticaseibacillus paracasei*, *Lactobacillus helveticus*, *Lactococcus lactis subsp. Lactis*, and *Leuconostoc mesenteroides* have been isolated from kefir grains and found to produce EPS (Hamet et al., 2015; Wang et al., 2021; You et al., 2020). In the last 10 years, several mesophilic LAB strains were investigated for their ability to ferment milk and produce EPS to improve the rheology of fermented milk-based foodstuffs (Hamet et al., 2015; Wang et al., 2021). Kefir-fermented non-dairy products also attract much attention. Nondairy kefir grains can have a wider range of microbial diversity than traditional kefir grains, mainly because such kefir grains are produced with a wide range of sugars and subtracts involved in the fermentation process. Many authors reported 45 bacteria and 23 yeasts (Egea et al., 2022). However, kefir fermentation is complex so the relative composition of bacteria and yeasts can vary and this is difficult to control for the production of homogeneous commercial products (Egea et al., 2022).

3.3 Bakery and EPS Use

LAB-made EPS possesses a thickening effect and could substitute polysaccharides used as additives in bakeries in sourdough. EPS-forming LAB in fermented sourdough enhances the techno-functional qualities of baked products (Păcularu-Burada et al., 2020; Sahin et al., 2019; Tinzl-Malang et al., 2020). In situ, EPS manufacturing improves dough handling and stability and bread texture and volume (Tang et al., 2018; Tinzl-Malang et al., 2020). The use of specific EPS-producing LAB strains in sourdoughs has been examined as a novel way to compensate for quality losses in reduced sugar products or to substitute baking fat (Sahin et al., 2019; Tang et al., 2018). Some studies have demonstrated that EPS-producing LAB preserves frozen dough products against cold and enhances quality (Zhang et al., 2021). Polysaccharides are now popular due to the growth of gluten-free products. Consumers and patients of celiac disease require enhanced gluten-free products from bakeries (Akobeng et al., 2020; Daba et al., 2021; Korcz & Varga, 2021). LAB-prepared EPS may interact with water and form a high-quality network with other dough components to improve the structure and volume of gluten-free or gluten-containing bread. This results in softer bread crumbs and a longer shelf life (Lynch et al., 2018; Xu et al., 2020). Recently, Zheng et al. (2020) used Limosi *Lactobacillus reuteri* to ferment in situ and produce EPS. It produced fructans and glucans for excellent gluten-free bread.

3.4 Meat Industry and EPS Application

Although EPS-producing LABs are widely used in dairy products, they are recently under exploration in meat processing (Loeffler et al., 2020). The origins of meat manufacturing and lactic acid bacteria have produced a great extent of the cultural heritage of world food. Through the formation of lactic or acetic acids, humans can regulate the food's safety, texture, color, etc., when they are taken raw (Korcz & Varga, 2021). Meat products contain protein, iron, zinc, vitamins, and many others, but high-fat meat might lead to diabetes or cardiovascular disease. The meat business needs to formulate new low-fat products to meet customer demand. This poses various challenges for the meat business since low-fat processed beef products might lack technological and sensory qualities (Korcz & Varga, 2021; Kumar, 2021; Loeffler et al., 2020). Hydrocolloids are commonly used food additives. They improve the moisture retention and gelation properties of meat proteins and the texture of low-fat meat products, but consumers refuse food preservatives on the label (Asioli et al., 2017). In situ LAB EPS production during the meat processing could, therefore, be an attractive option. Relatively few research have explored the in situ EPS synthesis by LAB in beef products. Sucuk and salami raw fermented sausages as well as cooked ham were assayed for in situ EPS production using LAB (Loeffler et al., 2020). For raw fermented sausages, conditions in fermentation proved to be more beneficial than those related to ham processing, which decreased the level of EPS formation (Loeffler et al., 2020). Hilbig et al. (Hilbig et al., 2019) showed that the selected LAB strains (*L. plantarum* and *L. curvatus*) were able to produce enough EPS during sausage fermentation to decrease the high-fat content of fermented raw sausages. EPS-producing LAB cultures increased the hardness, reduced stickiness, and made a Turkish fermented sausage harsher (Dertli et al., 2016).

4 Conclusion

EPS from lactic acid bacteria are generally recognized as safe immunomodulators to stimulate the mucosal immune system because of their "generally regarded as safe" classification. Because EPSs are nontoxic, they have various potential applications in food, medicine, and even pharmaceuticals. However, EPS has not been widely utilized for its immunomodulatory and other beneficial properties. Because current research shows that the beneficial effects of EPS present on the bacterial cell walls are highly dependent on their molecular structures, further understanding of the specific structure–function relationships of EPS of *lactobacilli* and other strains may aid in the formulation of specific EPS structures and propel their application as novel food additives or as therapeutics to promote intestinal health. Various bacterial strains can have somewhat varied EPS structures, which can have varying biological effects. It is sometimes difficult to ascertain the precise structure–function correlations because of the lack of information regarding the particular EPS structures that are employed in research. In presenting the biological activities of EPS, particular care should be given to include comprehensive information on the individual EPS structure or bacterial strain employed. However, the use of dietary EPS for particular individuals with varying health conditions is not without its difficulties, though. Our knowledge of EPS structures is still lacking, despite the fact that several research have clarified how variations in EPS structures affect health advantages. It is critically necessary to conduct further study on the chemical structures that provide certain health advantages. The small sample size of accessible EPS is one of the limitations that scientists now confront. Because of the limited quantities of EPS that were separated during extractions, alternative techniques for acquiring EPS were required. EPS might become more readily available as a result of synthesis. Modification of the structures of EPS or chemical production to produce specific side chains or single monosaccharide modifications that clarify or enhance certain EPS activities can also be very useful. It will be necessary to know more about the beneficial effects of EPS on the host before one could introduce the former into the health sector to promote certain health advantages.

References

Agus, A., Planchais, J., & Sokol, H. (2018). Gut microbiota regulation of tryptophan metabolism in health and disease. *Cell Host & Microbe, 23*(6), 716–724. https://doi.org/10.1016/j.chom.2018.05.003

Akobeng, A. K., Singh, P., Kumar, M., & Al Khodor, S. (2020). Role of the gut microbiota in the pathogenesis of coeliac disease and potential therapeutic implications. *European Journal of Nutrition, 59*(8), 3369–3390. https://doi.org/10.1007/s00394-020-02324-y

Asioli, D., Aschemann-Witzel, J., Caputo, V., Vecchio, R., Annunziata, A., Næs, T., & Varela, P. (2017). Making sense of the "clean label" trends: A review of consumer food choice behavior and discussion of industry implications. *Food Research International, 99*, 58–71. https://doi.org/10.1016/j.foodres.2017.07.022

Berthold-Pluta, A. M., Pluta, A. S., Garbowska, M., & Stasiak-Różańska, L. (2019). Exopolysaccharide-producing lactic acid bacteria—Health-promoting properties and application in the dairy industry. *Postępy Mikrobiologii – Advancements of Microbiology, 58*(2), 191–204. https://doi.org/10.21307/PM-2019.58.2.191

Castro-Bravo, N., Hidalgo-Cantabrana, C., Rodriguez-Carvajal, M. A., Ruas-Madiedo, P., & Margolles, A. (2017). Gene replacement and fluorescent labeling to study the functional role of exopolysaccharides in Bifidobacterium animalis subsp. lactis. *Frontiers in Microbiology, 8*, 1405. https://doi.org/10.3389/fmicb.2017.01405

Castro-Bravo, N., Wells, J. M., Margolles, A., & Ruas-Madiedo, P. (2018). Interactions of surface exopolysaccharides from Bifidobacterium and Lactobacillus within the intestinal environment. *Frontiers in Microbiology, 9*, 2426. https://doi.org/10.3389/fmicb.2018.02426

Chen, Y., Zhang, M., & Ren, F. (2019). A role of exopolysaccharide produced by Streptococcus thermophilus in the intestinal inflammation and mucosal barrier in Caco-2 monolayer and dextran sulphate sodium-induced experimental murine colitis. *Molecules, 24*(3), 513. https://doi.org/10.3390/molecules24030513

Daba, G. M., Elnahas, M. O., & Elkhateeb, W. A. (2021). Contributions of exopolysaccharides from lactic acid bacteria as biotechnological tools in food, pharmaceutical, and medical applications. *International Journal of Biological Macromolecules, 173*, 79–89. https://doi.org/10.1016/j.ijbiomac.2021.01.110

Del Piano, M., Balzarini, M., Carmagnola, S., Pagliarulo, M., Tari, R., Nicola, S., Deidda, F., & Pane, M. (2014). Assessment of the capability of a gelling complex made of tara gum and the exopolysaccharides produced by the microorganism Streptococcus thermophilus ST10 to prospectively restore the gut physiological barrier. *Journal of Clinical Gastroenterology, 48*(Supplement 1), S56–S61. https://doi.org/10.1097/MCG.0000000000000254

Dertli, E., Colquhoun, I. J., Gunning, A. P., Bongaerts, R. J., Le Gall, G., Bonev, B. B., Mayer, M. J., & Narbad, A. (2013). Structure and biosynthesis of two exopolysaccharides produced by Lactobacillus johnsonii FI9785. *Journal of Biological Chemistry, 288*(44), 31938–31951. https://doi.org/10.1074/jbc.M113.507418

Dertli, E., Mayer, M. J., & Narbad, A. (2015). Impact of the exopolysaccharide layer on biofilms, adhesion and resistance to stress in Lactobacillus johnsonii FI9785. *BMC Microbiology, 15*(1), 8. https://doi.org/10.1186/s12866-015-0347-2

Dertli, E., Toker, O. S., Durak, M. Z., Yilmaz, M. T., Tatlısu, N. B., Sagdic, O., & Cankurt, H. (2016). Development of a fermented ice-cream as influenced by in situ exopolysaccharide production: Rheological, molecular, microstructural and sensory characterization. *Carbohydrate Polymers, 136*, 427–440. https://doi.org/10.1016/j.carbpol.2015.08.047

Duranti, S., Gaiani, F., Mancabelli, L., Milani, C., Grandi, A., Bolchi, A., Santoni, A., Lugli, G. A., Ferrario, C., Mangifesta, M., Viappiani, A., Bertoni, S., Vivo, V., Serafini, F., Barbaro, M. R., Fugazza, A., Barbara, G., Gioiosa, L., Palanza, P., et al. (2016). Elucidating the gut microbiome of ulcerative colitis: Bifidobacteria as novel microbial biomarkers. *FEMS Microbiology Ecology, 92*(12), fiw191. https://doi.org/10.1093/femsec/fiw191

Egea, M. B., dos Santos, D. C., de Oliveira Filho, J. G., Ores, J. D. C., Takeuchi, K. P., & Lemes, A. C. (2022). A review of nondairy kefir products: Their characteristics and potential human health benefits. *Critical Reviews in Food Science and Nutrition, 62*(6), 1536–1552. https://doi.org/10.1080/10408398.2020.1844140

Ewaschuk, J. B., Diaz, H., Meddings, L., Diederichs, B., Dmytrash, A., Backer, J., Looijer-van Langen, M., & Madsen, K. L. (2008). Secreted bioactive factors from *Bifidobacterium infantis* enhance epithelial cell barrier function. *American Journal of Physiology-Gastrointestinal and Liver Physiology, 295*(5), G1025–G1034. https://doi.org/10.1152/ajpgi.90227.2008

Fanning, S., Hall, L. J., Cronin, M., Zomer, A., MacSharry, J., Goulding, D., O'Connell Motherway, M., Shanahan, F., Nally, K., Dougan, G., & van Sinderen, D. (2012). Bifidobacterial surface-exopolysaccharide facilitates commensal-host interaction through immune modulation and pathogen protection. *Proceedings of the National Academy of Sciences, 109*(6), 2108–2113. https://doi.org/10.1073/pnas.1115621109

García-Almendárez, B. E., Cann, I. K. O., Martin, S. E., Guerrero-Legarreta, I., & Regalado, C. (2008). Effect of Lactococcus lactis UQ2 and its bacteriocin on Listeria monocytogenes biofilms. *Food Control, 19*(7), 670–680. https://doi.org/10.1016/j.foodcont.2007.07.015

Gibson, G. R., Hutkins, R., Sanders, M. E., Prescott, S. L., Reimer, R. A., Salminen, S. J., Scott, K., Stanton, C., Swanson, K. S., Cani, P. D., Verbeke, K., & Reid, G. (2017). Expert consensus document: The International Scientific Association for Probiotics and Prebiotics (ISAPP) consensus statement on the definition and scope of prebiotics. *Nature Reviews Gastroenterology & Hepatology, 14*(8), 491–502. https://doi.org/10.1038/nrgastro.2017.75

Hamer, H. M., Jonkers, D., Venema, K., Vanhoutvin, S., Troost, F. J., & Brummer, R.-J. (2008). Review article: The role of butyrate on colonic function. *Alimentary Pharmacology & Therapeutics, 27*(2), 104–119. https://doi.org/10.1111/j.1365-2036.2007.03562.x

Hamet, M. F., Piermaria, J. A., & Abraham, A. G. (2015). Selection of EPS-producing Lactobacillus strains isolated from kefir grains and rheological characterization of the fermented milks. *LWT – Food Science and Technology, 63*(1), 129–135. https://doi.org/10.1016/j.lwt.2015.03.097

Han, X., Yang, Z., Jing, X., Yu, P., Zhang, Y., Yi, H., & Zhang, L. (2016). Improvement of the texture of yogurt by use of exopolysaccharide producing lactic acid bacteria. *BioMed Research International, 2016*, 1–6. https://doi.org/10.1155/2016/7945675

Harapanahalli, A. K., Younes, J. A., Allan, E., van der Mei, H. C., & Busscher, H. J. (2015). Chemical signals and mechanosensing in bacterial responses to their environment. *PLoS Pathogens, 11*(8), e1005057. https://doi.org/10.1371/journal.ppat.1005057

Hilbig, J., Loeffler, M., Herrmann, K., & Weiss, J. (2019). The influence of exopolysaccharide-producing lactic acid bacteria on reconstructed ham. *International Journal of Food Science & Technology, 54*(9), 2763–2769. https://doi.org/10.1111/ijfs.14188

Holscher, H. D. (2017). Dietary fiber and prebiotics and the gastrointestinal microbiota. *Gut Microbes, 8*(2), 172–184. https://doi.org/10.1080/19490976.2017.1290756

Jurášková, D., Ribeiro, S. C., & Silva, C. C. G. (2022). Exopolysaccharides produced by lactic acid bacteria: From biosynthesis to health-promoting properties. *Foods, 11*(2), 156. https://doi.org/10.3390/foods11020156

Kim, Y., Kim, S.-H., Whang, K.-Y., Kim, Y.-J., & Oh, S. (2008). Inhibition of Escherichia coli O157:H7 attachment by interactions between lactic acid bacteria and intestinal epithelial cells. *Journal of Microbiology and Biotechnology, 18*(7), 1278–1285.

Korakli, M., Ganzle, M. G., & Vogel, R. F. (2002). Metabolism by bifidobacteria and lactic acid bacteria of polysaccharides from wheat and rye, and exopolysaccharides produced by Lactobacillus sanfranciscensis. *Journal of Applied Microbiology, 92*(5), 958–965. https://doi.org/10.1046/j.1365-2672.2002.01607.x

Korcz, E., & Varga, L. (2021). Exopolysaccharides from lactic acid bacteria: Techno-functional application in the food industry. *Trends in Food Science & Technology, 110*, 375–384. https://doi.org/10.1016/j.tifs.2021.02.014

Kumar, Y. (2021). Development of low-fat/reduced-fat processed meat products using fat replacers and analogues. *Food Reviews International, 37*(3), 296–312. https://doi.org/10.1080/87559129.2019.1704001

Lebeer, S., Claes, I. J. J., Verhoeven, T. L. A., Vanderleyden, J., & De Keersmaecker, S. C. J. (2011). Exopolysaccharides of *Lactobacillus rhamnosus* GG form a protective shield against innate immune factors in the intestine. *Microbial Biotechnology, 4*(3), 368–374. https://doi.org/10.1111/j.1751-7915.2010.00199.x

Lee, I.-C., Caggianiello, G., van Swam, I. I., Taverne, N., Meijerink, M., Bron, P. A., Spano, G., & Kleerebezem, M. (2016). Strain-specific features of extracellular polysaccharides and their impact on Lactobacillus plantarum-host interactions. *Applied and Environmental Microbiology, 82*(13), 3959–3970. https://doi.org/10.1128/AEM.00306-16

Liu, H.-N., Wu, H., Chen, Y.-Z., Chen, Y.-J., Shen, X.-Z., & Liu, T.-T. (2017). Altered molecular signature of intestinal microbiota in irritable bowel syndrome patients compared with healthy controls: A systematic review and meta-analysis. *Digestive and Liver Disease, 49*(4), 331–337. https://doi.org/10.1016/j.dld.2017.01.142

Loeffler, M., Hilbig, J., Velasco, L., & Weiss, J. (2020). Usage of *in situ* exopolysaccharide-forming lactic acid bacteria in food production: Meat products—A new field of application? *Comprehensive Reviews in Food Science and Food Safety, 19*(6), 2932–2954. https://doi.org/10.1111/1541-4337.12615

London, L. E. E., Chaurin, V., Auty, M. A. E., Fenelon, M. A., Fitzgerald, G. F., Ross, R. P., & Stanton, C. (2015). Use of Lactobacillus mucosae DPC 6426, an exopolysaccharide-producing strain, positively influences the techno-functional properties of yoghurt. *International Dairy Journal, 40*, 33–38. https://doi.org/10.1016/j.idairyj.2014.08.011

Lynch, K. M., Coffey, A., & Arendt, E. K. (2018). Exopolysaccharide producing lactic acid bacteria: Their techno-functional role and potential application in gluten-free bread products. *Food Research International, 110*, 52–61. https://doi.org/10.1016/j.foodres.2017.03.012

Marchesi, J. R., & Ravel, J. (2015). The vocabulary of microbiome research: A proposal. *Microbiome, 3*(1), 31. https://doi.org/10.1186/s40168-015-0094-5

Milani, C., Duranti, S., Bottacini, F., Casey, E., Turroni, F., Mahony, J., Belzer, C., Delgado Palacio, S., Arboleya Montes, S., Mancabelli, L., Lugli, G. A., Rodriguez, J. M., Bode, L., de Vos, W., Gueimonde, M., Margolles, A., van Sinderen, D., & Ventura, M. (2017). The first microbial colonizers of the human gut: Composition, activities, and health implications of the infant gut microbiota. *Microbiology and Molecular Biology Reviews, 81*(4), e00036-17. https://doi.org/10.1128/MMBR.00036-17

Morrison, D. J., & Preston, T. (2016). Formation of short chain fatty acids by the gut microbiota and their impact on human metabolism. *Gut Microbes, 7*(3), 189–200. https://doi.org/10.1080/19490976.2015.1134082

Murofushi, Y., Villena, J., Morie, K., Kanmani, P., Tohno, M., Shimazu, T., Aso, H., Suda, Y., Hashiguchi, K., Saito, T., & Kitazawa, H. (2015). The toll-like receptor family protein RP105/MD1 complex is involved in the immunoregulatory effect of exopolysaccharides from Lactobacillus plantarum N14. *Molecular Immunology, 64*(1), 63–75. https://doi.org/10.1016/j.molimm.2014.10.027

Oerlemans, M. M. P., Akkerman, R., Ferrari, M., Walvoort, M. T. C., & de Vos, P. (2021). Benefits of bacteria-derived exopolysaccharides on gastrointestinal microbiota, immunity and health. *Journal of Functional Foods, 76*, 104289. https://doi.org/10.1016/j.jff.2020.104289

Oleksy, M., & Klewicka, E. (2018). Exopolysaccharides produced by *Lactobacillus* sp.: Biosynthesis and applications. *Critical Reviews in Food Science and Nutrition, 58*, 450–462. https://doi.org/10.1080/10408398.2016.1187112

Păcularu-Burada, B., Georgescu, L. A., & Bahrim, G.-E. (2020). Current approaches in sourdough production with valuable characteristics for technological and functional applications. *The Annals of the University Dunarea de Jos of Galati Fascicle VI – Food Technology, 44*(1), 132–148. https://doi.org/10.35219/foodtechnology.2020.1.08

Patel, A., & Prajapati, J. B. (2013). Food and health applications of exopolysaccharides produced by lactic acid bacteria. *Advances in Dairy Research, 1*(2), 107. https://doi.org/10.4172/2329-888X.1000107

Poole, J., Day, C. J., von Itzstein, M., Paton, J. C., & Jennings, M. P. (2018). Glycointeractions in bacterial pathogenesis. *Nature Reviews Microbiology, 16*(7), 440–452. https://doi.org/10.1038/s41579-018-0007-2

Proctor, L. M., Creasy, H. H., Fettweis, J. M., Lloyd-Price, J., Mahurkar, A., Zhou, W., Buck, G. A., Snyder, M. P., Strauss, J. F., Weinstock, G. M., White, O., & Huttenhower, C. (2019). The integrative human microbiome project. *Nature, 569*(7758), 641–648. https://doi.org/10.1038/s41586-019-1238-8

Püngel, D., Treveil, A., Dalby, M. J., Caim, S., Colquhoun, I. J., Booth, C., Ketskemety, J., Korcsmaros, T., van Sinderen, D., Lawson, M. A., & Hall, L. J. (2020). Bifidobacterium breve UCC2003 exopolysaccharide modulates the early life microbiota by acting as a potential dietary substrate. *Nutrients, 12*(4), 948. https://doi.org/10.3390/nu12040948

Rios-Covian, D., Cuesta, I., Alvarez-Buylla, J. R., Ruas-Madiedo, P., Gueimonde, M., & de los Reyes-Gavilán, C. G. (2016). Bacteroides fragilis metabolises exopolysaccharides produced by bifidobacteria. *BMC Microbiology, 16*(1), 150. https://doi.org/10.1186/s12866-016-0773-9

Rojo, D., Méndez-García, C., Raczkowska, B. A., Bargiela, R., Moya, A., Ferrer, M., & Barbas, C. (2017). Exploring the human microbiome from multiple perspectives: Factors altering its composition and function. *FEMS Microbiology Reviews, 41*(4), 453–478. https://doi.org/10.1093/femsre/fuw046

Rossi, O., Khan, M. T., Schwarzer, M., Hudcovic, T., Srutkova, D., Duncan, S. H., Stolte, E. H., Kozakova, H., Flint, H. J., Samsom, J. N., Harmsen, H. J. M., & Wells, J. M. (2015). Faecalibacterium prausnitzii strain HTF-F and its extracellular polymeric matrix attenuate clinical parameters in DSS-induced colitis. *PLoS One, 10*(4), e0123013. https://doi.org/10.1371/journal.pone.0123013

Ruas-Madiedo, P., Medrano, M., Salazar, N., De Los Reyes-Gavilán, C. G., Pérez, P. F., & Abraham, A. G. (2010). Exopolysaccharides produced by Lactobacillus and Bifidobacterium strains abrogate in vitro the cytotoxic effect of bacterial toxins on eukaryotic cells. *Journal of Applied Microbiology, 109*(6), 2079–2086. https://doi.org/10.1111/j.1365-2672.2010.04839.x

Ruijssenaars, H. J., Stingele, F., & Hartmans, S. (2000). Biodegradability of food-associated extracellular polysaccharides. *Current Microbiology, 40*(3), 194–199. https://doi.org/10.1007/s002849910039

Sahin, A. W., Zannini, E., Coffey, A., & Arendt, E. K. (2019). Sugar reduction in bakery products: Current strategies and sourdough technology as a potential novel approach. *Food Research International, 126*, 108583. https://doi.org/10.1016/j.foodres.2019.108583

Salazar, N., Gueimonde, M., Hernández-Barranco, A. M., Ruas-Madiedo, P., & de los Reyes-Gavilán, C. G. (2008). Exopolysaccharides produced by intestinal *Bifidobacterium* strains act as fermentable substrates for human intestinal bacteria. *Applied and Environmental Microbiology, 74*(15), 4737–4745. https://doi.org/10.1128/AEM.00325-08

Salazar, N., Ruas-Madiedo, P., Kolida, S., Collins, M., Rastall, R., Gibson, G., & de los Reyes-Gavilán, C. G. (2009). Exopolysaccharides produced by Bifidobacterium longum IPLA E44 and Bifidobacterium animalis subsp. lactis IPLA R1 modify the composition and metabolic activity of human faecal microbiota in pH-controlled batch cultures. *International Journal of Food Microbiology, 135*(3), 260–267. https://doi.org/10.1016/j.ijfoodmicro.2009.08.017

Sánchez, B., Delgado, S., Blanco-Míguez, A., Lourenço, A., Gueimonde, M., & Margolles, A. (2017). Probiotics, gut microbiota, and their influence on host health and disease. *Molecular Nutrition & Food Research, 61*(1), 1600240. https://doi.org/10.1002/mnfr.201600240

Staudacher, H. M., Lomer, M. C. E., Farquharson, F. M., Louis, P., Fava, F., Franciosi, E., Scholz, M., Tuohy, K. M., Lindsay, J. O., Irving, P. M., & Whelan, K. (2017). A diet low in FODMAPs reduces symptoms in patients with irritable bowel syndrome and a probiotic restores Bifidobacterium species: A randomized controlled trial. *Gastroenterology, 153*(4), 936–947. https://doi.org/10.1053/j.gastro.2017.06.010

Tahoun, A., Masutani, H., El-Sharkawy, H., Gillespie, T., Honda, R. P., Kuwata, K., Inagaki, M., Yabe, T., Nomura, I., & Suzuki, T. (2017). Capsular polysaccharide inhibits adhesion of Bifidobacterium longum 105-A to enterocyte-like Caco-2 cells and phagocytosis by macrophages. *Gut Pathogens, 9*(1), 27. https://doi.org/10.1186/s13099-017-0177-x

Tang, X., Liu, R., Huang, W., Zhang, B., Wu, Y., Zhuang, J., Omedi, J. O., Wang, F., & Zheng, J. (2018). Impact of in situ formed exopolysaccharides on dough performance and quality of Chinese steamed bread. *LWT, 96*, 519–525. https://doi.org/10.1016/j.lwt.2018.04.039

Tinzl-Malang, S. K., Grattepanche, F., Rast, P., Fischer, P., Sych, J., & Lacroix, C. (2020). Purified exopolysaccharides from Weissella confusa 11GU-1 and Propionibacterium freudenreichii JS15 act synergistically on bread structure to prevent staling. *LWT, 127*, 109375. https://doi.org/10.1016/j.lwt.2020.109375

Tiwari, S., Kavitake, D., Devi, P. B., & Halady Shetty, P. (2021). Bacterial exopolysaccharides for improvement of technological, functional and rheological properties of yoghurt. *International Journal of Biological Macromolecules, 183*, 1585–1595. https://doi.org/10.1016/j.ijbiomac.2021.05.140

Van Itallie, C. M., & Anderson, J. M. (2006). Claudins and epithelial paracellular transport. *Annual Review of Physiology, 68*(1), 403–429. https://doi.org/10.1146/annurev.physiol.68.040104.131404

Verma, R., Lee, C., Jeun, E.-J., Yi, J., Kim, K. S., Ghosh, A., Byun, S., Lee, C.-G., Kang, H.-J., Kim, G.-C., Jun, C.-D., Jan, G., Suh, C.-H., Jung, J.-Y., Sprent, J., Rudra, D., De Castro, C., Molinaro, A., Surh, C. D., & Im, S.-H. (2018). Cell surface polysaccharides of *Bifidobacterium bifidum* induce the generation of $Foxp3^+$ regulatory T cells. *Science Immunology, 3*(28), eaat6975. https://doi.org/10.1126/sciimmunol.aat6975

Wang, H., Sun, X., Song, X., & Guo, M. (2021). Effects of kefir grains from different origins on proteolysis and volatile profile of goat milk kefir. *Food Chemistry, 339*, 128099. https://doi.org/10.1016/j.foodchem.2020.128099

Welman, A. D., & Maddox, I. S. (2003). Exopolysaccharides from lactic acid bacteria: Perspectives and challenges. *Trends in Biotechnology, 21*(6), 269–274. https://doi.org/10.1016/S0167-7799(03)00107-0

Xu, D., Hu, Y., Wu, F., Jin, Y., Xu, X., & Gänzle, M. G. (2020). Comparison of the functionality of exopolysaccharides produced by sourdough lactic acid bacteria in bread and steamed bread. *Journal of Agricultural and Food Chemistry, 68*(33), 8907–8914. https://doi.org/10.1021/acs.jafc.0c02703

Yan, S., Yang, B., Zhao, J., Zhao, J., Stanton, C., Ross, R. P., Zhang, H., & Chen, W. (2019). A ropy exopolysaccharide producing strain *Bifidobacterium longum* subsp. *longum* YS108R alleviates DSS-induced colitis by maintenance of the mucosal barrier and gut microbiota modulation. *Food & Function, 10*(3), 1595–1608. https://doi.org/10.1039/C9FO00014C

You, X., Yang, L., Zhao, X., Ma, K., Chen, X., Zhang, C., Wang, G., Dong, M., Rui, X., Zhang, Q., & Li, W. (2020). Isolation, purification, characterization and immunostimulatory activity of an exopolysaccharide produced by Lactobacillus pentosus LZ-R-17 isolated from Tibetan kefir. *International Journal of Biological Macromolecules, 158*, 408–419. https://doi.org/10.1016/j.ijbiomac.2020.05.027

Zarour, K., Vieco, N., Pérez-Ramos, A., Nácher-Vázquez, M., Mohedano, M. L., & López, P. (2017). Food ingredients synthesized by lactic acid bacteria. In *Microbial production of food ingredients and additives* (pp. 89–124). Elsevier. https://doi.org/10.1016/B978-0-12-811520-6.00004-0

Zhang, B., Omedi, J. O., Zheng, J., Huang, W., Jia, C., Zhou, L., Zou, Q., Li, N., & Gao, T. (2021). Exopolysaccharides in sourdough fermented by Weissella confusa QS813 protected protein matrix and quality of frozen gluten-red bean dough during freeze-thaw cycles. *Food Bioscience, 43*, 101180. https://doi.org/10.1016/j.fbio.2021.101180

Zheng, J., Wittouck, S., Salvetti, E., Franz, C. M. A. P., Harris, H. M. B., Mattarelli, P., O'Toole, P. W., Pot, B., Vandamme, P., Walter, J., Watanabe, K., Wuyts, S., Felis, G. E., Gänzle, M. G., & Lebeer, S. (2020). A taxonomic note on the genus Lactobacillus: Description of 23 novel genera, emended description of the genus Lactobacillus Beijerinck 1901, and union of Lactobacillaceae and Leuconostocaceae. *International Journal of Systematic and Evolutionary Microbiology, 70*(4), 2782–2858. https://doi.org/10.1099/ijsem.0.004107

Zhou, X., Hong, T., Yu, Q., Nie, S., Gong, D., Xiong, T., & Xie, M. (2017). Exopolysaccharides from Lactobacillus plantarum NCU116 induce c-Jun dependent Fas/Fasl-mediated apoptosis via TLR2 in mouse intestinal epithelial cancer cells. *Scientific Reports, 7*(1), 14247. https://doi.org/10.1038/s41598-017-14178-2

Zhou, X., Qi, W., Hong, T., Xiong, T., Gong, D., Xie, M., & Nie, S. (2018). Exopolysaccharides from *Lactobacillus plantarum* NCU116 regulate intestinal barrier function via STAT3 signaling pathway. *Journal of Agricultural and Food Chemistry, 66*(37), 9719–9727. https://doi.org/10.1021/acs.jafc.8b03340

Živković, M., Miljković, M. S., Ruas-Madiedo, P., Markelić, M. B., Veljović, K., Tolinački, M., Soković, S., Korać, A., & Golić, N. (2016). EPS-SJ exopolisaccharide produced by the strain Lactobacillus paracasei subsp. paracasei BGSJ2-8 is involved in adhesion to epithelial intestinal cells and decrease on E. coli association to Caco-2 cells. *Frontiers in Microbiology, 7*, 286. https://doi.org/10.3389/fmicb.2016.00286

Exopolysaccharides as Prebiotics: Modulating Gut Microbiota and Host Metabolism

Sakshi Santosh Marathe

1 Introduction

The human gut microbiota comprises a large number of different microorganisms, which consists of bacteria, fungi, and viruses among other microorganisms. Understanding how an individual's gut microbiota can affect his well-being is a crucial part of implementing new and innovative approaches to getting a healthy microbiome. The gut microbiota is dependent on several factors, including prebiotics, that are considered non-digestive foods for beneficial bacteria within the gut. There are many different types of prebiotics with some being exopolysaccharides and have emerged as a novel group of biological compounds with important health benefits.

Exopolysaccharides are classified as high molecular weight polysaccharides that are produced by cells of microbial origins, which can include yeast and all forms of bacteria during various metabolic processes. As much as very little information regarding the facilities that promote the development of these polysaccharides remains, it's clear that these compounds encourage the cultivation of beneficial microorganisms in the gut. Because of the various means of controlling gut microbiota, exopolysaccharides have been associated with health benefits such as better digestion, enhanced immune response, and even improvement of mental health.

The chapter examines current literature on exopolysaccharides in a very detailed manner by reviewing their effects on gut microbiota modulating process and the mechanisms of how the structure and biosynthesis of exopolysaccharides promote prebiotic activities.

S. S. Marathe (✉)
Sathaye College, Mumbai, India

T. Sarkar et al. (eds.), *Extracellular Polysaccharides*,
https://doi.org/10.1007/978-3-031-94472-7_10

2 Exopolysaccharides: Definition, Classification, and Biosynthesis

EPS or exopolysaccharides are microorganisms secreted complex carbohydrates with high molecular weight produced by a number of microorganisms including bacteria, yeast, and fungi. These naturally occurring biopolymers are synthesized and enzymes secreted to the surrounding media where they provide various functions including biofilm production, environmental stress shielding, and host immune system interaction.

EPS classification depends on the composition and there are several fundamental typologies: homopolysaccharides and heteropolysaccharides. Homopolysaccharides are made of only a monocastrate monosaccharide, which comprise of glucans, fructans, and galactans. Common examples of homosaccharides are dextran, levan and pullulan. Heteropolysaccharides include xanthan, gellan, dextran, and other monosaccharides. Examples of heteropolysaccharides include xanthan gum, gellan gum, and succinoglycan.

Some of the factors that have been found to influence the production of EPS include specific carbon sources, nitrogen sources, and environmental factors. The biosynthesis of EPS begins with the reaction of precursor molecules involving a number of enzyme-catalyzed reactions, which then leads to the production of the final EPS end product. The biosynthetic pathways usually begin with the precursors m synthetic.

3 Chemical Composition of EPS

Exopolysaccharides are biopolymers produced exogenously by microorganisms such as bacteria, fungi, and algae. Their chemical backbone on which the sugar entities are incorporated into the chain consists largely of alternate arrangements of repetitive monosaccharides: glucose, galactose, mannose, rhamnose, xylose, or sugar derivatives such as glucuronic acid, fucose, or N-acetylglucosamine. The unique structural complexness of EPS arises from the variabilities of these monosaccharides in the selected arrangements in the polymer backbone and side chains. The glycosidic linkages of these monosaccharides, along with those of α or β-configurations and their branching modes, are critical factors through which the physical and biological properties of these polymeric materials are determined. For instance, highly branched EPS may have a better gel-forming capability or improved biological activity corresponding to immunomodulation or antioxidant actions. Apart from molecular weight, another significant factor influencing the functionality of any polysaccharide is exopolysaccharides. As mentioned in work up to October 2023, they have a molar mass that usually ranges from very poorly soluble polysaccharides with molar masses below a few kilodaltons (kDa) to highly soluble polysaccharides with molar masses above 1 megadalton (MDa). Generally, higher molar mass polysaccharides have a strong thickening or emulsifying capability and can be used as thickeners, emulsifiers, and stabilizers in the food and pharmaceutical industry. Their solubility is generally dictated by the structural factors of the sugar residue ratio, the presence of uronic acids, and the development of branches. Most of the EPS are hydrophilic and soluble in water, but there are a few, such as xanthan gum, that show partial solubility due to some hydrophobic substituents. The existence of functional groups such as OH, COOH, SO3H, and acetate groups is of great importance in determining the chemical reactivity and bioactivity of polysaccharides. The functional groups allow for the polysaccharide to complex with metal ions, proteins, or other biomolecules that contribute to antioxidant, antimicrobial, and immunomodulatory actions. In a specific example, sulfate groups reportedly increase the antiviral activities of EPS through their interactions with viral proteins; carboxyl groups, on the other hand, provide chelation of metal ions, assisting in detoxification and again acting as functions to combat microbial matter itself situationally. EPS has been recognized for their encouraging properties extending to the prebiotic effect on gut flora, antioxidant activity, and immune modulation with respect to host immunity. This, together with their versatility of applications, sees EPS as useful in fields spanning from healthcare to biotechnology.

4 Prebiotic Properties of Exopolysaccharides

Exopolysaccharides (EPS) are increasingly recognized as potential prebiotics since they selectively stimulate the growth and activity of beneficial gut microbiota, thus improving the overall health of the host. Prebiotics, in general, refer to non-digestible food components that resist digestion in the upper gastrointestinal (GI) tract. These components reach the colon intact, where they provide substrates for fermentation by commensal bacteria such as Bifidobacterium and Lactobacillus. Fermentation of EPS results in various bioactive metabolites, improving gut health through more efficient digestion, immune modulation, and the production of short-chain fatty acids (SCFAs). While probiotics involve the introduction of live beneficial microbes, prebiotics such as EPS catalyze the native microbiota to enhance resilience and balance within the gut ecosystem.

The special structure of EPS interacts with gut microbes and gives them prebiotic properties. They are fermented by certain microorganisms into SCFAs, including acetate, propionate, and butyrate, that lower colonic pH, suppress pathogenic bacteria, and encourage growth of beneficial microbes in the gut. EPS also induce microbial diversity in the gut, an important factor in preventing dysbiosis-associated disease including inflammatory bowel disease (IBD) and metabolic disorders. Furthermore, EPS are instrumental in maintenance of gut permeability through stimulation of mucin synthesis and tight junction protein assembly, both of which protect against translocation of pathogens and systemic inflammation. They even interchange with gut-associated lymphoid tissue (GALT) immune cells to promote anti-inflammatory cytokines and regulatory T cells, thereby promoting gut homeostasis and systemic health.

The potential of EPS as a prebiotic can also be demonstrated by examples. EPS produced by lactobacilli species such as *Lactobacillus plantarum*, *Lactobacillus rhamnosus*, and *Bifidobacterium bifidum* help beneficial microbes while exhibiting antioxidant and immunomodulatory effects. Those produced by cyanobacteria include examples such as *Spirulina platensis* and *Chlorella vulgaris*, which stimulate the growth of Bifidobacterium and decrease gut inflame.

Sinorhizobium meliloti and other related soil bacteria have been found to produce heteropolysaccharides that serve as protection for the gut lining against oxidative stress. Fungal-derived EPS, such as β-glucans from *Saccharomyces cerevisiae* and *Aspergillus* spp., work not only in promoting microbial growth but also in boosting the immune response. In summary, the polysaccharides demonstrate prebiotic properties via fostering microbial diversity, enhancing gut barrier function, and modulating the immune system. In this regard, EPS are undoubtedly promising candidates for functional foods and nutraceuticals owing to the synergistic benefits they confer to digestion and overall systemic health. The need of the hour seems to be to analyze structure–function relationships of EPS in more detail and develop robust processes that commercialize EPS production so that they may be put to full use in solving some of the most urgent global health issues.

5 Modulation of Gut Microbiota by Exopolysaccharides

Exopolysaccharides, or EPS, are high-molecular-weight polysaccharide polymers secreted by microorganisms that play essential roles in binding the gut microbiome together through affecting its composition, diversity, and function. These polymeric substances, which are non-digestible, remain unscathed while transiting through the stomach before reaching the colon and acting as substrates for microbial fermentation, selectively stimulating helpful microbes and suppressing opportunistic pathogenic ones. This results in an enrichment of helpful microbes, such as Bifidobacterium and Lactobacillus, in the gut, thereby providing a boost in microbial diversity, which is a critical aspect of maintaining gut stability and resilience against dysbiosis. Since EPS provide a unique carbon source for these microbial communities, this can allow both dominant commensals and less numerous beneficial species that have customarily been supported to gain strength.

Beneficial microbes will metabolize EPS to facilitate the production of short-chain fatty acids such as acetate, propionate, and butyrate; reduce colonic pH; inhibit pathogenic bacteria; and provide energy for colonocytes. This process of fermentation then intensifies the metabolic effects of Bifidobacterium and Lactobacillus that further ameliorate their actions: enhanced nutrient absorption, reduced inflammation, and strengthened immunity. EPS also support growing beneficial microbes such as *Akkermansia muciniphila*, which are needed for gut barrier function and metabolic health. Lastly, EPS contribute to reinforcing the gut barrier by facilitating mucin production, providing a stronger mucus layer of protection, and upregulating tight junction proteins that facilitate closure of the spaces between epithelial cells against pathogen invasion.

In addition to their direct involvement in physical gut barrier functions, EPS also exhibit several properties as immunomodulators, mainly through their interaction with immune cells in the gut-associated lymphoid tissue (GALT). The interaction resulting from this will increase the production of anti-inflammatory cytokines and regulatory T cells, thereby controlling the chronic inflammation and ensuring immune tolerance toward the commensal microbes. EPS additionally augment the immune response via macrophage activation, dendritic cell activation, and natural killer cell activation, thereby enhancing the guts' abilities to combat infections. Collectively, these effects allow EPS to function as a potential source for therapeutic modalities for gut disorders through microbial diversity enrichment, gut barrier protection, and immune modulation. Their ability to influence the gut-immune axis makes them critical ingredients for functional foods, pharmaceuticals, and personalized nutrition strategies that target gut health. As research progresses, knowledge on the links between EPS structure and function will create further opportunities to harness their application toward global health challenges.

6 Health Benefits of Exopolysaccharides

6.1 Health and Other Benefits of Exopolysaccharides

Health Benefits of Exopolysaccharides

Exopolysaccharides that are secreted by microorganisms such as bacteria, fungi, and cyanobacteria are regarded as the new bright bioactive compound facilitating a myriad of health benefits. Their non-digestibility and ability to interact with gut microbiota make them progressively influential in human health. EPS modify digestion, modulation of immune function, and exhibit anti-inflammatory, neuroprotective, and anticancer properties, establishing themselves as promising components in the health-promoting strategy through their special chemical structure and biological activities.

6.2 Better Digestion and Bowel Activity

EPS contribute toward digestion and bowel function improvement through a prebiotic substrate for the beneficial gut microbiota, e.g., Bifidobacterium and Lactobacillus. Upon fermentation of EPS in the colon, they give rise to SCFAs: acetate, propionate, and butyrate. SCFAs regulate bowel motility, enhance uptake of water and electrolytes, and lead to a decrease in colonic pH, which is not a suitable growth environment for pathogenic bacteria. They help feed on good bacteria, thereby boosting nutrient absorption in the course of relieving constipation, diarrhea, or irritable bowel syndrome symptoms. Mucin-stimulating properties further contribute to strengthening the intestinal mucus layer and thus promote the function of the gut barrier and prevention of leaky gut syndrome.

6.3 Enhanced Functions of the Immune System and Inhibition of Inflammation

Immune modulation is guided through EPS in interaction with gut-associated lymphoid tissue. EPS act on inducing the production of anti-inflammatory cytokines, activation of macrophages and dendric cells, and promotion of regulatory T-cell functions. This modulates inflammation through a decrease in systemic inflammation and an extension of the immune homeostasis. By enhancing the gut barrier and regulating microbial populations, EPS inhibits the translocation of toxic pathogens. Thus, keeping under control chronic inflammatory obstructions often associated with inflammatory bowel disease. Other notable activities displayed by EPS are stimulation activities on the autocrine and paracrine mechanisms of immune regulation through enhancing the performance of natural killer cells and cytotoxic T lymphocytes in the promotion of the defense of the body against infections and diseases.

6.4 Potential Benefits for Mental Health and Cognitive Function

The gut-brain axis, a bidirectional communication station from the gut to the brain, is modulated largely by the gut microbiota and their metabolites. EPS-induced modulation of gut microbiota may play a vital role in mental health and cognitive function. SCFAs produced from fermentation by EPS enhance the integrity of the blood–brain barrier and decrease neuroinflammation and synthesis of several neurotransmitters, including serotonin. These effects have been associated with benefits to mood, anxiety and depression symptoms, and cognitive performance. Emerging research indicates that EPS may defend against brain impairments common in neurodegenerative diseases such as Alzheimer's and Parkinson's by reducing oxidative stress and inflammation in the brain.

6.5 Anticancer and Anti-Inflammatory Properties

EPS exert very strong anticancer and anti-inflammatory effects via several mechanisms. They limit cancer cell proliferation, induce apoptosis, and modulate the immune response to target tumor cells effectively. For example, EPS from Lactobacillus and Bifidobacterium were found to exert antitumor activity on colorectal cancer models. More importantly, EPS are known for their ability to scavenge free radicals due to their antioxidant property, hence combat the oxidative stress—a key component in inflammation and carcinogenesis. Furthermore, these compounds modulate the pro-inflammatory pathways of the immune system and thus lower the risk of a variety of chronic inflammatory diseases and cancers.

The broad spectrum of these health benefits includes improvement of digestion and modulation of immune response, mental health boosting, and anticancer potentiality. Such compounds have great prospect for functional food, probiotic, and pharmaceutical applications within the area of personalized medicine, becoming a key factor in health promotion and customer wellness.

7 Food and Pharmaceutical Applications of Exopolysaccharides

7.1 Exopolysaccharides Have Found Considerable Place in Food and Pharmaceutical Applications

Exopolysaccharides, or EPS, are a group of microbes that are produced by bacteria, fungi, and algae, during the course of fermentation. They have attracted a great deal of attention for their various applications within the food and pharmaceutical industries. Their unique functional properties, bioactivity, and ability to modulate gut microbiota qualify them as bioactive ingredients, therapeutic agents, and additives in diverse formulation types. Remaining hurdles to large-scale production notwithstanding, further progress in biotechnology opens up new doors to utilitarian applications of EPS.

7.2 Use as Prebiotic Ingredients in Food Products

EPS have been known to possess prebiotic properties, which has now led to its incorporation in functional food products with benefits for gut health. They act as substrates for the beneficial gut microbes Bifidobacterium and Lactobacillus, promoting their growth and activity. As a result, a wide variety of EPS-enriched food products, such as yogurt, fermented milk, and infant formulas, have been developed. EPS-producing strains of Lactobacillus were utilized in the production of yogurt to improve its texture and shelf life, concurrently providing prebiotic benefits. Infant formulas that are EPS added could provide indigestible oligosaccharides similar to those found in human milk, thereby promoting a healthy gut microbiota during infancy.

Besides promoting gut health, EPS promote food quality through their action as natural stabilizers, emulsifiers, and thickeners. They add viscosity and water-holding capacity suitable in soups, sauces, and bakery products. Furthermore, their inherently low caloric count and resistance to enzymatic degradation thereby promote inclusion in dietary products that target weight-loss and metabolic health.

7.3 Potential Applications in Pharmaceuticals and Nutraceuticals

Bioactive properties of EPS that encompass antioxidant, anti-inflammatory, and immunomodulatory actions recognize their considerable promises as pharmaceuticals and nutraceuticals. EPS are considered for use as drug delivery agents due to their biocompatibility and ability to form hydrogels.

They can encapsulate drugs, probiotics, or bioactive compounds, ensuring targeted delivery and controlled release in the gastrointestinal tract. In the nutraceutical industry, EPS are incorporated into dietary supplements to promote gut health, boost immunity, and reduce inflammation. For instance, EPS derived from *Lactobacillus plantarum* are marketed as prebiotic supplements for improving digestion and preventing gastrointestinal disorders. EPS are also potentially the area of research regarding wound healing and tissue engineering because of their ability to form biofilms and stimulate cell proliferation. Their anti-cancer properties are also being explored for therapeutic interventions, mainly in colorectal cancer.

7.4 Challenges and Opportunities for Large-Scale Production and Commercialization

Despite their potential, EPS production and commercialization on a large scale face several challenges. Among these are high production costs, variability in yield, and the complexity of downstream processing. Optimization of fermentation processes for improved EPS yield against production costs is to be carried out. Besides, for achieving commercial success, ensuring batch-to-batch homogeneity in the properties and features of EPS becomes necessary.

However, advances in biotechnology can provide some solutions for these problems. Genetic modification of microbial strains would enable them to produce tailor-made EPS with much higher yield than conventional fermentations. Agro-industrial waste-based fermentation substrates would lower the production costs and help meet sustainability aspirations. Escalating consumer demand for naturally occurring health-promoting ingredients could provide an extensive opportunity for

EPS-enriched products. Thus, the application of EPS in food and pharmaceuticals holds great promise for human health and product quality improvements. Challenges remain; however, research and technological advances are shortly bringing EPS toward the realizable area for large-scale production and commercialization, making it a vital asset for the functional food and pharmaceutical industries.

8 Biosynthetic Pathways and Genetic Regulation of Exopolysaccharides

Exopolysaccharides (EPS) are produced in microorganisms through polymorphic biosynthetic pathways, thus demonstrating a higher degree of regulation and complexity. These pathways ensure that polysaccharides are formed by simple sugar precursors so as to incorporate structural diversity with functional specificity. Understanding the mechanisms of biosynthesis and gene regulation of EPS production is crucial for their manufacture, followed by the tailoring of their properties for incorporation in industrial and pharmaceutical endeavors and in the food industry.

8.1 Biosynthetic Pathways for EPS Production

Three major steps are involved: sugar uptake, precursor synthesis, and polymerization/export.

1. Sugar Uptake and Precursor Synthesis: Monosaccharide uptake by microorganisms such as glucose or fructose is facilitated through specific transport systems from the external environment. These sugars are then converted to activated sugar nucleotides, UDP-glucose, GDP-mannose, or dTDP-rhamnose, which serve as precursors in the formation of EPS.
2. Polymerization: The activated sugar precursors are catalytically linked together by glycosyltransferases to form an oligosaccharide chain. Such mixed linkage makes possible the desired arrangement of the associated sugars into EPS by curtailing linkage types upon forming glycosidic bonds.
3. Export and Assembly: The newly synthesized polysaccharide chains are actively exported into the extracellular space via the action of membrane-associated proteins. The transport and assembly process often involves the polysaccharide co-polymerase (PCP)-associated and outer membrane polysaccharide export (OPX)-associated proteins to secure proper assembly and secretion of EPS.

8.2 Role of Specific Genes and Regulatory Networks in EPS Synthesis

The genes participating in EPS biosynthesis are generally organized into operons or clusters, encoding enzymes for sugar metabolism, glycosyltransferases, and export proteins. For example, the eps and cps gene clusters in *Lactobacillus* and *Streptococcus* species, respectively, have been largely ascribed a crucial role in the production of EPS. Regulators located within these clusters modulate the expression of Eps in response to operon-specific environmental and physiological stimuli.

Some pivotal regulatory proteins include two-component systems and sigma factors that influence the levels of EPS produced. Two-component systems sense external stimuli, such as nutrient availability or environmental stress, and subsequently activate the transcription of EPS biosynthetic genes for their production. Additionally, cyclic di-GMP, a small intracellular effector, acts as a signal in regulating EPS production in biofilm formation and stress adaptation.

8.3 Genetic Engineering Approaches to Enhance EPS Yield and Functionality

The advent of genetic engineering techniques enabled the manipulation of polysaccharide-producing microorganisms to boost yields and tailor properties. Strategies utilized are:

1. Gene Overexpression: Boosting key biosynthetic genes, such as other glycosyltransferases or sugar nucleotide synthases involved in EPS production.
2. Knockout Mutations: Deletions from the genetic material so that undesirable by-products are not synthesized, guiding metabolic flux specifically toward EPS biosynthesis.
3. Pathway Engineering: Tweaking or adding entire biosynthetic pathways resulting in novel EPS with specific structures and functionalities. For instance, hybrid EPS with improved rheological or bioactive properties have been developed using synthetic biology approaches.
4. Regulatory Modulation: Target engineering of regulatory genes or promoters to ensure optimum expression of EPS biosynthetic pathways under appropriate growth conditions.

8.4 *Applications and Future Directions*

Genetic understanding of EPS biosynthesis allows for the industrial production of high-value polysaccharides for numerous applications, from food additives to pharmaceuticals. The emerging technologies CRISPR-Cas9- and omics-based approaches might speed up the discovery of new EPS-producing strains and pathways. It is anticipated that fusion between systems biology and metabolic engineering will offer an optimum glimpse into the full technological potential in EPS technology for industrial or therapeutic applications.

9 Role of Exopolysaccharides in Biofilm Formation

The true crux of biofilm formation—important for microbial attachment to supramolecular polymeric aggregates—is its role as that of a matrix produced outside the cell. The majority of the EPS matrix, which is also known as extracellular polymeric substances, is composed of other substances, including proteins, nucleic acids, and lipids. EPS accounts for a greater proportion of the biofilm matrix, commonly 50–90%, and is an integral part of its structure, function, and durability.

9.1 *Mechanisms of EPS-Mediated Biofilm Formation*

Biofilm formation occurs in five stages: initial attachment, irreversible attachment, EPS production, maturation, and dispersal. EPS plays an important role in the biofilm by promoting irreversible attachment. During this stage, it facilitates cell adhesion to surfaces and to one another. EPS molecules interact with the surface through hydration. They form a three dimensional hydrated gel like matrix, which stabilizes the biofilm and acts as a matrix for the incorporation of other cell types. EPS serves as a shield, insulating the biofilm community from unfavorable physical stresses, including desiccation, UV radiation, and antimicrobial agents. EPS is also known to mediate cell-to-cell signaling through quorum sensing, leading to correlated gene expression and the ability of cells to alter their phenotype in response to nutrient or other environmental changes. Differing EPS composition—involving charged and neutral sugars, proteins, and extracellular DNA—provides great diversity and functional versatility to biofilms.

9.2 *Importance of Biofilms in Natural Ecosystems and Industrial Processes*

Biofilm plays an important role in nutrient cycling, ecological stability, and symbiotic relationships in natural ecosystems. In aquatic ecosystems, for example, biofilms formed on rocks and sediments are very important for organic matter decomposition and nutrient liberation. In the rhizosphere, microbial biofilms improve plant growth by enhancing nutrient availability and protecting roots against pathogens.

Biofilms have favorable and harmful effects in industrial processes. Biofilms in wastewater treatment systems support degradation of organic pollutants and heavy metals. Similarly, in bioleaching, EPS-producing bacteria biofilms facilitate the extraction of valuable metals from ores. But biofilm formation on industrial equipment, pipelines, and medical devices often leads to biofouling, causing reduced efficiency, corrosion, and contamination. Understanding and utilizing EPS's role in biofilm formation is important to optimize its benefits, besides alleviating its negative impacts.

Implications of EPS in Pathogenic Biofilms and Strategies for Control

Pathogenic biofilms are a cause of concern in healthcare and industrial setups because of their increased resistance to antibiotics and immune responses. The biofilm formation and virulence of such bacteria as *Pseudomonas aeruginosa*, *Staphylococcus aureus*, and *Escherichia coli* depend largely on the EPS. EPS in pathogenic biofilms acts like a physical shield that limits penetration of antibiotics, providing protection to the acting bacteria from immune responses from the host body.

Strategies to control pathogenic biofilms include several approaches targeting EPS synthesis and disruption:

1. Enzymatic Degradation: Enzymes, such as polysaccharide lyases, can break down EPS and destabilize the biofilm matrix.
2. Chemical Inhibitors: Such chemicals can inhibit EPS production or quorum sensing to prevent biofilm formation.
3. Nanotechnology: Nanoparticles are now studied for their uterine penetration and EPS disruption capacity.
4. Probiotic and Phage Therapy: Beneficial microbes or bacteriophages targeting specific pathogens within biofilms.

EPS mediation in biofilm formation is a multifaceted process, with wide-ranging implications. Biofilms are critical in natural ecosystems and useful in certain industrial processes, but pathogenic biofilms present significant hurdles. Further research into mechanisms of EPS and innovative control strategies is paramount in biofilm management under different confines.

10 Environmental Applications of Exopolysaccharides

EPS is widely interoperable with multiple biological functions because of their unique physicochemical characteristics. Produced by various microorganisms, namely, bacteria, fungi, and microalgae, EPS are instrumental in bioremediation, soil health, promotion of plant growth, and wastewater treatment. These applications depict the potential of EPS in delivering sustainable solutions to environmental problems.

10.1 EPS in Heavy Metal and Pollutant Bioremediation

EPS are well known for their exceptional metal-binding capacity by virtue of the functional groups, like carboxyl, hydroxyl, and phosphate, present in their structure. The existence of these groups allows EPS to chelate heavy metals, such as Pb, Cd, Hg, and As, thereby allowing their removal from contaminated environments. EPS from both fungi and bacteria have been utilized for biosorption processes involving the immobilization of heavy metals, thus reducing their availability and bioavailability. EPS from Pseudomonas and Bacillus species have been reported to sequester metals in soil and polluted waters. Also, EPS can bind organic pollutants like polycyclic aromatic hydrocarbons (PAHs), enriching their biodegradation by microbial communities.

EPS applications in bioremediation are environmentally friendly and cost-effective alternatives to chemical precipitation and ion exchange. Scaling up EPS production and optimizing biosorption conditions are key areas of research to enhance their effectiveness in bioremediation actions on a broad scale.

10.2 Soil Health and Plant Growth Promotion

Being in the rhizosphere, EPS-producing microbes are contributing significantly to soil health and plant growth promotion. EPS improve the soil structure by binding soil particles together to form stable aggregates that improve water retention and aeration. This is especially relevant for arid and degraded soils, wherein EPS are able to mitigate erosion and improve fertility. EPS also promote nutrient cycling because they serve as substrates for microbial colonization and activate the enzymatic activities. Various nitrogen-fixing bacteria-like Rhizobium and phosphate-solubilizing bacteria-like Bacillus have relied on EPS for establishing symbiotic relationships with plant roots. Such association enhances nutrient availability to the plant, thereby promoting growth and yield. Additional protective barriers offered by EPS shield rhizosphere microorganisms and plant roots from several environmental stressors, including drought, salinity, and pathogen attacks. The application of

EPS-producing bioinoculants in sustainable agriculture is gaining momentum as part of improved crop productivity and resilience.

10.3 Water Treatment and Other Bio-Based Solutions: A Potential Perspective

EPS has good efficacy in water treatment applications, owing to their flocculation and bio-adsorption potential. In many wastewater treatment systems, EPS promote the agglomeration and settling of suspended particles, which renders treated water relatively clearer. These include aerobic and anaerobic systems for waste treatment, where soluble EPS in activated sludge systems are built up by organisms predominantly from the genera Zoogloea and Sphaerotilus.

Moreover, EPS themselves could also have the ability to adsorb and immobilize toxic compounds in industrial effluents such as dyes or heavy metals that would have adverse effects on the environment. Additionally, EPS adds value to biofilm formation in constructed wetlands because they contribute to degradation activities within them. The use of EPS toward creating bio-based materials for environmental applications is another promising area of particular interest. The production of EPS-based hydrogels and films, toward the objective of soil moisture retention, pollutant filtration, and packaged degradation, is also aligned with worldwide efforts in dealing with plastic waste and fostering circular economies. The use of EPS for their environmental applications provides an astonishing opportunity for meeting pressing ecological challenges sustainably and ecologically. In leveraging various properties of EPS, researchers and industries may well be able to tackle pollution, soil degradation, and water scarcity and ensure a better planet.

11 Industrial Production and Economic Feasibility of Exopolysaccharides

The industrial production of exopolysaccharides has recently garnered tremendous interest on account of their wide-ranging applications in foods, pharmaceuticals, cosmetics, and environmental management. The economical production of EPS in larger quantities will therefore decide the real value of it. Better fermentation technologies and economic assessments have made way for efficient EPS production.

Innovations for Industrial-Scale EPS Production: From Fortified Fermentation Techniques to Molecular Biology

Fermentation constitutes the basis of industrial production of EPS. The main microbial fermentation systems, submerged fermentation (SmF) and solid-state fermentation, use different microorganisms—bacteria, fungi, or microalgae—for cultivation in the production of EPS.

The commonly accepted method for EPS production is submerged fermentation because of the wide scale of operation and control of parameters like temperature, pH, and aeration. With the development of improved bioreactor designs, such as fed-batch and continuous culture, the EPS yield and productivity have been increased. These improved designs allow for controlled nutrient addition, thus helping to avoid substrate inhibition and growth limitation of the microbes being cultivated. Further improvements in EPS production have come from advances in bioprocess optimization, including using immobilized cell reactors and cocultures of synergistic microorganisms. The use of immobilized systems allows the stability of the microorganism, reduces contamination risks, and in few cases, achieves prolonged fermentation cycles. Genetic engineering of EPS-producing strains with modification of crucial pathways has also led to a very promising way of increasing EPS production with respect to yield, functionality, and reduced costs.

11.1 Cost–Benefit Analysis of EPS Production in Industrial Settings

Economic feasibility remains one of the key considerations for industrial-scale adoption of EPS. The economic feasibility of EPS production scales with the costs of raw materials, energy, and downstream processing. Carbon sources such as glucose, sucrose, and lactose contribute significantly to production costs. Compared to substitutional compounds, EPS has improved industrial production due to fermentation technology improvement and rationalization of the polysaccharide biosynthesis. Besides the same, sustainability issues get equated with renewable exploitation, ensuring economic profitability by EPS biogenesis process, among its conservationism. The key to the economy of scale will include low-cost, renewable substrates made from agro-industrial residues (molasses, whey, and lignocellulosic biomass), whereby these substrates will not only contribute to the lowering of costs but also embody circular economy principles by utilizing waste streams. The mission that will be a challenge but where maximum expenditure is appended to appropriate ultrafiltration, precipitation, and freeze-drying. But membrane filtration and hybrid energy-efficient dryers are presently studied to alleviate some of the points of the downstream process while being energy-efficient. EPS production has high upfront costs, but the payback period is significant in pharmaceuticals and food because of the strong increase in the demand for natural and bioactive polymers. Developing products with added value such as prebiotics, emulsifiers, and biofilm increases profitability and market potential.

11.2 Sustainability Aspects and the Use of Renewable Substrates

Sustainability in the EPS perspectives arises in line with challenges in the broader view of our environment and planetary health. For it to be equal and fed-back in terms of sustainability means this must include renewable and waste-derived substrates. These have included agricultural residues such as corn steep liquor, sugar cane bagasse, and dairy waste water. A perfect example of sustainable production therefore deals with the consumption of energy and organization of waste disposal processes during fermentation and downstream processes; thereby introducing some concept of renewable energy such as solar or biogas into the EPS plants further reduces the impact on the environment. A further future-oriented notion for producing EPS is of bioengineered microorganisms being able to utilize nonconventional carbon sources like carbon dioxide or methane. Some selected production strategies enhance the possibility of regular, dependable, and cost-effective production of EPS. The importance of sustainability in these was enhanced by a combination of careful environmental considerations on renewable resources along with sustainable measures in ensuring economic viability.

12 The Role of Exopolysaccharides in Metabolic Disorders

Metabolic disorders such as diabetes, obesity, and metabolic syndrome pose a serious health challenge globally. The recent exopolysaccharides (EPS), which are the natural polymers secreted by microorganisms and potent therapeutic agents intended for treating metabolic disorders, were very promising due to their bioactive properties including strengthening of immunity and reducing inflammation. They have also been used as modulators of gut microbiota, lipid metabolism, and glucose response, leaving them useful for application in these metabolic diseases.

12.1 Potential of EPS for the Treatment of Diabetes and Obesity

EPS represents an important opportunity in the management of diabetes due to their modulation of glucose metabolism and enhancement of insulin sensitivity. Many EPS possess prebiotic properties whereby they stimulate the growth of beneficial gut microorganisms such as Lactobacillus and Bifidobacterium, fermenting them into short-chain fatty acids (SCFAs) such as butyrate and propionate, which can modulate blood glucose homeostasis through GLP-1 secretion. This is a hormone that increases the release of insulin and decreases blood sugar levels. By forming viscous gels in the gut, some EPS slow carbohydrate absorption into the bloodstream and thus reduce spikes in blood glucose that typically occur after eating. For instance, in animal models, EPS of *Lactobacillus plantarum* and *Leuconostoc mesenteroides* have been shown to reduce glucose intake. In treating obesity, EPS enhance satiety and inflammation of adipose tissue. The EPS further act to promote gut barrier integrity so as to decrease the translocation of lipopolysaccharides (LPS) and, in effect, reduce the systemic inflammation further propagating obesity. Some other.

13 Clinical Studies and Human Trials on Exopolysaccharides

The therapeutic potential of exopolysaccharides (EPS) in promoting health has gained significant attention, provoking a rise in human clinical trials. These trials aim primarily to evaluate the efficacy, safety, and appropriate dosage of EPS for human consumption, and cover aspects including but not limited to gut health, immune modulation, metabolic disorders, and others.

13.1 Overview of Clinical Trials Evaluating EPS for Health Benefits

Clinical trials demonstrate the ability of EPS to support gut microbiota and improve digestive health. For instance, the prebiotic effects of EPS from Lactobacillus and Bifidobacterium species make them a source of good gut microbiota and short-chain fatty acids, SCFAs associated with good gut and metabolic health. Studies on EPS derived from Kefir have shown immunomodulatory and anti-inflammatory effects that reduce systemic inflammation in inflammatory bowel disease (IBD). In metabolic health, the role of EPS in managing diabetes and hypercholesterolemia has been investigated. For instance, in diabetic patients, EPS from *Leuconostoc mesenteroides* improved fasting blood glucose and lipid profiles, indicating potential consumption benefits in glycemic control and cardiovascular health.

13.2 Safety, Efficacy, and Dosage Considerations for Human Consumption

Safety in clinical applications stands first. Most EPS used in human trials are derived from GRAS microorganisms, such as lactic acid bacteria. Studies have proven on more than one occasion that EPS are safe and nontoxic even at varying doses. In addition, dosage and side effect evaluation is essential because excessive use can lead to gastrointestinal discomfort and bloating.

14 Challenges and Ethical Considerations in Exopolysaccharide Research

Exopolysaccharides (EPS) have grown in importance as valuable biopolymers in health, food, pharmaceuticals, and the environment. The investigation and exploitation of EPS come with challenges and ethical considerations. These problems include genetic modifications, intellectual property, and innovative sustainable balance. The solutions to addressing the challenges are imperative for the proper engraving of the chances that EPS present while making sure of responsible industrial and scientific practices.

14.1 Ethical Considerations Linked to Genetic Modification of EPS-Producing Microbes

Increased EPS yield and functionality typically occur through the incorporation of genetic engineering. Genetic modification creates tolerable strains, and it might include the alteration of metabolic pathways or the introduction of exogenous genes. While such changes may result in increased industrial productivity, they may raise ethical concerns. Genetic modification could have unforeseen ecological consequences if engineered strains escape from their cells into a natural ecosystem. For example, if transgenic DNA enters indigenous populations, it could threaten microbial diversity or even create an antibiotic resistance problem. In addition, public unease regarding the use of genetically modified organisms (GMOs) requires communication of safety and environmental tiers. Regulatory frameworks developed by the FDA and EFSA require stringent testing prior to the approval of genetically modified microbes; researchers and industry must abide by these guidelines for the ethical sake of scientific society's fear of GMOs.

14.2 Addressing Challenges in Intellectual Property and Patents

The commercial importance of EPS has led to an increasing fight for patent rights. Protecting patent rights is fundamental for innovation's drive; however, it is not without challenges. Disputes in patent ownership can grow into an ethical crisis when indigenous knowledge or natural microbial strains have been used. For example, microorganisms producing EPS may be sourced from biodiversity-rich regions, raising questions about benefit-sharing and respect for local communities' contributions. The Nagoya Protocol within the Convention on Biological Diversity emphasizes equitable sharing of benefits arising from the use of genetic resources. In such contexts, researchers and industries must navigate through these frameworks to ensure that their practices adhere to the principles of fair and just practice. Also, the overlapping of patents on production methods, formulations, and applications of EPS can stifle innovation. Open-source initiatives and collaborative research models are emerging as potential solutions to overcome these hurdles while enabling equitable access to EPS technology.

14.3 Balancing Innovation with Sustainability and Environmental Impact

Massive-scale EPS production requires extraordinary amounts of resources including the carbon sources and energy consuming processes. An overdependence on nonrenewable substrates or intensive fermentation processes can cause environmental degradation. The challenge lies in the sustainable establishment of production methods, such as utilizing agricultural waste or renewable feedstock to attain reduced carbon footprint. Moreover, great effort and dedication should go in to create those processes for EPS production that will not be in direct competition for food resources, especially when this involves crops like maize and sugarcane. Circular economy principles activating the integration of EPS production in the conventional agro-industrial waste streams can provide a sustainable solution. The balance between innovation with environmental stewardship also involves lifecycle assessments to evaluate the overall sustainability of EPS applications. In each case, the science, politics, industry, and development communities must collaborate toward the creation of incentives for sustainability while addressing the concerns of our global environment. The challenges and ethical considerations in EPS research point to the absolute necessity of a holistic approach, coupled with scientific advancement, ethical responsibility, and sustainability. Such a concerted effort should facilitate further progress toward fair and environmentally conscious solutions that allow full benefit of the EPS technologies across a wide range of applications.

15 Emerging Technologies and Future Perspectives in Exopolysaccharide (EPS) Research

The rapid evolution of emerging technologies is radically transforming the core policy in EPS research and enabling varied understandings of their granulation, properties, and applications. The integration of advanced tools, such as omics technologies, synthetic biology, and innovations in precision medicine, aims in delivering tailored EPS having specific functional benefits.

15.1 Emerging Omics Technologies in EPS Research

Omics technologies—genomics, proteomics, and metabolomics—remain at the forefront of unraveling the complexities of EPS biosynthesis. Genomics entails sequencing the genomes of EPS-producing microorganisms closely to unveil their genetic identity and physiological pathways responsible for EPS production. This information can lead to identification of new strains or modification of existing strains with the goal of higher production yields and functionality. Proteomics concentrates on the functional study of proteins involved with the biosynthesis and regulation of EPS production. Map-out functioning of enzymes makes for a better understanding of proteins that catalyze and control the production of a specific EPS. Metabolomics allows better understanding of metabolite flux within microbial systems such as discovering the intermediates and end-products salient for EPS production. Knowledge about the mentioned metabolite alterations allows the engineering of pathways to enhance EPS yield and property modulation. Together, this institution of technology provides a systemic approach enabling the advancement of strains designed for EPS productions.

15.2 Synthetic Biology Potential in EPS Design

Synthetic biology is fast being recognized as a disruptive technique in engineering microorganisms to make customizable EPS with functional properties. With synthetic gene circuit design and the composition of metabolic pathways, the goal is to engineer strains capable of producing EPS with improved viscosity, emulsification, or bioactivity. For instance, synthetic biology may allow the introduction of functional groups in EPS to augment their interaction with biological systems to foster applications in drug delivery or tissue engineering. Synthetic biology also permits the construction of microbial consortia of several microbial morphologies that work tropically in an effort to further optimize EPS. Also, CRISPR-Cas9 and other gene-editing techniques seem to boost the accuracy with which such pathways of metabolism can be modified, hence enhancing EPS production efficiency and economic viability. Whether some new class of chemicals or biochemicals, such as EPS, may be used to relieve the patient's condition is an important question for today. In precision medicine, EPS progress toward objects that interact specifically with suspicious tissue or modulate the microbiome in controlling and/or mitigating the chronic conditions can be engineered for drug provision or treatment agents. In personalized nutrition, EPS will go to fortifying functional food tailored to certain dietary habits that advance gut health and enhance immunity. These applications demand continued progression of omics technologies and synthetic biology to design EPSs of high specificity, safety, and efficacity. Omics technologies, synthetic biology, and personalization approaches pattern the EPS future. This innovation holds enormous potential for solving global challenges in health, food, and sustainability, carrying EPS from a basic investigation to transformative real-life applications.

16 Challenges and Ethical Considerations in Exopolysaccharide (EPS) Research

Thus far, it has been a short leap of rapid progress in the EPS field, especially the genetic modification of EPS-producing microbes for enhanced productivity. Such innovation raises ethical questions. A main concern is whether these modified organisms will accidentally affect the ecosystems into which they are released. These altered microbes may affect ecosystems in ways unforeseen and, thus, threaten microbial diversity and ecological balance. Horizontal gene transfer from modified microbes to native microbes might further seed engineered traits and increase ecological risks. It leads to the added risk of public skepticism toward GMOs: many are "wary" of GMOs and perceive possible hazards to health and the environment. To counter their influences, the researchers should promote transparency, rigorously assess the possible risk, and engage the public in the decision-making process. Stronger regulation also needs to create a balance between encouragement to create and ensure safety and accountability going forward. In addition to that, patents and intellectual property within EPS present yet another hurdle. Innovations in EPS research involve inventing proprietary genetic engineering techniques, fermentation processes, or microbial strains. Although patents are employed to secure such innovations in some way and provide incentives to spread that technology, it can also establish monopolization, closed access of EPS-related technology, especially among the low-income regions. As such, overlapping patents confer legal squabbles among researchers or companies, thus making the commercialization processes slower. Natural resources and indigenous knowledge used in the EPS research are usually not taken into consideration in patent applications, which poses moral dilemmas relating to benefit sharing. This calls for a more just IP regime giving emphasis to open-access research, public–private collaborations, and fair benefit sharing frameworks that credit the contributions of local communities. Balancing innovations garnered through EPS with sustainable operation within the ecosystem is crucial. Notably, large-scale production of EPS demands vast amounts of substrates, water, and energy, with notable depletion of natural resources or with possible pollution caused. This is part of the reason why researchers are currently looking into the use of agro-industrial waste, dairy wastewater, and renewable resources as substrates that go in line with the circular economy. Energy-efficient fermentation technologies and waste-reduction measures are also very important for sustainable production. We must consider the environmental impacts of EPS-based products, which include bioplastics, in order for such materials to be biodegradable and not contribute to microplastic pollution. Legal compliance in resource extraction and labor practices is also important; companies must, in this regard, put the needs of fair labor standards foremost in order to avoid community exploitation and reduced carbon footprints.

EPS research promises to be beneficial in several areas of an industrial and environmental nature. However, it is beset with several challenges and ethical issues. Genetic engineering, intellectual property issues, and sustainability challenges should be tackled through team players comprising researchers, policymakers, and stakeholders. Transparency, equity, and responsibility of dealing with the environmental impact should be among the main principles with which the society develops and utilizes EPS by then ensuring compliance for the ethical and responsible development of innovation programs.

17 Conclusions and Future Directions

Exopolysaccharides (EPS) are thus becoming the new frontier in biotechnology, with great potential in broad spectra, including healthcare, agri-and food technology, and the environment. These biopolymers produced from microbial sources are endowed with excellent properties such as biofilm formation, immunomodulation, and structural variations, making them beneficial for utilization in bioplastics, prebiotics, and bioremediation. As research continues to unravel the intricate roles of EPS, the importance of these polymers in combating daunting global challenges such as climate change, food security, and human health is taken center stage. In order to maximize this potential, a well-directed action plan must be developed to fill the voids in the scientific knowledge base, development of technologies, and commercializing them. Continuous EPS research and applications will require such a roadmap. Beginning with the improvement of microbial strains and production systems for yield, quality, and cost-effectiveness, advances in synthetic biology and metabolic engineering should greatly assist in the design of microbes for the production of a particular type of EPS with desired characteristics. Any plan toward industrial production must promote the sustainable production processes, with special attention given to renewable substrates and energy-efficient means of minimizing environmental burdens. In addition, the establishment of thorough safety assessments would provide adequate security to ensure that EPS-based products are safe for consumers and ecosystems alike. Equally important will be commercialization and scaling. Cooperation among academia, industry, and government can speed the transition of the laboratory findings into market-ready products. Investment in pilot production facilities and funding support for EPS seed ventures may assist in breaking many of the bottlenecks governing the path toward large-scale production. Public awareness campaigns and transparent communication on the beneficial and safe side of EPS-based technologies could help boost consumer acceptance and, consequently, the market potential. Interdisciplinary collaboration is simply the basis for sustainable EPS development. Chemists, biologists, engineers, and environmental scientists must unite to treat technical, ethical, and environmental issues. Coordination between the policymakers and industry stakeholders according to the principles of a circular economy and sustainable development will ensure that EPS innovations advocate a greener and fairer future.

In conclusion, the transformative power of EPS resides in addressing critical global challenges while catalyzing innovation. A strategic, interdisciplinary, sustainability-oriented approach can open up new avenues for EPS application. With this, we are called upon to act together, pledging our commitment toward the advancement of research, technology development, and practices into responsible and impactful use of EPS for a better life in the years to come.

Further Readings

Abdalla, A. K., Ayyash, M. M., Olaimat, A. N., Osaili, T. M., Al-Nabulsi, A. A., Shah, N. P., & Holley, R. (2021). Exopolysaccharides as antimicrobial agents: Mechanism and spectrum of activity. *Frontiers in Microbiology, 12*, 664395.

Angelin, J., & Kavitha, M. (2020). Exopolysaccharides from probiotic bacteria and their health potential. *International Journal of Biological Macromolecules, 162*, 853–865.

Bengoa, A. A., Dueñas, M. T., Prieto, A., Garrote, G. L., & Abraham, A. G. (2023). Exopolysaccharide-producing Lacticaseibacillus paracasei strains isolated from kefir as starter for functional dairy products. *Frontiers in Microbiology, 14*, 1110177.

Hidalgo-Cantabrana, C., Sánchez, B., Milani, C., Ventura, M., Margolles, A., & Ruas-Madiedo, P. (2014). Genomic overview and biological functions of exopolysaccharide biosynthesis in Bifidobacterium spp. *Applied and Environmental Microbiology, 80*(1), 9–18.

Hussein, M.-D. M., Ghaly, M. F., Osman, M. Y., Shalaby, A. S. G., & Helal, M. M. I. (2015). Production and prebiotic activity of exopolysaccharides derived from some probiotics. *Egyptian Pharmaceutical Journal, 14*(1), 1–9. https://doi.org/10.4103/1687-4315.154687

İspirli, H., Demirbaş, F., & Dertli, E. (2018). Glucan type exopolysaccharide (EPS) shows prebiotic effect and reduces syneresis in chocolate pudding. *Journal of Food Science and Technology, 55*(9), 3821–3826.

Jurášková, D., Ribeiro, S. C., & Silva, C. C. G. (2022). Exopolysaccharides produced by lactic acid bacteria: From biosynthesis to health-promoting properties. *Foods, 11*(2), 156.

Khan, R., Shah, M. D., Shah, L., Lee, P. C., & Khan, I. (2022). Bacterial polysaccharides – A big source for prebiotics and therapeutics. *Frontiers in Nutrition, 9*, 1031935.

Lee, M. G., Joeng, H., Shin, J., Kim, S., Lee, C., Song, Y., Lee, B. H., Park, H. G., Lee, T. H., Jiang, H. H., Han, Y. S., Lee, B. G., Lee, H. J., Park, M. J., Jun, Y. J., & Park, Y. S. (2022). Potential probiotic properties of exopolysaccharide-producing Lacticaseibacillus paracasei EPS DA-BACS and prebiotic activity of its exopolysaccharide. *Microorganisms, 10*(12), 2431.

Majee, S. B., Avlani, D., & Biswas, G. R. (2017). Rheological behaviour and pharmaceutical applications of bacterial exopolysaccharides. *Journal of Applied Pharmaceutical Science, 7*(9), 224–232.

Miyamoto, J., Shimizu, H., Hisa, K., Matsuzaki, C., Inuki, S., Ando, Y., Nishida, A., Izumi, A., Yamano, M., Ushiroda, C., Irie, J., Katayama, T., Ohno, H., Itoh, H., Yamamoto, K., & Kimura, I. (2023). Host metabolic benefits of prebiotic exopolysaccharides produced by Leuconostoc mesenteroides. *Gut Microbes, 15*(1), 2161271.

Mohd Nadzir, M., Nurhayati, R. W., Idris, F. N., & Nguyen, M. H. (2021). Biomedical applications of bacterial exopolysaccharides: A review. *Polymers, 13*(4), 530.

Netrusov, A. I., Liyaskina, E. V., Kurgaeva, I. V., Liyaskina, A. U., Yang, G., & Revin, V. V. (2023). Exopolysaccharides producing bacteria: A review. *Microorganisms, 11*(6), 1541.

Salazar, N., Gueimonde, M., de los Reyes-Gavilán, C. G., & Ruas-Madiedo, P. (2015). Exopolysaccharides produced by lactic acid bacteria and bifidobacteria as fermentable substrates by the intestinal microbiota. *Critical Reviews in Food Science and Nutrition, 56*(9), 1440–1453.

Salimi, F., & Farrokh, P. (2023). Recent advances in the biological activities of microbial exopolysaccharides. *World Journal of Microbiology & Biotechnology, 39*(8), 213.

Sanders, M. E., Merenstein, D. J., Reid, G., Gibson, G. R., & Rastall, R. A. (2019). Probiotics and prebiotics in intestinal health and disease: From biology to the clinic. *Nature Reviews. Gastroenterology & Hepatology, 16*(10), 605–616.

Schmid, J., Sieber, V., & Rehm, B. (2015). Bacterial exopolysaccharides: Biosynthesis pathways and engineering strategies. *Frontiers in Microbiology, 6*, 496.

Sheng, S., Fu, Y., Pan, N., Zhang, H., Xiu, L., Liang, Y., Liu, Y., Liu, B., Ma, C., Du, R., & Wang, X. (2022). Novel exopolysaccharide derived from probiotic Lactobacillus pantheris TCP102 strain with immune-enhancing and anticancer activities. *Frontiers in Microbiology, 13*, 1015270. https://doi.org/10.3389/fmicb.2022.1015270

Sims, I. M., Frese, S. A., Walter, J., Loach, D., Wilson, M., Appleyard, K., Eason, J., Livingston, M., Baird, M., Cook, G., & Tannock, G. W. (2011). Structure and functions of exopolysaccharide produced by gut commensal Lactobacillus reuteri 100-23. *The ISME Journal, 5*(7), 1115–1124.

Werning, M. L., Hernández-Alcántara, A. M., Ruiz, M. J., Soto, L. P., Dueñas, M. T., López, P., & Frizzo, L. S. (2022). Biological functions of exopolysaccharides from lactic acid bacteria and their potential benefits for humans and farmed animals. *Foods, 11*(9), 1284.

Wu, M., Pakroo, S., Nadai, C., Molinelli, Z., Speciale, I., De Castro, C., Tarrah, A., Yang, J., Giacomini, A., & Corich, V. (2024). Genomic and functional evaluation of exopolysaccharide produced by Liquorilactobacillus mali t6-52: Technological implications. *Microbial Cell Factories, 23*(1), 158.

Xu, X., Peng, Q., Zhang, Y., Tian, D., Zhang, P., Huang, Y., Ma, L., Qiao, Y., & Shi, B. (2020). A novel exopolysaccharide produced by Lactobacillus coryniformis NA-3 exhibits antioxidant and biofilm-inhibiting properties in vitro. *Food & Nutrition Research, 64*, 3744. https://doi.org/10.29219/fnr.v64.3744

Xu, M., Li, Z., Zhao, X., & Li, W. (2022). Prebiotic properties of exopolysaccharides from Lactobacillus helveticus LZ-R-5 and L. pentosus LZ-R-17 evaluated by in vitro simulated digestion and fermentation. *Foods, 11*(16), 2501.

Zaghloul, E. H., & Ibrahim, M. I. A. (2022). Production and characterization of exopolysaccharide from newly isolated marine probiotic Lactiplantibacillus plantarum EI6 with in vitro wound healing activity. *Frontiers in Microbiology, 13*, 903363.

Immunomodulatory Role of Microbial EPS in Food and Dairy Industry

Bidisha Chatterjee, Debdutta Ghosh, Ayesha Shamim, Prerana Das, Titir Choudhuri, Sharadia Dey, and Arup Kumar Mitra

1 Introduction

The carbohydrate group of molecules is the most crucial and abundant biomolecule. They can be categorized into three groups primarily depending on their size: mono-, oligo-, and polysaccharides. Monosaccharides consist of simple sugar moiety containing a single polyhydroxy ketone or aldehyde group. Oligosaccharides contain glycosidic linkage joining monosaccharides to form short chains. Polysaccharides (glycans) are long chain carbohydrates containing more than 20 monomers joined by glycosidic linkage (Porter & Martens, 2017). Polysaccharides can be further subdivided into homopolysaccharide (one type of monosaccharide) and heteropolysaccharide (two of more different monosaccharides). Polysaccharides are storage house for living cells that store various structural materials required to build the outer structure of the cell. They are also important for diverse cellular mechanism on the cell surface such as cell adhesion, cell–cell interaction, and molecular recognition (Chaisuwan et al., 2017).

Microorganisms produce diverse polysaccharide for various cellular functions. The polysaccharides secreted by the microbes are of two types: intracellular and

B. Chatterjee (✉) · D. Ghosh · A. Shamim · P. Das · T. Choudhuri
Department of Microbiology, St. Xavier's College, (Autonomous), Kolkata, West Bengal, India

S. Dey
Department of Environmental Studies, St. Xavier's College, (Autonomous), Kolkata, West Bengal, India

A. K. Mitra
Department of Microbiology, St. Xavier's College, (Autonomous), Kolkata, West Bengal, India

Department of Environmental Studies, St. Xavier's College, (Autonomous), Kolkata, West Bengal, India

T. Sarkar et al. (eds.), *Extracellular Polysaccharides*,
https://doi.org/10.1007/978-3-031-94472-7_11

extracellular polysaccharides. The extracellular polysaccharide secreted by the microbes helps them to survive in the environment by helping in cell adhesion, biofilm formation, and cell protection (Donot et al., 2012). However, they do not serve the purpose of carbon and energy sources. The extracellular polysaccharide can be further categorized into capsular polysaccharide and exopolysaccharide (EPS). These exopolysaccharides are excreted into the environment and is widely responsible for biofilm formation (Sajna et al., 2021). Microbial EPS includes both homopolysaccharide (e.g., cellulose, dextran, pullulan, levan, and curdlan) and heteropolysaccharide (e.g., alginate, mauran, succinoglycan, and haloglycan). These polymers have been known for their diverse properties and are currently used in different fields such as in drug delivery systems, food additives, thickeners, cosmetics, heavy metal removal, and emulsifiers (Freitas et al., 2011). EPS may contain different functional groups including carboxyl, phosphate, uronic acids, pyruvate, acetyl, and sulphate. These functional groups determine the physicochemical properties well as the biological properties of EPS (Sajna et al., 2021).

The bioactive EPS has been isolated from a diverse variety of microorganisms including bacteria, fungi, algae, microalgae, and archaea. These microbial EPS have revealed potential bioactive properties such as antimicrobial, anti-inflammatory, immunomodulatory, antioxidant, and anticancer activities (Chaisuwan et al., 2017). The immunity boosting EPS have been thoroughly studied since the polysaccharides showed potential immunity stimulation in both in vitro and in vivo cytotoxicity to the immune cells (Adebayo-Tayo et al., 2018; Surayot et al., 2014). The microbial EPS have been applied in the bone repair, tissue engineering, and numerous other medical fields like ophthalmology and orthopedic surgery. They have been utilized to release drug in the body at a slower pace. They have also been applied to treat inflammation and tumors by placing the drugs at the target area. Some EPS help to improve the gut health by maintaining the gut microflora (Razzaghi et al., 2021). The primary role of EPS is to protect the cells from unfavorable or toxic conditions like phagocytosis, desiccation, phage attacks, and antibiotics. This results in an improved microbial interaction in diverse environment. It also helps the cells to adhere to the substratum in the form of biofilm, thus forming a symbiotic community (Jaseera & Abdulla, 2021).

Microbial EPS have also gained popularity in the food industry as a result of its renewable potential. The EPS is naturally synthesized, cost-effective, nontoxic, biodegradable, and an eco-friendly alternative. Microbial EPS have been widely used as stabilizing and binding agent due to its hydrocolloid property. For example, xanthan gum can improve the egg white foaming property, can be applied in low fat food preparation, can improve gel physicochemical property when mixed with potato starch, can be applied as gluten replacement, and can be applied in the development of low glycemic food formulations. Curdlan can be applied as an ingredient for fat replacement in meat products as well as to set yoghurt formulations. Alginate have been utilized in low fat mayonnaise fabrication. Levan and dextran have been applied as microbial EPS in kefir beverage. Microbial EPS is also widely applied as a prebiotic. Antimicrobial agents such as plant extracts, metallic nanoparticles, essential oils, and bacteriocins can be incorporated in the microbial EPS to form an active ingredient for food packaging to maintain the food safety and food quality during storage (Arserim Ucar et al., 2021).

This chapter will elucidate the interaction of the immune system with the microbial EPS, their application in the drug delivery system and as an immunomodulator. It further explains the microbial EPS immunomodulators application in the food industry followed by the commercialization and cost-effectiveness of the methods.

2 Interaction of Immune System with the Microbial EPS

The structural variations provide unique physicochemical properties and functionalities to the exopolysaccharides (EPS) (Zhang & Edgar, 2014). The structural analysis of EPS generally includes an examination of several crucial aspects: the types and arrangements of monosaccharides, variations in molecular weight, the conformation of glycosidic linkages and the characteristics of repeating units, specific bonding locations among monosaccharides, and the overall three-dimensional structure. Together, these factors characterize the physical, chemical, and biological attributes of EPS (Wang et al., 2023). It is proposed that EPS characterized by a porous structure are more prone to demonstrate antioxidant activity, whereas those with immunomodulatory properties usually possess a triple-helical configuration. In-depth examinations of their monosaccharide makeup, molecular weight, structural arrangement, and biological functions can yield important insights into the relationships between structure and activity, revealing how these structural characteristics affect their biological roles (Wang et al., 2023).

2.1 Relationship Between Structure and Activity of Exopolysaccharides (EPS)

The makeup and arrangement of exopolysaccharides (EPS) significantly affect their microstructure and surface characteristics, which are essential in influencing their biological activity. Changes in these structural features can alter the interactions of EPS with biological systems, impacting attributes like bioavailability, functionality, and efficiency in particular applications (Kanamarlapudi & Muddada, 2017). The alteration of exopolysaccharides (EPS) via group substitution to change the structure of these polysaccharides and improve their specific biological functions has become a significant focus in studies related to their interactions with the immune system (El-Ghonemy, 2021; Larsen & Engelsen, 2015) Current techniques for modifying polysaccharides include acetylation, carboxymethylation, phosphorylation, and sulfonation. A case in point that demonstrates the addition of functional groups to the glucan chain is lasiodiplodia, a (1→6)-β-D-glucan obtained from fungi (Wang et al., 2018).

2.1.1 Acetylation Modification

Acetylation alters the spatial configuration of polysaccharides, thereby affecting their biological roles (Ren et al., 2020). The exopolysaccharides (EPS) generated by *Lactobacillus plantarum* 70810 showed antioxidant properties after the incorporation of a novel acetyl group (Wang et al., 2015).

2.1.2 Carboxymethylation Modification

Carboxymethylation involves adding carboxymethyl groups to polysaccharide chains via etherification reactions between polysaccharides and acids or their derivatives. This modification results in changes to the three-dimensional structure and water solubility of the polysaccharides, which in turn affects their biological functions (Ren et al., 2020). For example, polysaccharides obtained from the fermentation broth of Lachnum YM240 underwent carboxymethyl modification. The results showed that diabetic mice treated with carboxymethylated Lachnum polysaccharides displayed significantly lower fasting glucose and serum triglyceride levels, as well as notably enhanced insulin sensitivity (Wang et al., 2017b).

2.1.3 Phosphorylation Modification

Phosphorylation is an effective method for improving the bioactivity of exopolysaccharides (EPS) (Zhang et al., 2017). The exopolysaccharides (EPS) isolated from *Lactococcus lactis* subsp. Lactis underwent phosphorylation and exhibited antioxidant properties in both in vivo and in vitro (Guo et al., 2013).

2.1.4 Sulfonation Modification

Sulfonation is employed to modify the chain length and hydrophilicity of exopolysaccharides (EPS), impacting their biological effectiveness. Sulfonated EPS obtained from *Enterobacter cloacae* Z0206 safeguarded RAW264.7 mouse macrophages against oxidative damage induced by H_2O_2 and inhibited DNA fragmentation. These results suggest that sulfonation boosts antioxidant properties by enhancing water solubility and chain length, and offers cell protection by promoting increased hydrogen atom exchange (Jin et al., 2014).

2.2 Immunomodulation by LAB-Derived EPS and Antimicrobial Activity

Lactic acid bacteria (LAB) show their immunomodulatory effects via cell wall (peptydoglycan) and interact to the receptor of the target cell by their secreted EPS. The dendritic cells use their pattern recognition receptors (PRR) to interact with the gastrointestinal microbiota by sensing their microbe-associated molecular pattern (MAMP). Then the association between PRR and MAMP will stimulate different molecular activities such as antimicrobial activity, which involves cytokines. Glycan helps to alter ligand-receptor communication and differentiate among MAMPs of LAB, pathogen, and commensals. For ex., EPS secreted from *Leuconostoc mesenteroides* NTM048 stimulates Immunoglobulin A and this IgA captures antigen and prevent it from binding to the cell surface receptors. It helps to neutralize cholera toxin and reduces the motility of *Salmonella* (Abdalla et al., 2021). A *L. mesenteroides* strain (isolated from Italian Cheese) has bifidogenic, antimicrobial, antioxidant, and antibiofilm property. EPS-B3, produced from this bacterium shows antibacterial activity against some foodborne pathogens and induces the growth of Bifidobacteria (Probiotics) (Bisson et al., 2023).

The EPS, secreted by *L. rhamnosus* KL37 interacts with Murine peritoneal macrophages (Laiño et al., 2016). EPS stimulate the production of TNF-α, IL-6, and IL-12 and desensitize the macrophages at higher concentration than LPS. The pro-inflammatory effects can be analyzed by TNF-α/IL-10 production. After the restimulation with LPS or EPS, the production of TNF-α is reduced which indicates the state of hyperresponsiveness (induced by macrophages) (Ciszek-Lenda et al., 2011).

The EPS, secreted from *Bifidobacterium breve* UCC2003 interacts with Mice. It modulates immune cells recruitment and stimulate cytokine production (Laiño et al., 2016). The EPS produced by *B. breve* reduces the colonization of *Citrobacter rodentium* inside the murine gut (Fanning et al., 2012).

Lactobacillus bulgaricus OLL1073-R1 interacts with murine lymphocytes and murine macrophages through the pattern recognition receptors. This interaction

- Induces the phagocytic activity of macrophages (Laiño et al., 2016)
- Helps to kill tumor cells by improving macrophage cytotoxicity (Laiño et al., 2016)
- Increases murine splenocyte mytogenic activity (Laiño et al., 2016)
- Helps to increase cytokine production in macrophages (Laiño et al., 2016)

Lactobacillus plantarum N14 possesses anti-allergic property and immunostimulatory effects (Laiño et al., 2016). Enterotoxigenic *Escherichia coli* and LPS trigger the effect of *L. jensenii* TL2937 and induces inflammatory immune response inside the porcine intestinal epitheliocyte cell line (PIE cells). In the PIE cells, the inflammatory response mediated by toll-like receptors (TLR-4) indicates inflammation inside the gut of weaning piglets and releases cytokines (Shimazu et al., 2012). *L. plantarum* 70810 (produces c-EPS and r-EPS) and *L. plantarum* KX041 show natural antioxidant ability (Wang et al., 2014, 2017a). *Lactobacillus plantarum* JLAU103 shows immunostimulatory effect (Wang et al., 2020).

When EPS is of high molecular weight, it will give a negative impact on intestinal colonization like the *B. animalis* subsp. *Lactis* that reduces its capability to adhere to human IEC and inhibit Biofilm formation (Abdalla et al., 2021).

Lactobacillus confuses TISTR1498 has the immunomodulatory ability (Zhang et al., 2023). The EPS of this bacteria contains only D-glucose, connected by α-(1–6) glycosidic linkages. EPS was not able to activate RAW264.7 cells so that they cannot produce pro-inflammatory mediators such as cytokines and nitric oxides. But the EPS of low molecular weight ($\leq 70 \times 10^3$ g/mol) and partially hydrolyzed EPS can activate RAW264.7 cells and produces different cytokines and nitric oxide (Surayot et al., 2014).

Lactobacillus curvatus SJTUF62116 (isolated from fish, *Gymnocypris przewalskii* in Qinghai Lake) helps to prevent the growth of *E. coli*, *Streptococcus enteritidis*, and *Staphylococcus aureus* by using its secreted EPS (Zhu et al., 2022).

Lactobacillus lactis ssp. *Cremoris* possesses immunostimulatory ability (Zhang et al., 2023).

Lactobacillus kefiranofaciens enhances the secretion of cytokines and immunoglobulin at the intestinal mucosa level. As a result, the number of IgA+ cells are increased in the small and large intestines (Vinderola et al., 2006). The number of macrophages is also increased in lamina propria and peritoneal cavity (Medrano et al., 2011).

High serum cholesterol is a major cardiovascular risk factor. To reduce cholesterol level, different drugs are used nowadays but most of them have severe side effects. Exopolysaccharide of *L. paracasei* helps to reduce cholesterol level of serum (Bhat & Bajaj, 2019).

Lactobacillus fermentum YL-11 shows antitumor ability (Zhang et al., 2023). *Lactobacillus fermentum* ME-3 shows antimicrobial property against intestinal pathogen and also possess antioxidant property (Mikelsaar & Zilmer, 2009).

2.3 Interactions of Exopolysaccharides with Intestinal Microbiota

The gastrointestinal (GI) tract of human is a resident to a variety of microorganisms, predominated by bacteria, which plays an important role in enhancing human health. Bacterial exopolysaccharide (EPS) has played an essential role in human health by influencing the interaction of bacteria with the host-like adhesion to GIT epithelium leading to immune response. The gut microbiome contains a number of species that helps in food digestion, protection from pathogen, and regulating intestinal immunity (Oerlemans et al., 2021). The most dominant bacterial species that contribute to healthy gut environment are Firmicutes and Bacteroidetes along with Lactobacillus and Bifidobacterium (Thursby & Juge, 2017). Alteration in gut microbiota, which is also known as dysbiosis, is linked to various diseases. This emphasizes the importance of maintaining a healthy gut health (Rojo et al., 2017). Modification of the gut microbiome by various approaches like transfer of microbiota or addition of beneficial bacteria to the diet like Bifidobacterium or Lactobacillus through supplementation has shown good results for reducing some common diseases like irritable bowel syndrome, inflammatory bowel disease, and ulcerative colitis (Staudacher et al., 2017). These bacteria are not only responsible for healthy gut microbiome but also produce some essential molecules like short-chain fatty acid, indole derivatives, along with products of bacterial cell wall that are beneficial for human health (Agus et al., 2018). The intact prebiotic fibers reaches the colon and serves as the energy source for gut microbiota which will promote the growth of beneficial species. Recently, glycans in the microbial envelope have been identified as important molecules that promote intestinal health (Hidalgo-Cantabrana et al., 2012). If EPS is considered to be prebiotic it should resist digestion in the upper gastrointestinal tract and can be utilized by specific microbes in the colon. But it has been found that fructans and β-glucans are resistant to digestion and reaches the colon undigested. Fermentation of EPS by certain bacteria leads to the production of SCFAs, which play a very crucial role in providing energy to colonocytes, for maintaining the integrity of intestinal barrier and modulating host metabolism and immune response (Hamer et al., 2007). A crucial role of EPS is seen where it prevents both commensal and pathogenic bacteria to the intestinal epithelium. It has been found that EPS produced from *L. paracasei* subsp. paracarasei prevents *E. coli* adhesion to Caco-2 cells. In poultry, a LAB strain *L. johnsonii* F19785 reduces the colonization of *Clostridium perfringens* and *E. coli* in chicks (Dertli et al., 2018). EPS from two bacteria *B. fragilis and F. prausnitzii* have shown immunomodulatory effect in the gut (Rossi et al., 2015).

Probiotics are the good bacteria, which provide benefits to our body. They maintain healthy microbial flora in our gut and intestine and inhibit the growth of pathogenic bacteria. Bifidobacterium and Lactobacillus are two probiotic bacteria that keep our gut healthy. The viability of probiotic bacteria depends on different factors such as storage temperature, pH, and production of hydrogen peroxide. Microencapsulation increases their stability in adverse condition (Iurciuc et al., 2016).

2.4 Interaction of Immune System with Fungal Beta-Glucan EPS

Synthesis of cytokines such as IL-8 and IL-12 are stimulated after binding of fungal schizophyllan with receptors present on different types of immune cells such as leukocytes and polymorphonuclear leukocytes. In vitro and in vivo studies have reported that cytotoxic lymphokine-activated killer (LAK) cells and natural killer (NK) cells get progressively sensitized in the presence of IL-2 upon activation of macrophages by this fungi-derived beta-glucan (Kano et al., 1996). Comparative in vivo analysis in between tumor antigen-sensitive lymphokine-activated killer cells, which have been exposed to schizophyllan and those without prior exposure, resulted in better survivability in the former category of cells during experiments carried out in mice models (Kano et al., 1998). Thus, this fungal exopolysaccharide exhibits antitumoral properties (Zhong et al., 2013). Beta-D-glucans such as scleroglucan with its triple helix structure and optimal degree of branching of 0.20–0.33 have been reported to exhibit amplification of inhibition of tumors. In vitro, human monocytes are activated by these polymers coupled with the induction of TNF-alpha (Falch et al., 2000). It has also been reported that scleroglucan upon hydrolysis resulted in the formation of short chain lysates, which exhibited a stimulatory effect on the reinforcement of dendritic cells associated with the increase of CD40 and CD80/86 coupled with the induced secretion of pro-inflammatory cytokines during studies carried out in porcine systems (Bimczok et al., 2009).

3 Application of EPS in Drug Delivery System

Microbial exopolysaccharides (EPS) can be designated as complex carbohydrate moieties produced by microorganisms such as bacteria, algae, and fungi, which are extruded outside into the microenvironments to make up the extracellular matrix. As these polysaccharides are biologically synthesized by living organisms, they are regarded as biopolymers. Biopolymers should be extracellularly derived from non-pathogenic or recombinant microbial strains, which are categorized as GRAS (generally regarded as safe) variants, to be applicable for use in biomedical and pharmaceutical fields. These microbially derived biopolymers on account of their primarily intrinsic properties such as biocompatibility, noncytotoxic and biodegradable nature, which may be genetically and chemically augmented, if necessary, feature as nonnegotiable benefits for implementation in the preparation of drug delivery systems. Microbial exopolysaccharides can be fabricated to develop desirable polymer-based tablets, sprays and syrups, micro- and nano-systems, hydrogel, aerogel and nanogel beads, liposomes and micelles, and films and membranes for targeted, controlled, and sustained release of drugs, and this has already been documented in scientific literature. Microbial exopolysaccharides, which have been typically used as drug carriers, can be of bacterial origin such as xanthan, gellan, hyaluronic acid, dextran, bacterial cellulose, curdlan, and mauran. They can be of fungal origin such as pullulan, schizophyllan and scleroglucan. The polymer levan is derived from both bacteria and fungi. The exopolysaccharides, which are obtained from algae, include carrageenan and agar whereas another biopolymer alginate can be derived from both bacteria and algae (Razzaghi et al., 2021).

3.1 Alginate

It's a linear exopolysaccharide, which can be synthesized by both bacteria and algae. Alternating residues of beta-D mannuronic acid (M) and alpha-L-guluronic acid (G) bonded by (1,4) glycosidic linkages constitute the biopolymer characterized by selective O-acetylation, which can be acquired naturally in bacteria or by chemical modification as in the case of algal alginates (Lee & Mooney, 2012; Windhues & Borchard, 2003). Bacterial alginates are primarily derived from *Pseudomonas* and *Azotobacter* species, and among them, recombinant *Pseudomonas aeruginosa* PGN5 strain (Valentine et al., 2020) and *Azotobacter vinelandii* (Clementi et al., 1995) have been reported to exhibit promising results in commercial production of this polymer for pharmaceutical applications. Enhanced rheological stability due to selective O-acetylation, increased chain length, high water retention, and capacity to cross-link with divalent cations make it a suitable platform for biofabrication in several forms such as that of hydrogel beads assembled with magnetic nanocrystals of cellulose for pH-mediated delivery of drugs (Supramaniam et al., 2018), fibers for the sustained release of the drug CHG (Shi et al., 2020) and hybrid aerogels of dual porosity attained by expansion of carbon dioxide (Martins et al., 2015) among other approaches for controlled and targeted release of drugs in therapeutics.

3.2 Xanthan Gum

It's a microbial exopolysaccharide composed of glucose monomers with a trisaccharide side chain made up of mannose (beta-1,4), glucuronic acid (beta-1,2), and terminal mannose residue linked by alpha-1,3 glycosidic bond to alternate glucose monomeric units in the backbone (Patel et al., 2020), which is obtained by the fermentation of sugars by *Xanthomonas* species. Xanthan is remarkable for its temperature-dependent and ionic strength-dependent conformation in solution (Morris et al., 1977), which renders it highly susceptible to utilization in drug delivery systems in the form of hydrogels, which can be cross-linked with sodium Trimetaphosphate (Bejenariu et al., 2009) and with other polymers like chitosan (Faralli et al., 2019). It has been reported that gold nanoparticles are stabilized by xanthan and polyethylene glycol for the delivery of drugs in murine melanoma cells with implications in the treatment of disorders like cancer (Muddineti et al., 2016).

3.3 Gellan

Gellan, similar to xanthan is an anionic polymer synthesized by *Sphingomonas paucimobilis* which consists of a repeating linear tetrasaccharide unit comprising of two D-glucose residues linked to one D-glucuronic acid by beta-(1,4) glycosidic bond, which is bonded to one L-rhamnose residue in a specific arrangement with high susceptibility of the 1,3-D glucose residue to substitution by acetyl or glycerate groups, which is influential in dictating the mechanical and rheological properties of the biopolymer (Fialho et al., 1999). Enhanced mechanical stability can be attained by copolymerization with other polymers like hydroxypropyl methylcellulose, sodium alginate, and starch that impart improved muco-adhesiveness (De Oliveira Cardoso et al., 2017) for formulation of pharmaceuticals, which can be administered nasally (Salunke & Patil, 2016) or orally for alleviating gastrointestinal symptoms (Milivojevic et al., 2019).

3.4 Hyaluronic Acid

Microbial *hyaluronic acid* produced by *Streptococcus equi* (Kim et al., 1996) and other recombinant strains is a glycosaminoglycan, which consists of a repeating linear disaccharide comprising of beta-D-glucuronic acid and beta-D-N-acetylglucosamine residues bonded by alternate beta-1,3 and beta-1,4 glycosidic linkages. Hyaluronic acid is remarkable for its hygroscopicity and lubricating viscoelasticity (Chen, 2002), which render it rheologically highly capable of fabrication in the form of hydrogels and functional silica nanoparticles with mesoporosity (Chen et al., 2019) in the preparation of anticancer therapeutics. Biofabrication of hyaluronic acid has exhibited specificity in recognition of the CD44 receptor which is overexpressed in tumor cells (Bourguignon, 2014) with major implications in targeted drug therapy. As hyaluronic acid is a natural component of the vitreous humor of the eyes, it can be an important ingredient in ocular formulations to alleviate ophthalmic irritations (Kalam, 2016).

3.5 Dextran

It is a neutral exopolysaccharide obtained from *L. mesenteroides* strains, which consist of a linear chain of D-glucose monomers bonded by alpha-(1,6) glycosidic linkages and variable branching points linked via alpha-(1,4), alpha-(1,3) or alpha-(1,2) bonds marked by the occurrence of hydroxyl functional groups with strain-dependent variation in molecular weight and branching patterns influencing the solubility, viscosity, and structural stability of the polymer (Sarwat et al., 2008). The structural integrity of the biopolymer enables its application as suitable material to form hydrogels for organ-specific drug delivery (Hovgaard & Brøndsted, 1995), calcium alginate-associated interpenetrating polymeric networks (IPNs) (Pescosolido et al., 2011), nanoparticles, and paclitaxel loaded copolymerized dextran micelles (Liu et al., 2015) for anticancer therapy. In the biomedical field, in case of hypovolemic patients, dextran is used as a plasma expander (McCahon & Hardman, 2007) and in vascular operations, it exhibits antithrombotic activity (Abir et al., 2004).

3.6 Bacterial Cellulose (BC)

It is a linear homopolymer of beta-D-glucopyranose monomers joined together by beta-1,4 glycosidic linkage, which is obtained from *Komagataeibacter* sp. and this biopolymer is devoid of lignin, hemicellulose, pectin, and other lignocellulosic by-products with unique 3D structure advantageous for biomedical applications contrary to vegetal cellulose. High water absorbance, high tensile strength, good flexibility, and thermal stability possessed by bacterial cellulose films make it intrinsically suitable for use as coating agents in drug formulations (Amin et al., 2012). To develop wound dressing constructs, bacterial cellulose can be fabricated in the form of polymeric scaffolds in association with chitosan and alginate (Chang & Chen, 2016), chemically modified hydrogels, films, sponges and BC membrane-based transdermal diclofenac administration systems (Silva et al., 2014) for action of anti-inflammatory and antibacterial drugs.

3.7 Curdlan

Water-insoluble curdlan is an unbranched linear homopolysaccharide composed of beta-(1,3) glucan units, which assumes a rigid triple helical conformation subject to temperature-dependent reversible or irreversible gelling conformation (Zhang & Edgar, 2014) and this exopolysaccharide is synthesized by *Agrobacterium* sp. and *Alcaligenes faecalis* among other bacterial species. To develop bioactive and pharmaceutical formulations, curdlan can be subjected to diverse treatments including chemical modification of its hydroxyl groups in alkaline conditions (Yan et al., 2020) to increase solubility, augmentation with cellulose in cryogels (El-Naggar et al., 2017), development of complexed curdlan sulphate antivirals against HIV (Yan et al., 2017) and production of phosphorylated curdlan-based antibacterials along with other nanocarrier-based curdlan derivatives with further implications as anticancer therapeutics upon interaction with hepatoma cell lines (Na et al., 2000).

3.8 Levan

Water-soluble levan is a fructan homopolysaccharide made up of a core of D-fructofuranosyl monomers joined together by beta-(2,6) glycosidic bonds with beta-(1,2) bonds at branched points and a terminal D-glucosyl subunit, which impart the polymer with desirable tensile strength, concentration-dependent intrinsic viscosity in solutions, and good thermal and chemical stability for developing therapeutics (Han, 1990). Microbial levan can be obtained from bacteria and fungi with *Zymomonas mobilis* and *Bacillus subtilis* as the producer strains among other bacterial species. This biopolymer can be fabricated in the form of nano- and microcarriers coated with vancomycin (Sezer et al., 2017) and 5-ASA releasing complexed hydrogels susceptible to temperature sensitivity (Osman et al., 2017) with potential bactericidal activity. Levan exhibits promise in the development of antitumor drugs via specificity in recognition of GLUT5 overexpressed in cancerous cells (Kim et al., 2015) and it also acts as a blood plasma expander.

3.9 Mauran

It is an anionic exopolysaccharide which is sulphated and consists of recurring mannose, galactose, glucose, and glucuronic acid residues and this biopolymer can be synthesized by the halophile *Halomonas maura*, an extremophilic bacterium. The biopolymer is characterized by its thixotropic and pseudoplastic nature (Llamas et al., 2006), which make it rheologically stable and advantageous for application in pharmacology in the form of hybrid nanoparticles complexed with chitosan effective against adenocarcinoma cell lines (Raveendran et al., 2013). Isolation of the exopolymer from *Pseudomonas otitidis*, an endophytic bacterium found in association with the roots of the medicinal plant *Tribulus terrestris* L., followed by its characterization and subsequent conversion into nanocarriers for the drug quercetin have been reported to study the pharmacokinetics of the release of this drug and its activity against breast cancer cells in vitro in varied pH environments along with the determination of its antioxidant and ferric ion reducing potential (Kalimuthu et al., 2023).

3.10 Pullulan

It is a linear alpha-(1,4) linked D-glucose containing exopolysaccharide made up of repeating maltotriose units linked together by alpha-(1,6) glycosidic bonds, which can be synthesized by the strains of the fungus *Aureobasidium pullulans*. This biopolymer is physico-mechanically characterized as thermostable, chemically stable, highly adhesive, and flexible, which makes it a suitable platform to develop pharmaceuticals (Grigoras, 2019) in the form of films, anticancer drug conjugated with pH-sensitive nanoparticles (Li et al., 2014), electrospun copolymer fibers in the presence of calcium chloride, which turn out to be ultrafine and stable (Xiao & Lim, 2018), and Prostaglandin E1 bearing nanogels (Kobayashi et al., 2009) among others. Pullulan can be implemented in organ-specific targeted drug delivery systems owing to its notable affinity for lectin-like receptors (Singh et al., 2015).

3.11 Schizophyllan

Fungal beta-glucan schizophyllan is obtained from *Schizophyllum commune*, which exhibits antitumoral properties. Schizophyllan nanoparticles can be added with anticancer agent ellagic acid for the treatment of breast cancer cells (Pirzadeh-Naeeni et al., 2020). *Scleroglucan* is extracellularly obtained from *Sclerotium rolfsii*, which has exhibited its ability as a vehicle in the delivery of the drug theophylline in vitro (Viñarta et al., 2007). The backbone of these biopolymers consists of beta-(1→3)-D-glucopyranose residue with a single beta-(1→6)-linked branching at every third monomeric unit. *Lasiodiplodan* is another fungal exopolysaccharide, which can be obtained from *Lasiodiplodia theobromae* with anticancer therapeutic potential (Alves da Cunha et al., 2012). This exopolymer can be implemented in healing hydrogels with a stimulatory effect on reepithelialization (Nissola et al., 2021). *Sacchachitin* is another fungal exopolysaccharide, which has been reported to stimulate healing of wounds of corneal cells in animal models (Chen et al., 2012). About 40% chitin and 60% beta-1,3-glucan constitute the polymer sacchachitin. *Botryosphaeran* is an extracellularly obtained polysaccharide from the fungus *Botryosphaeria rhodina*, which has been reported to exhibit antidiabetic properties in in vivo diabetic rat models (Miranda-Nantes et al., 2011).

3.12 Carrageenan

Anionic carrageenan is a sulphated linear polysaccharide, which is extracellularly obtained from red seaweeds and it is composed of repeating disaccharide units containing beta-D-galactopyranose linked by alpha-(1,3) glycosidic bond and alpha-D-galactopyranose or 3,6-anhydro-alpha-D-galactopyranose linked by alpha-(1,4) glycosidic bond, which can be categorized into several types such as kappa, iota, and lambda among others (Campo et al., 2009). The structural organization of the biopolymer makes it physicochemically stable for application in drug delivery. They can be applied in several forms such as that of ibuprofen encapsulated beads (Sipahigil & Dortunc, 2001), hydrogels delivering ciprofloxacin (Padhi et al., 2016), anticancer therapeutic quercetin (Varghese et al., 2014), and also as novel electrospun fibers associated with the formation of polyelectrolyte complex with antioxidant potential (Ignatova et al., 2016).

3.13 Agar

Exopolysaccharide agar is linear and unbranched with alternating beta-D-galactose and 3,6-anhydro-alpha-L-galactose monomers bearing hydroxyl groups, which impart hydrophilicity and this polymer is primarily obtained from red algae, namely, *Gelidium* and *Gracilaria* species with prevalent widespread applications as a solidifying and gelling agent. This biopolymer harbors the potential for implementation in oral drug delivery systems owing to its property as a modifying agent. Comparative in vitro pharmacokinetic analysis in between agar-alginate cross-linked hydrogel beads and native alginate gel samples demonstrated the promise of the composite beads over the latter as a sustainably releasing vehicle for loaded drugs with evident pH-sensitivity (Yin et al., 2018).

4 Utilization of EPS as an Immunomodulator

Exopolysaccharides (EPS) are complex biopolymer, which are secreted mostly by gram-negative bacteria, which creates biofilm in moist environment. These polymers are made up of polysaccharide, lipids, and proteins (Chaisuwan et al., 2020). These polysaccharides are made up of more than 20 monomer units linked by glycosidic bond (Porter & Martens, 2017). Microorganisms secrete EPS for various functions outside the cell. EPS can be used for heavy-metal removal, drug delivery, and as bioactive compounds having functions like anticancer and antimicrobial. EPS can enhance innate immune system and modulate innate immunity potentially preventing cancer caused by immune deficiencies.

The LAB-EPS involves the initiation of tol-DCs, which increases proliferation of anti-inflammatory cytokines and expands suppressor T cells. This mechanism involves the inhibition of T17 cell differentiation. *Bifidobacterium animalis* subsp. *lactis* IPLA-R1 has immunomodulatory capabilities seen in mice. The Dendrobium officinale polysaccharide (DOP) decrease the production of C-Reactive Protein (CRP) and downregulates inflammatory cytokines (Zhou et al., 2023). An increase in production of IL-10 and regulatory T cells occurs when in vitro experiments are done with *B. bifidum* PRI1 (Sadeghi et al., 2024). The EPS, which are derived from LABs, have antiviral and anti-inflammatory effects through modulation of cytokine. LAB-EPS can decrease inflammation by interacting with TLR and RP105/MD1 complex (Laiño et al., 2016).

The macrophages are induced to produce cytokines, which include TNF- α, Interleukin-6, and Interleukin-12 but by reducing TNF- α, the amount of desensitization regarding macrophage occurs after secondary stimulation with EPS. A LAB *Lactobacilllus rhamnosus* KL37 is found, which can decrease *Citrobacter rodentium* colonization by modulating cytokine recruitment and synthesize immune cells when mice are treated with *B. breve* UCC2003. Phagocytic activity of macrophage and mitogenic activity of murine splenocytes are increased along with an enhancement in cytotoxicity toward tumor cells. Treatment with *L. bulgaricus* OLL 1073-R1 stimulates cytokine production in macrophage, induces tolerogenic dendritic cell, and expands regulatory Treg cells. In mice if we introduce *B. animalis* subsp. *lactis* IPLA-R1, then we can control T17cell differentiation. Larger relative molecular mass of exopolysaccharide derived among *Lacticaseibacillus rhamnosus* RW-9595M decreases proinflammatory mediators in macrophage and lymphocyte while an increase in IL-10 production is essential for immune response regulation. Different LAB-EPS have different immunomodulatory activities with phospho-polysaccharide, which enhances macrophage function (Divanovic et al., 2005).

The TLR4 signaling pathway of porcine intestinal epithelial cell involves dimerization and activation of proinflammatory response through MyD88-dependent and MyD88-independent pathways, which ultimately lead to activation of NFκB and AP1, which are the key regulators of inflammatory response (Wachi et al., 2014). Negative regulators such as TOLLIP and IRK-M play an important role in inhibiting TLR signaling for maintaining immune response balance and also maintained by regulating TLR signaling and negative regulation of porcine intestinal epithelial cells (PIE). Porcine epithelial cells explain interpretation and functionality regarding porcine single immunoglobulin interleukin-1-related receptor, toll-interacting protein, A-20, B cell lymphoma-3, mitogen-activated protein kinase-1, and Interleukin-1 receptor-associated kinase-M genes. Two LAB variants *L. plantarum* N14 and *L. delbrueckii* TUA4408L reduce the interpretation of cytokines that enhances inflammation in porcine intestinal epithelial cells by modulation of NFκB and MAPK pathways and induces negative regulators of TLR signaling. The EPS which are produced by these strains plays an important role in anti-inflammatory effects (Murofushi et al., 2015).

A LAB-EPS strain *L. plantarum* MM89 enhances phagocytosis, and cytokine production in RAW264.7 cells (Rajoka et al., 2022). The LAB having variable gene clusters of EPS like *L. plantarum* has immunomodulatory effect; a *L. plantarum* strain JLAU103 and *L. helveticus* strain R-5-EPS has an anti-inflammatory effect in macrophages (Wang et al., 2020). EPS which are rich in fucose can activate TLR-4 receptor and Dectin-1 receptor and it can suppress cytokine secretion that brings about physical change in an antigen presenting cell (DC). It also promotes yield of molecules which are at the surface of an antigen presenting cell i.e. dendritic cell (DC). EPS from *B. adolescentis* strain have high molecular weight and has been found to induce different cytokine profiles when it is co-cultured with macrophage RAW264.7 and mouse spleen. It exhibits anti-inflammatory effects with strain IF1–03 and with strain IF1–11 it shows pro-inflammatory effects (Alessandri et al., 2019).

Probiotic bacteria regulate the production of cytokine that is essential for defence mechanism also it induces the release of cytokine release, induces inflammation, lymphocyte differentiation and activation of immune system. *Bifidobacterium. adolescentis* explains immunomodulatory effect by reducing pro-inflammatory cytokines in mice and anti-inflammatory cytokines in colitis mice. There are certain strains of *Bifidobacterium longum* which stimulate immune system by secreting cytokines which are specific and are secreted from peripheral blood mononuclear cells. Macrophages are an important immune cell because it recognizes foreign pathogen which are activated by microbial metabolites like polysaccharide and kills that pathogen by phagocytosis or producing cytokine or by presenting it to T-helper cells. There are certain strains of bacteria which can influence cytokine production in macrophage with rise of cytokines that reduces inflammation like Interleukin-10 that also reduces yield of cytokines that heightens inflammation like Tumor necrosis factor-α. An immune response which is shown in the cells can vary depending upon the purity of EPS and if there is contamination with other bacterial components. Some LAB strains like *L. mesenteroides* and *L. lactis* sp. *cremoris* KVS20 produces bioactive EPS and creates different effect on host immunity. Studies have shown that Glucose rich EPS can stimulate immune system very well. The α-1,6 glycosylated dextran from *L. plantarum* DMDL 9010 promotes IgA production and enhances host immunity by metabolizing into SCFA. TL4 signaling is very high in PIE cells which leads to inflammation. In PIE cells TL4 signaling mimics the intestinal epithelial cells of human and mice which activates the pro-inflammatory response by MYD88 dependent and MYD88 independent pathways. The LAB strains *L. plantarum*N14 and *L. delbrueckii* TUA4408L regulates the TLR signaling in PIE cells by the upregulation of negative regulators and reduction of inflammatory mediators and the EPS which are produced from these strains modulates the NF-kB and MAPK pathways by interacting with TLR2, TLR4 and RP105 in porcine epithelial cells. The different compositions of EPS result in different effects on the expression of PRR and downstream signaling pathways. It has been found that there is a mechanism by which the EPS regulates immune system which is found in strains of Lactobacillus strains that induces tolerance in porcine intestinal cell and by interacting with RP105 and TLR4 receptor and upregulation if Interleukin-1 receptor associated kinase-M and mitogen-associated protein phosphatase-1 expression. The IRAK-M and MKP-1 act as a biomarker for selecting immunomodulatory EPS from Lactobacilli, which can be done by co-activation of RP105 and TLR4 pathways. The intestinal epithelial cell (IEC) and dendritic cell both interact with each other along with gastrointestinal microbiota by pattern recognition receptor (PRR) to detect microbe-associated molecular pattern (MAMP), which in turn controls the antimicrobial action and cytokine production (Labeer et al., 2010). Glycan plays an important role in distinguishing MAMP produced by different microbes (Abdalla et al., 2021). It is found that Bifidobacterium EPS can enhance the production of regulatory T cells and IL10 by naïve CD4 T cells through treatment of DC that shows its crucial role to tolerate and also maintain anti-inflammatory conditions. These indicate the potential application of Bifidobacterium EPS to cure immune-related disorder, which shows the path for future research for their interaction with dendritic cells.

5 Microbial EPS in the Food Industry

Extracellular polymeric substances (EPS) are the high molecular weight, natural polymeric substances secreted by different species of microorganisms, plants, and seaweeds. EPS is composed of monosaccharides and other compounds such as acetate, phosphate, and succinate. EPS can be classified into two groups on the basis of their type of monosaccharides such as homopolysaccharides and heteropolysaccharides. Homopolysaccharides contain only one type of monosaccharides, on the other hand heteropolysaccharides contain two or more different types (three to seven) of monosaccharides (glucose, galactose, mannose, N-acetylglucosamine, N-acetyl galactosamine, rhamnose) (Basiri, 2021).

5.1 Extraction and Purification of Microbial Polysaccharides

Different methods are used for the extraction of various types of polysaccharides from microorganisms. Centrifugation (Ultracentrifugation may be used for better results) is generally used as an initial step for the extraction of EPS having slime fornications. If EPS is thermally stable, then the separation between EPS and cell can also be achieved by heat treatment (Jindal & Khattar, 2018). The solubilization of polysaccharides is enhanced by boiling the microbial cell suspension at control temperature (Jindal et al., 2022). But capsular polysaccharides can be extracted in a different way. It depends on the nature of association between the cell and the EPS to be extracted (Jindal & Khattar, 2018). If the association is very strong between the cell and EPS to be extracted then more severe treatments (NaCl treatment, EDTA treatment, Alkaline treatment) are required (Huang et al., 2022). Sonication of the cell can also be used (Jindal & Khattar, 2018).

5.2 Applications in Food Industry

Microbial EPS is extensively used in food industry due to its structural diversity and different unique properties. The high viscosity of EPS in aqueous solution makes them a very good stabilizer and emulsifier in food industry. For example, in food industry, xanthan gum is extensively used as stabilizer and emulsifier and also forms edible film to maintain the quality of freshly cut pears. Cinnamic acid is an aromatic polyphenol, which is embedded in xanthan gum and acts as an active polyphenol oxidase inhibitor. As a result, the edible film is formed and delay the browning of fresh-cut pears. The synergic effect between xanthan gum and cinnamic acid has an important role as a protective layer against oxygen. Pullulan and Levan are used as edible food coating nowadays. Galactopol is also used as a food coating and packaging material. The antimicrobial and antioxidant properties of some EPS help to increase the shelf life of food. The antioxidant property helps to scavenge free radicals rapidly to inhibit oxidation, which leads to slowing down the browning of the fruits. The EPS of *Porphyridium cruentum* is a very good inhibitor of browning of freshly cut banana pieces at 4 °C (Widyaningrum & Meindrawan, 2020).

Microbial EPS helps to increase the flavor, texture, and quality of the food as a thickeners, stabilizers, texturizers, and gelling agent. It helps to control the texture of the food by inhibiting or inducing ice crystal formation in the foods and also maintains the color and appearance of the foods. It also helps to enhance the rheology, stability, mouth feel, and texture of the fermented milk products. During yogurt manufacturing process, low viscosity, high syneresis, and gel fracture are the major problems that can be easily overcome by using these microbial EPS. The adhesiveness, hardness, and chewiness of cheese can be maintained by fermentation using EPS producing *L. delbruekii* subsp. *bulgaricus* and *Streptococcus thermophilus* culture. As a result, the cheese becomes soft and moist due to the presence of EPS, on the other hand in absence of EPS the cheese becomes dry and granular (Jindal & Khattar, 2018). Homopolysaccharide can be divided into two class of (i) glucan and (ii) fructan.

5.2.1 Glucan

Glucan is a high molecular weight polymer, which contain glucose molecules that are connected by glycosidic bonds (Basiri, 2021).

5.2.1.1 Dextran

Dextran is also a high molecular weight polymer, which is obtained from sucrose by the help of dextransucrase enzyme by the bacteria (Basiri, 2021). Dextran is obtained by fermentation of sucrose by using the enzyme of leuconostoc or streptococcus. The pH of the fermentation should be 6.5–7. As soon as the fermentation has been completed, the culture must be cleared to remove all the viable microorganisms. Then the precipitation of the solubilized dextran is done by adding organic solvents such as ethanol, methanol, acetone, and 2-propanol. Again, reprecipitation is done to remove contaminants and the white powder of dextran is produced (Jindal & Khattar, 2018). It can be widely applied in the food industry as follows:

1. Dextran plays an important role as viscosifying, gelling, emulsifying, texturing agents in different food stuffs. It also increases the mouth feel, freshness, and shelf life of bakery products (Ahmed et al., 2016).
2. Dextran helps to increase softness, loaf volume, and crumb texture of bakery products (De Bellis et al., 2022).
3. It plays an important role as a stabilizer, inhibits crystallization, increases moisture retention and viscosity, and enhances flavor (Dahiya & Nigam, 2023).

4. Dextran also plays an important role in manufacturing milk beverages, soft drinks, and ice creams (Basiri, 2021).
5. Dextran also plays an important role in cheese making (reduced fat cheese) by increasing water binding and moisture content in the nonfat substances and increasing creaminess of dairy products (Ahmed et al., 2016).
6. Dextran also has an important role in prebiotics manufacturing. Foods containing prebiotics are associated with prevention of several diseases (diarrhea, bowl cancer, coronary heart disease, inflammatory bowel disease, cholesterol lamia, obesity, type 2 diabetes, and osteoporosis) (Ahmed et al., 2016).
7. Dextran is also used as a cryoprotectant in ice cream production (Ahmed et al., 2016).

5.2.1.2 Pullulan

Pullulan is a noncarcinogenic, non-mutagenic, and nontoxic water-soluble polysaccharide, which contains repeating units of maltotriose (Basiri, 2021). Pullulan can be obtained from sucrose by the enzyme of *A. pullulan* when both UDP and ATP are added in the reaction mixture (Prajapati et al., 2013). The enzyme known as glycosyltransferase is involved in the biosynthesis of pullulan by catalyzing the formation of $\alpha(1{\rightarrow}4)$ and $\alpha(1{\rightarrow}6)$ glycosidic linkages between glucose units (Ponnusami & Gunasekar, 2015). Pullulan can also be produced by fermentation. The media should contain peptone, phosphate, and basal salts. The pH during the fermentation must be 6.5, which falls during first 24 h to 3.5. Maximum growth is achieved within 75 h and maximum pullulan is obtained within 100 h (Prajapati et al., 2013). It can be widely applied in the food industry as follows:

1. It can be utilized as a substitute of starch in pasta and baked products in food industry (Basiri, 2021).
2. Pullulan can also form edible film to food ingredients, which helps in coating the food and packaging material (Oğuzhan & Yangılar, 2013).
3. It can also be used as flavoring in seasoning agents and spicing agents in food industry. Due to this, it is used in different soups, sauces, and beverages for their intensifier properties (Badhwar et al., 2019).
4. It can also be used as a binder denture adhesive and stabilizer in different food paste and helps to adhere nuts to the cookies. It also helps in maintaining the quality and texture of mayonnaise (Oğuzhan & Yangılar, 2013).
5. It is also used as a binding agent in tobacco, plant fertilizer, and seed coating (Oğuzhan & Yangılar, 2013).
6. Maltotriose syrup can be obtained by enzymatic hydrolysis of pullulan by the help of the enzyme pullulanase. This syrup has some excellent properties such as mild sweetness, low freezing point depression, less color formation compared with glucose, maltose, or sucrose syrup, helps retain moisture, and has good heat stability. Due to these properties, it is suitable for the food industries (Oğuzhan & Yangılar, 2013).

5.2.1.3 Curdlan

Curdlan is a tasteless, colorless, odorless, indigestible compound, which has wide applications in the food and medical industries. It is insoluble in water and forms two types of gels, low set gel (thermo-reversible gel) and high set gel (thermo-irreversible gel) (Basiri, 2021). It is primarily applied in the food industry in the following ways:

1. Curdlan forms gels upon heating, and do not convert into original form upon cooling. This property is very useful where stability is needed during processing and storage (Divyasri et al., 2014).
2. Curdlan is also utilized as dietary fiber and gel binder in meat and fish products due to its ability to retain water, which can enhance the texture and moisture of the food (Al-Rmedh et al., 2023).
3. Curdlan enhances the adhesiveness, chewiness, viscosity, and gumminess of emulsified meat products (Basiri, 2021).
4. Curdlan can act like fat in low-fat or fat-free products, which provides creamy taste without added calories (Aquinas et al., 2022).
5. Curdlan can also form biodegradable, edible film for packaging foods and it also shows antimicrobial effect (Basiri, 2021).
6. Curdlan helps to reduce the growth of lactic acid bacteria during milk storage, which converts lactose to lactic acid (Al-Rmedh et al., 2023).
7. Curdlan has been also used as a prebiotic, promoting gut health by supporting beneficial gut microbiota (Chaudhari et al., 2021).

5.2.1.4 Alternan

Alternan is a long chain homopolysaccharide, which is obtained from sucrose by using alternansucrase enzyme (Basiri, 2021). It is used as a prebiotic ingredient in some food products due to its stability under acidic condition and also in syrups where low glycemic index is required (Pilling & Frohberg, 2009).

5.2.1.5 Reuteran

Reuteran is water-soluble α-glucan, which is obtained from reuteransucrase. It helps to improve the quality of gluten-free sourdough and sorghum bread. Benefits include extended shelf life, softer crumb, prebiotic activity, and enhanced nutritional quality of bread (Basiri, 2021). Reuteran exhibits some antimicrobial property, along with it uses in food preservation (El-Ziney et al., 1999).

5.2.1.6 Cellulose

It is a complex and most abundant polysaccharide, which is obtained from microbial sources. It is used in the food industry in the following ways:

1. Cellulose is a nontoxic, biocompatible, easy to make, and renewable compound. So, it has a wide application in the food industry (Jhang et al., 2022).
2. Modified cellulose is utilized to make film coating in food packaging. For example, under certain conditions, sodium carboxymethyl cellulose and hydroxyethyl cellulose are cross-linked with citric acid to make a cellulose-based edible film (Singh et al., 2019).
3. Acetylated cellulose nanofiber (ACNF) is produced from cellulose nanofiber by grafting hydroxyl group with the acetic anhydride on the main chain of the cellulose. Poly lactic acid (PLA) and ACNF are utilized as a filler to form PLA/ACNF film. Due to its good mechanical property, good compatibility, and strong hydrophobicity, it is widely used in food industries (Jhang et al., 2022).
4. Cellulose is also used as a thickening agent along with being used as an emulsifier, stabilizing agent, and expander (Manan et al., 2021).

5.2.1.7 Scleroglucan

Scleroglucan, a water-soluble polysaccharide produced by Sclerotium fungi, has various applications in the food industry (Basiri, 2021). The applications of scleroglucan in food industries are as follows:

1. It is used to enhance the viscosity of different food products like sauces, soups, and beverages (Rahman et al., 2021).
2. By preventing separation in food products, scleroglucan helps to stabilize emulsions and suspensions (Castillo et al., 2015).
3. Scleroglucan is a nondigestible dietary fiber, which promotes gut health (Zeng et al., 2021).

5.2.2 Fructan

Fructans are produced from sucrose by the help of fructosyltransferase enzyme. Fructans are classified into two group: Levan and inulin (Basiri, 2021).

5.2.2.1 Levan

Levan is a nontoxic polymer, which is soluble in water and oil and insoluble in organic solvents (Basiri, 2021). Its applications in the food industry are as follows:

1. Levan is utilized for film coating in food packaging. Addition of plasticizer is required for pure levan film because pure levan is very much brittle in nature due to lack of long flexible moieties (Basiri, 2021).
2. Levan is used as a sweetener (as a fructose source), industrial gum, emulsifying agent, and a compound that maintains texture (Zimmermann et al., 2021).
3. It also enhances the flavor of the food and acts as a filler (bulking agents) (Tomulescu et al., 2016).
4. It is also used as a thickener and stabilizer in food industry (Shi et al., 2023).
5. Due to its probiotic nature, levan can promote the growth of beneficial bacteria in the intestine (Domżał-Kędzia et al., 2023).

5.2.2.2 Inulin

Inulin is a type of fructo-oligosaccharide, which has different applications in food industry (Basiri, 2021). Its applications in the food industry are as follows:

1. It can be utilized as a fat replacer in nonfat functional dairy food and sausages and used as a sugar replacer in chocolate (Basiri, 2021).
2. Inulin can also be utilized as prebiotics in animal and human food stuffs because it is fermented by intestinal bacteria and produce large amount of short fatty acid chain (Basiri, 2021).

5.2.3 Xanthan Gum

In 1960, xanthan gum was discovered and from 1970 it was commercially used. It is formed by aerobic fermentation of glucose and sucrose by using pure culture of *Xanthomonas campestris*. The fermentation is performed for approximate 3 days at 30 °C. As soon as the fermentation is completed, the broth must be sterilized to remove the viable microorganisms. The extracellular components are recovered by centrifugation and xanthan is obtained by precipitation using isopropyl alcohol. Then these are allowed to be dried and milled for packaging. The moisture content should be about 11% and ash content should be 6–9% (Jindal & Khattar, 2018).

Xanthan gum has the following important physical properties:

1. High solubility: Xanthan gum is highly soluble in hot and cold water but hardly soluble in organic solvents. The highest solubility is found in 25% phosphoric acid then acetic acid, nitric acid, sulfuric acid, and sodium hydroxide (Chaturvedi et al., 2021).
2. Emulsifying nature: Xanthan gum is a very good stabilizing, emulsifying, thickening agent. That's why Xanthan gum has a wide application in food industry (Chaturvedi et al., 2021).
3. Highly viscous: Xanthan gum is very much viscous even at very low concentration. The viscosity is determined by the amount of salt, temperature, pH, and the biopolymer concentration (Chaturvedi et al., 2021).

4. High resistance to pH: Xanthan gum can highly tolerate a wide range of pH (pH 2–12), which make it favorable for foods. This property is very important for acidic food presentation (Chaturvedi et al., 2021).
5. Pseudoplasticity: Xanthan gum has a pseudoplasticity property (viscosity inversely changes with shear rate of xanthan solution). This property increases the mouth feel, flavor release in food products (Chaturvedi et al., 2021).

Application in Food Industry

1. It plays an important role as a thickening agent in fruit juices and chocolate (Chaturvedi et al., 2021).
2. Xanthan gum can form a very stable emulsion. That's why it is used in the production of oil-based and non-oil-based sauces and ketchups (Mollakhalili Meybodi & Mohammadifar, 2015).
3. Xanthan gum can also improve texture, flavor, liquidity, viscidness, and rheological performance of foods (Chaturvedi et al., 2021).
4. It acts as a stabilizing agent in desserts and dairy products (ice-creams, milk shakes, and ice-milk) (Katzbauer et al., 1998).
5. Xanthan gum has a protecting effect against the heat shock and it helps to control the formation of ice-crystals (Mollakhalili Meybodi & Mohammadifar, 2015).
6. It is used as a bodying and dangling agent in different drinks (Chaturvedi et al., 2021).
7. It forms an edible coating film, which leads to delay the browning of freshly cut fruits and maintains the quality (Widyaningrum & Meindrawan, 2020).

5.2.4 Gellan

Gellan is an anionic polysaccharide of high molecular weight, which is formed of tetrasaccharide backbone containing 2 β-D glucose, L-rhamnose, D-glucuronic acid, and acyl (glyceryl and acetyl) substituents (Basiri, 2021). Gellan is produced by aerobic fermentation with *S. paucimobilis*. Media composition is very important for gellan production. Media should contain carbon source, nitrogen source, and inorganic salts. Carbohydrates such as glucose, fructose, maltose, and sucrose are utilized as carbon source. Different nitrogen sources are present (organic nitrogen source, corn extract and inorganic nitrogen source, ammonium nitrate, potassium nitrate), which stimulate or limit gellan production. The factors that affect gellan production are oxygen transfer capacity, optimum pH 6.5–7, and optimum temperature 30 °C (Iurciuc et al., 2016).

Application of Gellan in Food Industry

1. When hydrocolloid or gum is dispersed in water, it will form gel. That's why hydrocolloid has a wide application in food industry. They are utilized as emulsifying, gelling, and stabilizing agent in foodstuffs. Gellan also has a water-binding capacity, that's why it increases the stability of the foods (Iurciuc et al., 2016).

2. Gellan plays an important role as suspending agent and stabilizing agent in puddings, icing and glazes, dessert gel, and sauces (Fialho et al., 2008).
3. It can also be utilized as biopolymer support, which helps to encapsulate the bioactive compounds (nutraceutics, enzymes, and cells) to give protection and to obtain foods with different benefits and also to treat chronic diseases (Iurciuc et al., 2016).
4. Gellan prevents the loss of moisture of the foodstuffs and provides clarity to the gellatine desserts. Gellan keeps the foods soft and juicy because the melting temperature of gellan can be increased. That's why Gellan has a wide application in gellatine desserts and Jam production (Iurciuc et al., 2016).

5.2.5 Alginate

Alginate are naturally found, indigestible polysaccharide that are produced and refined from various microorganisms (Senturk Parreidt et al., 2018).

1. Due to its gelling property in the presence of calcium ions, alginate is widely used in the jellies and dessert making (Gheorghita Puscaselu et al., 2020).
2. Alginate enhances the shelf life of fruits, vegetables, and meat from microbial contamination (Pournaki et al., 2024).
3. To enhance texture and stability, alginate is used in the food industry, it also act as a thickener and stabilizer (Basiri, 2021).

5.2.6 Acetan

Acetan is produced by *Acetobacter xylinum* and it is an anionic heteropolysaccharide. It is utilized as gelling agents and viscosifier in sweet product (Basiri, 2021). It improves the texture and increases the stability of the foodstuffs (Gomes et al., 2018). It is also utilized for fermentation of different beverages and acts as an antioxidant (Yassunaka Hata et al., 2023).

5.2.7 Viili

Viili is a ropy fermented milk product and it contains glucose, rhamnose, galactose, and phosphate. It is used as a thickening agent in dairy products and reduces the syneresis of the fermented milk products (Basiri, 2021).

5.2.8 Kefiran

Kefiran, is a polysaccharide that's is derived from kefir grains (Tan et al., 2020).

Application

1. Due to its antimicrobial property, kefiran is used for enhancing the shelf life of food (Tan et al., 2020).
2. To improves the viscosity and texture of sauces and dairy product, kefiran is used extensively (Correia et al., 2022).
3. Kefiran also plays an important role to promote the growth of gut bacteria and acts as a prebiotic (Correia et al., 2022).
4. Kefiran is also used as an eco-friendly packaging solution due to its biodegradable nature (Correia et al., 2022).
5. Kefiran is also utilized as a gelling agent, thickening agent, and water-binding agent (Dailin et al., 2021).

6 Commercialization and Cost-Effectiveness of EPS

Microbial polysaccharides provide several advantages for industrial-scale bioactive material production, including better yields under controlled fermentation and being unaffected by geographic or seasonal fluctuations. Furthermore, they do not compete with agricultural land use and can use a variety of organic sources, unlike plants that rely entirely on solar energy (Donot et al., 2012). Plant-based polymers largely dominate the global biopolymer market, with microbial polysaccharides such as xanthan and gellan making up a smaller portion. Although they hold industrial significance, the extensive production and downstream processing expenses restrict the widespread commercialization of microbial polysaccharides (Sutherland, 2002; Finore et al., 2014). Interest in addressing the economic challenges of using microbial exopolysaccharide (EPS) for industrial purposes has sparked the creation of different approaches. These approaches encompass the utilization of economical fermentation feedstock, the enhancement of production conditions to increase yield, the use of genetic modification to generate superior strains, and the improvement of downstream processing methods to reduce expenses (Rehm, 2010; Kreyenschulte et al., 2014).

EPS production has been the subject of numerous studies, with only a few focusing on the utilization of low-cost media. Approximately, 30% of the total production cost for bioactive compounds is attributed to fermentation media. The use of costly synthetic media, which are abundant in substrate as in yeast extract, peptone (nitrogen source), sugar (carbon source), and other metals, poses economic challenges for EPS production. Bacteria efficiently utilize available carbon and nitrogen sources in the medium under specific environmental conditions to secrete polysaccharides into their surroundings (Nouha et al., 2016). The carbon source provides about 70% of the energy needed for exopolysaccharide biosynthesis (Shukla & Dave, 2018). Developing affordable fermentation media that is biodegradable and economical is critical for increasing production yields while managing expenses. Achieving this involves the development and utilization of cheaper alternative substrates for fermentation, such as agro-industrial by-products and different waste materials (Öner, 2013).

6.1 Optimization of Exopolysaccharide Production from Bacterial Strains Using Agricultural Wastes: A Response Surface Methodology Approach

The study involved the isolation of bacterial strains to produce exopolysaccharides (EPS) from water samples collected at Bonny Island. Bacteria forming mucoid colonies after serial dilution were chosen for further examination. The quantification of EPS was accomplished by introducing pure cultures into nutrient broth with 4% sucrose, followed by centrifugation and ethanol precipitation. Identification of the most effective EPS-producing isolate was conducted through morphological and biochemical tests based on Bergey's Manual. Organic carbon sources derived from plantains, sugarcane bagasse, and sweet potato peels were diluted, autoclaved, and filtered. Similarly, organic nitrogen sources such as bean bran and poultry droppings were processed in potassium sulphate solutions and autoclaved. The optimization for the production of EPS is studied by taking into account one factor at a time methodology to inspect different carbon sources, pH levels, synthetic media, nitrogen source, and the incubation criteria. The condition that would be optimal for the production of EPS was determined via the Response Surface Methodology (RSM). It focuses on the impact of organic nitrogen and carbon sources, the incubation time, and the pH. The crude EPS was finally characterized using Fourier transform infrared (FTIR) spectroscopy to analyze its functional groups (Nwosu et al., 2019).

The focus of this research was on enhancing the production of exopolysaccharides (EPS) from the bacteria strain *Shewanella chilikensis* WAS11, which was identified based on the formation of mucoid colonies. Different culture media were quantitatively evaluated, and it was found that medium C gave an elevated yield of exopolysaccharide (1.29 g/L) and cell growth (1.62 OD). Sucrose was identified as the most effective inorganic carbon source, producing 0.18 g/L of EPS, while the addition of pretreated extracts from sweet potato peels (SPP) resulted in an optimal EPS yield of 1.23 g/L. Among the synthetic nitrogen sources, 0.5% yeast extract demonstrated an abundant production of EPS (0.54 g/L) along with an optimal growth (0.71 OD). When organic nitrogen sources were tested, bean bran yielded the highest EPS (0.95 g/L). The impact of initial pH demonstrated a direct relationship with EPS production, with a pH of 5 resulting in the highest EPS yield (1.98 g/L) and biomass (2.02 OD). With increasing incubation time, the maximum EPS yield was observed on day 6 (1.16 g/L), while the peak cell density occurred on day 9 (1.06 OD). Using Response Surface Methodology (RSM), the optimal conditions for EPS production were determined, highlighting the interactions among carbon and nitrogen sources, pH, and incubation time (Nwosu et al., 2019).

6.2 Physical Mutagenesis for Enhanced EPS Production in Bacillus licheniformis

The optimization of bioprocess for EPS production has been previously documented in different microorganisms, such as *Pseudomonas* species (Raza et al., 2012; Maalej et al., 2015), *Lactobacillus* spp. (Wang et al., 2017a), *Bacillus* spp. (Moghannem et al., 2018), and *Hirsutella* spp. (He et al., 2014). EPS production has been optimized by adjusting various parameters. Traditional optimization methods have been used in many studies, involving the alteration of one factor while keeping the others constant.

B. licheniformis MS3 strain of bacteria was suspected to have achieved the highest yield of EPS, 4.6 g/L with a three times dilution and an hour of irradiation. The carbohydrate content of 83.56% and the reducing ends of 4.69% was indicated by both phenol-sulfuric acid and DNSA tests, respectively. To the contrary, the *B. licheniformis MS5* strain gave a reasonably low yield with two times dilution and irradiation for 120 min. The amount of reducing terminals indicates the existence of monosaccharides or oligosaccharides, which might have been produced during fermentation via depolymerization (Asgher et al., 2020). To further optimize *B. licheniformis MS3* for the exponential production of EPS, the RSM -CCD strategy was operated. The heteropolymeric character of the exopolysaccharide with glucose, mannose, and fructose was revealed by HPLC analysis of a fragment having the highest yield. An evenly, compact construction with very little porosity was observed via SEM analysis. The appearance of glycosidic bonds and COOH/OH functional moieties as their monomeric units was observed by FTIR spectroscopy. Further characterization revealed the hydrophilic nature of EPS and its water and oil holding ability (Asgher et al., 2020).

6.3 Mass Production of Microbial Exopolysaccharide

High production of microbial EPS can be obtained by utilizing various types of bioreactors such as airlift reactors, flow packed bed continuous reactors, and continuous stirred tank reactors. The choice of a suitable bioreactor relies on several parameters such as the correlation between the growth density of the cell and the EPS production, rheometric characteristics of the polymer, and the cohesiveness of the polymer, e.g., highly cohesive xanthan in comparison to lesser cohesive curdlan. Furthermore, the cohesiveness of a media may drastically change from nonviscous in its initial stages to exceedingly cohesive during the process of fermentation, greatly interfering with the transfer of oxygen and thereby altering the growth of biomass (Seviour et al., 2011).

A recently manufactured bioreactor of volume 10 L having a working capacity of 6 L supported with three impellers comprising a set of six-blade turbine impeller has been introduced for successful dissolution of air bubbles, again a spiral impeller which creates an uphill movement and a blade of six turbine impeller has been associated with a spiral that rotates at a very high velocity in a reverse direction. This placement assures a thorough blending of the media amid the EPS production of very cohesive methylene by *Methylobacterium organophilum*. To overcome the problem of viscosity in the bioreactor, the velocity of the stirring was gradually increased to certify the productive mixing and to overcome the restriction of dissolved oxygen. The lower turbine along with the spiral impellers were set at a lower rpm (350) whereas the upper turbine impellers were set at higher rpm (1000). Resulting, the content of methylan grew up to 20.7 g/L, and the cohesiveness of the fermentation media peaked up to 127 MPa, similar to xanthan solution of 68 g/L (Oh et al., 1997).

6.4 Challenges and Upcoming Research Approaches for the Commercialization of EPS

In an attempt to direct the inconsistency and complexities involved during EPS production, several studies were made to find the most cost-effective technique. The experimentation aimed to regulate the production and downstream processing of the exopolysaccharide (EPS) from lactobacillus species, focusing on finding an improved biosynthetic cycle and employing genetically modified stains to peak up the output and efficiency of the EPS (Bibi et al., 2021). Correspondingly, Xiong et al. (2023) aimed to analyze a link between the configuration and functional properties of the exopolysaccharide formed by LAB highlighting their bond for the production of important biopolymers. Continuous advancement is ongoing yet several hindrances invade their practical production. The various requirement of different strains of bacteria and their complicated biosynthetic pathways further makes the production of EPS on a large scale challenging. Moreover, expanding the productivity and keeping in check the cost effectiveness, along with recent methodologies, create remarkably technical challenges (Ibrahim et al., 2022). In an effort to tackle such challenges, further research is to be made to focus on resolving the practical and the controlling difficulties that impede the actual administration of the techniques employed for the production of EPS.

Discovering different methods ranging from latest designs of bioreactors to innovative techniques of fermentation can provide improved and reasonable production prospective. The decreased susceptibility of extremophiles to the contamination of mesophilic bacteria along with its potential to reach very high cell growth density makes it a favorable microbial yard for the production of PHA in a cost-effective manner. The present endeavor is booming so as to aim for the enhancement of biosynthetic pathways with the help of synthetic biology. It focuses on drastically lowering the costs related to the utilization of energy, the complicated downstream processing and the demand for freshwater and substrate (Możejko-Ciesielska et al., 2023). Furthermore, it is necessary to study and research about the extremophiles so as to explore different abilities of EPS ranging from surviving in harsh conditions and protection against harmful radiation. Moreover, it is necessary to discover their biological properties, which include anticancer, antimicrobial, and antioxidant attributes that could improve their potential applications of exopolysaccharide. Cutting-edge technologies such as 3D printing and electro spinning allows modern ways to embody exopolysaccharide into their functional moieties, which permits specialized characteristics, thereby allowing the expansion of the application in various industries (Mouro et al., 2024).

7 Conclusion

The main aspect of the study was the microbial exopolysaccharide. These exopolysaccharides are macromolecules, which vary in their physicochemical properties like thermal stability, cold endurance, salt tolerance depending on the composition of their monomeric subunits, the corresponding functional group attached to the monomer along with the diverse linkages found in their structure. These properties also participate in their potential to act as a drug delivery system, as an immunomodulator or in any other medical field. The production of the microbial EPS depends on various growth parameters like temperature, nutrient availability, salt concentration, pH, carbohydrate, and protein sources. However, these parameters can be optimized for maximum yield. But the use of microbial EPS is still widely restricted due to a number of reasons like side effects, prolonging the drug influence time, drug administration method, which would specifically administer the drug at the exact location and at the exact amount in the human system for optimum effect without any side effects. Further in vitro cytotoxicity tests are still required for their commercialization on a global basis. However, their application as an immunomodulators is being currently studied in the global medical field. They play an important role in immunomodulation by interacting with host immune system, thereby suppressing or stimulating the immune response as per the requirement of the host. These microbial EPS have high potential as alternative source of therapeutics in the medical and pharmaceutical field. Along with their diverse use in the medicinal world, these macromolecules are being also applied in the food industry as stabilizers, thickeners, and emulsifiers because of their eco-friendly, cost-effective, biodegradable, and multifunctional nature. The production complication of microbial EPS can be controlled to some extent with the use of agricultural waste as an alternative source for microbial nutrients. Their application in the modern world can be vastly utilized as a green technology favoring the environmental sustainability.

References

Abdalla, A. K., Ayyash, M. M., Olaimat, A. N., Osaili, T. M., Al-Nabulsi, A. A., Shah, N. P., & Holley, R. (2021). Exopolysaccharides as antimicrobial agents: Mechanism and spectrum of activity. *Frontiers in Microbiology, 12*, 664395.

Abir, F., Barkhordarian, S., & Sumpio, B. E. (2004). Efficacy of dextran solutions in vascular surgery. *Vascular and Endovascular Surgery, 38*(6), 483–491.

Adebayo-Tayo, B., Ishola, R., & Oyewunmi, T. (2018). Characterization, antioxidant and immunomodulatory potential on exopolysaccharide produced by wild type and mutant Weissella confusa strains. *Biotechnology Reports, 19*, e00271.

Agus, A., Planchais, J., & Sokol, H. (2018). Gut microbiota regulation of tryptophan metabolism in health and disease. *Cell Host & Microbe, 23*(6), 716–724. https://doi.org/10.1016/j.chom.2018.05.003

Ahmed, I., Lianfu, Z., & Mahdi, A. (2016). Preparation and application. *International Journal of Research Studies in Agricultural Sciences, 3*(5), 259–265.

Alessandri, G., Ossiprandi, M. C., MacSharry, J., van Sinderen, D., & Ventura, M. (2019). Bifidobacterial dialogue with its human host and consequent modulation of the immune system. *Front Immunology, 10*, 2348.

Al-Rmedh, Y. S. S., Ali, H. I., & Al-Sahlany, S. T. G. (2023). Curdlan gum, properties, benefits and applications. *IOP Conference Series: Earth and Environmental Science, 1158*(11), 112011. IOP Publishing.

Alves da Cunha, M. A., Turmina, J. A., Ivanov, R. C., Barroso, R. R., Marques, P. T., Fonseca, E. A., et al. (2012). Lasiodiplodan, an exocellular (1→6)-β-d-glucan from Lasiodiplodia theobromae MMPI: Production on glucose, fermentation kinetics, rheology and anti-proliferative activity. *Journal of Industrial Microbiology and Biotechnology, 39*(8), 1179–1188.

Amin, M. C. I. M., Abadi, A. G., Ahmad, N., Katas, H., & Jamal, J. A. (2012). Bacterial cellulose film coating as drug delivery system: Physicochemical, thermal and drug release properties. *Sains Malaysiana, 41*(5), 561–568.

Aquinas, N., Bhat, M. R., & Selvaraj, S. (2022). A review presenting production, characterization, and applications of biopolymer curdlan in food and pharmaceutical sectors. *Polymer Bulletin, 79*(9), 6905–6927.

Arserim Ucar, D. K., Konuk Takma, D., & Korel, F. (2021). Exopolysaccharides in food processing industrials. In A. K. Nadda, K. V. Sajna, & S. Sharma (Eds.), *Microbial exopolysaccharides as novel and significant biomaterials* (Springer series on polymer and composite materials). Springer. https://doi.org/10.1007/978-3-030-75289-7_8

Asgher, M., Urooj, Y., Qamar, S. A., & Khalid, N. (2020). Improved exopolysaccharide production from Bacillus licheniformis MS3: Optimization and structural/functional characterization. *International Journal of Biological Macromolecules, 151*, 984–992.

Badhwar, P., Kumar, P., & Dubey, K. K. (2019). Extractive fermentation for process integration and amplified pullulan production by A. pullulans in aqueous two phase systems. *Scientific Reports, 9*(1), 32.

Basiri, S. (2021). Applications of microbial exopolysaccharides in the food industry. *Avicenna Journal of Medical Biochemistry, 9*(2), 107–120.

Bejenariu, A., Popa, M., Dulong, V., Picton, L., & Le Cerf, D. (2009). Trisodium trimetaphosphate crosslinked xanthan networks: Synthesis, swelling, loading and releasing behaviour. *Polymer Bulletin, 62*, 525–538.

Bhat, B., & Bajaj, B. K. (2019). Hypocholesterolemic potential and bioactivity spectrum of an exopolysaccharide from a probiotic isolate Lactobacillus paracasei M7. *Bioactive Carbohydrates and Dietary Fibre, 19*, 100191.

Bibi, A., Xiong, Y., Rajoka, M. S. R., Mehwish, H. M., Radicetti, E., Umair, M., & Aadil, R. M. (2021). Recent advances in the production of exopolysaccharide (EPS) from Lactobacillus spp. and its application in the food industry: A review. *Sustainability, 13*(22), 12429.

Bimczok, D., Wrenger, J., Schirrmann, T., Rothkötter, H. J., Wray, V., & Rau, U. (2009). Short chain regioselectively hydrolyzed scleroglucans induce maturation of porcine dendritic cells. *Applied Microbiology and Biotechnology, 82*, 321–331.

Bisson, G., Comuzzi, C., Giordani, E., Poletti, D., Boaro, M., & Marino, M. (2023). An exopolysaccharide from Leuconostoc mesenteroides showing interesting bioactivities versus foodborne microbial targets. *Carbohydrate Polymers, 301*, 120363.

Bourguignon, L. Y. (2014). Matrix hyaluronan-activated CD44 signaling promotes keratinocyte activities and improves abnormal epidermal functions. *The American Journal of Pathology, 184*(7), 1912–1919.

Campo, V. L., Kawano, D. F., da Silva Jr, D. B., & Carvalho, I. (2009). Carrageenans: Biological properties, chemical modifications and structural analysis–A review. *Carbohydrate Polymers, 77*(2), 167–180.

Castillo, N. A., Valdez, A. L., & Fariña, J. I. (2015). Microbial production of scleroglucan and downstream processing. *Frontiers in Microbiology, 6*, 1106.

Chaisuwan, W., Jantanasakulwong, K., Wangtueai, S., Phimolsiripol, Y., Chaiyaso, T., Techapun, C., Phongthai, S., You, S., Regenstein, J. M., & Seesuriyachan, P. (2020). Microbial exopolysaccharides for immune enhancement: Fermentation, modifications and bioactivities. *Food Bioscience, 35*, 100564. https://doi.org/10.1016/j.fbio.2020.100564

Chang, W. S., & Chen, H. H. (2016). Physical properties of bacterial cellulose composites for wound dressings. *Food Hydrocolloids, 53*, 75–83.

Chaturvedi, S., Kulshrestha, S., Bhardwaj, K., & Jangir, R. (2021). A review on properties and applications of xanthan gum. In *Microbial polymers: Applications and ecological perspectives* (pp. 87–107). Springer.

Chaudhari, V., Buttar, H. S., Bagwe-Parab, S., Tuli, H. S., Vora, A., & Kaur, G. (2021). Therapeutic and industrial applications of curdlan with overview on its recent patents. *Frontiers in Nutrition, 8*, 646988.

Chen, W. J. (2002). Functions of hyaluronan in wound repair. In *Hyaluronan* (pp. 147–156). Elsevier.

Chen, R. N., Lee, L. W., Chen, L. C., Ho, H. O., Lui, S. C., Sheu, M. T., & Su, C. H. (2012). Wound-healing effect of micronized sacchachitin (mSC) nanogel on corneal epithelium. *International Journal of Nanomedicine, 7*, 4697–4706.

Chen, C., Tang, W., Jiang, D., Yang, G., Wang, X., Zhou, L., et al. (2019). Hyaluronic acid conjugated polydopamine functionalized mesoporous silica nanoparticles for synergistic targeted chemo-photothermal therapy. *Nanoscale, 11*(22), 11012–11024.

Ciszek-Lenda, M., Nowak, B., Śróttek, M., Gamian, A., & Marcinkiewicz, J. (2011). Immunoregulatory potential of exopolysaccharide from Lactobacillus rhamnosus KL37. Effects on the production of inflammatory mediators by mouse macrophages. *International Journal of Experimental Pathology, 92*(6), 382–391.

Clementi, F., Fantozzi, P., Mancini, F., & Moresi, M. (1995). Optimal conditions for alginate production by Azotobacter vinelandii. *Enzyme and Microbial Technology, 17*(11), 983–988.

Correia, S., Gonçalves, C., Oliveira, J. M., Radhouani, H., & Reis, R. L. (2022). Impact of kefiran exopolysaccharide extraction on its applicability for tissue engineering and regenerative medicine. *Pharmaceutics, 14*(8), 1713.

Dahiya, D., & Nigam, P. S. (2023). Dextran of diverse molecular-configurations used as a blood-plasma substitute, drug-delivery vehicle and food additive biosynthesized by Leuconostoc, Lactobacillus and Weissella. *Applied Sciences, 13*(22), 12526.

Dailin, D. J., Malek, R. A., Ting, T. L., Wehbe, R., Ramli, S., Elsayed, A. A., et al. (2021). Biosynthesis, production and application of kefiran in food industry: A review. *Bioscience Research, 18*(1), 102–119.

De Bellis, P., Ferrara, M., Bavaro, A. R., Linsalata, V., Di Biase, M., Musio, B., Gallo, V., Mulè, G., & Valerio, F. (2022). Characterization of dextran produced by the food-related strain Weissella cibaria C43-11 and of the relevant dextransucrase gene. *Foods (Basel, Switzerland), 11*(18), 2819.

De Oliveira Cardoso, V. M., Cury, B. S. F., Evangelista, R. C., & Gremião, M. P. D. (2017). Development and characterization of cross-linked gellan gum and retrograded starch blend hydrogels for drug delivery applications. *Journal of the Mechanical Behavior of Biomedical Materials, 65*, 317–333.

Dertli, E., Colquhoun, I. J., Côté, G. L., Le Gall, G., & Narbad, A. (2018). Structural analysis of the α-D-glucan produced by the sourdough isolate Lactobacillus brevis E25. *Food Chemistry, 242*, 45–52. https://doi.org/10.1016/j.foodchem.2017.09.017

Divanovic, S., Trompette, A., Atabani, S. F., Madan, R., Golenbock, D. T., Visintin, A., Finberg, R. W., Tarakhovsky, A., Vogel, S. N., Belkaid, Y., Kurt-Jones, E. A., & Karp, C. L. (2005). Negative regulation of Toll-like receptor 4 signaling by the Toll-like receptor homolog RP105. *Nature Immunology, 6*, 571–578.

Divyasri, D., Gunasekar, V., Benny, I. S., & Ponnusami, V. (2014). A review on industrial applications of curdlan. *International Journal of ChemTech Research, 6*(5), 3000–3003.

Domżał-Kędzia, M., Ostrowska, M., Lewińska, A., & Łukaszewicz, M. (2023). Recent developments and applications of microbial levan, a versatile polysaccharide-based biopolymer. *Molecules, 28*(14), 5407.

Donot, F., Fontana, A., Baccou, J. C., & Schorr-Galindo, S. (2012). Microbial exopolysaccharides: Main examples of synthesis, excretion, genetics and extraction. *Carbohydrate Polymers, 87*(2), 951–962.

El-Ghonemy, D. H. (2021). Antioxidant and antimicrobial activities of exopolysaccharides produced by a novel Aspergillus sp. DHE6 under optimized submerged fermentation conditions. *Biocatalysis and Agricultural Biotechnology, 36*, 102150.

El-Naggar, M. E., Abdelgawad, A. M., Tripathi, A., & Rojas, O. J. (2017). Curdlan cryogels reinforced with cellulose nanofibrils for controlled release. *Journal of Environmental Chemical Engineering, 5*(6), 5754–5761.

El-Ziney, M. G., Van Den Tempel, T., Debevere, J., & Jakobsen, M. (1999). Application of reuterin produced by Lactobacillus reuteri 12002 for meat decontamination and preservation. *Journal of Food Protection, 62*(3), 257–261.

Falch, B. H., Espevik, T., Ryan, L., & Stokke, B. T. (2000). The cytokine stimulating activity of (1→3)-β-D-glucans is dependent on the triple helix conformation. *Carbohydrate Research, 329*(3), 587–596.

Fanning, S., Hall, L. J., Cronin, M., Zomer, A., MacSharry, J., Goulding, D., et al. (2012). Bifidobacterial surface-exopolysaccharide facilitates commensal-host interaction through immune modulation and pathogen protection. *Proceedings of the National Academy of Sciences, 109*(6), 2108–2113.

Faralli, A., Shekarforoush, E., Ajalloueian, F., Mendes, A. C., & Chronakis, I. S. (2019). In vitro permeability enhancement of curcumin across Caco-2 cells monolayers using electrospun xanthan-chitosan nanofibers. *Carbohydrate Polymers, 206*, 38–47.

Fialho, A. M., Martins, L. O., Donval, M. L., Leitão, J. H., Ridout, M. J., Jay, A. J., et al. (1999). Structures and properties of gellan polymers produced by Sphingomonas paucimobilis ATCC 31461 from lactose compared with those produced from glucose and from cheese whey. *Applied and Environmental Microbiology, 65*(6), 2485–2491.

Fialho, A. M., Moreira, L. M., Granja, A. T., Popescu, A. O., Hoffmann, K., & Sá-Correia, I. (2008). Occurrence, production, and applications of gellan: Current state and perspectives. *Applied Microbiology and Biotechnology, 79*, 889–900.

Finore, I., Di Donato, P., Mastascusa, V., Nicolaus, B., & Poli, A. (2014). Fermentation technologies for the optimization of marine microbial exopolysaccharide production. *Marine Drugs, 12*(5), 3005–3024.

Freitas, F., Alves, V. D., & Reis, M. A. (2011). Advances in bacterial exopolysaccharides: From production to biotechnological applications. *Trends in Biotechnology, 29*(8), 388–398.

Gheorghita Puscaselu, R., Lobiuc, A., Dimian, M., & Covasa, M. (2020). Alginate: From food industry to biomedical applications and management of metabolic disorders. *Polymers, 12*(10), 2417.

Gomes, R. J., Borges, M. F., Rosa, M. F., Castro-Gómez, R. J. H., & Spinosa, W. A. (2018). Acetic acid bacteria in the food industry: Systematics, characteristics and applications. *Food Technology and Biotechnology, 56*(2), 139–151.

Grigoras, A. G. (2019). Drug delivery systems using pullulan, a biocompatible polysaccharide produced by fungal fermentation of starch. *Environmental Chemistry Letters, 17*(3), 1209–1223.

Guo, Y., Pan, D., Sun, Y., Xin, L., Li, H., & Zeng, X. (2013). Antioxidant activity of phosphorylated exopolysaccharide produced by Lactococcus lactis subsp. lactis. *Carbohydrate Polymers, 97*(2), 849–854.

Hamer, H. M., Jonkers, D., Venema, K., Vanhoutvin, S., Troost, F. J., & Brummer, R. J. (2007). The role of butyrate on colonic function. *Alimentary Pharmacology and Therapeutics, 27*(2), 104–119. https://doi.org/10.1111/j.1365-2036.2007.03562.x

Han, Y. W. (1990). Microbial levan. *Advances in Applied Microbiology, 35*, 171–194.

He, L., Cheng, J. W., Wang, Y. B., Li, H. B., Qian, H., Li, W. Q., & Ren, X. Y. (2014). Statistics-based optimization of extracellular polysaccharide production from Hirsutella sinensis using a fermentation process and in vitro immunomodulatory activity. *Food Science and Biotechnology, 23*, 451–457.

Hidalgo-Cantabrana, C., López, P., Gueimonde, M., de Los Reyes-Gavilán, C. G., Suárez, A., Margolles, A., & Ruas-Madiedo, P. (2012). Immune modulation capability of exopolysaccharides synthesised by lactic acid bacteria and bifidobacteria. *Probiotics and Antimicrobial Proteins, 4*(4), 227–237.

Hovgaard, L., & Brøndsted, H. (1995). Dextran hydrogels for colon-specific drug delivery. *Journal of Controlled Release, 36*(1–2), 159–166.

Huang, X., Ai, C., Yao, H., Zhao, C., Xiang, C., Hong, T., & Xiao, J. (2022). Guideline for the extraction, isolation, purification, and structural characterization of polysaccharides from natural resources. *eFood, 3*(6), e37.

Ibrahim, H. A., Abou Elhassayeb, H. E., & El-Sayed, W. M. (2022). Potential functions and applications of diverse microbial exopolysaccharides in marine environments. *Journal of Genetic Engineering and Biotechnology, 20*(1), 151.

Ignatova, M., Manolova, N., Rashkov, I., & Markova, N. (2016). Quaternized chitosan/κ--carrageenan/caffeic acid–coated poly (3-hydroxybutyrate) fibrous materials: Preparation, antibacterial and antioxidant activity. *International Journal of Pharmaceutics, 513*(1–2), 528–537.

Iurciuc, C., Savin, A., Lungu, C., Martin, P., & Popa, M. (2016). Gellan food applications. *Cellulose Chemistry and Technology, 50*, 1–13.

Jaseera, K. V., & Abdulla, T. (2021). Microbial EPS as immunomodulatory agents. In A. K. Nadda, K. V. Sajna, & S. Sharma (Eds.), *Microbial exopolysaccharides as novel and significant biomaterials* (Springer series on polymer and composite materials). Springer. https://doi.org/10.1007/978-3-030-75289-7_9

Jhang, J. C, Lin, J. H, Lou, C, W, Chen, Y. S. Biodegradable and conductive PVA/CNT nanofibrous membranes used in nerve conduit applications. *Journal of Industrial Textiles*. 2021;51(1_suppl):1048S–1065S. https://doi.org/10.1177/15280837211032086

Jin, M., Wang, Y., Huang, M., Lu, Z., & Wang, Y. (2014). Sulphation can enhance the antioxidant activity of polysaccharides produced by Enterobacter cloacae Z0206. *Carbohydrate Polymers, 99*, 624–629.

Jindal, N., & Khattar, J. S. (2018). Microbial polysaccharides in food industry. In *Biopolymers for food design* (pp. 95–123). Academic Press.

Jindal, N., Khattar, J. S., & Singh, D. P. (2022). Isolation of microbial polysaccharides. In *Polysaccharides of microbial origin: Biomedical applications* (pp. 439–452). Springer International Publishing.

Kalam, M. A. (2016). Development of chitosan nanoparticles coated with hyaluronic acid for topical ocular delivery of dexamethasone. *International Journal of Biological Macromolecules, 89*, 127–136.

Kalimuthu, A. K., Pandian, S. R. K., Pavadai, P., Panneerselvam, T., Kabilan, S. J., Sankaranarayanan, M., et al. (2023). Drug delivery applications of exopolysaccharides from endophytic bacteria Pseudomonas otitidis from Tribulus terrestris L. *Journal of Polymers and the Environment, 31*(8), 3632–3649.

Kanamarlapudi, S. L. R. K., & Muddada, S. (2017). Characterization of exopolysaccharide produced by Streptococcus thermophilus CC30. *BioMed Research International, 2017*(1), 4201809.

Kano, Y., Kakuta, H., & Hashimoto, J. (1996). Augmentation of antitumor effect by combined administration with interleukin-2 and sizofiran, a single glucan, on murine EL-4 lymphoma. *Biotherapy, 9*, 241–247.

Kano, Y., Kakuta, H., & Hashimoto, J. (1998). Augmentation of the antitumor effect of adoptive immunotherapy by in vivo sensitization of EL-4 lymphoma and pre-treatment with sizofiran. *Biotherapy, 11*, 1–6.

KATZBAUER, B. (1998). Properties and applications of xanthan gum. Polymer degradation and stability, 59(1–3), 81–84.

Kim, J. H., Yoo, S. J., Oh, D. K., Kweon, Y. G., Park, D. W., Lee, C. H., & Gil, G. H. (1996). Selection of a Streptococcus equi mutant and optimization of culture conditions for the production of high molecular weight hyaluronic acid. *Enzyme and Microbial Technology, 19*(6), 440–445.

Kim, S. J., Bae, P. K., & Chung, B. H. (2015). Self-assembled levan nanoparticles for targeted breast cancer imaging. *Chemical Communications, 51*(1), 107–110.

Kobayashi, H., Katakura, O., Morimoto, N., Akiyoshi, K., & Kasugai, S. (2009). Effects of cholesterol-bearing pullulan (CHP)-nanogels in combination with prostaglandin E1 on wound healing. *Journal of Biomedical Materials Research Part B: Applied Biomaterials, 91*(1), 55–60.

Kreyenschulte, D., Krull, R., & Margaritis, A. (2014). Recent advances in microbial biopolymer production and purification. *Critical Reviews in Biotechnology, 34*(1), 1–15.

Labeer, S., Vanderleyden, J., & De Keersmaecker, S. C. J. (2010). Host interactions of probiotic bacterial surface molecules: Comparison with commensals and pathogens. *Nature Reviews Microbiology, 8*(3), 171–184.

Laiño, J., Villena, J., Kanmani, P., & Kitazawa, H. (2016). Immunoregulatory effects triggered by lactic acid bacteria exopolysaccharides: New insights into molecular interactions with host cells. *Microorganisms, 4*(3), 27.

Larsen, F. H., & Engelsen, S. B. (2015). Insight into the functionality of microbial exopolysaccharides by NMR spectroscopy and molecular modeling. *Frontiers in Microbiology, 6*, 1374.

Lee, K. Y., & Mooney, D. J. (2012). Alginate: Properties and biomedical applications. *Progress in Polymer Science, 37*(1), 106–126.

Li, H., Bian, S., Huang, Y., Liang, J., Fan, Y., & Zhang, X. (2014). High drug loading pH-sensitive pullulan-DOX conjugate nanoparticles for hepatic targeting. *Journal of Biomedical Materials Research Part A, 102*(1), 150–159.

Liu, P., Situ, J. Q., Li, W. S., Shan, C. L., You, J., Yuan, H., et al. (2015). High tolerated paclitaxel nano-formulation delivered by poly (lactic-co-glycolic acid)-g-dextran micelles to efficient cancer therapy. *Nanomedicine: Nanotechnology, Biology and Medicine, 11*(4), 855–866.

Llamas, I., Moral, A. D., Martínez-Checa, F., Arco, Y., Arias, S., & Quesada, E. (2006). Halomonas maura is a physiologically versatile bacterium of both ecological and biotechnological interest. *Antonie Van Leeuwenhoek, 89*, 395–403.

Maalej, H., Hmidet, N., Boisset, C., Buon, L., Heyraud, A., & Nasri, M. (2015). Optimization of exopolysaccharide production from Pseudomonas stutzeri AS22 and examination of its metal-binding abilities. *Journal of Applied Microbiology, 118*(2), 356–367.

Manan, S., Shahzad, A., Ul-Islam, M., Ullah, M. W., & Yang, G. (2021). Synthesis routes and applications of cellulose in food industry. In *Bacterial cellulose* (pp. 115–143). CRC Press.

Martins, M., Barros, A. A., Quraishi, S., Gurikov, P., Raman, S. P., Smirnova, I., et al. (2015). Preparation of macroporous alginate-based aerogels for biomedical applications. *The Journal of Supercritical Fluids, 106*, 152–159.

McCahon, R., & Hardman, J. (2007). Pharmacology of plasma expanders. *Anaesthesia & Intensive Care Medicine, 8*(2), 79–81.

Medrano, M., Racedo, S. M., Rolny, I. S., Abraham, A. G., & Pérez, P. F. (2011). Oral administration of kefiran induces changes in the balance of immune cells in a murine model. *Journal of Agricultural and Food Chemistry, 59*(10), 5299–5304.

Mikelsaar, M., & Zilmer, M. (2009). Lactobacillus fermentum ME-3 – An antimicrobial and antioxidative probiotic. *Microbial Ecology in Health and Disease, 21*(1), 1–27.

Milivojevic, M., Pajic-Lijakovic, I., Bugarski, B., Nayak, A. K., & Hasnain, M. S. (2019). Gellan gum in drug delivery applications. In *Natural polysaccharides in drug delivery and biomedical applications* (pp. 145–186). Academic Press.

Miranda-Nantes, C. C., Fonseca, E. A., Zaia, C. T., Dekker, R. F., Khaper, N., Castro, I. A., & Barbosa, A. M. (2011). Hypoglycemic and hypocholesterolemic effects of botryosphaeran from Botryosphaeria rhodina MAMB-05 in diabetes-induced and hyperlipidemia conditions in rats. *Mycobiology, 39*(3), 187–193.

Moghannem, S. A., Farag, M., Shehab, A. M., & Azab, M. S. (2018). Exopolysaccharide production from Bacillus velezensis KY471306 using statistical experimental design. *Brazilian Journal of Microbiology, 49*, 452–462.

Mollakhalili Meybodi, N., & Mohammadifar, M. A. (2015). Microbial exopolysaccharides: A review of their function and application in food sciences. *Journal of Food Quality and Hazards Control, 2*(4), 112–117.

Morris, E. R., Rees, D. A., Young, G., Walkinshaw, M. D., & Darke, A. (1977). Order-disorder transition for a bacterial polysaccharide in solution. A role for polysaccharide conformation in recognition between Xanthomonas pathogen and its plant host. *Journal of Molecular Biology, 110*(1), 1–16.

Mouro, C., Gomes, A. P., & Gouveia, I. C. (2024). Microbial exopolysaccharides: Structure, diversity, applications, and future frontiers in sustainable functional materials. *Polysaccharides, 5*(3), 241–287.

Możejko-Ciesielska, J., Ray, S., & Sankhyan, S. (2023). Recent challenges and trends of polyhydroxyalkanoate production by extremophilic bacteria using renewable feedstocks. *Polymers, 15*(22), 4385.

Muddineti, O. S., Kumari, P., Ajjarapu, S., Lakhani, P. M., Bahl, R., Ghosh, B., & Biswas, S. (2016). Xanthan gum stabilized PEGylated gold nanoparticles for improved delivery of curcumin in cancer. *Nanotechnology, 27*(32), 325101.

Murofushi, Y., Villena, J., Morie, K., Kanmani, P., Tohno, M., Shimazu, T., Aso, H., Suda, Y., Hashiguchi, K., Saito, T., & Kitazawa, H. (2015). The toll-like receptor family protein RP105/MD1 complex is involved in the immunoregulatory effect of exopolysaccharides from Lactobacillus plantarum N14. *Molecular Immunology, 64*(1), 63–75. https://doi.org/10.1016/j.molimm.2014.10.027

Na, K., Park, K. H., Kim, S. W., & Bae, Y. H. (2000). Self-assembled hydrogel nanoparticles from curdlan derivatives: Characterization, anti-cancer drug release and interaction with a hepatoma cell line (HepG2). *Journal of Controlled Release, 69*(2), 225–236.

Nissola, C., Marchioro, M. L. K., Mello, E. V. D. S. L., Guidi, A. C., de Medeiros, D. C., da Silva, C. G., et al. (2021). Hydrogel containing (1→6)-β-D-glucan (lasiodiplodan) effectively promotes dermal wound healing. *International Journal of Biological Macromolecules, 183*, 316–330.

Nouha, K., Rd, T., & Ry, S. (2016). EPS producing microorganisms from municipal wastewater activated sludge. *Journal of Petroleum & Environmental Biotechnology, 7*(1), 1000255.

Nwosu, I. G., Abu, G. O., & Agwa, K. O. (2019). Production of microbial exopolysaccharide by cost-effective medium opimization method. *Journal of Advanced Microbiology, 19*, 1–13.

Oerlemans, M., Akkerman, R., Ferrari, M., Walvoort, M. T. C., & de Vos, P. (2021). Benefits of bacteria-derived exopolysaccharides on gastrointestinal microbiota, immunity and health. *Journal of Functional Foods, 76*, 104289. https://doi.org/10.1016/j.jff.2020.104289

Oğuzhan, P., & Yangılar, F. (2013). Pullulan: Production and usage in food ındustry. *African Journal of Food Science and Technology, 4*(3), 2141–5455.

Oh, D. K., Kim, J. H., & Yoshida, T. (1997). Production of a high viscosity polysaccharide, methylan, in a novel bioreactor. *Biotechnology and Bioengineering, 54*(2), 115–121.

Öner, E. T. (2013). Microbial production of extracellular polysaccharides from biomass. In *Pretreatment techniques for biofuels and biorefineries* (pp. 35–56). Springer.

Osman, A., Oner, E. T., & Eroglu, M. S. (2017). Novel levan and pNIPA temperature sensitive hydrogels for 5-ASA controlled release. *Carbohydrate Polymers, 165*, 61–70.

Padhi, J. R., Nayak, D., Nanda, A., Rauta, P. R., Ashe, S., & Nayak, B. (2016). Development of highly biocompatible gelatin & i-carrageenan based composite hydrogels: In depth physiochemical analysis for biomedical applications. *Carbohydrate Polymers, 153*, 292–301.

Patel, J., Maji, B., Moorthy, N. H. N., & Maiti, S. (2020). Xanthan gum derivatives: Review of synthesis, properties and diverse applications. *RSC Advances, 10*(45), 27103–27136.

Pescosolido, L., Vermonden, T., Malda, J., Censi, R., Dhert, W. J., Alhaique, F., et al. (2011). In situ forming IPN hydrogels of calcium alginate and dextran-HEMA for biomedical applications. *Acta Biomaterialia, 7*(4), 1627–1633.

Pilling, J., & Frohberg, C. (2009, March 6). *Use of alternan as a thickening agent and thickening agent compositions contining altrnan and another thickening agent*. WIPO Patent WO200910408A1.

Pirzadeh-Naeeni, S., Mozdianfard, M. R., Shojaosadati, S. A., Khorasani, A. C., & Saleh, T. (2020). A comparative study on schizophyllan and chitin nanoparticles for ellagic acid delivery in treating breast cancer. *International Journal of Biological Macromolecules, 144*, 380–388.

Ponnusami, V., & Gunasekar, V. (2015). Production of pullulan by microbial fermentation. In *Polysaccharides* (pp. 581–596). Springer.

Porter, N. T., & Martens, E. C. (2017). The critical roles of polysaccharides in gut microbial ecology and physiology. *Annual Review of Microbiology, 71*, 349–369.

Pournaki, S. K., Aleman, R. S., Hasani-Azhdari, M., Marcia, J., Yadav, A., & Moncada, M. (2024). Current review: Alginate in the food applications. *J, 7*(3), 281–301.

Prajapati, V. D., Jani, G. K., & Khanda, S. M. (2013). Pullulan: An exopolysaccharide and its various applications. *Carbohydrate Polymers, 95*(1), 540–549.

Rahman, M. M., & Lamsal, B. P. (2021). Ultrasound-assisted extraction and modification of plant-based proteins: Impact on physicochemical, functional, and nutritional properties. Comprehensive reviews in food science and food safety, 20(2), 1457–1480. https://doi.org/10.1111/1541-4337.12709

Rajoka, M., Mehwish, H. M., Kitazawa, H., Barba, F. J., Berthelot, L., Umair, M., Zhu, Q., He, Z., & Zhao, L. (2022). Techno-functional properties and immunomodulatory potential of exopolysaccharide from Lactiplantibacillus plantarum MM89 isolated from human breast milk. *Food Chemistry, 377*, 131954.

Raveendran, S., Poulose, A. C., Yoshida, Y., Maekawa, T., & Kumar, D. S. (2013). Bacterial exopolysaccharide-based nanoparticles for sustained drug delivery, cancer chemotherapy and bioimaging. *Carbohydrate Polymers, 91*(1), 22–32.

Raza, W., Yang, W., Jun, Y., Shakoor, F., Huang, Q., & Shen, Q. (2012). Optimization and characterization of a polysaccharide produced by Pseudomonas fluorescens WR-1 and its antioxidant activity. *Carbohydrate Polymers, 90*(2), 921–929.

Razzaghi, M., Navvabi, A., Homaee, M. B., Sani, R., Michaud, P., & Homaei, A. (2021). Exopolysaccharides in drug delivery systems. In A. K. Nadda, K. V. Sajna, & S. Sharma (Eds.), *Microbial exopolysaccharides as novel and significant biomaterials* (Springer series on polymer and composite materials). Springer. https://doi.org/10.1007/978-3-030-75289-7_7

Rehm, B. H. (2010). Bacterial polymers: Biosynthesis, modifications and applications. *Nature Reviews Microbiology, 8*(8), 578–592.

Ren, Y. Y., Sun, P. P., Ji, Y. P., Wang, X. T., Dai, S. H., & Zhu, Z. Y. (2020). Carboxymethylation and acetylation of the polysaccharide from Cordyceps militaris and their α-glucosidase inhibitory activities. *Natural Product Research, 34*(3), 369–377.

Rojo, D., Méndez-García, C., Raczkowska, B. A., Bargiela, R., Moya, A., Ferrer, M., & Barbas, C. (2017). Exploring the human microbiome from multiple perspectives: Factors altering its composition and function. *FEMS Microbiology Reviews, 41*, 453–478. https://doi.org/10.1093/femsre/fuw046

Rossi, O., Khan, M. T., Schwarzer, M., Hudcovic, T., Srutkova, D., Duncan, S. H., Stolte, E. H., Kozakova, H., Flint, H. J., Samsom, J. N., Harmsen, H. J. M., & Wells, J. M. (2015). Faecalibacterium prausnitzii strain HTF-F and its extracellular polymeric matrix attenuate clinical parameters in DSS-induced colitis. *PLoS One, 10*(4), 1–17. https://doi.org/10.1371/journal.pone.0123013

Sadeghi, M., Haghshenas, B., & Nami, Y. (2024). Bifidobacterium exopolysaccharides: New insights into engineering strategies, physicochemical functions, and immunomodulatory effects on host health. *Food Microbiology, 15*, 1396308.

Sajna, K. V., Sharma, S., & Nadda, A. K. (2021). Microbial exopolysaccharides: An introduction. In A. K. Nadda, K. V. Sajna, & S. Sharma (Eds.), *Microbial exopolysaccharides as novel and significant biomaterials* (Springer series on polymer and composite materials). Springer. https://doi.org/10.1007/978-3-030-75289-7_1

Salunke, S. R., & Patil, S. B. (2016). Ion activated in situ gel of gellan gum containing salbutamol sulphate for nasal administration. *International Journal of Biological Macromolecules, 87*, 41–47.

Sarwat, F., Qader, S. A. U., Aman, A., & Ahmed, N. (2008). Production & characterization of a unique dextran from an indigenous Leuconostoc mesenteroides CMG713. *International Journal of Biological Sciences, 4*(6), 379.

Senturk Parreidt, T., Müller, K., & Schmid, M. (2018). Alginate-based edible films and coatings for food packaging applications. *Foods, 7*(10), 170.

Seviour, R. J., McNeil, B., Fazenda, M. L., & Harvey, L. M. (2011). Operating bioreactors for microbial exopolysaccharide production. *Critical Reviews in Biotechnology, 31*(2), 170–185.

Sezer, A. D., Kazak Sarılmışer, H., Rayaman, E., Çevikbaş, A., Öner, E. T., & Akbuğa, J. (2017). Development and characterization of vancomycin-loaded levan-based microparticular system for drug delivery. *Pharmaceutical Development and Technology, 22*(5), 627–634.

Shi, S. S., Lei, S. J., & Fu, C. X. (2020). Studies of the properties of CHG-loaded alginate fibers for medical application. *Polymer Testing, 83*, 106141.

Shi, Y., Si, D., Zhang, X., Chen, D., & Han, Z. (2023). Plant fructans: Recent advances in metabolism, evolution aspects and applications for human health. *Current Research in Food Science, 7*, 100595.

Shimazu, T., Villena, J., Tohno, M., Fujie, H., Hosoya, S., Shimosato, T., et al. (2012). Immunobiotic Lactobacillus jensenii elicits anti-inflammatory activity in porcine intestinal epithelial cells by modulating negative regulators of the Toll-like receptor signaling pathway. *Infection and Immunity, 80*(1), 276–288.

Shukla, P. J., & Dave, B. P. (2018). Screening and molecular identification of potential exopolysaccharides (EPS) producing marine bacteria from the Bhavnagar coast, Gujarat. *International Journal of Pharmaceutical Sciences and Research, 9*(7), 2973–2981.

Silva, N. H., Rodrigues, A. F., Almeida, I. F., Costa, P. C., Rosado, C., Neto, C. P., et al. (2014). Bacterial cellulose membranes as transdermal delivery systems for diclofenac: In vitro dissolution and permeation studies. *Carbohydrate Polymers, 106*, 264–269.

Singh, R. S., Kaur, N., & Kennedy, J. F. (2015). Pullulan and pullulan derivatives as promising biomolecules for drug and gene targeting. *Carbohydrate Polymers, 123*, 190–207.

Singh, P., Magalhães, S., Alves, L., Antunes, F., Miguel, M., Lindman, B., & Medronho, B. (2019). Cellulose-based edible films for probiotic entrapment. *Food Hydrocolloids, 88*, 68–74. https://doi.org/10.1016/j.foodhyd.2018.08.057

Sipahigil, O., & Dortunc, B. (2001). Preparation and in vitro evaluation of verapamil HCl and ibuprofen containing carrageenan beads. *International Journal of Pharmaceutics, 228*(1–2), 119–128.

Staudacher, H., Lomer, M. C. E., Farquharson, F. M., Louis, P., Fava, F., Franciosi, E., Scholz, M., Tuohy, K. M., Lindsay, J. O., Irving, P. M., & Whelan, K. (2017). A diet low in FODMAPs reduces symptoms in patients with irritable bowel syndrome and A probiotic restores

Bifidobacterium species: A randomized controlled trial. *Gastroenterology, 153*(4), 936–947. https://doi.org/10.1053/j.gastro.2017.06.010

Supramaniam, J., Adnan, R., Kaus, N. H. M., & Bushra, R. (2018). Magnetic nanocellulose alginate hydrogel beads as potential drug delivery system. *International Journal of Biological Macromolecules, 118*, 640–648.

Surayot, U., Wang, J., Seesuriyachan, P., Kuntiya, A., Tabarsa, M., Lee, Y., Kim, J. K., Park, W., & You, S. (2014). Exopolysaccharides from lactic acid bacteria: Structural analysis, molecular weight effect on immunomodulation. *International Journal of Biological Macromolecules, 68*, 233–240.

Sutherland, I. (2002). A sticky business. Microbial polysaccharides: Current products and future trends. *Microbiology Today, 29*, 70–71.

Tan, K. X., Chamundeswari, V. N., & Loo, S. C. J. (2020). Prospects of kefiran as a food-derived biopolymer for agri-food and biomedical applications. *RSC Advances, 10*(42), 25339–25351.

Thursby, E., & Juge, N. (2017). Introduction to the human gut microbiota. *Biochemical Journal, 474*(2), 1823–1836. https://doi.org/10.1042/BCJ20160510

Tomulescu, C., Stoica, R., Sevcenco, C., Căşărică, A., Moscovici, M., & Vamanu, A. (2016). Levan – A mini review. *Scientific Bulletin Series F. Biotechnologies, 20*, 309–317.

Valentine, M. E., Kirby, B. D., Withers, T. R., Johnson, S. L., Long, T. E., Hao, Y., et al. (2020). Generation of a highly attenuated strain of Pseudomonas aeruginosa for commercial production of alginate. *Microbial Biotechnology, 13*(1), 162–175.

Varghese, J. S., Chellappa, N., & Fathima, N. N. (2014). Gelatin–carrageenan hydrogels: Role of pore size distribution on drug delivery process. *Colloids and Surfaces B: Biointerfaces, 113*, 346–351.

Viñarta, S. C., François, N. J., Daraio, M. E., Figueroa, L. I., & Fariña, J. I. (2007). Sclerotium rolfsii scleroglucan: The promising behavior of a natural polysaccharide as a drug delivery vehicle, suspension stabilizer and emulsifier. *International Journal of Biological Macromolecules, 41*(3), 314–323.

Vinderola, G., Perdigón, G., Duarte, J., Farnworth, E., & Matar, C. (2006). Effects of the oral administration of the exopolysaccharide produced by Lactobacillus kefiranofaciens on the gut mucosal immunity. *Cytokine, 36*(5–6), 254–260.

Wachi, S., Kanmani, P., Tomosada, Y., Kobayashi, H., Yuri, T., Egusa, S., Shimazu, T., Suda, Y., Aso, H., Sugawara, M., Saito, T., Mishima, T., Villena, J., & Kitazawa, H. (2014). Lactobacillus delbrueckii TUA4408L and its extracellular polysaccharides attenuate enterotoxigenic Escherichia coli-induced inflammatory response in porcine intestinal epitheliocytes via Toll-like receptor-2 and 4. *Molecular Nutrition & Food Research, 58*(10), 2080–2093. https://doi.org/10.1002/mnfr.201400218

Wang, K., Li, W., Rui, X., Chen, X., Jiang, M., & Dong, M. (2014). Characterization of a novel exopolysaccharide with antitumor activity from Lactobacillus plantarum 70810. *International Journal of Biological Macromolecules, 63*, 133–139.

Wang, K., Li, W., Rui, X., Li, T., Chen, X., Jiang, M., & Dong, M. (2015). Chemical modification, characterization and bioactivity of a released exopolysaccharide (r-EPS1) from Lactobacillus plantarum 70810. *Glycoconjugate Journal, 32*, 17–27.

Wang, X., Shao, C., Liu, L., Guo, X., Xu, Y., & Lü, X. (2017a). Optimization, partial characterization and antioxidant activity of an exopolysaccharide from Lactobacillus plantarum KX041. *International Journal of Biological Macromolecules, 103*, 1173–1184.

Wang, Y., Su, N., Hou, G., Li, J., & Ye, M. (2017b). Hypoglycemic and hypolipidemic effects of a polysaccharide from Lachnum YM240 and its derivatives in mice, induced by a high fat diet and low dose STZ. *MedChemComm, 8*(5), 964–974.

Wang, Z., Xie, J., Shen, M., Nie, S., & Xie, M. (2018). Sulfated modification of polysaccharides: Synthesis, characterization and bioactivities. *Trends in Food Science & Technology, 74*, 147–157.

Wang, J., Fang, X., Wu, T., Fang, L., Liu, C., & Min, W. (2020). In vitro immunomodulatory effects of acidic exopolysaccharide produced by Lactobacillus planetarium JLAU103 on RAW264.7 macrophages. *International Journal of Biological Macromolecules, 156*, 1308–1315.

Wang, W., Ju, Y., Liu, N., Shi, S., & Hao, L. (2023). Structural characteristics of microbial exopolysaccharides in association with their biological activities: A review. *Chemical and Biological Technologies in Agriculture, 10*(1), 137.

Widyaningrum, D., & Meindrawan, B. (2020). The application of microbial extracellular polymeric substances in food industry. *IOP Conference Series: Earth and Environmental Science, 426*(1), 012181. IOP Publishing.

Windhues, T., & Borchard, W. (2003). Effect of acetylation on physico-chemical properties of bacterial and algal alginates in physiological sodium chloride solutions investigated with light scattering techniques. *Carbohydrate Polymers, 52*(1), 47–52.

Xiao, Q., & Lim, L. T. (2018). Pullulan-alginate fibers produced using free surface electrospinning. *International Journal of Biological Macromolecules, 112*, 809–817.

Xiong, J., Liu, D. M., & Huang, Y. Y. (2023). Exopolysaccharides from Lactiplantibacillus plantarum: Isolation, purification, structure–function relationship, and application. *European Food Research and Technology, 249*(6), 1431–1448.

Yan, J. K., Wang, Y. Y., Qiu, W. Y., & Wu, J. Y. (2017). Construction and characterization of nanosized curdlan sulfate/chitosan polyelectrolyte complex toward drug release of zidovudine. *Carbohydrate Polymers, 174*, 209–216.

Yan, J. K., Cai, W. D., Wang, C., Yu, Y. B., Zhang, H. N., Yang, Y., & Wang, W. H. (2020). Macromolecular behavior, structural characteristics and rheological properties of alkali-neutralization curdlan at different concentrations. *Food Hydrocolloids, 105*, 105785.

Yassunaka Hata, N. N., Surek, M., Sartori, D., Vassoler Serrato, R., & Aparecida Spinosa, W. (2023). Role of acetic acid bacteria in food and beverages. *Food Technology and Biotechnology, 61*(1), 85–103.

Yin, Z. C., Wang, Y. L., & Wang, K. (2018). A pH-responsive composite hydrogel beads based on agar and alginate for oral drug delivery. *Journal of Drug Delivery Science and Technology, 43*, 12–18.

Zeng, W., Wang, J., Shan, X., Yu, S., & Zhou, J. (2021). Efficient production of scleroglucan by Sclerotium rolfsii and insights into molecular weight modification by high-pressure homogenization. *Frontiers in Bioengineering and Biotechnology, 9*, 748213.

Zhang, R., & Edgar, K. J. (2014). Properties, chemistry, and applications of the bioactive polysaccharide curdlan. *Biomacromolecules, 15*(4), 1079–1096.

Zhang, M., Su, N., Huang, Q., Zhang, Q., Wang, Y., Li, J., & Ye, M. (2017). Phosphorylation and antiaging activity of polysaccharide from Trichosanthes peel. *Journal of Food and Drug Analysis, 25*(4), 976–983.

Zhang, J., Xiao, Y., Wang, H., Zhang, H., Chen, W., & Lu, W. (2023). Lactic acid bacteria-derived exopolysaccharide: Formation, immunomodulatory ability, health effects, and structure-function relationship. *Microbiological Research, 274*, 127432.

Zhong, K., Liu, L., Tong, L., Zhong, X., Wang, Q., & Zhou, S. (2013). Rheological properties and antitumor activity of schizophyllan produced with solid-state fermentation. *International Journal of Biological Macromolecules, 62*, 13–17.

Zhou, W., Tao, W., Wang, M., Liu, W., Xing, J., & Yang, Y. (2023). Dendrobium officinale Xianhu 2 polysaccharide helps forming a healthy gut microbiota and improving host immune system: An in vitro and in vivo study. *Food Chemistry, 401*, 134211.

Zhu, J., Tan, Z., Zhang, Z., & Shi, X. (2022). Characterization on structure and bioactivities of an exopolysaccharide from Lactobacillus curvatus SJTUF 62116. *International Journal of Biological Macromolecules, 210*, 504–517.

Zimmermann, C., & Wagner, A. E. (2021). Impact of Food-Derived Bioactive Compounds on Intestinal Immunity. Biomolecules, 11(12), 1901. https://doi.org/10.3390/biom11121901

Concealed Prebiosis in Human Gut Microflora

Sanjana Ghosh, Teesta Bhowmick, Jagrita Paul, Aniksha Mazumder, Sushmita Shaw, Arup Kumar Mitra, Sudeshna Shyam Choudhury, and Ayan Chandra

1 Introduction

1.1 What Are Extracellular Polymer Substances?

EPS are extracellular polymeric substances which are produced by bacteria in order to form biofilms. Biofilms are defined as aggregates of microbes which are embedded into the matrix of self-produced extracellular polymeric substances. These aggregates are formed by cell-to-cell contact which are either attached to a surface or they can be mobile biofilms. The organization of EPS molecules within the matrix depends on intramolecular interaction that occur between the EPS molecules, this interaction tells us about the mechanical strength and properties of the biofilm and the activity if microbes within the biofilm. EPS results in the formation of biofilm architecture and produces spatial organization in which the cells within the biofilm form micro-colonies. The bacteria within the matrix or EPS are different from the microbes that live individually. EPS molecules help to trap nutrients as well as another molecule. Therefore, EPS mainly increases the strength and stability of the biofilm architecture, forms pores or channel to provide nutrient to the microbes present within the biofilm, helps them to survive them in stressful as well as desiccation, etc., the biofilm production by microbes is mainly their survival strategy) (Flemming et al., 2016).

S. Ghosh (✉) · T. Bhowmick · J. Paul · A. Mazumder · S. Shaw · A. K. Mitra · S. S. Choudhury
Department of Microbiology, St. Xavier's College (Autonomous), Kolkata, West Bengal, India
e-mail: sanjanaghosh@sxccal.edu

A. Chandra
Department of Statistics, St. Xavier's College (Autonomous), Kolkata, West Bengal, India

T. Sarkar et al. (eds.), *Extracellular Polysaccharides*,
https://doi.org/10.1007/978-3-031-94472-7_12

EPS are the microbial biosynthetic compound or we can say the **biopolymers**. EPS can also be called as "exopolysaccharide," "extracellular polysaccharide," "exopolymers," etc., the major constituent of EPS are the polysaccharide along with other constituents.

By the colorimetric analysis of EPS we found that EPS is composed of the following components, i.e., proteins (serum albumin, bovine, and cytochrome C), saccharides, monosaccharides, phenolic compounds, and uronic acids. The protein content was estimated by Lowry's test and BCA assay (absorbance was measured at 562 nm), saccharide is determined via phenol sulfuric acid method (OD was measured between 480 and 490 nm) and uronic acid was determined by the carbazole sulfuric acid method (OD measured at 750 nm). All these tests were done over the 96 well plates and the OD were measured by ELISA (Felz et al., 2019). EPS synthesis is a common property of all the microbes occurring in both prokaryotes and eukaryotes such as bacteria, fungi, algae, and archaea.

EPS results in the formation of 3D, gel like, hydrates mesh like matrix which can sometimes be charged. The microbes are immobilized within the matrix. EPS concentration in the biofilm can be between 50% and 90%. EPS are formed by the polymerization various polymer molecules like polysaccharide. EPS also contains non-polymeric substances like proteins, which change the physiochemical properties of the matrix (Zhou et al., 2019; Nie et al., 2018). The matrix is produced by various biochemical substances, produced by microbes, release of products as a consequence of cell lysis or from the organic compounds present in the medium. EPS production is present in both eukaryotic and prokaryotic organisms both. The concentration of EPS and chemical characterization are two important factors that determine the viscosity of biofilm. Microbial EPS can be either capsular or it can form slimy layer. If the EPS is covalently bound to the cell surface of the bacteria, then it forms a capsule and if it is very loosely associated with the cell surface, then it forms a slimy layer (Flemming, 2016). Figure 1 represents the steps of bacterial biofilm formation.

1.2 Chemical Characterization of EPS

The EPS majorly consist of protein, carbohydrate, nucleic acid, organics material, etc., but the concentration of this composition varies with their environment and form. The dominant constituents of EPS are carbohydrate and protein. It may also consist constituents' protein derivatives such as glycoprotein, lipoprotein, and lipopolysaccharide.

1. Carbohydrate: It is the major constituent of EPS. The EPS majorly contains hexose sugar and rarely pentose sugar and also uranic acid. EPS can be homopolysaccharide or heteropolysaccharide. The homopolysaccharide contains a repeated unit of single sugar residue which is either L-fructose or D-Glucose. The homopolysaccharides are divided into mainly three groups. The first group

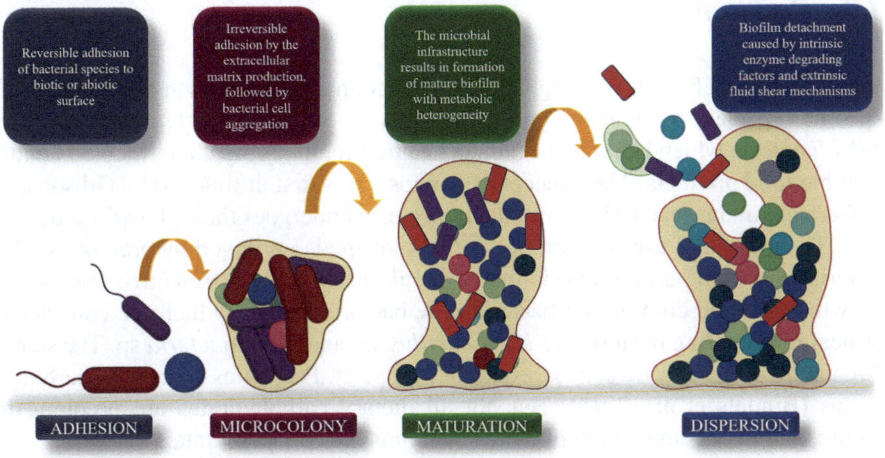

Fig. 1 Steps of bacterial biofilm formation

consists of α-D glucans, the second group consists of β-D-glucan, and the third group consists of fructans. Examples of homopolysaccharide produced by microbes are dextran and cellulose.

Heteropolysaccharide EPS consist of more than one sugar residue in a repeated manner. Heteropolysaccharide are made up of D-glucose, D-Glucuronic acid, D-galactose, L-rhamnose, L-fructose, etc. These EPS are produced by lactic acid bacteria. Examples of heteropolysaccharide include alginate, Xanthan gum, and gellan gum. The most well studied heteropolysaccharide EPS is that of Pseudomonas aeruginosa and Azotobacter, which acts as the opportunistic pathogen.

2. Protein: EPS contain protein as either enzymes or as structural protein. These enzymes are present in the EPS matrix and a few of them are involved in degradation of EPS. Under the stress condition, these enzymes are secreted by microbes which result in the degradation of EPS component. These enzymes can either degrade the EPS of their own species or their own EPS. The structural protein is involved in the formation of the EPS matrix and results in the formation of a link between the bacterial surface and the substrate surface.
3. Nucleic acid: The EPS of various microbes contains extracellular DNA present in their EPS. The EPS of *Staphylococcus aureus* contains nucleic acid or extracellular DNA as a major constituent, also a minor constituent of EPS in *S. epidermidis*. They are also present in the EPS of *Bacillus cereus* and *Streptococcus mutans*.
4. Lipid: The matrix also contains lipids and their derivatives. One such lipid derivative is lipopolysaccharide which helps in the adhesion of the cell to the surface. Lipids also act as biosurfactants which help the microbial cells to get detached from the oil droplets as well as for their attachment (More et al., 2014).

1.3 Prebiosis

Prebiotics are defined as the ingredients which enhance the growth and function of specific species of human gut microbiota specially *Lactobacillus* sp. and *Bifidobacteria* which benefits the human health. Prebiotics are the fermented nondigestible food products. The concept of prebiosis was first introduced by Gibson and Roberfroid in the year 1955. From them this term undergoes through various definition. The growth of useful microbes in the intestine is improved by prebiotics. The criteria to become a prebiotic is that it should be specific or selective toward the growth beneficial commensal bacteria. The bacterial groups which are considered as health promoting bacteria are *Lactobacillus* sp. and *Bifidobacteria* sp. The selective nature of a prebiotic is its key qualifier property. Prebiotics are mostly carbohydrates (Bindels et al., 2015). But not all dietary carbohydrates are prebiotic in nature, and to become a prebiotic the key points have to be maintain.

1. They must be resistant toward gastric acids.
2. Fermentable substrate to intestinal microflora only.
3. Selective proliferation of only those species of gut flora that is responsible for human health wellbeing (Markowiak & Ślizewska, 2017).

The most common types of prebiotics which are used for benefiting the human health are galacto-oligosaccharide, trans galactosidase, and fructo-oligosaccharide. In addition to galacto and fructo-oligosaccharide other carbohydrates which are considered as prebiotics are gluco-oligosaccharides, mannan-oligosaccharide, and lactulose. These ingredients are fermented by specific gut microbiota by which they receive their survival energy for effective proliferation and also enhancing their physiological activity inside the human body such as metabolism and enhancing digestion process, also they can effectively suppress human pathogens by showing antagonistic effects (Valcheva & Dieleman, 2016).

1.4 Mechanism of Action of Prebiotics

Our gastrointestinal tract consists of both beneficial as well as detrimental microbes. This pathogenic microbe produces various toxic substances such as ammonia, amines, and H2S. The growth of these detrimental microbes are inhibited by beneficial microbes such as *Bifidobacteria* and *Lactobacilli* which are enhanced by prebiotics. As it has been found that intestinal microbes are involved in pathogenesis of GI tract, scientists had increased their interest over dietary fibers to regulate this microbe and for that reason scientists have come across the concept of prebiotics. They ferment this polysaccharide and produce SCFAs (Short chain fatty acids). According to recent research and study, it has been found that EPS have a profound effect on the GI tract, on mental health, cardiovascular system, bones, and on metabolic activity of our body. The prebiotics can strongly influence the composition of gut flora. As humans lacks the enzymes which can hydrolyze prebiotics, they become resistant toward the intestinal digestion and reach the colon where they serve as fermentable substrate for *Bifidobacteria* and *Lactobacilli* (Guarino et al., 2020).

On fermenting these prebiotics by gut microflora produces SCFAs (Short chain fatty acids) which are butyric acid, propionic acid, and lactic acid. These SCFAs has diverse effect on human health including decreasing the PH of the colon, stimulating the immune system, intestinal epithelial cell development and various other physiological advantages. Therefore, prebiotics can only show their effect if they are degraded by these microbes. SCFAs can effectively diffuse through the blood circulatory system and have their effect on not only GI but also on various organs. The SCFAs also have immunomodulators effect which effectively kills the pathogenic microbes (Davani-Davari et al., 2019).

1.5 EPS as Prebiotics

Bacterial EPS are unique biomolecules because of their structure, composition, purity, and their 3D structure. In recent days, EPS has provided an excellent property in a variety of food products. It plays a vital role in immunomodulation, production of yoghurt, parabiosis property, etc. In this review, we will specifically focus toward the prebiotic property of EPS. EPS are the bioactive compounds effectively used in fermented food items which enhance the human digestive system and other metabolic activities (Korcz et al., 2018). Nowadays they are frequently used in various fermented food products such as yoghurt which enhances its firmness and texture. From studies, it has been found that capsular and ropy EPS can improve the physical structure of yoghurts by increasing its firmness and texture. Also, they can be used as a prebiotic in dairy industry as it enhances their production. EPS addition increases the microbiological, physical, and chemical properties of the yoghurt and also increases its viscosity (Bibi et al., 2021).

The LAB has been used in various fermented and dairy food product such as cheese, milk, and buttermilk. LAB is a group of probiotic bacteria. Probiotics are the live microbes which when consumed in sufficient amount, provides health benefit to the host. Among this LAB, some Lactic acid bacteria produce EPS which possess the prebiotic properties. These EPS are very well resistant toward the gastric region and easily reaches the distal region of the colon. They stimulate the growth of specific intestinal microbes in the gut. Also, it has been found that some LAB strain can potentially degrade the EPS produced from other LAB strains. Therefore, EPS from LAB serves as fermentable substrates for other microbes (Prete et al., 2021). EPS of LAB are of two types based on their units: Homopolysaccharide EPS (HoPS) and Heteropolysaccharide EPS (HePS).

Homopolysaccharide EPS consist of only one monosaccharide unit, i.e., either glucose or fructose. These monosaccharide units are polymerized into a single chain. The homopolysaccharide EPS are made from sucrose using either of the two enzymes, i.e., levaosucrase or glucanosucrase.

Heteropolysaccharide EPS are made from more than one monosaccharide unit. It composed of repeating units of three to eight different sugar residues. Its synthesis process is different from HoPS and involves four steps: transportation of sugar, synthesis of sugar nucleotides, synthesis of repeating units, and lastly their polymerization. Studies have shown that both homo- and heteropolysaccharides EPS can be used as prebiotics produced by *Lactobacillus* and *Bifidobacteria*. These EPS are quite resistant toward gastrointestinal enzymes and can also be effectively fermented. The structure of both HoPS and HePs are quite different but they serve as easy fermentable substrates for the gut microbiota and act as effective prebiotics. As the HoPS have the simplest primary structure, they can be easily degraded by specific gut microbes' results in production of oligosaccharides which are then further utilized by other microbes in the gut. As Heteropolysaccharide EPS possess a complex structure, they are less easily degraded by gut microbiota compared to HoPS.

As a result, both HoPS and HePS can be used as a fermentable substrate but their degree of bio degradation by gut microbes are highly influenced by the physiochemical property of the polymers of EPS and also the number of hydrolytic enzymes the microbe possess. This EPS can effectively rearrange the microbes within the gut flora and enhance the gut health. Once these EPS are degraded, they release small mono or oligosaccharides which are then effectively fermented by microbes (Salazar et al., 2016).

The EPS from LAB mostly HoPS have similar chemical and structural composition as that of prebiotic carbohydrates, but the only difference between them is that they differ in their size and molecular weight. Studies have shown that, the Levan HoPS synthesized from *Lactobacillus sanfranciscensis* extracted from sourdough effectively proliferates Bifidiobacteria in the human colon. Also, the fructan type HoPS can also effectively proliferates *Bifidobacteria*.

The HePS are less likely fermented by both human and animal microbiota.

Bifidobacterium strains can also produce various type of EPS which benefited the human health. According to study, *Bifidobacterium* can produce approx. 11 different EPS molecules which can act as substrate for fermentation by gut microbes (Kaur & Dey, 2023).

In recent studies, the EPS from various probiotics have been examined, and the following examples were shown: Hongpattarakere et al., 2012 reported that EPS from the bacteria of the genera *Lactobacillus* sp., *Weisella* sp., *Pediococcus* sp. can be effectively utilized by *Bifidobacterium bifidum*. Russo et al. also reported that the HoPS β-glucan has potential effect on three probiotics of the genus *Lactobacillus* sp., produced by *Pediococcus parvulus*. Also, Grosu-Tudor et al. studied the prebotic effect of EPS produced from the strains of LAB. They extracted 6 EPS from LAB and find their probiotic potential over the bacterium *Lactobacillus lactis* and *Bifidobacterium pseudocatenulatum* effectively (Helal et al., 2015).

The EPS was derived from a new *Lactobacilli* strain YY-112. The strain was named as *Lactobacillus pentosus*. The EPS that they derived from this strain has a molecular weight of approx. 5.9 × 104 Da. On administrating this EPS within a gastrointestinal system, they found that, this EPS increases the ratio of Bacteroides and Firmicutes and decreases the level of Proteobacteria specially *Shigella* sp. and *Escherichia* sp. This EPS also reduces the ratio of acetic to propionic acid which has a significant effect on weight loss and is also successful in preventing diseases associated with cardiovascular system. This EPS also increases the butyrate production which helps to prevent many colons disease. Thus, it was concluded that EPS derived from the new strain *Lactobacillus pentosus* have effective prebiotic activity (Wang et al., 2021).

Biofilm Formation by Gut Microbiota and Their Role as Prebiotics

Exopolysaccharides (EPS), a kind of carbohydrate polymers, envelop a diverse array of microorganisms. Accordingly, a number of EPS can be produced by specific Lactic acid bacterial (LAB) strains like *Lactococcus* sp., *Leuconostoc* sp., *Lactobacillus* sp., *Pediococcus* sp., *Weissella* sp., and Streptococcus spp. and are exported into the extracellular environment. Furthermore, different species and strains of EPS may have different molecular properties and different effects on host health or microbial functions. High molecular weight sugar polymers known as homopolysaccharides and heteropolysaccharides make up EPS from LAB. Homopolysaccharides are made of a single kind of monosaccharide, either glucose or fructose, and are produced by extracellular or cell wall-anchored glycanases that employ sucrose as a substrate.

Although most prebiotics are sourced from plants, there is evidence that EPS from microbial origin have analogous structures which execute similar physiological effects (Lindstrom et al., 2012).

Prebiotic Properties of EPS

For EPS to be used as a prebiotic food supplement, they should be used as substrates which are utilized selectively by microorganisms of host origin, thereby conferring health benefits (Gibson et al., 2017). For this, EPS should meet three criteria. It should not be absorbed or digested in the upper GI tract by mammalian enzymes and reach the colon. There are no human enzymes in the colon for degradation of fructans and β-glucans and thus the EPS in the form of homopolysaccharides reaches the colon undigested. The indigestibility of EPS was proven by using a β-2,6 linked fructan which is a heteropolysaccharide EPS derived from *Lactobacillus sanfranciscensis* which is stable at low pH on heat treatment (Korakli et al., 2002). Heteropolysaccharide form of EPS has been hypothesized to be more resistant to digestion compared to the homopolysaccharide form due to its complex nature. However, this has not been experimentally determined in vitro or in vivo till date (Castro-Bravo et al., 2018). Secondly, it should undergo fermentation by intestinal microbes. EPS should be used by only specific microorganisms of the gut microbiota in the colon. The structures of various homopolysaccharide forms of EPS have been degraded by enzymes derived from the microbiota into products which are beneficial for the host. Fermentability by *Lactobacilli* and *Bifidobacteria* of high molecular weight homopolysaccharides produced by *Lactobacillus sanfranciscensis* was studied which included 99.3% fructose and 0.7% glucose. Capability of fermentation was based on the type of microbe, for example, the EPS of *Lactobacillus sanfranciscensis* could be fermented by *Bifidobacteria*, like *B. bifidum, B. adolescentis, B. breve,* and *B. infantis* to produce lactate, formate, and acetate but *Lactobacilli* were not able to do the same.

EPS has varying abilities of being resistant to biodegradation while moving through the GI tract. Some EPS get readily degraded while some remain integrated during their passage through the GI tract. Prebiotic properties and varying biodegradability abilities depend on the complexity of the structures of various EPS. Irrespective of all differences, EPS on reaching the lumen of colon, should be able to selectively stimulate the proliferation of colonic bacteria, which imparts benefits to host, which is the third criteria for it to serve as a prebiotic. For example, Streptococcus thermophilus SFi12 produces heteropolysaccharide comprising of glucose, galactose, and rhamnose which is readily degraded by gut microbiota in adults. It consists of a single side chain consisting of β-galactosyl residue. On the other hand, EPS isolated from *Lactobacillus sakei, Lactobacillus helveticus* Lh59, and *Lactococcus lactis* ssp. *cremoris*, were resistant to degradation due to their complex structure and presence of one charged and one uncharged side chain. *Lactobacillus sakei* had one glucose side chain and the other comprised of rhamnose, glycerol, and phosphate. *Lactococcus lactis* ssp. *cremoris* had a rhamnose side chain and the other was composed of phosphate and galactose. This shows that the structure of EPS plays a vital role in determining its biodegradation ability.

It was reported that EPS from *Lactobacillus plantarum* showed considerable prebiotic properties due to their increased resistance to gastric and intestinal digestion and their ability to selectively enhance proliferation of beneficial gut microflora, specially *Bifidobacteria* (Hongpattarakere et al., 2012). Also, digestibility of this EPS by artificial gastric juices was low and exhibited prebiotic properties in vitro. EPS isolated from L. plantarum DM5 showed low digestibility by gastric juices, α-amylase, and intestinal juices. EPS of Lb. plantarum stimulated the growth of *Bifidobacterium infantis* and *Lactobacillus acidophilus* which display probiotic properties; on the other hand, it demotes the growth of a non-probiotic microbe, Enterobacteriaceae (Das et al., 2014). *Bifidobacteria* synthesize EPS in the intestines which can be used as substrates for fermentation readily by the gut microbes (Kaur et al., 2022).

Bifidobacteria, one of the most significant bacteria in the human gut, has the ability to lower blood cholesterol, treat diarrheal illnesses, alleviate lactose intolerance or maldigestion, guard against infections, and prevent cancer (O'Callaghan & van Sinderen, 2016). Compared to inulin, a conventional prebiotic carbohydrate, the pure EPS generated by *Bifidobacterium bifidum* EPS DA-LAIM had a better prebiotic effect. Compared to inulin, it exhibits growth-promoting action on probiotic strains, particularly lactic acid bacteria.

Production of SCFAs

Short chain fatty acids (SCFAs) are formed by the fermentation of EPS by various bacteria, thereby imparting beneficial aspects to host which include maintaining the intestinal barrier, modulating host metabolism, supplying energy to colonocytes, and regulating immune responses (Maciaet al., 2015). These SCFAs are a part of biofilm formation of various gut bacteria. Heteropolysaccharides isolated from *Bifidobacteria* like *Bifidobacterium longum* and *Bifidobacterium animalis*, which mainly comprised of glucose, galactose, and occasionally rhamnose, were readily fermented by microorganisms belonging to the human gut microbiota, thereby increasing levels of short chain fatty acids. Acetate which is produced by *Bifidobacterium adolescentis*, is the most abundant short chain fatty acid in circulation, also transports through the blood-brain barrier (Frost et al., 2014). They have also been demonstrated to lower cholesterol production rates, which in turn lowers cholesterol concentrations in plasma as seen in humans and rodents. Furthermore, high levels of SCFA in the gut lumen lower pH in colon where the proximal colon has a pH of 5.5–6.5 where fermentation is highest, compared to the distal colon which has noticeably higher pH levels of 6.5–7.0, which inhibits the growth of Gram-negative pathogenic Enterobacteriaceae, such as *Salmonella* spp. and *Escherichia coli*. This establishes a significant correlation between the composition of the microbiota and SCFA levels.

Certain EPS structures can directly affect the development of SCFA. Since different *Bifidobacteria* strains produce different heteropolysaccharide structures, it was discovered that the SCFA production during the fermentation of heteropolysaccharide from different strains of the bacteria produce different ratios of acetic acid and propionic acid. This suggests that the SCFA production is dependent on the particular EPS composition rather than the species producing the EPS itself. This study also discovered variations in butyric acid synthesis that are based on the structure of EPS. This suggests that various conformations of EPS may be used to induce the production of particular metabolites and, consequently, impart certain health benefits (Salazar et al., 2008).

Using dextransucrase, a kind of glycosyltransferase, *Leuconostoc mesenteroides* breaks down sucrose into glucose and fructose, using fructose as an energy source to generate dextran which is an exopolysaccharide, from glucose. This EPS is an indigestible polysaccharide, which on fermentation by gut microbiota, produces SCFAs and changes gut microbial composition. By controlling the production of gut hormones, such as the incretin hormone GLP-1, via the SCFA receptors GPR41 and GPR43, EPS considerably reduced the rise in blood glucose levels after glucose delivery. Constant EPS consumption significantly raised the levels of SCFAs, especially propionate, which is crucial for the positive effects on the host. It was also experimentally found that *Bacteroides* species produced higher levels of SCFAs such as acetate and propionate in mice after intake of EPS from *L. mesenteroides*. This proves that EPS of *L. mesenteroides* promotes *Bacteroides* in gut (Miyamoto et al., 2023).

Leuconostoc pseudomesenteroides XG5 synthesizes Exopolysaccharide (EPS) which is a linear dextran that leads to a 1.5-fold increase in the short chain fatty acids, especially acetic acid, and butyric acid. Since acetic acid serves as one of the end products of metabolic pathways of *Bifidobacterium* and LABs, their proliferation is indirectly facilitated by *L. pseudomesenteroides* XG5. It has been reported that acetic acid improves the function of the intestinal epithelial cell barrier (Pan et al., 2020).

Role of EPS in Infants
One of the main bacterial genera in the early life gut microbiome comprises of members of the genus *Bifidobacterium*; greater concentrations of this group are linked to better newborn health. *Bifidobacterium breve* UCC2003 produce EPS which establishes various cross-feeding networks during different growth phases of infants. Degradation of EPS produced by *B. breve* UCC2003 by *Bacteroides* results in acetate formation which further facilitates the growth of other members of the microbiota in later growth stages of infants, such as *Faecalibacterium* which are essential producers of butyrate. Also, EPS of *B. breve* increases proliferation of *Tyzzerella* which produces formate which in turn is metabolized by *Faecalibacterium*. Presence of EPS of *B. breve* is also positively correlated with the short chain fatty acids including butyrate, propionate and formate and also increased proliferation of *Clostridioides*. Thus, during early life transitions, EPS serves as a source of nutrients for some microbiota members, which in turn promotes community reorganization. By assisting and fostering the establishment of a healthy newborn microbiome, *Bifidobacterium* strains may improve the wellbeing of infant (Püngel et al., 2020).

Role of EPS Against Pathogenic Bacteria in Gut
The biological activities of EPS are significantly correlated with their physicochemical properties, for instance, the electric charge on the EPS polymer. For example, higher inhibitory effect against Gram-positive than Gram-negative pathogens was demonstrated by negatively charged EPS from the *Lactococcus lactis* F-mou strain, the highest inhibition being exhibited by *Bacillus cereus* ATCC 10702. Due to the negative charge of the sulphate group, it was proposed that the negatively charged EPS could interact more effectively with Gram-positive bacteria because of their larger positive charge on the cell wall (Nehal et al., 2019).

As of right now, no clear mechanism has been reported to account for the antibacterial activity of the EPS generated by the LAB against both Gram-positive and Gram-negative bacteria. Investigations into a possible mechanism or mechanisms to account for the detected antibacterial activity are still ongoing. It was believed that EPS may operate as an inhibitory mechanism since it has the ability to alter the bacterial cell envelope's structure, particularly the peptidoglycan layer. Additionally, it was noted that EPS might encourage the build-up of secondary metabolites in the growth media, thereby having a deleterious impact on both Gram-positive and Gram-negative bacteria. Another hypothesis is that functional groups in the EPS structure are most likely to interact with bacterial cell envelopes in some way to bring about their antibacterial potential.

Humans harbor *Clostridium difficile*, a toxin-producing bacterium, through feces or oral route. Spores resistant to heat, acids, and antibiotics disseminate its infection. Enterotoxins A and B are the two pathogenic forms produced by *C. difficile*. It is the main causative agent of pseudomembranous colitis, colonizes the intestines, and induces diarrhea brought on by antibiotics. The prebiotic effect of EPS aids LAB in gut colonization; it protects commensal microorganisms from the host's adaptive immune system; it increases their competition against the pathogenic bacteria; it inhibits the growth of pathogenic microorganisms and interferes with their adhesion to the intestinal epithelium; and it prevents or reduces the formation of biofilm by the pathogenic strains. These mechanisms together account for the antimicrobial activity of EPS in vivo. Bacteriocins, which are synthesized by *Lactobacilli*, aid in defending the host against a variety of diseases, including those caused by *Vibrio parahaemolyticus* and *Clostridium difficile* (Schoster et al., 2013).

The EPS generated by *L. plantarum* YW32 was capable of inhibiting the formation of biofilms by both Gram-positive and Gram-negative pathogens. It was reported that EPS may disrupt the formation of biofilms by altering the surfaces of bacteria, preventing the initial adherence of bacteria to the surface, or by functioning as a signalling molecule to downregulate the expression of genes necessary for the synthesis of biofilms (Wang et al., 2015). Blocking the process of quorum sensing, is a good strategy to inhibit the early adhesion and subsequent growth of mature structures, thereby preventing or eradicating biofilm formation (Paluch et al., 2020).

It was reported that heteropolysaccharides produced by *L. rhamnosus* and *L. plantarum* WLPL04 strongly inhibited the development of biofilms formed by harmful bacteria, such as *Salmonella enterica serovar typhimurium, Pseudomonas aeruginosa, Escherichia coli* O157:H7, and *Staphylococcus aureus* (Rajoka et al., 2018). Furthermore, the highest disruptive effect of EPS was observed when *Loigolactobacillus coryniformis* NA-3 produced it on pre-formed biofilms of *S. typhimurium* and *Bacillus cereus* (Liu et al., 2017). The EPS of LAB may have an antibacterial impact by interfering with the signalling molecules that these pathogens use to create biofilms on their surface. This interaction would prevent the pathogens from forming biofilms and result in an antimicrobial effect (Spanò et al., 2016).

The heteropolysaccharide EPS derived from *Lactobacillus gasseri* was evaluated for its antimicrobial activity using various common food borne pathogens. It exhibited maximum inhibitory effect on *Listeria monocytogenes* MTCC 657 and also prevented its biofilm formation (Parveen et al., 2018).

Another potential method of stopping pathogenic bacteria from developing biofilms is the generation of exopolysaccharides by some biofilm-producing, probiotic LAB. By inhibiting the colonization of bacterial pathogens, probiotic LAB biofilms can improve their own colonization and stability in the host mucosa. Given that probiotic LAB may effectively compete with bacterial pathogens for resources and space, biofilm development by these bacteria is acknowledged as a promising method to impede the production of biofilms by harmful bacteria (Barzegari et al., 2020).

Additionally, the increased pathogen-EPS interaction may help retain the pathogen in the intestinal lumen and aid in its eventual clearance. It has been proposed that the branching polysaccharide, kefiran serves as a masking agent, keeping bacterial cells and their toxins from attaching to the EPS rather than the receptors on the surface of the enterocyte (Lynch et al., 2018).

The principal components the monosaccharide content of EPS produced by the *Lacticaseibacillus paracasei* EPS DA-BACS are glucose and mannose. Glucomannans are non-ionic, water-soluble hydrocolloidal dietary fibers that have been shown to have anti-inflammatory, prebiotic, anti-diabetic, and anti-obesity properties. It has also been suggested that EPS synthesized by *L. paracasei* EPS DA-BACS has prebiotic and anti-inflammatory properties. They have strong resistance to pancreatin, bile salts, and acids. It gives the cells acid tolerance and increases the cells' survival rate in gastrointestinal environments. Tolerance to gastrointestinal stress is promoted by EPS generated by *L. paracasei* EPS DA-BACS. In addition to showing strong growth inhibitory action against *Staphylococcus aureus* and *Escherichia coli*, EPS fully prevented the growth of *Pseudomonas aeruginosa, Bacillus subtilis,* and *Aspergillus brasiliensis*. Proliferation of bacteria with high prebiotic activity such as *Bifidobacterium bifidum*, *Lactobacillus gasseri*, *B. faecale,* and *B. animalis* was facilitated by purified EPS. There are no virulence factors or antibiotic resistance genes present in *L. paracasei*. As a result, *L. paracasei* EPS DA-BACS has the potential to be a probiotic because to its strong gut adhesion ability, gastrointestinal tolerance, and anti-inflammatory and antibacterial properties (Lee et al., 2022).

Bactericidal and bacteriostatic activities exhibited by the EPS produced by *Lactobacillus kefiranofaciens* DN1 was reported against *S. enterica serovar Enteritidis* and *Listeria monocytogenes*. The efficacy of the EPS was proportional to its concentration (Jeong et al., 2017). Moreover, EPS isolated from *Lactobacillus* spp., exhibited significant antibacterial activity against *Micrococcus luteus* Ca6 and *S. enterica* ATCC 43972 (Trabelsi et al., 2017). Through surface hydrophobicity and auto-aggregation, EPS extracted from *Lactobacillus johnsonii* FI9785 was reported to promote the competitive inhibition of pathogens (Dertli et al., 2015). According to reports, *Lactobacillus*-derived EPSs include a number of functional groups, including hydroxyl, phosphate, and carbonyl groups. These groups are thought to be crucial in enabling the antibacterial and antioxidant properties of EPS (Riaz et al., 2020). These polymers inhibited Gram-positive and Gram-negative microorganisms with varying degrees of resistance, either in vitro or in vivo.

Role of EPS Against Pathogenic Fungi in Gut

Antifungal action against *Candida albicans* was exhibited by the negatively charged EPS generated by *Lactobacillus cereus* F-mou strain. It proposed that negatively charged EPS may have superior pathogen interaction due to its sulphate group (Nehal et al., 2019). 70% of the biofilm produced by *Candida albicans* SC5314 was suppressed by the dextran produced by the probiotic bacteria *Weissella* confuse, which was isolated from Romanian yoghurt. Through cell-to-cell contacts and exometabolite excretion, *Lactobacillus rhamnosus* has shown the ability to reduce the induction of hyphae of C. albicans as well as the formation of biofilms (Matsubara et al., 2016). Additionally, by inhibiting fungal adherence and lowering the amount of nutrients accessible, *L. rhamnosus* GG shielded the oral epithelial cells from damage brought about by *C. albicans* (Mailänder-Sánchez et al., 2017). Antifungal activity against *C. pelliculosa* was also exhibited by EPS produced by various strains of *Lactobacillus* (Abdalla et al., 2021).

Antiviral Activity of EPS in Gut

Strong anti-rotavirus activity was demonstrated by EPS from *L. plantarum* LRCC5310 in vitro, especially against extracellular rotaviruses. It also demonstrated a high adhesion rate in vitro, which hindered the rotavirus's ability to adhere to MA104 cells. In vivo, *L. plantarum* LRCC5310's EPS shortened the duration of diarrhea, inhibited rotavirus proliferation in the colon, and sped up recovery as seen in young mice (Kim et al., 2009). Additionally, it has been proposed to lessen shedding and other damage by viral infection, thereby safeguarding the gut mucosal barrier. The probiotic strain, *L. rhamnosus* GG, exhibited positive efficacy against rotavirus, thus preventing gastroenteritis in hospitalized children and infants (Sindhu et al., 2014). *Lactobacillus plantarum* produces EPS which exhibit antiviral potential against transmissible gastroenteric corona virus as tested in swine models (Yang et al., 2017).

Role of EPS in Heavy Metal Detoxification

An essential part of heavy metal biosorption is played by microbial EPS. Particularly, compared to non-EPS-producing species, EPS-producing LAB have been shown to exhibit better absorption, tolerance, and effective reduction of the toxicity of heavy metals in vitro and/or in vivo. The negatively charged acidic groups and distinctive steric structure on the surface of EPS are the primary factors in the mechanisms supporting EPS-metal interaction. (Wang et al., 2024) The process has been represented in Fig. 2.

Gut Microflora and the Mechanism of Action

The human Gut Microbiota has remained as a major complex ecosystem due to its dynamic and spatial structure composed by lots of microorganisms that are highly different in taxonomic levels, the main inhabitants are bacteria, fungi, archaea, and viruses inside our body interacting among them, but also with the host. The genetic

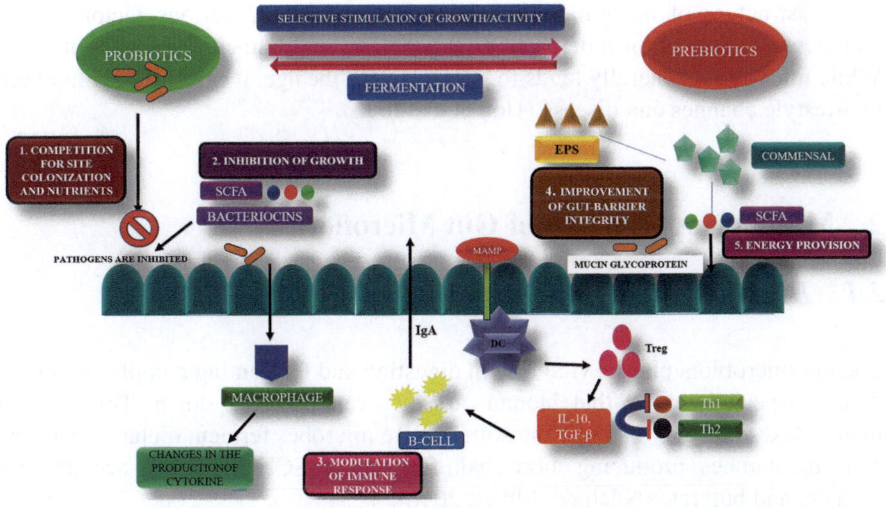

Fig. 2 Improvement of intestinal microbiota homeostasis. [MAMP = Microorganism associated molecular pattern; DC = Dendritic cell; Treg = Regulatory T cell; IL-10 = Interleukin-10, TGF-β = Transforming growth factor β]

repertoire responsible of the whole gut microbial genes together is at least one order of magnitude higher to that of Human genome. To some extent, it is also regarded as an "essential organ" of the human body. The gut microbiota is primarily composed of four main groups: Firmicutes, Bacteroidetes, Actinobacteria, and Proteobacteria (Chen et al., 2021).

1. Composition of Gut Microflora

The gut microbiota is composed of thousands of microbial species and is primarily dominating by the two bacterial phyla: Firmicutes and Bacteroidetes. In addition, other significant groups include Actinobacteria, Proteobacteria, and Verrucomicrobia.

- Firmicutes: Gram-positive bacteria like *Clostridium*, *Lactobacillus*, and *Faecalibacterium* make this phylum. These microbes are essential breaking down dietary fibre into short-chain fatty acids (SCFAs), which provide the health benefits for host health (Hou et al., 2022).
- Bacteroidetes: This phylum is comprised of Gram-negative bacteria like *Bacteroides* and *Prevotella*, this group is responsible in the fermentation of complex carbohydrates and aiding in energy extraction from food (Flint et al., 2012).
- Actinobacteria: This phylum includes notable members such as *Bifidobacterium*, which contribute to immune modulation and supports to maintain the integrity of the gut barrier.
- Proteobacteria: This phylum includes potential pathogens like *Escherichia coli*; however, they are typically the part of the normal gut flora. Their overgrowth can result into dysbiosis and inflammation.

The structure of the gut microbiota is determined by various factors which includes the mode of birth delivery, diet, genetics, environment, and antibiotic use. While microbiota generally tends to stabilize with the age, it can still be influenced by lifestyle changes and disease (Hou et al., 2022).

2 Mechanism of Action of Gut Microflora

2.1 Digestion and Fermentation

The gut microbiota plays a vital role in digesting and fermenting complex carbohydrates, especially those that human enzymes cannot break down. This process mainly takes place in the large intestine, where microbes ferment dietary fibers and resistant starches, producing short-chain fatty acids (SCFAs) such as acetate, propionate, and butyrate (Ndeh & Gilbert, 2018).

2.1.1 Breakdown of Complex Carbohydrates

When dietary fibers, primarily composed of polysaccharides, reach the colon undigested, specific microbial enzymes facilitate their breakdown. Human enzymes are incapable of digesting many complex carbohydrates (e.g., cellulose, hemicellulose, and inulin), but gut microbes, especially those in the phyla Firmicutes and Bacteroidetes, possess hydrolases and glycosidases that enable fermentation of these substrates (Ndeh & Gilbert, 2018).

The key steps include the following:

- Hydrolysis of polysaccharides: Microbial enzymes break down complex polysaccharides into simpler oligosaccharides and monosaccharides.
- Fermentation: These sugars are then anaerobically fermented by gut bacteria into organic acids, gases (such as carbon dioxide, methane, and hydrogen), and SCFAs.

SCFAs, the primary by-products of fermentation, provide numerous health benefits to the host (Oliphant & Allen-Vercoe, 2019).

2.1.2 Production of Short-Chain Fatty Acids (SCFAs)

Short-chain fatty acids (SCFAs) are produced when gut microbes ferment dietary fibers, playing a vital role in human health. There are three primary SCFAs and they are: acetate, propionate, and butyrate, each perform unique functions, with their production closely tied to the metabolic activities of the gut microbiota (Fusco et al., 2023).

Acetate (60% of total SCFAs): Acetate is the most abundant SCFA in the colon, it has effects that extends beyond the gut. Once the absorption is done into the bloodstream, it is distributed to peripheral tissues, where it serves as an important energy source. Acetate is also known to play a role in lipid metabolism, helps in regulating appetite and fat storage (Morrison & Preston, 2016).

Propionate (25% of total SCFAs): Propionate is a volatile fatty acid and is primarily absorbed by the liver. It also plays a major role in gluconeogenesis. In addition, propionate lowers the cholesterol synthesis in the liver, which results into lowering blood cholesterol levels, and thus it supports the overall metabolic health (Chambers et al., 2015).

Butyrate (15% of total SCFAs): Butyrate plays a significant role in maintaining the gut health. It serves as the primary source of energy for colonocytes, the cells lining the colon, and is important for maintaining the integrity of the gut barrier. By supporting the tight junctions between cells, butyrate helps to reduce the intestinal permeability, thus helps to prevent "leaky gut," which can lead to inflammation and disease (Elce et al., 2017).

2.1.3 Health Benefits of Short-Chain Fatty Acids (SCFAs)

SCFAs play several important roles in maintaining host health:

Energy Source: SCFAs, mainly butyrate, serve as an important primary energy source for colonocytes, which helps to maintain the integrity of the gut lining. When butyrate levels are insufficient, it can weaken epithelial cells, thus resulting into increased gut permeability, which is commonly referred to as "leaky gut" (Yan & Ajuwon, 2017).

Anti-inflammatory Properties: The anti-inflammatory effects on butyrate are significant. By inhibiting histone deacetylase (HDAC) activity, it can also promote the development of regulatory T cells (Tregs) with the ability to maintain the immune response. These Tregs play a significant role in increasing immune tolerance and restricting immune-related tissue damage (Facchin et al., 2024).

Metabolic Regulation: SCFAs, particularly propionate, play a role in energy balance by decreasing fat synthesis (lipogenesis) and lowering cholesterol. Both acetate and propionate also influence the release of hormones such as glucagon-like peptide-1 (GLP-1) and peptide YY (PYY), which help regulate appetite and glucose metabolism (Facchin et al., 2024).

Gut Barrier Enhancement: Butyrate supports the intestinal barrier by enhancing the formation of tight junctions between epithelial cells, preventing harmful pathogens and toxins from entering the bloodstream (Karahanian, 2021).

2.2 Immune Modulation

The gut microbiota plays a crucial role in shaping and regulating the immune system. By interacting with microbial antigens, immune cells learn to differentiate between harmful and harmless stimuli. This process therefore helps to prevent the improper immune reactions against the food antigens and beneficial gut bacteria, and by that it promotes the immune tolerance.

2.2.1 Induction of Regulatory T Cells (Tregs)

Polysaccharide A which is produced by the gut bacteria, such as *Bacteroides fragilis*, acts on CD4+ T cells to help in the differentiation of Treg. These Tregs produce IL-10, an anti-inflammatory cytokine which has immune promoting effects and negative inflammatory responses toward gut microenvironment, thereby maintaining the gut homeostasis. This process is facilitated by the activation of TLR2 signalling pathways, which acquires an immune tolerance in body without excessive triggering of the immune responses (Sittipo et al., 2018).

2.2.2 Mucosal Immunity

The gut microbiota enhances mucosal immunity by promoting the secretion of secretory immunoglobulin A (sIgA) in the gut lumen. Secretory IgA acts as a first line of defense, which neutralizes the pathogens and toxins without inducing inflammation, preserves the intestinal integrity and prevents from infections. Maintaining immune balance and preventing the autoimmune responses depends critically on the interaction between microbiota and immune cells (Di Vincenzo et al., 2024).

3 Gut Barrier Integrity

The gut barrier plays an important defense mechanism which protects the body's internal environment from harmful external factors in the intestines. It plays a dual purpose by allowing the nutrient absorption while preventing the entry of pathogens, toxins, and other harmful substances entering the bloodstream. The structural integrity of this barrier is maintained by the balanced gut microbiota. The following section provides a detailed overview of how the gut microbiota supports the gut barrier's structure and function (Turner, 2009):

Key Components of the Gut Barrier

1. *Epithelial Cells*: A continuous lining is formed on the intestinal surface by these cells, which acts as a physical shield. The barrier's permeability is regulated, and the passage of harmful substances is prevented by the tight junctions which links these cells.
2. *Mucus Layer*: These protective layers are produced by specialized goblet cells, which cover the epithelial surface, preventing it from getting contact with the harmful microorganisms and the gut contents.
3. *Immune Cells*: These immune cells are positioned beneath the epithelial layer, these cells monitor for potential threats that may penetrate the epithelial lining, which ensures prompt immune responses to eliminate them.

Gut Microbiota: The protective microbes in the gut directly contribute to fortifying the barrier by enhancing the epithelial cell function and supporting the immune regulation (Turner, 2009).

3.1 Microbiota's Role in Maintaining Gut Barrier Integrity

The gut barrier is maintained by the gut microbiota through production of the metabolites, especially the *short-chain fatty acids (SCFAs)* like butyrate, acetate, and propionate. These SCFAs are essential as they preserve the integrity of the epithelial tight junctions, which prevent the excessive permeability of the gut barrier.

Among them, *butyrate* is particularly important as it has been found to upregulate the key proteins, like occludin and claudins, which reinforce tight junctions, and in return help in preventing the condition commonly referred to as "leaky gut" (Monolayers et al., 2009).

Mechanism: Mucin which is produced by the goblet cells is enhanced by the butyrate, strengthening the protective mucus layer covered by the epithelial cells. It is also used as an energy source for colonic cells and promotes the expression of tight junction proteins. This formation helps in preventing the gaps between cells which could allow the harmful substances, including pathogens, to cross into the bloodstream (Monolayers et al., 2009).

3.2 Modulation of Immune Response

The immune system communicates with the gut bacteria through pattern recognition receptors (PRRs), like Toll-like receptors (TLRs). This interaction helps the immune system to distinguish between harmless microbes and harmful pathogens, which promotes the tolerance toward beneficial gut microbes while preparing a defence against potential threats. Properly balanced microbial interactions help to minimize excessive inflammatory responses that could otherwise compromise the integrity of the gut barrier (Littman,.).

Mechanism: Gut bacteria stimulate the production of anti-inflammatory molecules, such as cytokines (e.g., IL-10), and promote the activation of regulatory T cells (Tregs). These Tregs are critical in maintaining immune balance, ensuring that inflammation remains controlled and does not damage the gut barrier, thus preserving immune homeostasis (Littman, 2012).

3.3 Prevention of Pathogen Colonization

Useful bacteria present in the intestine plays a crucial role in occupying the available space and utilizing nutrients. This process is known as competitive exclusion which reduces the chances of infections targeted at the gut lining and overall degradation in the intestinal health (Hooper, 2012).

3.4 Regulation of Intestinal Permeability

Maintaining a balanced gut microbiota is crucial for proper intestinal permeability. It strengthens the epithelial barrier, blocking the harmful substances, such as pathogens, toxins, or undigested food particles, from passing through the intestinal lining. However, when the gut microbiota balance is disturbed, it leads to a condition known as dysbiosis, where intestinal permeability is increased. This leads to the entry of the harmful agents into the bloodstream, triggering inflammation and systemic immune responses that can cause tissue damage and organs (Chelakkot et al., 2018).

4 Neurotransmitter Production by Gut Microbiota

The communication with the central nervous system (CNS) plays a critical role by the gut microbiota through the gut-brain axis. This interaction goes in both the direction of communication system and includes both the central and enteric nervous systems, which is regulated by hormonal, immune, and neural pathways. The primary way by which gut microbiota influences brain function is through the production of neurotransmitters, chemical messengers responsible for signal transmission in the brain and nervous system. Below is a closer look at how intestinal bacteria contribute to neurotransmitter production and how these interactions affect both the mental health and neurological conditions:

4.1 Serotonin Production

Serotonin, the neurotransmitter associated with mood regulation or emotion is often referred as the "feel-good" hormone. The *enterochromaffin cells* produce 90% of the body's serotonin in the gut and that gut bacteria plays a major role in this process.

Enterococcus, *Streptococcus*, *Escherichia*, and *Lactobacillus* are certain gut bacteria that either produce serotonin directly or modulate its production by promoting the synthesis of its *tryptophan* which is its precursor. Tryptophan which is derived from dietary proteins, and with the help of gut microbiota it gets converted into serotonin. The serotonin which gets synthesized in the gut not only have an effect on the enteric nervous system which is responsible for the gut motility, but also communicates with the brain through the gut-brain axis, and thus it affects the mood, behavior, and conditions such as anxiety and depression (Yano et al., 2015).

4.2 Gamma-Aminobutyric Acid (GABA)

GABA or Gamma-Aminobutyric Acid is a primary inhibitory neurotransmitter that reduces the excitability of the neurons and plays a vital role by increasing relaxation, decreasing anxiety, and providing better sleep quality.

GABA production has been found in certain gut bacteria, like *Lactobacillus* and *Bifidobacterium*. By modulating GABA levels, these advantageous microbes can regulate anxiety, stress responses, and depressive symptoms. This GABA produced by microbes focuses the significant relation between gut health and mental health which further points out the role of the gut-brain axis in controlling emotional and mental health (Yunes et al., 2016).

4.3 Dopamine Production

Dopamine functions as the leading neurotransmitter which is associated with the regulation of pleasure, motivation, and reward. This influences the dopamine levels by directly producing the dopamine or by affecting the precursors of dopamine, like *tyrosine*.

Some gut bacteria, such as *Bacillus* species, are known for its capability to produce dopamine in the gut. This gut derived dopamine can affect not only gut motility but also communicate with the brain through the vagus nerve, potentially have an effect on the brain's reward and pleasure pathways. This connects the link between gut health and neurological functions such as motivation and reward behaviors (Strandwitz, 2018).

4.4 Acetylcholine Production

Acetylcholine is a vital neurotransmitter which is crucial for functions such as memory, attention, and muscle control. The production or synthesis of acetylcholine can be influenced by certain the gut microbes, thus impacting both cognition and the autonomic nervous system.

Specific genus of probiotics, *Lactobacillus* species, are known to influence acetylcholine levels, which may impact on both neural signalling as well as gut motility. This interaction between gut bacteria and acetylcholine points out the essential role of the gut microbiota in maintaining both mental and physical health (Furness et al., 2013).

4.5 Influence on the Gut-Brain Axis

The gut microbiota interacts with the central nervous system (CNS) through several interlinked pathways:

4.5.1 Vagus Nerve

This major nerve connects the gut to the brain, enabling the direct communication between the gut and the brain. This nerve transmits signals that are influenced by neurotransmitters produced in the gut, affecting mood, cognition, and emotional responses (Breit et al., 2018).

4.5.2 Immune System

The gut microbiota plays a significant role in modulating immune responses. By regulating inflammatory markers such as cytokines, the microbiota can influence brain function, thereby affecting the mental health and the cognitive processes (Belkaid & Hand, 2014).

4.5.3 Hypothalamic-Pituitary-Adrenal (HPA) Axis

The gut microbiota impacts the stress responses by modulating the HPA axis, which regulates the release of cortisol, a hormone which plays a significant role in both stress and mood regulation. An imbalanced microbiota can contribute to disruption of the HPA axis, leading to increased stress and mood disorders (Dinan & Cryan, 2012).

4.6 Impact on Mental Health and Neurological Conditions

4.6.1 Mood Disorder

Alternations in the gut microbiota are often seen in the individuals with depression or anxiety, which indicates its close relationship with neurotransmitter production and mental health regulation. Gut dysbiosis is an imbalance in gut bacteria which can impair the production of key neurotransmitters targeted for mood stabilization, like serotonin and GABA. Serotonin, which is also known as the "feel-good" neurotransmitter, is primarily produced in the gut. Reduced level in serotonin can lead to mood disorders like depression. In addition, reduced level in GABA, which is an inhibitory neurotransmitter that helps to calm the neural activity, can result in high risk of anxiety and stress (Kelly et al., 2016).

4.6.2 Neurodevelopment Disorder

Increasing evidence indicates that the gut microbiota plays a major role in neurodevelopmental disorders like autism spectrum disorder (ASD). The brain development is crucially affected by the gut brain axis, and interference in the gut microbiota during early life stages can affect the central nervous system (CNS) through mechanisms such as immune system signalling and neurotransmitter regulation. Variations in gut microbiota are often observed in children with ASD compared to neurotypical children (Fattorusso et al., 2019).

4.6.3 Neurodegenerative Disease

Gut microbiota imbalances are increasingly associated with neurodegenerative diseases such as Parkinson's disease and Alzheimer's disease. Variations in the gut microbial composition can affect the production of neurotransmitters, activate chronic inflammation, and further helps in release of inflammatory molecules such as cytokines, which may cross the blood-brain barrier and affect the brain activities. In Parkinson's disease, the misfolding of protein alpha-synuclein, which is a critical factor in disease progression believed to be influenced by the gut microbiota. Likewise, the development of Alzheimer's disease has been linked to chronic gut inflammation, where increase in inflammation can critically affects the brain health and cognitive function (Houser & Tansey, 2017).

5 Dysbiosis and Disease

5.1 Inflammatory Bowel Disease (IBD)

Conditions like Crohn's disease and ulcerative colitis Crohn's disease, both forms Inflammatory Bowel Disease (IBD) are significantly associated with the composition of gut microbiota. The abundance of beneficial bacteria, such as *Faecalibacterium prausnitzii*, which are known for producing the short-chain fatty acids (SCFAs) such as butyrate, are reduced. These SCFAs play a crucial role in minimizing inflammation and maintaining the integrity of the intestinal wall. With the decreasing number of these protective bacteria, the immune regulation of the gut is affected, allowing the pathogenic bacteria like *Escherichia coli* to grow rapidly. This gives rise to chronic intestinal inflammation, which is characterized by typical IBD symptoms like abdominal pain, diarrhea, and weight loss (Gustafson, 2019).

5.2 *Obesity and Metabolic Syndrome*

Research demonstrates that the balance between two dominant bacterial groups, i.e., Firmicutes and Bacteroidetes, is disruptive in individuals with obesity. Complex carbohydrates are competently broken into sugars by Firmicutes, which leads to greater calorie absorption. This excess energy, together with microbials effects on fat metabolism, which is responsible for promoting the fat storage and increases the risk of obesity. Moreover, dysbiosis can be activated by low-grade inflammation through the stimulation of the immune cells and producing inflammatory compounds, which leads to insulin resistance and the onset of metabolic syndrome (Leopold, 2020).

5.3 *Mental Health Disorders (Gut-Brain Axis)*

The gut-brain axis serves as a communication route linking the gut and brain through neural, hormonal, and immune signals. Dysbiosis can disrupt the synthesis of neurotransmitter serotonin, which is mainly produced in the gut and plays a significant role in mood regulation. Also, gut microbes affects the immune system, and dysbiosis can drive an increased systemic inflammation, which is associated with mental health conditions like depression, anxiety, and stress. Moreover, research also suggests that microbial imbalances may contribute to some neurodevelopmental disorders, which includes autism spectrum disorders (Michael et al., 2018).

The gut microbiota is extremely important for human health, affecting a wide range of processes, which includes digestion, immune system regulation, mood, and metabolic equilibrium. The interactions between the microbiota and the host occur through various mechanisms, like fermentation, immune modulation, and the production of neurochemicals. Preserving a diverse and balanced microbial community is essential for overall well-being, as imbalances, or dysbiosis, can result into various diseases. The improved understanding of the mechanisms through which the gut microbiota impacts the health enables the development of new therapeutic strategies to modulate the microbiota, providing potential opportunities for disease prevention and treatment.

6 Conclusion

Exopolysaccharides (EPS) derived from probiotics have a wide domain to function, they are extracellularly secreted out of microorganisms and are majorly effective in cosmetology, wastewater treatment, textiles, industries, etc., which exacerbates that EPS-producing probiotic strains have a potential role in protecting the environment by detoxifying the harmful wastewater generated by anthropogenic activities, besides being used as a potent strain in industries for generating profits based upon the products produced. Along with being industrially advanced, EPS-producing probiotics are also an advantageous member of the food industries where they are used as gelling agents, emulsifiers, stabilizers, and viscosifiers, the prebiotic effects in it can change the texture and nutritional value of food, which can eventually increase the market value of a product simultaneously enhancing the quality content of the food (Pourjafar et al., 2023).

In addition to all these benefits, it poses significant positive health effects in humans, one of those important benefits is EPS produced from probiotic strains allows them to adhere in the gut and promotes gut health which maintains the gut-brain axis, a crucial route of communication between the digestive system and the brain of the human body, promoting gut health will lower the risk of digestion problems and will lead to easier digestion and proper working of metabolic activities. Improved gut health will assist in the effective functioning of the human body as it helps colonize the gut microflora excluding the harmful pathogens that attack the enteric region of the body. However, improving gut health also intensifies the human immune system's activity to counter foreign antigens by increasing immunogenicity. Kefir, a type of fermented milk helps in reducing the cholesterol levels in the blood, manages immunomodulatory tasks, and is also proven to be efficient in anti-tumor and anti-microbial functions. Studies have reported that EPS-producing strains have coupled roles to manage our immune system, especially the macrophages, which are one of the antigen-presenting cells, without any harmful effects (Bibi et al., 2021).

The most common strain of improving gut health is the EPS-producing *Lactobacillus* which comes under the realm of lactic acid bacteria. Many *Lactobacillus* strains can produce exopolysaccharides, and they are also a capable candidate for probiotics which can be readily available in some of the common preparations like starter cultures of yogurt and cheese. A recent study has revealed that six *Lactobacillus plantarum* strains produce ropy exopolysaccharides and hence, they can readily improve gut health. Tarhana, a fermented food made of grains, possesses remarkable probiotic properties to improve gut health (Bibi et al., 2021). EPS produced by *L. plantarum* in a recent study also exposed its anti-diabetic activity by inhibiting α-amylase. So, this reveals the health-promoting aspect of *L. plantarum* which produces exopolysaccharides (Angelin & Kavitha, 2020).

Lactobacillus adjusts the human intestine in such a way as to grant some health benefits. *Lactobacillus* sp. has been found to generate various metabolites secreted out in the log phase such as acetic acids, lactic acids, fatty acids, and antibiotics secreted out in the stationary phase, which are all necessary for the food industry as *Lactobacillus* is regarded as GRAS (Generally Regarded As Safe). So, exopolysaccharides produced from *Lactobacillus* are bioproducts with diversities of biological function and activities. Some important strains include *L. plantarum, L. johnsonii, L. helveticus, L. curvatus, L. rhamnosus, L. acidophilus, L.delbrueckii subsp bulgaricus, L.casei* which apart from enriching the gut microflora also function in resisting themselves from the environmental stress (Bibi et al., 2021).

Probiotics are antagonistic to antibiotics, antibiotics suppress the immune system when taken at a regular dose, whereas probiotics replenish the microflora enhancing their colonization in the gut and hence increasing the immune system to fight against potent pathogens. Probiotics with EPS are an added advantage to treating human disorders mostly related to gastrointestinal diseases such as inflammatory bowel disease, gastric ulcers, cardiovascular diseases, colon cancer, and some autoimmune diseases. EPS is used as a protecting barrier by probiotic strains because EPS has some charming abilities like water retention ability, water binding capacity, gelling, and swelling potential. EPS like alginate, cellulose, and xanthan promote biofilm formation for the adherence of bacterial cells to the gut and it also protects the good microbes from a highly acidic stomach environment due to the secretion of hydrochloric acid. EPS is widely effective as an anti-thrombotic, anti-allergic, and anti-coagulant (Angelin & Kavitha, 2020).

The benefits of EPS-producing probiotic strains are exemplary, but the sources in which they can be found are dairy-based and non-dairy-based. Dairy-based source such as milk is a highly potent source of lactose and probiotic strains, but a lactose-intolerant individual cannot take the benefits of it whereas non-dairy-based sources include fruits and vegetables (Angelin & Kavitha, 2020). EPS serves as an added advantage along with the benefits of probiotics. There are varieties of EPS-producing probiotics that safeguard our gut from pathogenic invasion and therefore can be used in pharmacological and food industries.

Enterococcus faecium WEFA23 an exopolysaccharide-producing probiotic strain sourced from healthy infant's feces has an antioxidant activity that restricts the oxidation of nucleic acid (DNA), proteins, and lipids by stopping further oxidative chain reactions. It scavenges the free radicals and prevents further formation of free radicals and therefore prevents the cell from damage. Apart from exhibiting antioxidant activity, these strains also prohibit the adherence of *Listeria monocytogenes* CMCC54007 on HT-29 cells (Angelin & Kavitha, 2020).

In a study by (Ayyash et al., 2020), it was found that *L. plantarum* an EPS synthesizing probiotic strain EPS-C70 isolated from camel's milk has cytotoxic activity against the colon cancer cell line and breast cancer cell line. Also, it can act against food-borne infections caused by *Salmonella typhimurium, E. coli* O157:H7, *Staphylococcus aureus*, and *Listeria monocytogenes* (Bibi et al., 2021). In addition, *Lactobacillus plantarum* KX041 obtained from Chinese Paocai has a strong activity to stimulate the immune system and radical scavenging activity by DPPH. Another EPS producing strain *Lactobacillus plantarum* MTCC9510 isolated from curd exhibits immunomodulatory and anti-tumorigenic activity. *Lactobacillus plantarum* JK0142 obtained from fermented tofu shows raised IgA content, cytokines, interleukin-2, and tumor necrosis factor-α eventually stimulating the immune system.

Lactobacillus paracasei M7, also isolated from human breast milk, helps establish the gut microflora of infants, acts as an antioxidant, helps in biofilm formation, and reduces cholesterol through its hypercholesterolemic activity. The presence of these strains in healthy infant feces has health-promoting aspects such as its antioxidant activity, which was experimented with by delaying aging in mice and enhancing the antioxidant property to lower oxidative damage. *Lactobacillus paracasei*

ZFM54 protects against food-borne pathogens by inhibiting certain vital pathogenic bacteria such as *Salmonella typhimurium, Micrococcus luteus,* and *Listeria monocytogenes*. So, this probiotic strain apart from producing exopolysaccharide has remarkable gut-promoting benefits (Bengoa et al., 2023).

Humans widely consume meats for protein, fats, and carbohydrates preferably including chicken, its gastrointestinal tract comprises *Lactobacillus* sp.ca6 which is an exopolysaccharide-producing probiotic strain, in addition, this strain could produce anti-microbial activity and is itself sensitive to several antibiotics. Whereas, *Lactobacillus acidophilus* harbors in infant feces which promotes the gut microbiota of the infant and has antioxidant activity by suppressing the nitric oxide and malondialdehyde levels in the serum. However, this strain also has antitumorigenic activity against Ehrlich ascites carcinoma (EAC) cells. EAC cells are experimentally non-differentiated mouse cell lines that can express tumors (Angelin & Kavitha, 2020).

Exopolysaccharides are classified into homopolysaccharides and heteropolysaccharides. The Homopolysaccharides are mostly made up of glucose and fructose and are further categorized into α-D glucan and β-D glucan whereas heteropolysaccharides are made up of D- glucose, D-galactose, and some sugars such as L-rhamnose, arabinose, fucose, and mannose. Some homopolysaccharides-producing probiotic strains such *as Lactobacillus casei, Lactobacillus sakei, Lactobacillus parabuchneri* help in producing Dextran (EPS) and they all play a major role as adjuvant, stabilizer, emulsifier, and carrier, these are widely used in pharmaceutical industries, matrix of column chromatography, as anticoagulant or can be also used for oil recovery and biomaterials. These probiotic strains exceptionally bear a smart strategy to be industrially efficient as well as provide some good influence to the intestine.

Homopolysaccharides from probiotic strains such as *Streptococcus mutans, Streptococcus salivarius,* and *Bacillus subtilis* produce Levan (EPS), whose monomer is made up of glucose, these all strains contribute in producing prebiotic, showing anti-tumorigenic and hypocholesterolemic activity (which reduces cholesterol), as bio-thickener in food industries (Angelin & Kavitha, 2020).

Streptococcus thermophilus CRL1190 produces EPS, which has the potential to protect the gastrointestinal tract and it could be a potential candidate to be used as a food-grade additive for specific causes such as good water, oil holding capacities, thermal stability, as anti-oxidant and flocculating activities. Streptococcus thermophilus is most commonly present in the starter culture of the yogurt and helps in its formation. Through this, we are assured of the benefits incurred from *S. thermophilus* while simultaneously being advantageous to the gut microbiota (Angelin & Kavitha, 2020).

Lactobacillus paracasei strain CIDCA 83123 and CIDCA 83124 sourced from the kefir grain of Argentina can adhere, invade, and colonize the intestine. The ability to adhere to epithelial intestinal cells protects against *Salmonella* infection, and immunomodulates and protects the gastric epithelial cells. Due to their presence, T cells increase CD4+ and CD8+ count which eventually leads to a lot of mucin production (Bengoa et al., 2023).

Other added advantageous homopolysaccharide-producing probiotic strains are *Streptococcus mutans* and *Lactobacillus reuteri* which produce insulin-type exopolysaccharides, that provide prebiotic like property to nourish the gut mucosal cell for maintaining the gut microbiota with good bacteria that influences the gut-brain axis to function properly.

Now comes the supremacy of heteropolysaccharides, they consist of many strains that contain the advantage for the well-being of human beings by maintaining the gut habitat. Some of the mentioned strains are mentioned below.

Lactobacillus plantarum JLK0142, composed of basic sugars like glucose and galactose, are used as food adjuncts, which means they are used as additives replacing an essential component. These probiotic strains are also used as clinical immunomodulatory substances in medicines that stimulate or suppress the immune system helping the body fight any infection, specific immunomodulatory substances help manufacture vaccines, cytokines, etc. (Angelin & Kavitha, 2020).

Lactobacillus johnsonii 142 is also a probiotic strain that produces heteropolysaccharides whose sugar composition is D-galactose and D- glucose, this strain is profoundly used in stimulating the anti-inflammatory cytokine IL-10 activating the CD206+ macrophages that helps in mitigating colitis, creates anti-inflammatory substances to counter *S. typhimurium* it also assists in doing the probiotic treatment, which treats hay fever, respiratory tract diseases, diarrhea, constipation (Angelin & Kavitha, 2020).

Additionally, *Enterococcus faecium* MC13 and *Streptococcus thermophilus* CC30 are also heteropolysaccharides producing probiotic strains. *Enterococcus faecium* MC13 is used as a strong emulsifier and flocculating agent which further can be utilized as an emulsifying agent and flocculating agent-producing industries whereas *S. thermophilus* can only act as a strong emulsifier not as a flocculating agent and hence can be used in food industries, while both being composed of galactose and glucose sugars (Angelin & Kavitha, 2020).

Furthermore, another probiotic-producing polysaccharide strain is *Lactobacillus plantarum* C70, which has a sugar composition of arabinose, mannose, glucose, and galactose that helps in improving the surface properties so that it can be used in various food systems.

Heteropolysaccharides are widely known for their diverse functions because some strains have beneficial roles in providing vital reasons for a good gut. So, the probiotic strains have proved with evidence that exopolysaccharides produced by them are profound candidates for food, pharmaceutical, and other manufacturing industries.

Lactobacillus gasseri FR4 contains glucose, mannose, fructose, rhamnose, and some small amount of fucose which makes it a potential strain to be used in food industries as an antioxidant, viscosifying, and antimicrobial agent.

As we can witness the genus *Lactobacillus* has a lot of ways of producing exopolysaccharides that benefit the gut intestine, another such strain is *Lactobacillus casei* WDX030 contains glucose, glucosamine, and mannose to be used as an adjuvant for creating subunit vaccines (Angelin & Kavitha, 2020). Vaccines have a great market industry nowadays, be it intrarenal, intramuscular, or through any other route they boost the immunity of the person by producing antibodies prior to the entry of the pathogen and remain as memory cells, when the infective dose of the same pathogen increases then these memory cells (B cells) produce more antibodies to counter the pathogen. This strategic immunogenic response shown by our body has a wise connection with this probiotic strain that helps to produce subunit vaccines.

Exopolysaccharides upon secretion have lot more ways to improve other targets apart from the human gut-intestine, which includes lowering the cholesterol activity that eventually reduces the risk of heart attack and death, they have remarkable anti-viral activity that lowers down the potency of the virus which may target the human by binding its receptors of any organs, they have anti-cancer property which suppresses tumorous growth in potential regions, they have anti-microbial activity which bring down the effect of the harmful pathogens on target organs, they have anti-inflammatory activity that reduces swelling and inflammation by inducing cytokines to fight for the pathogen, they have anti-biofilm activity that prevents the adherence of pathogenic bacteria to their receptors which is present on specific sites of human organs, they too have anti-oxidant activity that kills reactive oxygen species (ROS) produced by superoxide radicals and hydrogen peroxide that damages the cells, creates mutation in the DNA, oxidizes membrane phospholipids due the oxidative stress produced, however lactic acid bacterial strains prevent the cells from being damaged by producing superoxide dismutase, and glutathione as a free radical scavenger. Due to this antioxidant property, it enrolls the probiotic strains into the food industry as food supplements.

Although synthetic antioxidants like butyl-hydroxylated toluene and n-propyl gallate are available in the market, they possess bad aftermath as it could lead to cancer due to incomplete elimination of the pathogen. To avoid such circumstances, Exopolysaccharides naturally produced by the probiotic strains are safe to be used. One such experimental witness is that *Lactobacillus plantarum* YML009 proved that it has an antioxidant activity that scavenges reactive oxygen species (ROS) (Angelin & Kavitha, 2020).

Exopolysaccharides too have a role as immunomodulators that regulate and stimulate the immune system of the body, both innate and adaptive immunity. The function of antibody-driven elimination of the pathogen is regulated by the B cells which are responsible for antigen recognition and elimination, so immunomodulation of certain molecules like the cytokines increases the proliferation of T-lymphocytes, B-lymphocytes, Natural Killer cells (NK cells), mononuclear cells phagocytic activity and many more. This is capable of reducing the risk of daily infections like inflammatory bowel disease (IBD) and prevents gastrointestinal cancers like peptic ulcers, gastric cancer, and colon cancer (Angelin & Kavitha, 2020).

An experimental evidence that proves that exopolysaccharides produced by probiotic strain increase immune activity is that exopolysaccharide produced by *Lactobacillus casei* was subcutaneously injected into mice (experimental model), the observation produced was quite exciting as it elicited the humoral and cellular immunity by inducing the T-cell proliferation, enhanced expression of the cytokines and matured dendritic cells have produced that act as antigen presenting cells (APC).

Furthermore, pieces of evidence suggest that *Lactobacillus paracasei* CIDCA 8339 and another strain *Lactobacillus paracasei* CIDCA 83124 have specialized roles in metabolizing the fecal microbiota that modulates the fecal microflora profile which leads to the formation of short-chain fatty acids (SCFA), propionic acid (PA), and butyric acid (BA). The role of PA and BA inhibits the entry of pathogenic factors into the body by acting as a gut barrier and lowering high cholesterol levels, cancer, etc. The EPS produced by this strain also helps inhibit the biofilm-forming pathogenic bacteria and prevents infectious disease (Bengoa et al., 2023).

Bacteriocins are chemical compounds secreted by bacteria that kill or inhibit the growth of other or the same bacterial species, similarly, these compounds are also secreted by probiotic strains that produce exopolysaccharides such as *Lactobacillus paracasei* FX6 which produce bacteriocin that has a wide antimicrobial spectrum for killing diverse pathogenic organisms (Angelin & Kavitha, 2020).

In a recent study, it was found that exopolysaccharides produced by the *Lactobacillus rhamnosus* ZFM231 strain are profoundly evident to mitigate the inflammatory bowel disease and colitis (IBD) induced by dextran sulfate solution in a mice model (Wan et al., 2022).

A report suggests that other exopolysaccharide-producing probiotic strains, *Lactobacillus delbrueckii* and *Lactocaseibacillus rhamnosus* have positive effects on gut microbiota, i.e., they can promote *Faecalibacterium prausnitzii* and *Ruminococcus bromii* which can metabolize carbohydrate by their metabolic pathways and can also produce short chain fatty acids (Tarique et al., 2023).

Anti-cancerous activity is also exhibited by exopolysaccharide-producing probiotic strain which is remarkably shown in an experiment where *Lactobacillus kefiri* when administered on HT29 cancer cells, it showed increased apoptotic activity by increased regulation of caspase-3, caspase-8, caspase-9, cytochrome c, BAX, BAD. Also, *Lactobacillus* strain SB27 which produces acidic exopolysaccharide shows upregulation in caspase-3 activity finally resulting in apoptosis, arresting the G0 and G1 phase of the cell cycle which overall leads to inhibiting the proliferation of the HT29 cancer cells (Angelin & Kavitha, 2020).

Additionally, exopolysaccharides produced by probiotic strains are also a potential candidate to show anti-biofilm activity the evidence lies in *Lactobacillus acidophilus* A4, which stopped the biofilm formation of pathogenic enterohemorrhagic *E. coli* strain O157:H7. Dextran formed by probiotic strain *Weissella confusa* sourced from Romanian yogurt showed maximum (70%) anti-biofilm activity against *Candida albicans* SC5314. *Lactobacillus fermentum* LB-69 shows anti-biofilm activity against spore-forming *Bacillus cereus* RSKK863 (Angelin & Kavitha, 2020).

Recently in 2020, the entire world survived the deadly coronavirus pandemic, excitingly exopolysaccharide producing *Lactobacillus plantarum* LRCC5310 decreases rotavirus replication and *Lactobacillus plantarum* inhibits the proliferation of gastroenteritis coronavirus experimented on the cell line of epithelial swine testicle. This entirely shows the anti-viral activity of this probiotic strain (Angelin & Kavitha, 2020).

Exopolysaccharides produced from probiotic strains which are most commonly present in the human gut to reestablish the gut microflora are preferably the genus *Lactobacillus*, which has numerous dominating characteristics in every field of industries be it food, pharmaceutical, or vaccine-producing industries. The production of EPS through these strains requires low-cost and affordable substrates to produce high-quality polysaccharides which are mostly declared safe by the USFDA like xanthan which can be used in food industries. EPS producing strains exhibit maximum output to the gut health maintaining the human gut-brain axis. Current research focuses on modulating the structure of EPS to discover some therapeutics. The advantages of EPS show that they have wholesome potential to be used for commercial purposes, however, research till now has only been done on in vivo mice models, and on cell lines but more exploration is needed in this regard. From this book chapter, we can conclude that more research is needed to unlock the prospects of exopolysaccharides produced by probiotic strains, for extracting out the benefits from it and implementing it for the betterment of society.

7 Future Prospects

Exopolysaccharides of probiotic strains isolated from human gut microflora possess bright hopes due to their advantageous properties in each field of industry apart from providing good gut health. Microorganisms produce EPS as natural substances and hence EPS contains diverse physicochemical and rheological possessions. Due to this, the application of the exopolysaccharides is wide in bakery products, processed meat products, dairy, and nondairy packaged food products, and delivery methods that contain probiotics. We already know about the features that exopolysaccharides uphold, such as immunomodulation, anti-inflammatory activity, antibiofilm activity, and anti-viral activity, and their applications with some experimental shreds of evidence. Through all this experimental evidence, the application of naturally obtained exopolysaccharides or biopolymers without using synthetic polysaccharides in industries holds diverse potential for commercial purposes, as synthetic exopolysaccharides are unsafe to use. Now, we shall delve into the prospects of the exopolysaccharides produced by the probiotic strains in the sectors, to inculcate the knowledge among us for using them in future grounds.

Of late, we all are accustomed to consuming food products that are readily available in the market but we will be wondering about the application of EPS in the food industry. EPS produced from non-lactic acid bacteria (LAB) such as dextran, cellulose, xanthan, gellan, and pullulan is used to package food products because naturally obtained biopolymers do not impose any significant harm to the food product but synthetic polymer does. Additionally, Lactic acid bacteria (LAB) produce exopolysaccharides such as kefirans, dextrans, hyaluronic acid, and levans are also used for manufacturing food packets replacing the water-insoluble polymers which are considerably bad for human health (Moradi et al., 2021). In the future, research should more focus on the extensive use of biopolymers with a higher yield instead of synthetic polysaccharides.

Exopolysaccharides exhibit a non-Newtonian behavior of liquid in an aqueous medium due to this they possess qualities like viscosity, gelation, and firmness these properties are strategically utilized further by the food industries to provide texture, thickness, good structure, and shininess. Furthermore, exopolysaccharides have water-retaining properties which makes it a potential candidate to enhance flavor, retain moisture, and prevent sugar crystallization. EPS obtained from probiotic strains such as *Leuconostoc mesenteroides*-TMS, *Leuconostoc citreum*-BMS, *Bacillus tequilensis*-GM, *Pediococcus pentosaceus*-DPS are also used to make gel formulations through electrostatic interaction mediated complex formation. The accurate protein-to-polysaccharide ratio is suitable for such a formulation. For example, during gastric digestion, whey (protein) and xanthan (polysaccharide) have a higher probability of gel formation. So, food products that contain these ingredients are probably to be a gel. Milk is also such example where the milk proteins interact with EPS (negatively charged) at optimal pH and form curd, yogurt. Instead of using direct gelling agents like gelatin, pectin, and alginate. EPS is produced by incubating the strains of starter cultures, which leads to the production of natural low-fat, low-sugar, and low-calorie products that potentially deal with health-conscious consumers (Muninathan et al., 2022).

According to the American Cancer Society dataset, about 106,590 people are diagnosed with colon cancer in the United States, so we need a solution to this inevitable problem, EPS from probiotic strain *Lactobacillus plantarum* C70 has promising and futuristic role in battling colon as well as breast cancer. EPS-C70 has a sugar composition of glucose, galactose, arabinose, and mannose, they have experimentally proven that at a concentration of 10 mg/ml it showed 73.1% and 88.1% against colon cancer and breast cancer cell lines. Moreover, EPS-C70 is also capable of inhibiting some food-borne pathogens like *Salmonella typhimurium*, Pathogenic strain *E. coli* OH157:H7, *Staphylococcus aureus* ATCC and *Listeria monocytogenes* (Bibi et al., 2021). This suggests that *Lactobacillus* strains which are harbored in the human gut and contribute to a healthy microflora have encouraging and hopeful applications in the food industry as well as pharmaceutical industries.

Bacterial exopolysaccharides manifest some advantageous features to the human gut by dwelling inside them. We in our daily life consume so much prebiotic (dietary fibers and carbohydrates which are indigestible), probiotic (live beneficial microorganisms), and symbiotic (contains both pre and probiotic). These all contribute to a healthy gut composition and help in re-establishing the gut microflora, the EPS produced by the probiotic strains shorten the long polysaccharides to ferment short-chain fatty acids (SCFA) like some vitamins, butyrate, and propionate which would not have been produced as the host lacks specific enzymes to do this (Pérez-Ramos et al., 2015).

EPS-producing probiotic strains are also futuristic about gastrointestinal digestion. Dietary EPS has two main features (i) being indigestible to the gastric and digestive components, (ii) being fermentable by the gut microflora. For example, exopolysaccharide *from Lactobacillus plantarum* DM5 shows low digestibility against gastric juice (hydrochloric acid), α-amylase than inulin. Further, *Lactobacillus plantarum* A3, *Weissella cibaria* A2, *Weissella confuse* A9, *Pediococcus pentosaceus* S4 produced exopolysaccharide in MRS broth and exhibited resistance to the gastric acid and pancreatic juice also *Leuconostoc pseudomesenteroides* showed less digestibility (Muninathan et al., 2022). This suggests that EPS associated with the bacterial surface in the gastrointestinal tract protects from low pH, gastric, and pancreatic enzymes that are known to create an unfavorable environment in the tract.

When the sourdough is fermented by exopolysaccharides produced by probiotic strains like Lactobacillus sakei LS81, *Leuconostoc mesenteroides*, and *Weissella cibaria* they could reduce the hardness of the bread crumbs and can be used in bread industries as bread improvers. *Lactobacillus paracasei* shows that exopolysaccharides too act as emulsifying agents as it showed a high emulsification index in toluene and benzene-supplemented medium in comparison with coconut oil. This proposes the idea to use exopolysaccharides as emulsifiers and stabilizers in food products (Bibi et al., 2021).

We often witness the availability of meat products in the market. But very few of us know that the tenderness and juiciness of meat products could be preserved by using exopolysaccharides produced by the probiotic strain. *Lactobacillus plantarum, Lactobacillus curvatus*, and *Lactobacillus sakei* are used during the tumbling process of ham production.

Lactobacillus plantarum S12 is used as a natural preservative, as a starter culture, and as antimicrobial agent against *Listeria monocytogenes* ATCC–7644, *Staphylococcus aureus* ATCC-6538, *Pseudomonas aeruginosa* ATCC-27853, and *Enterococcus faecalis* ATCC-29212.

Hence, naturally produced exopolysaccharides could be used as additives for preservation purposes as the *Lactobacillus* genus is regarded as generally regarded as safe (GRAS) (Bibi et al., 2021).

There are still many roles of exopolysaccharides produced by the probiotic strains that are unexplored and waiting to be discovered. Summing up the futuristic advantages of probiotics producing EPS, which could be as thickeners, texturizers for enhancing the viscosity of the food product, lowering syneresis, stabilizers, flavor and texture enhancers, preservatives, improves rheological features and structure, texture improvers, as a gelling agent, as a shelf-life enhancer, immunomodulator, antioxidant, emulsifier, stabilizer, and many more. However, the problem still exists that the exopolysaccharides produced by the probiotic strains are less in quantity to be used in industries for commercial purposes. So, the solution to this particular problem is that the production parameters should be carefully analyzed and optimized, selecting the right voltage is a good option. EPS production should be simultaneous with optimal growth conditions for specific production processes concerning different strains in the industries. Using exopolysaccharides in pharmaceutical industries as immunomodulators would serve as a greater benefit to activate the human immune system. Awareness is to be created among the industries to use commonly available strains that produce exopolysaccharides for food and medicinal purposes.

References

Abdalla, A. K., Ayyash, M. M., Olaimat, A. N., Osaili, T. M., Al-Nabulsi, A. A., Shah, N. P., & Holley, R. (2021). Exopolysaccharides as antimicrobial agents: mechanism and spectrum of activity. *Frontiers in Microbiology*, (12), 664395. https://doi.org/10.3389/fmicb.2021.664395. PMID: 34093478; PMCID: PMC8170130.

Angelin, J., & Kavitha, M. (2020). Exopolysaccharides from probiotic bacteria and their health potential. *International Journal of Biological Macromolecules, 162*, 853–865.

Ayyash, M., Abu-Jdayil, B., Itsaranuwat, P., Galiwango, E., Tamiello-Rosa, C., Abdullah, H., et al. (2020). Characterization, bioactivities, and rheological properties of exopolysaccharide produced by novel probiotic Lactobacillus plantarum C70 isolated from camel milk. *International Journal of Biological Macromolecules, 144*, 938–946.

Barzegari, A., Kheyrolahzadeh, K., Khatibi, S. M. H., Sharifi, S., Memar, M. Y., & Vahed, S. Z. (2020). The battle of probiotics and their derivatives against biofilms. *Infection and Drug Resistance, 13*, 659. https://doi.org/10.2147/idr.s232982

Belkaid, Y., & Hand, T. W. (2014). Review role of the microbiota in immunity and inflammation. *Cell, 157*(1), 121–141. https://doi.org/10.1016/j.cell.2014.03.011

Bengoa, A. A., Dueñas, M. T., Prieto, A., Garrote, G. L., & Abraham, A. G. (2023). Exopolysaccharide-producing Lacticaseibacillus paracasei strains isolated from kefir as starter for functional dairy products. *Frontiers in Microbiology, 14*, 1110177.

Bibi, A., Xiong, Y., Riaz Rajoka, M. S., Mehwish, H. M., Radicetti, E., Umair, M., Shoukat, M., & Khan, M. (2021). Recent advances in the production of exopolysaccharide (EPS) from lactobacillus spp. and its application in the food industry: A review. *Sustainability, 13*, 12429. https://doi.org/10.3390/su132212429

Bindels, L. B., Delzenne, N. M., Cani, P. D., & Walter, J. (2015). Towards a more comprehensive concept for prebiotics. *Nature Reviews Gastroenterology & Hepatology, 12*(5), 303–310. https://doi.org/10.1038/nrgastro.2015.47

Breit, S., Kupferberg, A., Rogler, G., Hasler, G., & Hasler, G. (2018). Vagus nerve as modulator of the brain–gut Axis in psychiatric and inflammatory disorders. *Frontiers in Psychiatry, 9*. https://doi.org/10.3389/fpsyt.2018.00044

Chambers, E. S., Viardot, A., Psichas, A., Morrison, D. J., Murphy, K. G., Zac-Varghese, S. E. K., MacDougall, K., Preston, T., Tedford, C., Finlayson, G. S., Blundell, J. E., Bell, J. D., Thomas, E. L., Mt-Isa, S., Ashby, D., Gibson, G. R., Kolida, S., Dhillo, W. S., Bloom, S. R., et al. (2015). Effects of targeted delivery of propionate to the human colon on appetite regulation, body weight maintenance and adiposity in overweight adults. *Gut, 64*(11), 1744–1754. https://doi.org/10.1136/gutjnl-2014-307913

Chelakkot, C., Ghim, J., & Ryu, S. H. (2018). Mechanisms regulating intestinal barrier integrity and its pathological implications. *Experimental & Molecular Medicine, 50*, 1. https://doi.org/10.1038/s12276-018-0126-x

Chen, Y., Zhou, J., & Wang, L. (2021). Role and mechanism of gut microbiota in human disease. In *Frontiers in cellular and infection microbiology* (Vol. 11). Frontiers Media S.A. https://doi.org/10.3389/fcimb.2021.625913

Castro-Bravo N, Wells J. M, Margolles A, Ruas-Madiedo P. Interactions of Surface Exopolysaccharides From Bifidobacterium and Lactobacillus Within the *Intestinal*

Environment. Front Microbiol. 2018 Oct 11;9:2426. https://doi.org/10.3389/fmicb.2018.02426. PMID: 30364185; PMCID: PMC6193118

Das, D., Baruah, R., & Goyal, A. (2014). A food additive with prebiotic properties of an α-d-glucan from lactobacillus plantarum DM5. *International Journal of Biological Macromolecules, 69*, 20–26.

Davani-Davari, D., Negahdaripour, M., Karimzadeh, I., Seifan, M., Mohkam, M., Masoumi, S. J., Berenjian, A., & Ghasemi, Y. (2019). Prebiotics: Definition, types, sources, mechanisms, and clinical applications. *Foods, 8*(3) MDPI Multidisciplinary Digital Publishing Institute. https://doi.org/10.3390/foods8030092

Dertli, E., Mayer, M. J., & Narbad, A. (2015). Impact of the exopolysaccharide layer on biofilms, adhesion and resistance to stress in Lactobacillus johnsonii FI9785. *BMC Microbiology, 15*, 8. https://doi.org/10.1186/s12866-015-0347-2

Di Vincenzo, F., Del Gaudio, A., Petito, V., Riccardo, L., & Franco, L. (2024). Gut microbiota, intestinal permeability, and systemic inflammation: a narrative review. *Internal and Emergency Medicine, 19*(2), 275–293. https://doi.org/10.1007/s11739-023-03374-w

Dinan, T. G., & Cryan, J. F. (2012). Regulation of the stress response by the gut microbiota: Implications for psychoneuroendocrinology. *Psychoneuroendocrinology, 37*(9), 1369–1378. https://doi.org/10.1016/j.psyneuen.2012.03.007

Elce, A., Amato, F., Zarrilli, F., Calignano, A., Troncone, R., Castaldo, G., & Canani, R. B. (2017). Butyrate modulating effects on pro-inflammatory pathways in human intestinal epithelial cells. *Beneficial Microbes, 8*(5), 841–847. https://doi.org/10.3920/BM2016.0197

Facchin, S., Bertin, L., Bonazzi, E., Lorenzon, G., De Barba, C., Barberio, B., Zingone, F., Maniero, D., Scarpa, M., Ruffolo, C., Angriman, I., & Savarino, E. V. (2024). Short-chain fatty acids and human health: From metabolic pathways to current therapeutic implications. *Life (Basel), 14*(5), 559.

Fattorusso, A., Di Genova, L., Battista, G., Isola, D., Mencaroni, E., & Esposito, S. (2019). Autism spectrum disorders and the gut microbiota. *Nutrients, 11*, 521. https://doi.org/10.3390/nu11030521

Felz, S., Vermeulen, P., van Loosdrecht, M. C. M., & Lin, Y. M. (2019). Chemical characterization methods for the analysis of structural extracellular polymeric substances (EPS). *Water Research, 157*, 201–208. https://doi.org/10.1016/j.watres.2019.03.068

Flemming, H. C. (2016). Eps—Then and now. *Microorganisms, 4*(4) MDPI AG. https://doi.org/10.3390/microorganisms4040041

Flemming, H.-C., Wingender, J., Szewzyk, U., Steinberg, P., Rice, S. A., & Kjelleberg, S. (2016). Biofilms: An emergent form of bacterial life. *Nature Reviews Microbiology, 14*(9), 563–575. https://doi.org/10.1038/nrmicro.2016.94

Flint, H. J., Scott, K. P., Louis, P., & Duncan, S. H. (2012). The role of the gut microbiota in nutrition and health. *Nature Reviews Gastroenterology and Hepatology, 9*(10), 577–589. https://doi.org/10.1038/nrgastro.2012.156

Frost, G., Sleeth, M. L., Sahuri-Arisoylu, M., Lizarbe, B., Cerdan, S., Brody, L., Anastasovska, J., Ghourab, S., Hankir, M., Zhang, S., Carling, D., Swann, J. R., Gibson, G., Viardot, A., Morrison, D., Thomas, E. L., & Bell, J. D. (2014). The short-chain fatty acid acetate reduces appetite via a central homeostatic mechanism. *Nature Communications, 5*, 3611.

Furness, J. B., Rivera, L. R., Cho, H., Bravo, D. M., & Callaghan, B. (2013). The gut as a sensory organ. *Nature Reviews Gastroenterology & Hepatology, 10*, 729. https://doi.org/10.1038/nrgastro.2013.180

Fusco, W., Lorenzo, M. B., Cintoni, M., Porcari, S., Rinninella, E., Kaitsas, F., Lener, E., Mele, M. C., Gasbarrini, A., Collado, M. C., Cammarota, G., & Ianiro, G. (2023). Short-chain fatty-acid-producing bacteria: Key components of the human gut microbiota. *Nutrients, 15*(9). https://doi.org/10.3390/nu15092211

Guarino, M. P. L., Altomare, A., Emerenziani, S., Di Rosa, C., Ribolsi, M., Balestrieri, P., Iovino, P., Rocchi, G., & Cicala, M. (2020). Mechanisms of action of prebiotics and their effects on gastro-intestinal disorders in adults. *Nutrients, 12*(4). https://doi.org/10.3390/nu12041037

Gustafson, C. (2019). Pete Myers, MD: Epigenetics and Endocrine Disruption.

Gibson, G., Hutkins, R., Sanders, M. et al. Expert consensus document: The International Scientific Association for Probiotics and Prebiotics (ISAPP) consensus statement on the definition and scope of prebiotics. Nat Rev Gastroenterol Hepatol 14, 491–502 (2017). https://doi.org/10.1038/nrgastro.2017.75

Helal, M., Hussein, M.-D., Osman, M., Shalaby, A. S., & Ghaly, M. (2015). Production and prebiotic activity of exopolysaccharides derived from some probiotics. *Egyptian Pharmaceutical Journal, 14*, 1. https://doi.org/10.4103/1687-4315.154687

Hongpattarakere, T., Cherntong, N., Wichienchot, S., Kolida, S., & Rastall, R. A. (2012). *Carbohydrate Polymers, 87*, 846–852.

Hooper, L. V. (2012). Interactions between the microbiota and the immune system. *Science, 336*, 1268. https://doi.org/10.1126/science.1223490

Hou, K., Wu, Z. X., Chen, X. Y., Wang, J. Q., Zhang, D., Xiao, C., Zhu, D., Koya, J. B., Wei, L., Li, J., & Chen, Z. S. (2022). Microbiota in health and diseases. *Signal Transduction and Targeted Therapy, 7*(1), 135. https://doi.org/10.1038/s41392-022-00974-4

Houser, M. C., & Tansey, M. G. (2017). The gut-brain axis: Is intestinal in fl ammation a silent driver of Parkinson's disease pathogenesis? *NPJ Parkinson's Disease, 3*, 3. https://doi.org/10.1038/s41531-016-0002-0

Jeong, D., Kim, D.-H., Kang, I.-B., Kim, H., Song, K.-Y., Kim, H.-S., & Seo, K.-H. (2017). Characterization and antibacterial activity of a novel exopolysaccharide produced by *Lactobacillus kefiranofaciens* DN1 isolated from kefir. *Food Control, 78*, 436–442.

Karahanian, E. (2021). Use of short-chain fatty acids for the recovery of the intestinal epithelial barrier affected by bacterial toxins. *Frontiers in Physiology, 12*(May), 1–8. https://doi.org/10.3389/fphys.2021.650313

Kaur, N., & Dey, P. (2023). Bacterial exopolysaccharides as emerging bioactive macromolecules: From fundamentals to applications. *Research in Microbiology, 174*(4), 104024. https://doi.org/10.1016/j.resmic.2022.104024

Kelly, J. R., Borre, Y., Brien, C. O., Patterson, E., Aidy, S., Kennedy, P. J., Beers, S., Scott, K., Moloney, G., Hoban, A. E., Scott, L., Fitzgerald, P., Ross, P., Stanton, C., Clarke, G., Cryan, F., & Dinan, T. G. (2016). Transferring the blues: Depression-associated gut microbiota induces neurobehavioural changes in the rat. *Journal of Psychiatric Research, 82*, 109. https://doi.org/10.1016/j.jpsychires.2016.07.019

Kim, Y., Oh, S., & Kim, S. H. (2009). Released exopolysaccharide (r-EPS) produced from probiotic bacteria reduce biofilm formation of enterohemorrhagic Escherichia coli O157:H7. *Biochemical and Biophysical Research Communications, 379*, 324–329. https://doi.org/10.1016/j.bbrc.2008.12.053

Korcz, E., Kerényi, Z., & Varga, L. (2018). Dietary fibers, prebiotics, and exopolysaccharides produced by lactic acid bacteria: Potential health benefits with special regard to cholesterol-lowering effects. *Food and function, 9*(6), 3057–3068. https://doi.org/10.1039/c8fo00118a

Korakli M, Gänzle M. G, Vogel R. F. Metabolism by bifidobacteria and lactic acid bacteria of polysaccharides from wheat and rye, and exopolysaccharides produced by Lactobacillus sanfranciscensis. *J Appl Microbiol.* 2002;92(5):958-65. https://doi.org/10.1046/j.1365-2672.2002.01607.x PMID: 11972702

Kaur, Navneet & Dey, Priyankar. (2022). Bacterial Exopolysaccharides as Emerging Bioactive Macromolecules: From Fundamentals to Applications. *Research in Microbiology*. 174. 104024. https://doi.org/10.1016/j.resmic.2022.104024

Leopold, S. S. (2020). A conversation with … Charles L. Bosk PhD, expert on surgical education and medical error, and author of forgive and remember: Managing medical failure. *Clinical Orthopaedics and Related Research, 478*, 1147–1151. https://doi.org/10.1097/CORR.0000000000001184

Littman, D. R. (2012). The microbiome in infectious disease and inflammation. *Annual Review of Immunology, 30*, 759. https://doi.org/10.1146/annurev-immunol-020711-074937

Liu, Z., Zhang, Z., Qiu, L., Zhang, F., Xu, X., Wei, H., & Tao, X. (2017). Characterization and bioactivities of the exopolysaccharide from a probiotic strain of Lactobacillus plantarum WLPL04. *Journal of Dairy Science, 100*, 6895–6905.

Lynch, K. M., Zannini, E., Coffey, A., & Arendt, E. K. (2018). Lactic acid bacteria exopolysaccharides in foods and beverages: Isolation, properties, characterization, and health benefits. *Annual Review of Food Science and Technology, 9*, 155–176. https://doi.org/10.1146/annurev-food-030117-012537

Lindström C, Holst O, Nilsson L, Oste R, Andersson KE. Effects of Pediococcus parvulus 2.6 and its exopolysaccharide on plasma cholesterol levels and inflammatory markers in mice. *AMB Express*. 2012 Dec 13;2(1):66. https://doi.org/10.1186/2191-0855-2-66 PMID: 23234432; PMCID: PMC3560120.

Lee M. G, Joeng H, Shin J, Kim S, Lee C, Song Y, Lee BH, Park H. G, Lee T. H, Jiang HH, Han Y. S, Lee B. G, Lee H. J, Park M. J, Jun Y. J, Park Y. S. Potential Probiotic Properties of Exopolysaccharide-Producing Lacticaseibacillus paracasei EPS DA-BACS and Prebiotic Activity of Its Exopolysaccharide. *Microorganisms*. 2022 Dec 8;10(12):2431. https://doi.org/10.3390/microorganisms10122431. PMID: 36557684; PMCID: PMC9787920

Mailänder-Sánchez, D., Braunsdorf, C., Grumaz, C., Müller, C., Lorenz, S., Stevens, P., et al. (2017). Antifungal defense of probiotic lactobacillus rhamnosus GG is mediated by blocking adhesion and nutrient depletion. *PLoS One, 12*, e0184438. https://doi.org/10.1371/journal.pone.0184438

Markowiak, P., & Ślizewska, K. (2017). Effects of probiotics, prebiotics, and synbiotics on human health. *Nutrients., 9*(9). https://doi.org/10.3390/nu9091021

Matsubara, V. H., Wang, Y., Bandara, H., Mayer, M. P. A., & Samaranayake, L. P. (2016). Probiotic lactobacilli inhibit early stages of Candida albicans biofilm development by reducing their growth, cell adhesion, and filamentation. *Applied Microbiology and Biotechnology, 100*, 6415–6426. https://doi.org/10.1007/s00253-016-7527-3

Michael, H., Thornton, S., Xie, M., & Tian, L. (2018). Exact inference on the random-effects model for meta-analyses with few † studies †. *Biometrics, 75*, 485. https://doi.org/10.1111/biom.12998

Miyamoto, J., Shimizu, H., Hisa, K., Matsuzaki, C., Inuki, S., Ando, Y., et al. (2023). Host metabolic benefits of prebiotic exopolysaccharides produced by Leuconostoc mesenteroides. *Gut Microbes, 15*(1). https://doi.org/10.1080/19490976.2022.2161271

Monolayers, C.-C., Peng, L., Li, Z., Green, R. S., Holzman, I. R., & Lin, J. (2009). Butyrate enhances the intestinal barrier by facilitating tight junction assembly via activation of AMP-activated protein kinase. *Journal of Nutrition, 139*, 1619–1625. https://doi.org/10.3945/jn.109.104638.1619

Moradi, M., Guimarães, J. T., & Sahin, S. (2021). Current applications of exopolysaccharides from lactic acid bacteria in the development of food active edible packaging. *Current Opinion in Food Science, 40*, 33–39.

More, T. T., Yadav, J. S. S., Yan, S., Tyagi, R. D., & Surampalli, R. Y. (2014). Extracellular polymeric substances of bacteria and their potential environmental applications. *Journal of Environmental Management, 144*, 1–25. https://doi.org/10.1016/j.jenvman.2014.05.010

Morrison, D. J., & Preston, T. (2016). Formation of short chain fatty acids by the gut microbiota and their impact on human metabolism. *Gut Microbes, 7*(3), 189–200. https://doi.org/10.1080/19490976.2015.1134082

Muninathan, C., Guruchandran, S., Viswanath Kalyan, A. J., & Ganesan, N. D. (2022). Microbial exopolysaccharides: Role in functional food engineering and gut-health management. *International Journal of Food Science and Technology, 57*, 27–34. https://doi.org/10.1111/ijfs.15334

Macia, L., Tan, J., Vieira, A. et al. Metabolite-sensing receptors GPR43 and GPR109A facilitate dietary fibreinduced gut homeostasis through regulation of the inflammasome. *Nat Commun 6*, 6734 (2015). https://doi.org/10.1038/ncomms7734

Ndeh, D., & Gilbert, H. J. (2018). Biochemistry of complex glycan depolymerisation by the human gut microbiota. *FEMS Microbiology Reviews, 42*(2), 146–164. https://doi.org/10.1093/femsre/fuy002

Nehal, F., Sahnoun, M., Smaoui, S., Jaouadi, B., Bejar, S., & Mohammed, S. (2019). Characterization, high production and antimicrobial activity of exopolysaccharides from Lactococcus lactis F-mou. *Microbial Pathogenesis, 132*, 10–19. https://doi.org/10.1016/j.micpath.2019.04.018

Nie, Y., Luo, F., & Lin, Q. (2018). Dietary nutrition and gut microflora: A promising target for treating diseases. *Trends in Food Science & Technology, 75*, 72–80. https://doi.org/10.1016/j.tifs.2018.03.002

O'Callaghan, A., & van Sinderen, D. (2016). Bifidobacteria and their role as members of the human gut microbiota. *Frontiers in Microbiology, 7*, 925.

Oliphant, K., & Allen-Vercoe, E. (2019). Macronutrient metabolism by the human gut microbiome: Major fermentation by-products and their impact on host health. *Microbiome, 7*(1), 1–15. https://doi.org/10.1186/s40168-019-0704-8

Paluch, E., Rewak-Soroczyñska, J., Jêdrusik, I., Mazurkiewicz, E., & Jermakow, K. (2020). Prevention of biofilm formation by quorum quenching. *Applied Microbiology and Biotechnology, 104*, 1871–1881. https://doi.org/10.1007/s00253-020-10349-w

Pan, L., Han, Y., & Zhou, Z. (2020). In vitro prebiotic activities of exopolysaccharide from Leuconostoc pseudomesenteroides XG5 and its effect on the gut microbiota of mice. *Journal of Functional Foods, 67*, 103853. https://doi.org/10.1016/j.jff.2020.103853. ISSN 1756-4646.

Parveen, R., Anandharaj, M., & David, A. (2018). International journal of biological macromolecules characterization of a novel exopolysaccharide produced by Lactobacillus gasseri FR4 and demonstration of its in vitro biological properties. *International Journal of Biological Macromolecules, 109*, 772–783. https://doi.org/10.1016/j.ijbiomac.2017.11.062

Pérez-Ramos, A., Nácher-Vázquez, M., Notararigo, S., López, P., & Mohedano, M. L. (2015). Current and future applications of bacterial extracellular polysaccharides. In *Probiotics, prebiotics, and synbiotics* (pp. 329–344). Elsevier.

Pourjafar, H., Ansari, F., Sadeghi, A., Samakkhah, S. A., & Jafari, S. M. (2023). Functional and health-promoting properties of probiotics' exopolysaccharides; isolation, characterization, and applications in the food industry. *Critical Reviews in Food Science and Nutrition, 63*(26), 8194–8225.

Prete, R., Alam, M. K., Perpetuini, G., Perla, C., Pittia, P., & Corsetti, A. (2021). Lactic acid bacteria exopolysaccharides producers: A sustainable tool for functional foods. *Foods, 10*(7). https://doi.org/10.3390/foods10071653. MDPI.

Püngel, D., Treveil, A., Dalby, M. J., Caim, S., Colquhoun, I. J., Booth, C., Ketskemety, J., Korcsmaros, T., van Sinderen, D., Lawson, M. A., & Hall, L. J. (2020). Bifidobacterium breve UCC2003 exopolysaccharide modulates the early life microbiota by acting as a potential dietary substrate. *Nutrients, 12*(4), 948. https://doi.org/10.3390/nu12040948. PMID: 32235410; PMCID: PMC7231044.

Rajoka, M. S. R., Jin, M., Haobin, Z., Li, Q., Shao, D., Jiang, C., Huang, Q., Yang, H., Shi, J., & Hussain, N. (2018). Functional characterization and biotechnological potential of exopolysaccharide produced by Lactobacillus rhamnosus strains isolated from human breast milk. *LWT – Food Science and Technology, 89*, 638–647. [CrossRef].

Riaz Rajoka, M. S., Wu, Y., Mehwish, H. M., Bansal, M., & Zhao, L. (2020). Lactobacillus exopolysaccharides: New perspectives on engineering strategies, physiochemical functions, and immunomodulatory effects on host health. *Trends in Food Science and Technology, 103*, 36–48. https://doi.org/10.1016/j.tifs.2020.06.003

Salazar, N., Gueimonde, M., Hernández-Barranco, A. M., Ruas-Madiedo, P., De Los Reyes-Gavilán, C. G., de los Reyes-Gavilan, C. G., et al. (2008). Exopolysaccharides produced by intestinal Bifidobacterium strains act as fermentable substrates for human intestinal bacteria. *Applied and Environmental Microbiology, 74*(15), 4737–4745. https://doi.org/10.1128/AEM.00325-08

Salazar, N., Gueimonde, M., de los Reyes-Gavilán, C. G., & Ruas-Madiedo, P. (2016). Exopolysaccharides produced by lactic acid bacteria and Bifidobacteria as fermentable substrates by the intestinal microbiota. *Critical Reviews in Food Science and Nutrition, 56*(9), 1440–1453. https://doi.org/10.1080/10408398.2013.770728

Schoster, A., Kokotovic, B., Permin, A., Pedersen, P., Dal Bello, F., & Guardabassi, L. (2013). In vitro inhibition of Clostridium difficile and Clostridium perfringens by commercial probiotic strains. *Anaerobe, 20*, 36–41.

Sindhu, K. N., Sowmyanarayanan, T. V., Paul, A., Babji, S., Ajjampur, S. S., Priyadarshini, S., Sarkar, R., Balasubramanian, K. A., Wanke, C. A., Ward, H. D., & Kang, G. (2014). Immune response and intestinal permeability in children with acute gastroenteritis treated with Lactobacillus rhamnosus GG: a randomized, double-blind, placebo-controlled trial. *Clinical Infectious Diseases, 58*(8), 1107–1115. https://doi.org/10.1093/cid/ciu065. Epub 2014 Feb 5. PMID: 24501384; PMCID: PMC3967829.

Sittipo, P., Lobionda, S., Choi, K., Sari, I. N., & Kwon, H. Y. (2018). Toll-Like Receptor 2-mediated suppression of colorectal cancer pathogenesis by polysaccharide a from bacteroides fragilis. *Frontiers in Microbiology, 9*(July), 1588. https://doi.org/10.3389/fmicb.2018.01588

Spanò, A., Laganà, P., Visalli, G., Maugeri, T. L., & Gugliandolo, C. (2016). In vitro antibiofilm activity of an exopolysaccharide from the marine thermophilic Bacillus licheniformis T14. *Current Microbiology, 72*, 518–528.

Strandwitz, P. (2018). Neurotransmitter modulation by the gut microbiota. *Brain Research, 1693*, 128–133. https://doi.org/10.1016/j.brainres.2018.03.015

Turner, J. R. (2009). Intestinal mucosal barrier function in health and disease. *Nature Reviews Immunology, 9*(11), 799–809. https://doi.org/10.1038/nri2653

Trabelsi I, Ktari N, Ben Slima S, Triki M, Bardaa S, Mnif H, Ben Salah R. Evaluation of dermal wound healing activity and in vitro antibacterial and antioxidant activities of a new exopolysaccharide produced by Lactobacillus sp.Ca6. Int J Biol Macromol. 2017 Oct;103:194-201. https://doi.org/10.1016/j.ijbiomac.2017.05.017 Epub 2017 May 8. PMID: 28495632.

Tarique, M., Ali, A. H., Kizhakkayil, J., Gan, R. Y., Liu, S. Q., Kamal-Eldin, A., & Ayyash, M. (2023). Investigating the biological activities and prebiotic potential of exopolysaccharides Produced by Lactobacillus delbrueckii and Lacticaseibacillus rhamnosus: Implications for gut microbiota modulation and rheological properties in fermented milk. Food Hydrocolloids for Health, 4, Article 100162. https://doi.org/10.1016/j.fhfh.2023.100162

Valcheva, R., & Dieleman, L. A. (2016). Prebiotics: Definition and protective mechanisms. *Best Practice & Research Clinical Gastroenterology, 30*(1), 27–37. https://doi.org/10.1016/j.bpg.2016.02.008

Wang, J., Zhao, X., Yang, Y., Zhao, A., & Yang, Z. (2015). Characterization and bioactivities of an exopolysaccharide produced by *Lactobacillus plantarum* YW32. *International Journal of Biological Macromolecules, 74*, 119–126. https://doi.org/10.1016/j.ijbiomac.2014.12.006

Wang, M., Zhou, W., Yang, Y., Xing, J., Xu, X., & Lin, Y. (2021). Potential prebiotic properties of exopolysaccharides produced by a novel Lactobacillus strain, Lactobacillus pentosus YY-112. *Food & Function, 12*(19), 9456–9465. https://doi.org/10.1039/D1FO01261D

Wang, Y., Han, J., Ren, Q., et al. (2024). The involvement of lactic acid bacteria and their exopolysaccharides in the biosorption and detoxication of heavy metals in the gut. *Biological Trace Element Research, 202*, 671–684. https://doi.org/10.1007/s12011-023-03693-1

Wan, C., Qian, W. W., Liu, W., Pi, X., Tang, M. T., Wang, X. L., Gu, Q., Li, P., Zhou ,T., Exopolysaccharide from Lactobacillus rhamnosus ZFM231 alleviates DSS-induced colitis in mice by regulating gut microbiota. *J Sci Food Agric*. 2022 Dec;102(15):7087-7097. https://doi.org/10.1002/jsfa.12070 Epub 2022 Jul 2. PMID: 35707876

Yan, H., & Ajuwon, K. M. (2017). Butyrate modifies intestinal barrier function in IPEC-J2 cells through a selective upregulation of tight junction proteins and activation of the Akt signaling pathway. *PLoS One, 12*(6), e0179586.

Yang, Y., Song, H., Wang, L., Dong, W., Yang, Z., Yuan, P., et al. (2017). Antiviral effects of a probiotic metabolic products against transmissible gastroenteritis coronavirus. *Journal of Probiotics & Health, 05*(03). https://doi.org/10.4172/2329-8901.1000184

Yano, J. M., Yu, K., Mazmanian, S. K., Hsiao, E. Y., Yano, J. M., Yu, K., Donaldson, G. P., Shastri, G. G., Ann, P., Ma, L., Nagler, C. R., Ismagilov, R. F., Mazmanian, S. K., & Hsiao, E. Y. (2015). Indigenous bacteria from the gut microbiota regulate host serotonin biosynthesis article indigenous bacteria from the gut microbiota regulate host serotonin biosynthesis. *Cell, 161*(2), 264–276. https://doi.org/10.1016/j.cell.2015.02.047

Yunes, R. A., Poluektova, E. U., Dyachkova, M. S., Klimina, K. M., Kovtun, A. S., Averina, O. V., Orlova, V. S., & Danilenko, V. N. (2016). GABA production and structure of gadB/gadC genes in Lactobacillus and Bifidobacterium strains from human microbiota. *Anaerobe, 42*, 197. https://doi.org/10.1016/j.anaerobe.2016.10.011

Zhou, G., Li, N., Rene, E. R., Liu, Q., Dai, M., & Kong, Q. (2019). Chemical composition of extracellular polymeric substances and evolution of microbial community in activated sludge exposed to ibuprofen. *Journal of Environmental Management, 246*, 267–274. https://doi.org/10.1016/j.jenvman.2019.05.044

Exopolysaccharides (EPS): Exploring Their Prebiotic Functions and Effects on Gut Health

Rajani S. Bansode

1 Introduction

In 1982, Sutherland first used the word "Exopolysaccharide." to describe high molecular weight carbohydrate polymers that microorganisms exude into their surroundings. Exopolysaccharides known as extracellular polymeric substances account for polysaccharides between 40% and 95% of extracellular polymeric substances (Rana & Upadhyay, 2020). Exopolysaccharides (EPS) are extracellular polymers of carbohydrates, which are produced during the growth and metabolism of various microbes, such as bacteria, yeasts, fungi, and microalgae. Microbial cells release EPSs to enable them to bind other cells.

The chief component of EPSs is carbohydrates, along with other components such as acetate, pyruvate, succinate, and phosphate, among other non-carbohydrate substituents. The EPS structure contains variety of monosaccharides, which include fructose, glucose, arabinose, mannose, rhamnose, and xylose (Kaur & Dey, 2023). Nondigestible oligosaccharides known as prebiotics promote the growth of probiotic bacteria, especially bifidobacteria and lactobacilli. While there is some debate regarding the exact definition of prebiotics, the following standards are commonly used to classify substances as such: they must be unabsorbed in the gastrointestinal tract, resistant to the stomach's acidic environment and host enzyme hydrolysis, fermentable by the resident intestinal bacteria, able to selectively stimulate the growth and activity of the intestinal microbiota, and beneficial to the host's health and well-being (Pujari & Banerjee, 2020).

R. S. Bansode (✉)
Department of Stem Cell and Regenerative Medicine and Medical Biotechnology, Centre of Interdisciplinary Research, D. Y. Patil Education Society (Deemed to be University), Kolhapur, Maharashtra, India
e-mail: rajani_thanekar@yahoo.com

T. Sarkar et al. (eds.), *Extracellular Polysaccharides*,
https://doi.org/10.1007/978-3-031-94472-7_13

2 Types of Exopolysaccharides

EPS are classified as homopolysaccharides or heteropolysaccharides depending on their nature. Homopolysaccharides are made up of the same monosaccharide, such as curdlan, while heteropolysaccharides are made up of a mixture of different monosaccharides. In addition to the frequently observed sugars like glucose, galactose, and fructose, heteropolysaccharides also include certain uncommon sugars, viz., rhamnose, xylose, fucose, and mannose, as well as uronic acids, and amino sugars like xanthan.

The formation of heteropolysaccharides (HePS) requires the intracellular scope, which contains repeating units. Lipids like isoprenoid glycosyl may also put in to the creation of HePS. The repeating units are moved via the membrane outside of the cell before the polymerization. The conditions under which EPS is fermented determine its quantity and kind, and the different stages of microorganism growth influence the production and secretion of EPS by microbes (Silva et al., 2019).

Linear polysaccharides are polymer chains, which consist of straight chains of monosaccharides, such polypullan. Branching polysaccharides (EPS) are EPS that have arms and bends, such as EPS-W1 that was isolated from *Lactobacillus plantarum* W1 (Wang et al., 2023). EPSs offer microenvironments, which provide protection to bacteria from extreme conditions, help bacterial colonization and pathogenicity, and facilitate genes and metabolites exchange within bacterial communities. Microbial EPSs can be formed in a less time in high amounts under controlled conditions (Salimi & Farrokh, 2023).

Homopolysaccharides and heteropolysaccharides differ not only in their linkage bonds and chemical composition but also in the enzymes involved in their synthesis and the specific sites where this synthesis occurs (Rana & Upadhyay, 2020). The synthesis of homopolysaccharides occurs outside the cell, where a specific substrate, such as sucrose, is converted into the polymer through the action of enzymes that are secreted by the cell. In contrast, the formation of the precursor repeating units for heteropolysaccharides occurs inside the cell. These units are then transported across the membrane with the assistance of isoprenoid glycosyl carrier lipids, which facilitate their polymerization outside the cell (Kaur & Dey, 2023).

3 Mechanism of Synthesis of Exoploysaccharides

The biosynthesis of microbial exopolysaccharides (EPS) is a complex process involving numerous enzymes and regulatory proteins. Microorganisms that produce EPS rely on sugars as their energy source, while ammonium salts and amino acids serve as nitrogen sources. The production of EPS is influenced by the availability of carbon and nitrogen in the environment. A higher exopolysaccharide yield is often associated with lower nitrogen concentrations. Optimal EPS production typically occurs at a carbon-to-nitrogen (C/N) ratio of approximately 10:1. For instance, in a study by Zhang et al., the elemental composition of the EPS produced by *Alteromonas* sp. JL2810 showed that carbon comprised 33.84% of the total mass, while nitrogen contributed only 0.88%. The C/N ratio of the EPS was found to be around 38 (Rana & Upadhyay, 2020).

4 Microorganisms Producing Exopolysaccharides

Bacteria that produce exopolysaccharides (EPS) are classified into varying range of phylogenetic groups, including both Gram-negative and Gram-positive species. Among the Gram-negative bacteria, the genera such as Acetobacter, Gluconobacter, Gluconacetobacter, Komagataeibacter, Kozakia, Neoasaia, Agrobacterium, Rhizobium, and *Zymomonas* are included in the Alpha proteobacteria class. The Betaproteobacteria class comprises of genera like Alcaligenes and Achromobacter. The genera belonging to the Gammaproteobacteria class, prominently include *Azotobacter*, *Pseudomonas*, *Enterobacter*, *Alteromonas*, *Pseudoalteromonas*, *Xanthomonas*, *Halomonas*, *Erwinia*, *Vibrio*, and *Klebsiella*. On the Gram-positive side, the Bacilli class contains genera such as *Bacillus*, *Paenibacillus*, *Lactobacillus*, *Leuconostoc*, and *Streptococcus*. The *Sarcina* species belong to the class *Clostridia*. Actinomycetia class contain genera such as Bifidobacterium and Rhodococcus, among others. All of these diverse bacterial species are known to produce extracellular polymeric substances , which are crucial for a number of biological and commercial processes (Netrusov et al., 2023). Table 1 explains the exopolysaccharides produced by various microorganisms and their application.

5 Prebiotic Effect of Microbial Exopolysaccharides

5.1 Dextran

Louis Pasteur first discovered dextran exopolysaccharide as a microbial product of wine. Scheilber in 1974 named the exopolysaccharides as Dextran. Although their physiological functions in the microbial host are still unclear, EPS plays a part in quorum sensing, pathogenicity, biofilm formation, and defense against dehydration. The bacterial cell's cellular diffusion properties may be significantly impacted by a dextran layer surrounding it. Applications of dextran in the food, pharmaceutical, biomaterial, photofilm, and fine chemical industries have made it more significant. It is known that a wide variety of lactic acid bacteria (LAB) make dextran. Due to the difference in the degree of branching in their glucose backbone, dextran polymers exhibit a remarkable diversity in both chain length and physicochemical properties. Since dextran is a neutral polysaccharide, variations in pH or salt concentration have little effect on its viscosity. Dextrans that have more than 43% branching

Table 1 Exopolysaccharides produced by various microorganisms and their application

Sr. no.	Exopolysaccharides	Microorganisms	Application	References
1.	Dextran	*Leuconostoc mesenteroides*, *Streptococcus*, *Lactobacillus*, *Gluconobacter*	In food, used as emulsifier, viscosifier, stabilizer, gelling agent	Osemwegie et al. (2020)
2.	Xanthan	*Xanthomonas campestris* and *Sphingomonas paucimobilis*	Thickening agent in food, improve texture, blending agent, enhanced oil recovery, improve flowablity of fungicides, pesticides, thickener in latex paints, stabilizers in pharmaceutical preparation	Salimi and Farrokh (2023), Palaniraj and Jayaraman (2011)
3.	Gellan	*Sphingomonas paucimobilis*	Provide structure and texture in confectionary, bulking agent, solidifying agent in pet food, in electrophoresis, agar, microencapsulation	Bajaj et al. (2007)
4.	Levan	*P. syringae pv. Glycinea*, *Bacillus subtilis*, *B. megaterium* and *B. pumilis*	Industrial gum, blood plasma extender, sweetener, stabilizers, thickener, formulation aid	Sabra and Zeng (2009), Han (1990)
5.	Pullulan	*Aureobasidium pullulans*	Blood plasma extender conjugates in vaccines, as a carrier in targeted gene therapy, edible packaging, stabilizer, binder, electrophoresis, and filtration	Cheng et al. (2011), Sharma et al. (2021)
6.	Alginate	*A. vinelandii*	Encapsulation, edible film, wound dressings, foam dressing and hydrogel dressing, tissue engineering	Sabra and Zeng (2009), Puscaselu et al. (2020)
7.	Hyaluronan	*Streptococcus (Pa) belonging to group A*, *Bacillus cereus*, *Lactococcus lactis*	HA-derivatives for acceleration of wound-healing, treatment of cancer and vascular disease, and in anti-arthritis therapy	Vaishnav and Choudhary (2021), Volpi et al. (2009)
8.	Curdlan	*Alcaligenes faecalis var. myxogenes 10C3 strain*, *Agrobacterium*	Stabilizers, immobilzer, unique rheological application used in food industry, as thickener, as fat mimetic in meat products	Zhang and Edgar (2014), Kumar et al. (2020)
9.	Carrageenan	Seaweeds of the class Rhodophyceae	Used in food industry for gelling, thickening, and emulsifying	Nadda and Sharma (2021)

(continued)

Table 1 (continued)

Sr. no.	Exopolysaccharides	Microorganisms	Application	References
10.	Wellan	Sphingomonas sp. ATCC 31555, (referred also as Alcaligenes sp. ATCC 31555)	Improves texture of food, enhances water holding capacity while cooking and storage, drug release	Kaur et al. (2014)
11.	Kefiran	*Lactobacillus kefiranofaciens*	Bioactive agent, film coating	Nadda and Sharma (2021)
12.	Sclerogan	Sclerotium glucanicum NRRL 3006	Used in enhanced oil recovery, the preparation of adhesives, water colors, printing inks, and liquid animal feed composition, and thickeners in pesticides, as a stabilizer in food	Wang and McNeil (1995), Schmid et al. (2011)
13.	Alternan	*Leuconostoc mesenteroides*, *L. citreum*	As commercial gum Arabic, artificial sweetener	Wangpaiboo et al. (2018), Zikmanis et al. (2020), Vaishnav and Choudhary (2021)
14.	Cellulose	*Agrobacterium* spp., *Acetobacter* spp., *Azotobacter*, *Rhizobium* spp., *Sarcina*, *Alcaligenes*, and *Pseudomonas*	Bioadhesive, mucoadhesive, coating, gelling agent, thickening and stabilizing agent	Shokri and Khosro (2016), Lahiri et al. (2021)

through α-(1→3) links are insoluble in water. The molecular weight of dextrans is between 3 and 500,000 kDa. Dextran is more rod-like at lower molecular weights (<2000 kDa) and exhibits the characteristics of an extensible coil at molecular weights between 2000 and 10,000 kDa. In clinical applications, dextrans with low molecular weights (40, 60, and 70 kDa) are typically recommended (Kothari et al., 2015). Similar fermentation characteristics are seen in the highly branched 0.5 and 1 kDa dextran, which are fermented by Bifidobacterium species, *Lactobacillus*/*Enterococcus* species, and *Bacteroides*/*Prevotella* species.

Due to the difference in molecular weight, the branched 0.5 kDa dextran's rate of fermentation and gas generation was noticeably higher than that of 1 kDa dextran. Therefore, the 1 kDa dextran delivers greater advantage as it stays longer than 0.5 kDa dextran in the gastrointestinal system thus stimulating advantageous saccharolytic fermentation toward the distal region of the colon. The highly branched 1 kDa dextran preferentially stimulated *Bifidobacterium* sp. and *Roseburia*, generated substantial levels of SCFA, in particular acetate and propionate in continuous fermentation system while it lingered in the distal sections of the model. The third vessel of the system, which mimics the distal colon, showed a notable increase in

Bifidobacterium and Roseburia concentrations. This suggests that the dextran may be active in distal colonic regions, supporting the prebiotic effect of dextran (Sarbini et al., 2013).

5.2 Xanthan

One of the most studied heteropolysaccharides is xanthan, produced by *X. campestris*. Two d-glucoses, two d-mannoses, and a d-glucuronic acid combine to produce xanthan. Depending on the bacterial genus and species, this EPS often has a relatively high molecular weight ranging between 500 and 2000 kDa. Units of two to eight monomers repeat to make xanthan. Repeating pentasaccharide units are assembled to begin the production of xanthan. The macromolecule is subsequently created by polymerizing these components. Acetyl-CoA and phosphoenolpyruvate are two of the monosaccharides that are sequentially added to generate the repeating xanthan units (Kaur & Dey, 2023). Xanthan has outstanding suspending qualities, such as strong freeze-thaw stability, and thermal stability even in acidic environments, hence finds several applications as a viscosifier and stabilizer in syrups, sauces, dressings, baked goods, soft cheese, restructured meat, and more. Xanthan gum is used to increase volume and texture (particularly in gluten-free breads), decrease syneresis in creams and fruit fillings, increase flavor release, and improve freeze-thaw stability of chilled doughs. It also improves water binding during baking and the shelf life of baked foods. Xanthan gum gives dressings, sauces, and syrups the desired body, texture, and pourability as well as enhanced flavor release. It also helps the emulsion remain resilient to acid and salt and maintain a stable viscosity over a broad temperature range. Excellent consistency, viscosity, and freeze-thaw stability are provided by xanthan in buttered syrups and chocolate toppings. Because it is soluble and stable at low pH and enhances the suspension of insoluble particles (such as in fruit juices) as well as the body and mouthfeel of products, xanthan is also a useful stabilizer and bodying agent in cream cheese, where it improves flavor, shelf life, heat-shock protection, and decreases syneresis (Giavasis, 2013).

Khalid A. Alsaleem et al. showed that at lower concentrations (0.2–0.4%), xanthan gum (XG) acts as an efficient stabilizer in low-fat yoghurt (LFY), improving stiffness and overall texture while having no negative effect on microbiological stability. A 0.8% XG concentration enhanced the yoghurt's color, texture, and flavor. The XG-added samples of the LFY samples (0.2–0.6%) had the largest number of *Bifidobacterium bifidum* and maintained the greater numbers of *B. bifidum* throughout storage as compared to the control, suggesting that XG offers a favorable environment for *B. bifidum* development. Thus, their work shows the dual role of xanthan gum as both a stabilizer and a possible prebiotic, adding favor to the yogurt's quality and hypotheses on the viability of using probiotic cultures during storage (Alsaleem & Hamouda, 2024).

In addition to probiotics (*B. longum BB536*), xanthan gum and its hydrolyzed forms (0.1%, 0.5%, and 1% [w/v]) were used as prebiotics to create synbiotic yogurt. The viable count of *B. longum BB536* was maintained above 6 Log CFU/g during storage with a maximum count of 7.898 Log CFU/g in yogurt containing 1% of enzyme hydrolyzed xanthan gum (EHXG). The inclusion of hydrolyzed xanthan gum improved the survivability of yogurt culture and probiotics. The highest level of sensory acceptance and resistance to probiotic (*B. longum 536*) and yogurt culture bacteria (*Streptococcus thermophiles and L. bulgaricus*) was demonstrated by the 0.5% addition of EHXG. Thus, hydrolyzed xanthan gum may be the ideal option for prebiotics added to yogurt (Khalid et al., 2022). Scientists F. Liu et al. showed addition of xanthan gum in digestive biscuit proved to have prebiotic effect (Liu et al., 2023).

5.3 Gellan

Gellan gum's many functional qualities make it one of the exopolysaccharides with industrial applications. Members of the bacterium genus *Sphingomonas* excrete this sphingan group of heteropolysaccharides. L-rhamnose (L-Rha), D-glucuronic acid (D-GlcA), and b-D-glucose (D-Glc) make up the repeating unit of gellan polysaccharide. The approximate composition is 60% glucose, 20% rhamnose, and 20% glucuronic acid . Additionally, gellan gum contains a significant proportion of non-polysaccharide substances. The molecular mass of gellan gum is around 500,000 Da (Bajaj et al., 2007).

The benefits of gellan gum in avoiding Non alcoholic fatty liver disease (NAFLD) and encouraging the development of good bacteria were described by Moon et al. Gellan gum (GG) shown significant promise as a prebiotic by encouraging the development of probiotic strains of Bifidobacterium and Lactiplantibacillus. Furthermore, through enhanced gut microbial composition, decreased detrimental enzyme activity, and increased Short Chain Fatty Acid (SCFA) concentration, GG ameliorated HFD-induced hepatic inflammation and fat deposition, potentially through an effect on the gut–liver axis (Do et al., 2023). Xu et al. demonstrated the prebiotic qualities of gellan oligosaccharides in vitro through the cultivation of human fecal bacteria, *Bifidobacterium* sp., and *Lactobacillus* sp. Gellan Gum oligosaccharides GGOS had the best prebiotic effects by enhancing the butyric acid-producing bacteria and beneficial bacteroides (Xu et al., 2021).

5.4 Levan

Several species of *Bacillus*, *Zymomonas*, *Halomonas*, *Pseudomonas*, *Areobacter*, *Streptococcus*, and *Microbacterium* produce the exopolysaccharides levan (Rana & Upadhyay, 2020). Levan is a naturally occurring homopolysaccharide that has a

D-glucosyl residue at the end of the chain and is made up of D-fructo-furanosyl residues connected by β-(2,6) and β-(2,1) links in the core and branch chains, respectively. About 11,000 Da is the low molecular weight levan, and about 1,800,000 Da is the high molecular weight levan (Srikanth et al., 2015). Levan-type fructan significantly ($p < 0.05$) enhanced the content of both acetate and butyrate while specifically stimulating the growth of the Bifidobacterium and *Eubacterium rectale* and *Clostridium coccoides* group in all fermenter vessels. In comparison to the models treated with inulin HP (0.62–0.7 log cell/mL), the models treated with levan-type fructan showed significantly ($p<0.05$) greater increases in *Bifidobacterium* population (0.8–1.24 log cell/mL). This suggests that levan-type fructan has a stronger bifidogenic effect and a longer persistence in the colon because of its higher DP (Liu et al., 2020). According to structural analysis, the fructan Plev from *Paenibacillus* sp. FP01 is a linear levan-type fructan with a molecular mass of 3.11×10^6 Da. Plev has exceptional prebiotic properties , it control gut microbiota by producing SCFA, and enhancing intestinal metabolism. . Plev is a biopolymer that shows promise for widespread use in the food, feed, and pharmaceutical sectors (Cheng et al., 2021).

5.5 Pullulan

Aureobasidium pullulans' culture broth was the source of pullulan's initial isolation. α-(1→6)-linked (1→4)-α-d-triglucosides, which have the chemical formula $C_6H_{10}O_5$, make up the polysaccharide pullulan, a maltotriose trimer. The coexistence of α-(1→4) and α-(1→6) in a single compound makes this polysaccharide an intermediary between the structures of amylose and dextran. Pullulan is highly soluble in water and flexible due to its special structure. The polysaccharide is very soluble in water but insoluble in organic solvents. Pullulan dry powder has no taste or odor and is white in appearance. The pullulan is edible, biodegradable, nontoxic, nonmutagenic, and noncarcinogenic. Phthalyl pullulan nanoparticles, or PPNs, are a novel prebiotic formulation that boosts growth of probiotics'and show antibacterial activity. Following the coupling of hydrophobic phthalic anhydride to hydroxyl groups in pullulan via hydrophobic contacts, the PPNs were created by self-assembled nanoparticles. Esterification is thought to be the mechanism by which the pullulan's main hydroxyl groups react with the carboxylic acids of phthalic anhydride. Additionally, because of the increased hydrophobic contacts between the phthalic moieties in the PPNs, lower PPN sizes were the result of higher conjugation of phthalic groups (Hong et al., 2019). Pullulan-enriched breads were more resistant to AGHJ hydrolysis than the regular bread samples, according to the in vitro digestibility and fermentation tests. Pullulan outperformed other polysaccharides based on the Prebiotic Activity Score (PAS), and pullulan-enriched breads outperformed regular bread in terms of prebiotic efficacy. Pullulan's higher PAS suggests that it selectively promotes the growth of probiotic organisms while suppressing the growth of pathogenic organisms, meeting all the necessary conditions for an effective prebiotic (NithyaBalaSundari et al., 2020).

5.6 Alginate

The opportunistic pathogen *Pseudomonas aeruginosa* was the first known bacterium to produce alginate. Three nonpathogenic *Pseudomonas* species, including *P. mendocina*, *P. putida*, and *P. fluorescens*, were later found to produce alginate as well. For the formation of bacterial alginates, soil bacteria (*Azotobacter vinelandii*) are ideal. *Sargassum* spp., *Macrocystis pyrifera*, *Ascophyllum nodosum*, *Ecklonia maxima*, *Saccharina japonica* (formerly *Laminaria japonica*), *Durvillea antarctica*, *Lessonia nigrescens*, *Laminaria hyperborea*, and *Laminaria digitata* are used to create commercial alginates. It's estimated that approximately 85,000 tons of algae are used to create 23,000 tons of alginate (Wilkerson, 2022).

Enzymatic hydrolysis of alginate polymer was used to investigate the prebiotic potential of alginate oligosaccharides (AOS). Wang et al. observed that AOS affected the growth of bacteria. When cultivated in the medium supplemented with AOS, in vitro tests revealed that the quantities of *lactobacilli* or *bifidobacteria* were higher than those in the control media. Compared to fructo-oligosaccharides (FOS), alginate oligosaccharides considerably increased the development of *B. bifidum* ATCC 29521 and *B. longum SMU 27001*. While, AOS selectively boosted the male Wistar rat's fecal and cecal microbiota, according to in vivo investigations. Rats given a diet supplemented with 2.5% AOS for 2 weeks saw a 13-fold increase in the number of fecal bifidobacteria (Wang et al., 2006). In a review by Bi et al., studies supporting the prebiotic effect of alginate are described such as in a type 2 diabetic mouse model; *Sargassum fusiforme* alginate can dramatically reduce harmful bacteria like *Helicobacter* and *Turicibacter* while greatly increasing native bacteria like *Lactobacillus*, *Bacteroides*, *Akkermansia* *Alloprevotella*, *Weissella*, and *Enterorhabdus*. Alginate has been shown to have a beneficial effect on type 2 diabetes by considerably reducing the levels of aromatic amino acids (AAAs) and branched-chain amino acids (BCAAs) in the colons of type 2 diabetic mice (Bi et al., 2022).

5.7 Hyaluronan

Hyaluronan are produced from genus *Streptococcus* sp. Hyaluronan (HA) is a member of the linear polymers of repeating glucuronic acid and N-acetylglucosamine that make up the family of mucopolysaccharides, often known as glycosaminoglycans. The primary component of extracellular matrix (ECM), which keeps cells together and creates a channel for the flow of oxygen and nutrients to individual cells, is HA. It functions as a structural element in several organs and tissues. For instance, it maintains the elastoviscosity of liquid connective tissues, lubricates the synovial fluid in joints, gives the eye a jelly-like consistency, and regulates water transport and tissue hydration (1,2). Additionally, HA can function as signaling molecules through a variety of processes, including cell transformation, inflammation, wound healing, and morphogenesis. HA may raise the relative abundance of some helpful bacteria, such as those that produce SCFAs (Huang et al., 2022).

5.8 Curdlan

Curdlan is a bacterial polysaccharide that has garnered a lot of attention lately because of its intrinsic bioactivity and intriguing and useful rheological characteristics. Compared to many other prevalent natural polysaccharides, curdlan's simple (1→3)-β-glucan homopolymeric, unbranched structure promotes improved solubility, offering options for processing the polymer into desired shapes and formulations (Zhang & Edgar, 2014). High molecular weight glucose serves as curdlan's fundamental building block. In alkaline solution (0.3 M), the molecular weight of bacterial curdlan varies from 5.3 × 104 to 5.8 × 106 according to viscosity measurements, whereas the molecular weight of curdlan derived from fungal species ranges from 2.06×10^4 to 5.0×10^6. β-(1→3, 1→2)-glucose provides the curdlan polymer's basic bond, most likely with a small number of interchain 1,6-links. Curdlan has the chemical formula ($C_6H_{10}O_5$).

Due to its high insoluble dietary fiber content, curdlan is used to make low-calorie foods. The prebiotic effects of curdlan (1→3)-β-d-glucan oligosaccharides (GOS) was investigated. With the exception of *L. brevis* NRRL B-4527, GOS enhanced the growth of all examined Lactobacillus (L.) strains. GOS as the sole carbon source did not support the growth of *Escherichia coli*. Human fecal microbiota in vitro batch fermentation revealed that GOS markedly boosted the population of *Lactobacillus* sp., *Bifidobacterium* sp., and *Bacteroides* sp. As L. strains grew on GOS, the pH of the culture medium dropped and lactic, acetic, and propionic acids were generated. Representative strains of L. used GOS in a strain-dependent manner (Shi et al., 2018). Curdlan's prebiotic qualities would benefit the host by encouraging the development of a healthy gut flora and so lowering the incidence of gastrointestinal illnesses. The production of prebiotics to treat a variety of gastrointestinal dysbiosis issues can thus use curdlan (Chaudhari et al., 2021).

5.9 Carrageenan

Carrageenans are a group of biomolecules composed of linear polysaccharide chains with sulphate half-esters attached to the sugar unit. These properties allow carrageenans to dissolve in water, form highly viscous solutions, and remain stable over a wide pH range. There are three general forms (kappa, lambda, and iota), each with their own gelling property. Kappa carrageenan today is almost exclusively obtained from farmed *Kappaphycus alvarezii* and iota from farmed *Eucheuma denticulatum*. In preclinical settings, agarose-oligosaccharides show intriguing prebiotic potential by promoting the production of SCFAs, including butyrate and butyrate-producing bacteria, bifidobacteria, and lactobacilli. Although there are less studies evaluating carrageenans' prebiotic potential, they likewise boosted the creation of SCFA (Gotteland et al., 2020).

5.10 Wellan

Welan are produced by *Alcaligenes* (Ventorino et al., 2019). Among the exopolysaccharides, welan gum, a microbial product, is significant because of its unique qualities and possible uses (Kaur et al., 2014). Glycerol was used as the only carbon source during fermentation to produce low-molecular-weight welan gum (LMW-WG) utilizing adaptive laboratory evolution (ALE) of *Sphingomonas* sp. ATCC 31555. To assess LMW-WG's possible prebiotic qualities as dietary fiber, an in vitro modeling colon reactor was employed, which showed favorable effect on *Bifidobacterium* (Wang et al., 2024).

5.11 Kefiran

Kefiran is a hetero-polysaccharide biopolymer that contains galactose and glucose. It is a water-soluble prebiotic exopolysaccharide. It is either made from kefir grains, pure cultures of *L. kefiranofaciens* grown in aerobic environments, or mixed cultures of *L. kefiranofaciens* and *Saccharomyces cerevisiae* grown in anaerobic environments (Hasheminya & Dehghannya, 2020). The effect of glugalactan kefiran on human intestinal microbiota's and the biological impact of metabolites produced such as butyrate and others was investigated using , the feces of ten (10) children's , ranging in age from 4 months to 3 years, in an in vitro fermentation model, which demonstrated the formation of lactic acid and short-chain fatty acids (SCFA) (Medrano et al., 2020).

6 Prebiotic Effect of Mannan, Cellulose, and Polyglutamic Acid

Lactobacilli efficiently fermented Mannan oligosaccharides (MOS) to produce the primary short-chain fatty acids, propionate and acetate. Growing on Locust bean galactomannan (LBG) hydrolysate, lactobacilli generated branched-chain amino acids, including leucine, isoleucine, and valine. MOS-containing partially hydrolyzed mannans have prebiotic potential and could be investigated as a useful intervention to support probiotic gut microbiota and address health issues (Suryawanshi & Kango, 2021).

Dietary fibers that are difficult for the body to digest and absorb include cellulose. In addition to the health benefits of cellulose consumption in relation to weight control, if its solubility can be improved, this fiber may also offer prebiotic action. In order to increase cellulose's solubility, short-chain compounds were produced by hydrolyzing banana peel cellulose. Because enzymatic hydrolysis produced more soluble fibers of cellodextrins and water-soluble cellulose without producing any degraded byproducts, it was superior than diluted-acid hydrolysis. Probiotic growth was facilitated by the cellodextrins. While the water-soluble cellulose did not promote the growth of harmful bacteria, it did considerably boost the growth of *L. plantarum* TISTR2075, which met the requirements for prebiotic fiber. Additionally, compared to the commercial prebiotic, the water-soluble cellulose surprisingly offered a greater prebiotic activity index (Phirom-on & Apiraksakorn, 2021). A prebiotic called poly-gamma-glutamate (γ-PGA) was tested for its impact on the gut flora and organ distributions of mice. Oral treatment of γ-PGA raised the abundance of Lactobacillales while decreasing the abundance of *Clostridiales* in the stomach of mice, according to pyrosequencing study for the 16S rRNA genes of bacteria (Gut, 2017).

References

Alsaleem, K. A., & Hamouda, M. E. A. (2024). Enhancing low-fat probiotic yogurt: The role of xanthan gum in functionality and microbiological quality. *Processes, 12*(5), 1–12.

Bajaj, I. B., Survase, S. A., Saudagar, P. S., & Singhal, R. S. (2007). Gellan gum: Fermentative production, downstream processing and applications. *Food Technology and Biotechnology, 45*(4), 341–354.

Bi, D., et al. (2022). Potential food and nutraceutical applications of alginate: A review. *Marine Drugs, 20*(9), 1–15.

Chaudhari, V., et al. (2021). Therapeutic and industrial applications of curdlan with overview on its recent patents. *Frontiers in Nutrition, 8*, 646988.

Cheng, K. C., Demirci, A., & Catchmark, J. M. (2011). Pullulan: Biosynthesis, production, and applications. *Applied Microbiology and Biotechnology, 92*(1), 29–44.

Cheng, R., et al. (2021). Biosynthesis and prebiotic activity of a linear levan from a new Paenibacillus isolate. *Applied Microbiology and Biotechnology, 105*(2), 769–787.

Do, M. H., et al. (2023). Gellan gum prevents non-alcoholic fatty liver disease by modulating the gut microbiota and metabolites. *Food Chemistry, 400*, 134038. https://doi.org/10.1016/j.foodchem.2022.134038

Giavasis, I. (2013). Production of microbial polysaccharides for use in food. In B. McNeil, D. Archer, I. Giavasis, & L. Harvey (Eds.), *Microbial production of food ingredients, enzymes and nutraceuticals* (Woodhead publishing series in food science, technology and nutrition) (pp. 413–468). Woodhead Publishing. https://doi.org/10.1533/9780857093547.2.413

Gotteland, M., Riveros, K., Gasaly, N., & Carcamo, C. (2020). The pros and cons of using algal polysaccharides as prebiotics. *Frontiers in Nutrition, 7*, 1–15.

Gut, M. (2017). Prebiotic effects of poly-gamma-glutamate on bacterial flora in murine gut. *Journal of Microbiology and Biotechnology, 27*, 412–415.

Han, Y. W. (1990). Microbial levan. *Advances in Applied Microbiology, 35*, 171–194.

Hasheminya, S.-m., & Dehghannya, J. (2020). Novel ultrasound-assisted extraction of kefiran biomaterial, a prebiotic exopolysaccharide, and investigation of its physicochemical, antioxidant and antimicrobial properties. *Materials Chemistry and Physics, 243*, 122645. https://doi.org/10.1016/j.matchemphys.2020.122645

Hong, L., et al. (2019). Pullulan nanoparticles as prebiotics enhance the antibacterial properties of Lactobacillus plantarum through the induction of mild stress in probiotics. *Frontiers in Microbiology, 10*, 1–12.

Huang, G., et al. (2022). The prebiotic and anti-fatigue effects of hyaluronan. *Frontiers in Nutrition, 9*, 1–12.

Kaur, N., & Dey, P. (2023). Bacterial exopolysaccharides as emerging bioactive macromolecules: From fundamentals to applications. *Research in Microbiology, 174*(4), 104024. https://doi.org/10.1016/j.resmic.2022.104024

Kaur, V., et al. (2014). Welan gum: Microbial production, characterization, and applications. *International Journal of Biological Macromolecules, 65*, 454–461. https://doi.org/10.1016/j.ijbiomac.2014.01.061

Khalid, N., et al. (2022). Exploring the prebiotic potential of xanthan gum and its modified forms for the production of synbiotic yogurt. *Journal of Food Processing and Preservation, 46*, 1–12.

Kothari, D., Das, D., Patel, S., & Goyal, A. (2015). Dextran and food application. In K. G. Ramawat & J.-M. Mérillon (Eds.), *Polysaccharides: Bioactivity and biotechnology* (pp. 735–752). Springer International Publishing. https://doi.org/10.1007/978-3-319-16298-0_66

Kumar, D., et al. (2020). Chemistry and microbial sources of curdlan with potential application and safety regulations as prebiotic in food and health. *Food Research International, 133*, 109136. https://doi.org/10.1016/j.foodres.2020.109136

Lahiri, D., et al. (2021). Bacterial cellulose: Production, characterization, and application as antimicrobial agent. *International Journal of Molecular Sciences, 22*, 1–18.

Liu, C., Kolida, S., Charalampopoulos, D., & Rastall, R. A. (2020). An evaluation of the prebiotic potential of microbial levans from Erwinia sp. 10119. *Journal of Functional Foods, 64*, 103668.

Liu, F., et al. (2023). Effects of xanthan gum, konjac glucomannan, and arabinogalactan on the in vitro digestion and fermentation characteristics of biscuits. *Food & Function, 14*(13), 6036–6048. https://doi.org/10.1039/D3FO00703K

Medrano, M., Virginia, M., Simonelli, N., & Abraham, G. (2020). Bioactive carbohydrates and dietary fibre kefiran fermentation by human faecal microbiota: Organic acids production and in vitro biological activity. *Bioactive Carbohydrates and Dietary Fibre, 24*, 100229. https://doi.org/10.1016/j.bcdf.2020.100229

Nadda, A. K., & Sharma, S. (2021). *Microbial exopolysaccharides as novel and significant biomaterials*. Springer.

Netrusov, A. I., et al. (2023). Exopolysaccharides producing bacteria: A review. *Microorganisms, 11*, 1541.

NithyaBalaSundari, S., et al. (2020). Characterization of microbial polysaccharides and prebiotic enrichment of wheat bread with pullulan. *LWT, 122*, 109002. https://doi.org/10.1016/j.lwt.2019.109002

Osemwegie, O. O., et al. (2020). Exopolysaccharides from bacteria and fungi: Current status and perspectives in Africa. *Heliyon, 6*(6), e04205. https://doi.org/10.1016/j.heliyon.2020.e04205

Palaniraj, A., & Jayaraman, V. (2011). Production, recovery and applications of xanthan gum by Xanthomonas campestris. *Journal of Food Engineering, 106*(1), 1–12.

Phirom-on, K., & Apiraksakorn, J. (2021). Development of cellulose-based prebiotic fiber from banana peel by enzymatic hydrolysis. *Food Bioscience, 41*, 101083. https://doi.org/10.1016/j.fbio.2021.101083

Pujari, R., & Banerjee, G. (2020). Impact of prebiotics on immune response: From the bench to the clinic. *Immunology & Cell Biology, 99*(3), 1–19.

Puscaselu, R. G., Lobiuc, A., Dimian, M., & Covasa, M. (2020). Alginate: From food industry to biomedical applications and management of metabolic disorders. *Polymers, 12*(10), 1–30.

Rana, S., & Upadhyay, L. S. B. (2020). Microbial exopolysaccharides: Synthesis pathways, types and their commercial applications. *International Journal of Biological Macromolecules, 157*, 577–583. https://doi.org/10.1016/j.ijbiomac.2020.04.084

Sabra, W., & Zeng, A. P. (2009). Microbial production of alginates: Physiology and process aspects. In *Alginates: Biology and applications. microbiology monographs*. Springer.

Salimi, F., & Farrokh, P. (2023). Recent advances in the biological activities of microbial exopolysaccharides. *World Journal of Microbiology and Biotechnology, 39*(8), 1–22. https://doi.org/10.1007/s11274-023-03660-x

Sarbini, S. R., et al. (2013). The prebiotic effect of α-1,2 branched, low molecular weight dextran in the batch and continuous faecal fermentation system. *Journal of Functional Foods, 5*(4), 1938–1946. https://doi.org/10.1016/j.jff.2013.09.015

Schmid, J., Meyer, V., & Sieber, V. (2011). Scleroglucan: Biosynthesis, production and application of a versatile hydrocolloid. *Applied Microbiology and Biotechnology, 91*, 937–947.

Sharma, H., Pal, J., & Neelam, D. K. (2021). Bacterial extracellular polymers: A review. *Journal of Pure and Applied Microbiology, 15*(3), 1072–1082.

Shi, Y., et al. (2018). In vitro digestibility and prebiotic potential of curdlan (1→3)-β-d-glucan oligosaccharides in lactobacillus species. *Carbohydrate Polymers, 188*, 17–26. https://www.sciencedirect.com/science/article/pii/S0144861718301036

Shokri, J., & Khosro, A. (2016). Application of cellulose and cellulose derivatives in pharmaceutical industries chapter. *INTECH, 11*(tourism), 13. https://www.intechopen.com/books/advanced-biometric-technologies/liveness-detection-in-biometrics

Silva, L. A., Neto, J. H. P. L., & Cardarelli, H. R. (2019). Exopolysaccharides produced by Lactobacillus plantarum: Technological properties, biological activity, and potential application in the food industry. *Annals of Microbiology, 69*(4), 321–328.

Srikanth, R., et al. (2015). Review on production, characterization and applications of microbial levan. *Carbohydrate Polymers, 120*, 102–114. https://doi.org/10.1016/j.carbpol.2014.12.003

Suryawanshi, R. K., & Kango, N. (2021). Production of mannooligosaccharides from various mannans and evaluation of their prebiotic potential. *Food Chemistry, 334*, 127428. https://doi.org/10.1016/j.foodchem.2020.127428

Vaishnav, A., & Choudhary, D. K. (2021). *Microbial polymers*. Springer.

Ventorino, V., et al. (2019). Bioprospecting of exopolysaccharide-producing bacteria from different natural ecosystems for biopolymer synthesis from vinasse. *Chemical and Biological Technologies in Agriculture, 6*(1), 1–9. https://doi.org/10.1186/s40538-019-0154-3

Volpi, N., Schiller, J., Stern, R., & Soltes, L. (2009). Role, metabolism, chemical modifications and applications of hyaluronan. *Current Medicinal Chemistry, 16*(14), 1718–1745.

Wang, Y., & McNeil, B. (1995). Production of the fungal exopolysaccharide scleroglucan by cultivation of Sclerotium glucanicum in an airlift reactor with an external loop. *Journal of Chemical Technology & Biotechnology, 63*, 215–222.

Wang, Y., et al. (2006). In vivo prebiotic properties of alginate oligosaccharides prepared through enzymatic hydrolysis of alginate. *Nutrition Research, 26*, 597–603.

Wang, W., et al. (2023). Structural characteristics of microbial exopolysaccharides in association with their biological activities: A review. *Chemical and Biological Technologies in Agriculture, 10*(1), 1–16. https://doi.org/10.1186/s40538-023-00515-3

Wang, Y., et al. (2024). Glycerol-driven adaptive evolution for the production of low-molecular-weight Welan gum: Characterization and activity evaluation. *Carbohydrate Polymers, 339*, 122292.

Wangpaiboo, K., Padungros, P., & Nakapong, S. (2018). An α-1,6-and α-1,3-linked glucan produced by Leuconostoc citreum ABK-1 alternansucrase with nanoparticle and film-forming properties. *Scientific Reports, 8*, 1–10.

Wilkerson, M. Y. (2022). *Properties and applications of alginates* (pp. 1–181). InTech.

Xu, J., et al. (2021). In vitro assessment of prebiotic properties of oligosaccharides derived from four microbial polysaccharides. *LWT, 147*, 111544. https://www.sciencedirect.com/science/article/pii/S0023643821006976

Zhang, R., & Edgar, K. J. (2014). Properties, chemistry, and applications of the bioactive polysaccharide curdlan. *Biomacromolecules, 15*, 1079–1096.

Zikmanis, P., Brants, K., Kolesovs, S., & Semjonovs, P. (2020). Extracellular polysaccharides produced by bacteria of the Leuconostoc genus. *World Journal of Microbiology and Biotechnology, 9*, 1–18. https://doi.org/10.1007/s11274-020-02937-9

Extracellular Polysaccharides: An Innovative Prebiotic with Potential Health Benefits Targeting Gut Flora

Mousumi Ray, M. Ashwini, N. Bhargava, and Keshab Chandra Mondal

1 Introduction

The human gastrointestinal tract (GIT) is a complex ecosystem consisting of more than 1000 various species of broad-spectrum microorganisms comprising viruses, protozoa, archaea, fungi, and bacteria. These microbes are associated in symbiosis with their host thereby promoting health benefits (Proctor et al., 2019). Bacteria being the major population provide beneficial effects to the host in several ways such as improving intestinal health, preventing invasive pathogens, and supporting food digestion (Shi et al., 2017). Any dysbiosis in these beneficial bacterial inhabitants leads to various disorders, emphasizing the importance of healthy microflora for the host's well-being (Rojo et al., 2017). The rising occurrence of cancer, metabolic disorders, and neurodegenerative diseases causes a solemn risk to healthcare divisions worldwide challenging novel preventative and curative approaches.

A diet with probiotics and prebiotic fibres can help reduce disease symptoms by reestablishing precise vital intestinal microflora and preventing dysbiosis. The undigested prebiotic fibres, reach to the colon intact and being the food for human microflora that promote the growth of beneficial species in the intestine (Holscher, 2017). A bioactive compound that acts as a prebiotic is the functional polysaccharides, which have been a promising field of research as they are the components that improve intestinal health. They function as carbohydrate reserves and/or structural elements, playing critical roles in complex physiological processes in plants and

M. Ray · K. C. Mondal (✉)
Department of Microbiology, Vidyasagar University, Midnapore, West Bengal, India

M. Ashwini
Department of Developmental Biology and Genetics, Indian Institute of Science, Bengaluru, Karnataka, India

N. Bhargava
Department of MBA, Acharya Institute of Graduate Studies, Bengaluru, Karnataka, India

T. Sarkar et al. (eds.), *Extracellular Polysaccharides*,
https://doi.org/10.1007/978-3-031-94472-7_14

other organisms. Reserve polysaccharides are primarily found within the cytoplasm, while structural polysaccharides are generally located in the primary and secondary cell walls. Both types contribute to carbohydrate sources, provide dietary fibre for humans and animals, and influence physical function and health (Gan et al., 2022). Recently, scientists have identified glycans on the microbial cell wall as important molecules that play a key role in supporting intestinal health. Glycans are categorized into two main types: (1) capsular polysaccharides, which are tightly bound to the cell surface, and (2) exopolysaccharides (EPS), which are loosely associated with the extracellular surface or secreted into the surrounding environment. Interest in EPS has grown due to their significant role in interactions with the human host (Poole et al., 2018).

Polysaccharides are mostly ingested orally through food and move through the intestines for further utilization, offering significant benefits for gut health (Ho Do et al., 2021). Humans and animals can directly digest only simple sugars and specific types of starch, so a substantial portion of polysaccharides (such as fibre) reaches the hindgut largely intact, in the colon where it is then fermented by the gut microbiota (Murga-Garrido et al., 2021; Tanes et al., 2021). Also polysaccharides have demonstrated positive effects on the gut microbiota by supporting their growth and offering additional health benefits. Additionally, a deficiency in dietary fibre alters the gut microbiome and can result in gut dysbiosis, a condition linked to various diseases, particularly metabolic disorders (Agus et al., 2020).

The human gastrointestinal tract hosts hundreds of trillions of microbial cells and participates in numerous health-supporting and essential physiological processes that are essential for sustaining host health (Hooper & Macpherson, 2010). Dietary elements are the key components that directly affect the lining and functionality of intestinal microbiota (Shondelmyer et al., 2018). The microbiome plays a crucial role in supporting the host's health through various metabolic processes that include the synthesis of vital nutrients like short-chain fatty acids (SCFAs) and vitamins (Uebanso et al., 2020). Furthermore, it fosters the immune system and prevents the colonization of pathogens. The intestinal microbiota also digests complex carbohydrates into dietary fibres that are undigested by human enzymes (Myeong et al., 2023). During this carbohydrate metabolism by the microflora, the end product SCFAs produced are essential for immunological responses, regulation of inflammatory response, and prevention of metabolic disorders like obesity (Ratajczak et al., 2019; Aho et al., 2021; Cait et al., 2018; Ray et al., 2017; De la Cuesta-Zuluaga et al., 2018).

Thus, reinstating intestinal health through polysaccharide intermediation may influence metabolic ailments via modifying the gut microflora, interacting with the gut immune response and intestinal epithelial barrier (Chakaroun et al., 2020). Furthermore, the impacts of polysaccharides on the GIT and metabolic issues extend beyond mechanisms associated with intestinal microflora alone. This chapter suggests a comprehensive review of the structure-related functions, therapeutic potential, prebiotic favourable, and several uses of bacterial exopolysaccharides (EPS) in food and other fields. Moreover, it highlights the current research on how these polysaccharides modify intestinal microbiota to build host immunity, discovering their importance in promoting health.

2 Classification of Exopolysaccharides

The extracellular polymeric substances (EPSs) which are a layer of slimy produced by the microbes are discharged out of the cells and will promote cell adhesion. EPSs are typically long-chain homopolysaccharides or heteropolysaccharides composed by the repetition of sugar units, either through or deprived of chemical moieties (Fig. 1) (Muninathan et al., 2022). The molecular weights of homopolysaccharides ranges from 4* 10^4–6*10^6 Da, and they are produced in larger quantities in bacteria than heteropolysaccharides (Oleksy & Klewicka, 2018). The production of homopolysaccharides occurs through either intracellular or extracellular pathways, whereas that of heteropolysaccharides occurs through extracellular pathways.

Homopolysaccharides consist of a single type of repeating monosaccharide with a molecular mass of approximately 10^7 Da. They are either unbranched or branched and composed of either D-fructose or D-glucose. Polysaccharides that contain D-fructose frequently have β- glycosidic linkages. Nonetheless, in D-glucose homopolysaccharides is typically α-glycosidic or β-glycosidic linkage resulting α-type or in β-type fructans (Muninathan et al., 2022). The type of linkage helps classify the polysaccharide as glucan or fructan. Homopolysaccharides extracellular biosynthesis involves the action of fructosyl and glucosyl transferases during the polymerization process with sucrose or activated UDP-glucose as building blocks. Examples of D-glucose or D-fructose include pullulan, cellulose, levan, dextran, curdlan, mutan, and alternan.

Heteropolysaccharides have complicated structures and comprise two or more monosaccharide units with different branching patterns, linking locations, repetition units, and monosaccharide modifications. D-glucose, D-galactose, and D-rhamnose are the main monosaccharides in the backbone of heteropolysaccharides, with modifications involving D-mannose, D-glucuronic acid, and N-acetylated monosaccharides (Oerlemans et al., 2021). Different enzymes, proteins, and carriers are involved in the extracellular secretion of heteropolysaccharides. Hyaluronic acid (HA), xanthan gum, alginate (ALG), kefiran, and gellan are examples of heteropolysaccharides composed of many monosaccharides (Mohd Nadzir et al., 2021).

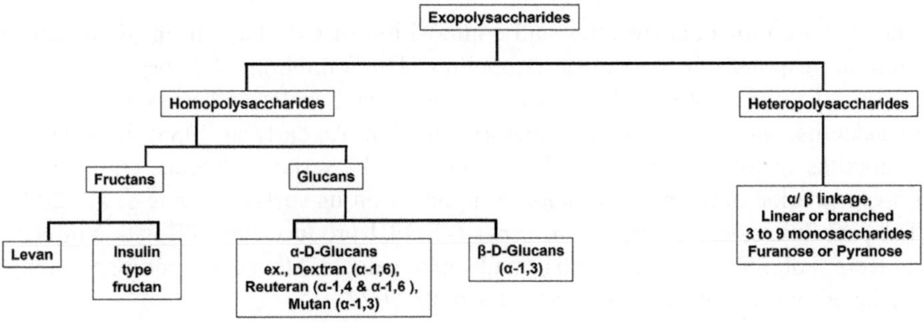

Fig. 1 Classification of microbial polysaccharides

2.1 Structure-Specific Function of Exopolysaccharides

EPS exhibits two distinct extracellular secretion states, slime polysaccharides, partially or entirely released into the surrounding environment to create slime, and capsular polysaccharides, which stick to the microbial cell wall to form a capsule. Monosaccharide composition and conformation, molecular weight range, glycosidic bond conformation, repeating units, linkage sites, and spatial structures are typically included in the structural description of EPS (Barcelos et al., 2020). The exterior morphology and microstructure of an EPS's are regulated by their structure and composition, that influences their biological activity (Wang et al., 2023). The major areas defined are the antibacterial, immunomodulatory, hypoglycemic, antioxidant, anticancer, and gut microbial-modulating characteristics of microbial EPS.

2.1.1 Antibacterial Activity of EPS

The functional groups, such as phosphate, hydroxyl, and carbonyl groups, are present in EPS. These functional groups interact with the cell membrane of bacteria that have antibacterial activity (Rajoka et al., 2020). The Aspergillus species DHE6 yield EPS with the chief functional groups -C=C, OH, -CH, and C-O-C. These EPS apply effective antibacterial activity against human health issues. The composition with rhamnose and glucose in EPS possess antibacterial activity, according to studies on EPS produced from *Lactococcus garvieae* C47, *Streptococcus thermophilus* GST-6, and *Lactobacillus gasseri* FR4 (Wang et al., 2023). Likewise, EPS, produced by *Pedipoccus pentosaceus* SSC-12, includes components of monosaccharide with rhamnose and glucose and has an effective antibacterial activity (Fan et al., 2021). However, *Lactobacillus crispatus,* yields an EPS that has a molecular weight of 53,887 Da and consists glucose and mannose. These EPS have antibacterial activity, that can proficiently limit the bacterial translocation and enhance the population of *Lactobacillus* and *Bifidobacterium* (Ding et al., 2022).

2.1.2 Immunomodulatory Activity of EPS

The specific molecular weights and composition of EPS have been implicated in immune responses in several investigations. The immunomodulatory activity and structural features of an EPS indicate a significant correlation between the content of galactose and its immunomodulatory function. As early as 1988, the galactose antibodies of rabbits were found to respond to bacterial pathogens composed of polysaccharide antigens as a major component on its surface (Wang et al., 2023). The probiotic starin *Enterococcus hirae* WEHI01 produces an EPS with a molecular weight of 2.59×10^3 Da which is made up of galactose that enhances macrophage-mediated immunological responses (Jia et al., 2022).

The structure-activity relationship analysis revealed a substantial correlation between EPS's immunomodulatory function and molecular weight. Higher molecular weight EPS may suppress the immune system (Wang et al., 2023). In general, the biological activity of EPS is much enhanced when it degrades from a higher molecular weight to a lower molecular weight.

Surayot et al. examined the effect of *Lactobacillus confusus* TISTR 1498's EPS, which only contained glucose with a high molecular weight of 65,000–506,000 kDa, affected immunomodulatory activity. They found that RAW264.7 cells were unable to produce the pro-inflammatory factors nitric oxide and cytokines. Its molecular weight was less than 70 kDa following partial acid hydrolysis, and it could strongly activate macrophages and trigger the synthesis of nitric oxide and cytokines like TNF-α, IL-1β, IL-6, and IL-10 (Surayot et al., 2014).

2.1.3 Hypoglycemic Activity of EPS

EPS can lower blood sugar through the inhibition of digestive enzymes (Huang et al., 2020). EPS's molecular weight, branching structure, and high-order structure are all strongly correlated with its hypoglycemic action. The right molecular weight is necessary for EPS to function at its best. Low molecular weight EPS typically has superior hypoglycemic action (Yang et al., 2021). A new Cordyceps polysaccharide was reported to have a 40.01% α-d-glucosidase inhibition rate is also known to have a low molecular weight of 28 kDa and was created by acid hydrolysis (Zhu et al., 2016).

The EPS's hypoglycemic activity is influenced by the glycosidic linkages (Mansel ct al., 2020).

These linkages 1 → 6, 1 → 3, and 1 → 4 were found in most of the EPS that have a hypoglycemic effect (Yang et al., 2021). For example, the leading galactose 1 → 4 linkage in the *Cordyceps militaris* EPS backbone effectively inhibits the activity of glucosidase and aids mice to regain their glucose tolerance (Sun et al., 2021). Analysis on *Pleurotus citrinopileatus* and *Cordyceps militaris* has encouraged the idea that the EPS with a helical shape is more likely to possess hypoglycemic effects (Sun et al., 2021). Additionally, sulfated EPS revealed better hypoglycemic ability than natural EPS (Wang et al., 2023).

2.1.4 Antioxidant Properties of EPS

Studies display that EPS has effective antioxidant properties. The main feature of EPS produced by bacteria is its ability to contribute hydrogen, a mechanism alike to that of any other polysaccharide sources. Some of the factors that affect antioxidant activity of EPS are monosaccharide content, glycosidic bond types, and its branching forms. The ratio of monosaccharide concentration and EPS component reveal a significant effect on antioxidant activity of EPS. For example, EPS with repeating units of glucose exhibits strong superoxide and 2,

2-diphenyl-1picrylhydrazyl-hydrate (DPPH) scavenging action and weak hydroxyl radical scavenging action, which is alike to the ascorbic acid standard (Wang et al., 2023). Moreover, the three main monosaccharides, i.e. rhamnose, galactose, and glucose, in EPS have antioxidant properties. EPS investigations on *Enterobacter ludwigeii*, *Porphyridium cruentum*, and *Bacillus* sp. S-1, are reported to have antioxidant activity (Paikra et al., 2022).

2.1.5 Anti-tumour Activity of EPS

The main chain composition, spatial structure, molecular chain conformation, degree of branching, helical conformation, flexibility of EPS is most important for its anticancer effect than its primary structure. More clearly, the EPS structure with high order is important for its anticancer effect than its primary structure. The anticancer activities of an EPS rely on β-1, 3 glycosidic bond on glucose chains and β-1, 6 glycosidic linkages on branched chains (Wang et al., 2023). The active component of several polysaccharides is isolated from *Auricularia auricula-judae* are β-(1, 3)-D-glucan with (1 → 6) branched chains that are reported to have potent antitumour activity (Tang et al., 2024).

The efficacy of EPS's anticancer activity is influenced by the chain conformation and morphological characteristics of EPS (Tang et al., 2020). The polysaccharides with triple-helical structures have the effects of anticancer activity. It has been confirmed that chain conformation increases the anticancer characteristics of EPS and enables polysaccharide to interact with the immune system. The other important factor for EPS anticancer activity is distinct viscosity. A suitable characteristic viscosity enables adhesion of polysaccharide to tumour cells. Sulfated polysaccharides showed higher activity of anticancer than non-sulfated polysaccharides. *Lactobacillus plantarum* 70,810 EPS (such as r-EPS1 and r-EPS2) inhibited tumour cell growth more quickly than r-EPS1 did. The authors postulated that the sulfate group and β-glycosidic bond composition may be closely linked to r-EPS2's strong antitumor activity (Wang et al., 2023).

2.1.6 Gut Microbial-Modulating Properties of Microbial EPS

Additionally, EPS can control the gut microbiota's composition and activity (Wan et al., 2022). The primary components of *Cordyceps sinensis* polysaccharides (CSPs) are phenolic chemicals, glucose, galactose, mannose, galacturonic acid, arabinose, and trace proteins. 1, 4-glucose and 1, 4-galactose, which have a molecular weight of about 28 kDa, make up the backbone of CSPs (Wang et al., 2017). The harmful bacteria (*Flexispira* and *Clostridium*) is decreased by CSPs and the number of probiotic bacteria such as *Bacteroides*, *Bifidobacterium*, and *Lactobacillus* are increased (Ying et al., 2020). Glucuronic acid, glucose, fucose, galactose/N-acetyl glucosamine, arabinose, xylose, and ribose make up *E. coli* EPS (EPS-m2) in a molar ratio of roughly 77:44:29:28:2:1:1. *Alstipes, Acinetobacter, Alloprevotella,*

Howardella, and *Oxalobacter* are more abundant when EPS-m2 is present, and GC detection shows that EPSm2 promotes the synthesis of SCFAs (Li et al., 2021). The dextran (LM742) is formed by *Leuconostoc mesenteroides* SPCL742 with a molecular weight of 1.3×10^6 Da and contains α-1, 3 and α-1, 6 glycosidic linkages in a 1:26.11 ratio. In human gastrointestinal disorders, the LM742 glucan exhibits resistance to digesting enzymes (Kim et al., 2022). EPS from *Paecilomyces cicadae* TJJ1213 also boosts probiotic abundance and controls metabolism and gut microbiota (Tian et al., 2023).

The method by which EPS regulates gut microbiota is still unclear, though, as there has been slight research on the subject and several limitations in the ones that have been published. The intrinsic mechanism of EPS regulation of the gut microbiota should thus be revealed, and future research on EPS structure and its relationship to gut microbiota regulation should be strengthened.

3 Therapeutic Potential of Exopolysaccharides

Microbial exopolysaccharides protect the host cell from a variety of threats, including pathogens, drought, antibiotics, and more. Their anthropogenic benefits have led to their increasing appeal in the food and pharmaceutical industries since they were originally studied in the 1880s. The chemical makeup of EPS determines its biological potential, which is impacted by the fermentation conditions used to cultivate probiotic bacteria (Li & Wang, 2012; Angelin & Kavitha, 2020). Monosaccharide content, glycosidic linkage, and chemical changes, including the particular conformation, extended chain, molecular weight, etc., all have an impact on EPS from various sources (Angelin & Kavitha, 2020). Because of their structural variety and broad range of applications, EPSs are now recognized as effective antibacterial, antioxidant, anticancer, and antitumor agents, the EPS generated by probiotic bacteria has become increasingly significant in therapeutic applications. They are also used as prebiotics (fermented milk products like kefir, villi, and taetea), nano drug delivery, and bioactive chemical encapsulation (Angelin & Kavitha, 2020).

Lactic acid bacteria are powerful makers of exopolysaccharides with various uses that are mostly isolated from dairy products such as yogurts, kefirs, fermented milk, cheese, etc. EPSs from LAB that have prebiotic qualities encourage the growth of good bacteria in the gut. The creation of biofilms, which facilitate colonization and population maintenance in the tough environment of the human gastrointestinal tract (GIT), is one of the advantageous properties of probiotic bacteria (Juraskova et al., 2022). Therefore, it has been demonstrated that probiotic LAB's attachment to GIT epithelial cells inhibits the colonization of pathogenic organisms, boosts the host immune system, and shields epithelial cells from harmful substances (such as carcinogens and toxic metal ions). Furthermore, LAB-derived EPS can protect against infectious illnesses by reducing or inhibiting the production of biofilms by dangerous bacteria (Oerlemans et al., 2021; Abdalla et al., 2021). According to other research, these EPS may have an impact on the gastrointestinal system by shielding intestinal cells from toxins and reducing cholesterol via increasing bile acid secretion (Juraskova et al., 2022). Extremophiles such as marine halophiles, thermophiles, psychrophiles, and EPSs from lactic acid bacteria (LAB) are important contributors to the contemporary pharmaceutical industry with proven pharmacological uses. These exopolymers also aid in therapeutic diagnostics, drug transport, wound healing, and other processes to directly prevent disease (Santra & Banerjee, 2021). Some of the therapeutic activities of EPS produced by microbial species are listed in Table 1.

Microorganisms are thought to be shielded by EPS against harmful substances, bacteriophages, antibiotics, and physical adversities. Exopolysaccharides generated by microorganisms have several benefits over other natural agents in terms of industrial and medicinal applications. The lactic acid bacterial group of microorganisms includes both probiotic and postbiotic members that produce health benefits such as EPS. These EPS have exceptional useful qualities and can be used to substitute plant or animal polysaccharides (Angelin & Kavitha, 2020). In recent years, non-dairy sources containing probiotic bacteria have been studied for their health benefits and rheological qualities. The EPS produced by the probiotics also operates as a physical barrier to stop toxins from interacting with eukaryotic cells by inhibiting toxin receptors on the cell surface or acting as toxin-scavengers (Gezginç et al., 2022). The anticancer, antioxidant, antibacterial, anti-inflammatory, and immunomodulatory properties of LAB EPS have been investigated. According to studies, LAB species may synthesize structurally diverse EPSs with various chemical substituents, linkages, and functional groups, allowing these compounds to have varying degrees of biological activity (Gezginç et al., 2022; Angelin & Kavitha, 2020). Metabolic engineering research uses knowledge of EPS genetics, biological variations, biosynthesis routes, and metabolic models to alter EPS composition and production. However, research on the fundamentals of using metabolic engineering to produce various EPSs from LAB is still ongoing, and it is likely to grow in scope and diversity. Given the previously mentioned bioactivities, LAB EPSs have the potential to be applied in a variety of industries, such as drug delivery, medicine, and food. Further research could yield ready-to-consume LAB EPSs with exceptional biological activities, and EPSs might be consistently present in specific product formulations in the food and medical industries.

4 Exopolysaccharide and Human Health/Effects on GI Tract

EPS play a key role in supporting both innate and adaptive immune cells. Since nearly 80% of the body's immune cells are found in the gut (Lewis & Blutt, 2019), the interactions between EPS and the immune system in this area have a big impact on our overall health (Shi et al., 2017). A deficiency in dietary polysaccharides can lead to decrease in gut microbiota's composition, diversity, and richness, which may ultimately contribute to the development of various diseases. Studies have shown that polysaccharides can be absorbed into the circulatory system following oral

Table 1 EPS-producing lactic acid bacteria species with its therapeutic activities

Bacterial species	Therapeutic application	References
Lactobacillus plantarum L-14	Anti-inflammatory	Kwon et al. (2020)
Lactobacillus plantarum RJF4	Antioxidant, cholesterol-lowering, antiproliferative, antiproliferative, antioxidant	Dilna et al. (2015)
Lactobacillus plantarum LRCC5310	Anti-viral	Kim et al. (2018)
Lactobacillus plantarum SKT109	Antioxidant	Wang et al. (2018)
Lactobacillus plantarum, Lactobacillus casei	Anti-tumour	Deepak et al. (2016a)
Lactobacillus casei 01	Anti-proliferative	Liu et al. (2011)
Lactobacillus kefiranofaciens DN1	Anti-bacterial	Jeong et al. (2017)
Lactobacillus reuteri DSM17938, *Lactobacillus reuteri* L26	Anti-gastrointestinal	Kšonžeková et al. (2016)
Lactobacillus delbureckii	Anti-bacterial	Adebayo-Tayo and Fashogbon (2020)
Lactobacillus sanfranciscensis	Antioxidant	Zhang et al. (2019)
Lactobacillus acidophilus	Antioxidant	Deepak et al. (2016b)
Streptococcus mutans MTCC 497	Anti-inflammatory	Buddana et al. (2015)
Weissella cibaria 27 (W27)	Anti-bacterial	Yu et al. (2018)
Pediococcus acidilactici NCDC 252	Antioxidant, anti-cancer	Kumar et al. (2020)
Enterococcus faecium WEFA23	Antioxidant	Jia et al. (2019)
Lactobacillus sp. Ca6	Antimicrobial activity	Trabelsi et al. (2015)
Pediococcus pentosaceus M41	Antitumor, antioxidant, antimicrobial	Ayyash et al. (2020b)
Lactobacillus strains	Cholesterol reduction, antioxidant, antimicrobial	Khalil et al. (2018)
Bacillus tequilensis FR9	Antioxidant	Li et al. (2014)
Lactobacillus plantarum C70	Antioxidant and cytotoxic activity	Ayyash et al. (2020a)
Lactobacillus plantarum OF101 and *YO175*	Antioxidant	Adesulu-Dahunsi et al. (2018)
Lactobacillus acidophilus (LA1)	Antioxidant	Abd El Ghany et al. (2015)
Lactobacillus plantarum MTCC9510	Antitumour activity and immunomodulatory activities	Ismail and Nampoothiri (2013)
Lactobacillus plantarum WLPL04	Antitumor activity	Liu et al. (2017)
Lactobacillus paracasei M7	Antioxidant, anti-biofilm	Bhat and Bajaj (2019)

(continued)

Table 1 (continued)

Bacterial species	Therapeutic application	References
Lactobacillus plantarum HY	Antioxidant activity	Kim et al. (2018)
Lactobacillus plantarum 86, *Weisellaconfusa* AI10, *Pediococcusparvulus* AI1, *Weisellacibaria* 142	Antibacterial activity	Patel et al. (2012)

administration, even when they possess high molecular weights and complex structures (Zhang et al., 2021). A functional intestine with an intact barrier that allows nutrient transport from the lumen to the bloodstream, while simultaneously restricting the passage of potentially harmful microorganisms and toxins, is essential for maintaining human health (Massier et al., 2021). Studies also reported that polysaccharides offer significant biological benefits for bowel health, including anti-inflammatory effects, protection of the gut epithelial barrier, and immune modulation through both microbiota-dependent and microbiota-independent mechanisms (Ho Do et al., 2021). Due to their water-retention capacity, polysaccharides in the large intestine can attract water and add bulk to the digesta, which enhances intestinal peristalsis and softens the stool. This helps dilute toxin concentrations, increase defecation frequency, and prevent constipation and related issues, such as haemorrhoids (Ho Do et al., 2021; Jeddou et al., 2016). Since the gut microbiota primarily relies on polysaccharides as their nutrient source, the lack of these nutrients in the diet influences the microbiota to shift towards utilizing indigenous host substrates like glycans. This modification turns into elevated pathogenic organisms with lowered gut commensals, probiotics, and their byproducts. In presence of low concentration or in absence of dietary polysaccharides that aids in inflammation and upregulated gut permeability. This inflammation and permeability is responsible for pathogen invasion and ultimately leads to the development of obesity and other metabolic disorders (Araujo et al., 2017). Consumption of dietary polysaccharides helps to improve and maintain the overall gastrointestinal health via gut microbiota through its function and stability. A study explored that polysaccharides derived from Lentinula edodes promote the growth of Bacteroides acidifaciens (Xu & Zhang, 2015).

Glucorhamna, a polysaccharide derived from the *Ruminococcus gnavus* shows gut health benefits through controlling inflammation within the gut (Henke et al., 2019). Furthermore, short-chain fatty acids (SCFAs) secrete as microbial byproducts in the intestine, perform as direct energy suppliers not only for gut commensals but also for colonocytes (Oerlemans et al., 2021), helps in cell proliferation, and support the function and integrity of the gut epithelial barrier (de la Cuesta-Zuluaga et al., 2018; Parada Venegas et al., 2019; Holscher, 2017). Short-chain fatty acids (SCFAs) synthesized by the microbial fermentation of exopolysaccharides (EPS), serve various health benefits through alteration of host metabolism and stimulate immune responses (Vieira et al., 2015). Recent study explained that EPSs derived from *Weissiella* and *Pediococcus* encourage to established of gut microbes like species belonging to the genus *Bifidobacterium* that are favourable to health (Khan et al., 2022). A recent analysis explored the effects of EPSs (EPS116) extracted from *Lactobacillus plantarum* NCU116 on gut epithelial regeneration. The study also investigated the interference of EPS116 in the regulation of gut homeostasis by altering the composition of gut commensals and affecting the proliferation and differentiation of intestinal stem cells (Zhou et al., 2021). Another study discovered that the increased production of the health beneficial SCFAs like acetate, propionate, and butyrate from the fermentation of EPS from *Bifidobacterium breve* in an infant colon with faeces from healthy, full-term, breast-fed infants (Püngel et al., 2020). A different report by Korcz et al., clarified SCFA production depends on particular EPS fermentation (Korcz et al., 2018). Researchers also demonstrated that the SCFA formation depends on the utilization of various hetero-exopolysaccharides (HePS) from different *Bifidobacterium* strains, turns into varying ratios of acetic acid-to-propionic acid.

Various studies also hypothesized that EPS can adversely influence bacterial adherence to the gut epithelium via shielding the bacterial cell wall and covering surface proteins involved in cell adhesion. Though, this shielding can reduce microbial adhesion efficacy and inhibits other surface proteins that activate the host's immune system. Moreover, EPS shows a vital function for bacterial survival in the gut and help bacteria to evade the host immune system (Oerlemans et al., 2021). The role of EPS helps in bacterial epithelial attachment and makes it especially useful for preventing harmful bacterial adherence. Various reports supported this outcome, for instance, Zivkovi'c et al. (2016) explored that EPS by *Lactobacillus paracasei* subsp. *paracasei* BGSJ2 effectively inhibits *Escherichia coli* from adhering to Caco-2 cells. Moreover, EPS from bacteria influences gut epithelial barrier integrity via altering tight junctions that cover epithelial cells. Claudins and occludin are two main proteins in these junctions, with some claudins decreasing permeability (1, 4, 5, 7, 8, 14) and others increasing it (2, 16) (Oerlemans et al., 2021). Scientists have explained, HePS from *Bifidobacterium animalis*, *Bifidobacterium longum*, and *Lactobacillus rhamnosus* GG increases barrier integrity.

The investigation on extensively studied indigestible polysaccharide beta-glucans, highlighted on their beneficial effects on colonic microbes such as *Lactobacillus acidophilus*, *Bifidobacterium animalis* subsp. *lactis*, and *Lactobacillus casei* in both natural and experimental settings (Eker & Mutlu, 2022). Researchers performing analysis on mice have explained that beta-glucan can influence and establish the number of *Lactobacilli* and *Bifidobacteria* (Eker & Mutlu, 2022) in the colon. Several gut microbes vary in their mechanism of digesting dietary glycans, such as *Bifidobacterium* depends on *Bacteroides* to break down complex carbohydrates, such as pectin, into functional oligosaccharides. The mutual interaction between *Bifidobacterium* and *Bacteroides* conserves the steadiness of gut microbiota, where *Bacteroides* act as primary degraders of complex glycans and *Bifidobacterium* functions as secondary degraders of simpler glycans (Nishiyama et al., 2021).

5 Application of Exopolysaccharides in Food Industries

Polysaccharides, a group of hydrophilic compounds called hydrocolloids, are extensively used in food and dairy industries as functional additives. Food enriched with polysaccharides are responsible to increase organoleptic properties and extend shelf-life of the product. For their thickening and gelling properties, polysaccharides are able to form viscous suspensions or gels when dissolved in water, thereby developing the texture, stability, and overall attributes of final products (Rana & Upadhyay, 2020). Moreover, other important hydrocolloid polysaccharides, such as celluloses, hemicelluloses, pectins, and exuded gums, are obtained from plants, and others like alginate, agar, and carrageenan are extracted from seaweeds, there are also commercially significant bacterial-isolated polysaccharides, including xanthan gum and gellan gum. These bacterial polymers are derived from Gram-negative bacteria: xanthan gum is synthesized by *Xanthomonas campestris* (Becker et al., 1998), while gellan gum is produced by *Sphingomonas paucimobilis* (Khan et al., 2022).

After the United States Food and Drug Administration (FDA) has sanctioned xanthan gum for use in human food, where it serves as a key stabilizer and thickening agent, improving viscosity, texture, mouthfeel, flavour release, appearance, and water-binding capacities, xanthan gum is regularly used in bakery goods, dairy products, beverages, dressings, sauces, gravies, syrups, and toppings (Palaniraj & Jayaraman, 2011). Now a days, Xanthan gum represents the most significant bacterial exopolysaccharide (EPS) in the hydrocolloid market worldwide, accounting for approximately 6% of the total market value (Freitas et al., 2011).

Gellan gum has also gained attention and found increasing application in the food industry, largely due to its functional properties, such as excellent thermal and acid stability, adjustable gel elasticity and rigidity, high transparency, and effective flavour release. Gellan gum is used as a stabilizer, suspending agent, and versatile gelling agent in a wide range of food applications, including bakery fillings, confections, dairy products, dessert gels, frostings, icings, jams, jellies, low-fat spreads, microwavable foods, puddings, sauces, and structured foods (Prajapati et al., 2013).

Current advancements have focused the modification of particular plant-based polymers with microbial alternatives, for instance bacterial alginate (Freitas et al., 2011) and cellulose (Shi et al., 2014). Bacterial derived cellulose offers several benefits over plant-based cellulose, with increased purity exclusive of contaminants, excluding the requirement for abrasive chemical application for isolation and purification. Moreover, bacterial cellulose is FDA-authorized, harmless, and crystalline in nature, allowing its production in a variety of forms and textures suitable for various food applications (Shi et al., 2014).

Dextran is used in bakery industry to improve moisture retention capability, enhance viscosity, and evade sugar crystallization of the food products (Rana & Upadhyay, 2020). Probiotics have gained importance due to proven health benefits, including normalization of intestinal flora, anticancer effects, cholesterol reduction, alleviation of lactose malabsorption, and allergy reduction. They also impact brain-gut signaling, reducing anxiety and depression without cognitive impairment (Saldías et al., 2015). Curdlan is used to increase creaminess, stability, and viscosity in food (Osmałek et al., 2014).

In cosmetics, marine-derived Hyanify boosts hyaluronic acid synthesis, helping reduce skin aging effects such as dehydration and volume loss (D'Arrigo et al., 2014). Pullulan, with stable viscosity under heat and pH changes, is used in lotions and creams (Prezotti et al., 2014). Xanthan gum enhances stability in toothpaste, creams, and shampoos, facilitating ingredient suspension (Rana & Upadhyay, 2020). Dextran is used in moisturizers for its emulsifying and thickening properties (Li et al., 2017).

Studies have identified a range of microbial exopolysaccharides, though only a few have significant commercial value. Agar has been used as a gelling agent since 1892, though some strains, like *Gemmatimonas aurantiaca*, prefer gellan gum as a growth medium (Rana & Upadhyay, 2020). Exopolysaccharide acrylamide (EPAA) has emerged as a viable agar substitute without affecting microbial growth or properties while maintaining protective habitats (Patel et al., 2010). Pullulan, a linear homopolysaccharide, is used in bone tissue engineering (Rana & Upadhyay, 2020). Exopolysaccharides also treat heavy metal-contaminated water, acting as bio-adsorbents for metals like cadmium, mercury, cobalt, and lead due to selective bio-sorption (Tiwari et al., 2017; Eddouaouda et al., 2012).

5.1 Application of Exopolysaccharides in Other Commercial Sectors

Exopolysaccharides such as gellan and curdlan are used as soil conditioners to improve water retention (Rana & Upadhyay, 2020). Microbial exopolysaccharides have various medical applications as well. Dextran is used as a plasma expander to control wound shock (Delattre et al., 2015), and gellan is employed in drug delivery systems, either as a disintegration agent in tablets or in sustained release formulations (Rana & Upadhyay, 2020). Various studies have explained that gellan-based hydrogels and hyaluronic acid-coated nanoparticles efficiently direct drugs delivery such as paclitaxel, prednisolone, and vorinostat respectively (Rana & Upadhyay, 2020; Sachdev & Cameotra, 2013). Levan shows antiviral properties against avian influenza and HIV (Sachdev & Cameotra, 2013). Hydrogels of xanthan gum rich with curcumin plays anti-inflammatory properties, effective in treating skin cancer and psoriasis (Koop et al., 2015). Xanthan gum derived exopolysaccharide, Xanthouronan, has explored antioxidant activity (Rana & Upadhyay, 2020). Xanthan gum is also used in oil production, fracturing, and pipeline cleanup. Microbial Enhanced Oil Recovery (MEOR) uses microorganisms like Enterobacter cloacae to produce several by-products (e.g., CO2, CH4, acetic acid, ethanol) that decrease oil viscosity, enabling increased oil recovery (Sun et al., 2011).

6 Future Aspects and Conclusion

As prebiotic the function of extracellular polysaccharides (EPS) is very promising, with optimistic impacts on gut health, microbiome targeted therapies, and tailored medicine. EPS from health beneficial bacteria like *Lactobacillus* and *Bifidobacterium* may be personalized to promote specific gut microbes for enhanced immunity and digestion (Angelin & Kavitha, 2020). Investigation suggests that EPS may support mental healthiness via the gut-brain axis, enrich immunity and metabolic condition (Singh et al., 2024). Enhancing EPS bioavailability and immovability in supplements could induce gastrointestinal efficacy, while uniting EPS with probiotics and postbiotics might improve curative outcomes (Oerlemans et al., 2021). EPS rich food supports the regulation of several gut diseases, improves regulatory conditions and enhances community alertness, ensuring its secure and effective integration into health treatments (Mouro et al., 2024). Analysis is yet exploring the effects of EPS on the gut-brain axis, actively facilitating mental health through neurotransmitter support (Singh et al., 2024). Due to the production of diverse health beneficial EPS by different bacterial strains, in depth research and a detailed understanding of EPS arrangements of EPS are key to attaining future breakthroughs. Detailed reporting on the bacterial strain and EPS structure is essential, as even minor variations can alter effects (Oerlemans et al., 2021). Careful extraction and purification also help ensure reliable results by removing impurities and preventing EPS degradation. Comprehensive details on EPS's chemical makeup, linkage types, modifications, molecular weight, and purity are critical for meaningful insights.

To integrate EPS into healthcare for targeted benefits, more research on its effects on the host, especially its interactions with immune receptors and its safety, is crucial. Identifying specific EPS structures linked to these effects may open up therapeutic possibilities. With rising antibiotic resistance, EPS could aid in reducing infections, boosting immune response, and supporting recovery from gut imbalances, potentially easing the antibiotic burden.

Declarations **Funding**: No external funding was used in the preparation of this article.

Conflict of Interest MR, AM, and KCM declare that there are no potential conflicts of interest that might be relevant to the contents of this chapter.

Ethics Approval: Not applicable.

Consent to Participate: Not applicable.

Consent for Publication: Not applicable.

Availability of Data and Material: The manuscript includes all the generated or analysed data during this study.

Code Availability: Not applicable.

Author's Contributions: MR, AM, and KCM produced the idea for the review article; MR, AM, and BN performed the literature survey; MR and AM critically revised and drafted the chapter. All authors reviewed the manuscript.

References

Abd El Ghany, K., Hamouda, R., Abd Elhafez, E., Mahrous, H., Salem-Bekhit, M., & Hamza, H. A. (2015). A potential role of Lactobacillus acidophilus LA1 and its exopolysaccharides on cancer cells in male albino mice. *Biotechnology & Biotechnological Equipment, 29*(5), 977–983.

Abdalla, A. K., Ayyash, M. M., Olaimat, A. N., Osaili, T. M., Al-Nabulsi, A. A., Shah, N. P., & Holley, R. (2021). Exopolysaccharides as antimicrobial agents: Mechanism and spectrum of activity. *Frontiers in Microbiology, 12*, 664395.

Adebayo-Tayo, B., & Fashogbon, R. (2020). In vitro antioxidant, antibacterial, in vivo immunomodulatory, antitumor and hematological potential of exopolysaccharide produced by wild type and mutant Lactobacillus delbureckii subsp. bulgaricus. *Heliyon, 6*(2), e03268.

Adesulu-Dahunsi, A. T., Jeyaram, K., Sanni, A. I., & Banwo, K. (2018). Production of exopolysaccharide by strains of *Lactobacillus plantarum* YO175 and OF101 isolated from traditional fermented cereal beverage. *PeerJ, 6*, e5326.

Agus, A., Clément, K., & Sokol, H. (2020). Gut microbiota-derived metabolites as central regulators in metabolic disorders. *Gut, 70*, 1174–1182.

Aho, V. T. E., Houser, M. C., Pereira, P. A. B., Chang, J., Rudi, K., Paulin, L., Hertzberg, V., Auvinen, P., Tansey, M. G., & Scheperjans, F. (2021). Relationships of gut microbiota, short-chain fatty acids, inflammation, and the gut barrier in Parkinson's disease. *Molecular Neurodegeneration, 16*(1), 6.

Angelin, J., & Kavitha, M. (2020). Exopolysaccharides from probiotic bacteria and their health potential. *International Journal of Biological Macromolecules, 162*, 853–865.

Araujo, J. R., Tomas, J., Brenner, C., & Sansonetti, P. J. (2017). Impact of high-fat diet on the intestinal microbiota and small intestinal physiology before and after the onset of obesity. *Biochimie, 141*, 97–106.

Ayyash, M., Abu-Jdayil, B., Itsaranuwat, P., Galiwango, E., Tamiello-Rosa, C., Abdullah, H., et al. (2020a). Characterization, bioactivities, and rheological properties of exopolysaccharide produced by novel probiotic Lactobacillus plantarum C70 isolated from camel milk. *International Journal of Biological Macromolecules, 144*, 938–946.

Ayyash, M., Abu-Jdayil, B., Olaimat, A., Esposito, G., Itsaranuwat, P., Osaili, T., et al. (2020b). Physicochemical, bioactive and rheological properties of an exopolysaccharide produced by a probiotic Pediococcus pentosaceus M41. *Carbohydrate Polymers, 229*, 115462.

Barcelos, M. C., Vespermann, K. A., Pelissari, F. M., & Molina, G. (2020). Current status of biotechnological production and applications of microbial exopolysaccharides. *Critical Reviews in Food Science and Nutrition, 60*(9), 1475–1495.

Becker, A., Katzen, F., Puhler, A., & Ielpi, L. (1998). Xanthan gum biosynthesis and application: A biochemical/genetic perspective. *Applied Microbiology and Biotechnology, 50*, 145–152.

Bhat, B., & Bajaj, B. K. (2019). Hypocholesterolemic potential and bioactivity spectrum of an exopolysaccharide from a probiotic isolate Lactobacillus paracasei M7. *Bioactive Carbohydrates and Dietary Fibre, 19*, 100191.

Buddana, S. K., Varanasi, Y. V. N., & Shetty, P. R. (2015). Fibrinolytic, anti-inflammatory and antimicrobial properties of α-(1-3)-glucans produced from Streptococcus mutans (MTCC 497). *Carbohydrate Polymers, 115*, 152–159.

Cait, A., Hughes, M. R., Antignano, F., Cait, J., Dimitriu, P. A., Maas, K. R., Reynolds, L. A., Hacker, L., Mohr, J., Finlay, B. B., Zaph, C., McNagny, K. M., & Mohn, W. W. (2018). Microbiome-driven allergic lung inflammation is ameliorated by short-chain fatty acids. *Mucosal Immunology, 11*(3), 785–795.

Chakaroun, R. M., Massier, L., & Kovacs, P. (2020). Gut microbiome, intestinal permeability, and tissue bacteria in metabolic disease: Perpetrators or bystanders? *Nutrients, 12*, 1082.

D'Arrigo, G., Navarro, G., Di Meo, C., Matricardi, P., & Torchilin, V. (2014). Gellan gum nanohydrogel containing anti-inflammatory and anti-cancer drugs: A multi-drug delivery system for a combination therapy in cancer treatment. *European Journal of Pharmaceutics and Biopharmaceutics, 87*(1), 208–216.

De la Cuesta-Zuluaga, J., Mueller, N. T., Álvarez-Quintero, R., Velásquez-Mejía, E. P., Sierra, J. A., Corrales-Agudelo, V., Carmona, J. A., Abad, J. M., & Escobar, J. S. (2018). Higher fecal short-chain fatty acid levels are associated with gut microbiome dysbiosis, obesity, hypertension and cardiometabolic disease risk factors. *Nutrients, 11*(1), 51.

Deepak, V., Ram Kumar Pandian, S., Sivasubramaniam, S. D., Nellaiah, H., & Sundar, K. (2016a). Optimization of anticancer exopolysaccharide production from probiotic Lactobacillus acidophilus by response surface methodology. *Preparative Biochemistry and Biotechnology, 46*(3), 288–297.

Deepak, V., Ramachandran, S., Balahmar, R. M., Pandian, S. R. K., Sivasubramaniam, S. D., Nellaiah, H., & Sundar, K. (2016b). In vitro evaluation of anticancer properties of exopolysaccharides from lactobacillus acidophilus in colon cancer cell lines. *In Vitro Cellular & Developmental Biology-Animal, 52*, 163–173.

Delattre, C., Pierre, G., Gardarin, C., Traïkia, M., Elboutachfaiti, R., Isogai, A., & Michaud, P. (2015). Antioxidant activities of a polyglucuronic acid sodium salt obtained from TEMPO-mediated oxidation of xanthan. *Carbohydrate Polymers, 116*, 34–41.

Dilna, S. V., Surya, H., Aswathy, R. G., Varsha, K. K., Sakthikumar, D. N., Pandey, A., & Nampoothiri, K. M. (2015). Characterization of an exopolysaccharide with potential health-benefit properties from a probiotic Lactobacillus plantarum RJF4. *LWT-Food Science and Technology, 64*(2), 1179–1186.

Ding, C., Wu, H., Cao, X., Gao, Z., Tang, Z., Fan, W., et al. (2022). Lactobacillus crispatus-derived exopolysaccharides with antibacterial activity limit Salmonella typhimurium invasion by inhibiting inflammasome-mediated pyroptosis. *Food & Function, 13*(20), 10501–10515.

Eddouaouda, K., Mnif, S., Badis, A., Younes, S. B., Cherif, S., Ferhat, S., et al. (2012). Characterization of a novel biosurfactant produced by Staphylococcus sp. strain 1E with potential application on hydrocarbon bioremediation. *Journal of Basic Microbiology, 52*(4), 408–418.

Eker, F. D., & Mutlu, A. (2022). Oat, buckwheat and whole Brown Rice flours as a potential prebiotic for Lactobacillus acidophilus (LA5), Lactobacillus casei and Bifidobacterium animalis subsp. lactis (BB-12). *Harran Üniversitesi Mühendislik Dergisi, 7*(2), 91–98.

Fan, Y., Li, X., Tian, R., Tang, R., & Zhang, J. (2021). Characterization and biological activity of a novel exopolysaccharide produced by Pediococcus pentosaceus SSC–12 from silage. *Microorganisms, 10*(1), 18.

Freitas, F., Alves, V. D., & Reis, M. A. (2011). Advances in bacterial exopolysaccharides: From production to biotechnological applications. *Trends in Biotechnology, 29*, 388–398.

Gan, L., Wang, J., & Guo, Y. (2022). Polysaccharides influence human health via microbiota-dependent and -independent pathways. *Frontiers in Nutrition, 9*, 1030063.

Gezginç, Y., Karabekmez-erdem, T., Tatar, H. D., Ayman, S., Ganiyusufoğlu, E., & Dayısoylu, K. S. (2022). Health promoting benefits of postbiotics produced by lactic acid bacteria: Exopolysaccharide. *Biotech Studies, 31*(2), 61–70.

Henke, M. T., Kenny, D. J., Cassilly, C. D., Vlamakis, H., Xavier, R. J., & Clardy, J. (2019). Ruminococcus gnavus, a member of the human gut microbiome associated with Crohn's disease, produces an inflammatory polysaccharide. *Proceedings of the National Academy of Sciences, 116*(26), 12672–12677.

Ho Do, M., Seo, Y. S., & Park, H. Y. (2021). Polysaccharides: Bowel health and gut microbiota. *Critical Reviews in Food Science and Nutrition, 61*, 1212–1224.

Holscher, H. D. (2017). Dietary fiber and prebiotics and the gastrointestinal microbiota. *Gut Microbes, 8*(2), 172–184.

Hooper, L. V., & Macpherson, A. J. (2010). Immune adaptations that maintain homeostasis with the intestinal microbiota. *Nature Reviews Immunology, 10*(3), 159–169.

Huang, Z., Lin, F., Zhu, X., Zhang, C., Jiang, M., & Lu, Z. (2020). An exopolysaccharide from Lactobacillus plantarum H31 in pickled cabbage inhibits pancreas α-amylase and regulating metabolic markers in HepG2 cells by AMPK/PI3K/Akt pathway. *International Journal of Biological Macromolecules, 143*, 775–784.

Ismail, B., & Nampoothiri, K. M. (2013). Exposition of antitumor activity of a chemically characterized exopolysaccharide from a probiotic Lactobacillus plantarum MTCC 9510. *Biologia, 68*, 1041–1047.

Jeddou, K. B., Chaari, F., Maktouf, S., Nouri-Ellouz, O., Helbert, C. B., & Ghorbel, R. E. (2016). Structural, functional, and antioxidant properties of water-soluble polysaccharides from potatoes peels. *Food Chemistry, 205*, 97–105.

Jeong, D., Kim, D. H., Kang, I. B., Kim, H., Song, K. Y., Kim, H. S., & Seo, K. H. (2017). Characterization and antibacterial activity of a novel exopolysaccharide produced by Lactobacillus kefiranofaciens DN1 isolated from kefir. *Food Control, 78*, 436–442.

Jia, K., Tao, X., Liu, Z., Zhan, H., He, W., Zhang, Z., et al. (2019). Characterization of novel exopolysaccharide of enterococcus faecium WEFA23 from infant and demonstration of its in vitro biological properties. *International Journal of Biological Macromolecules, 128*, 710–717.

Jia, K., Wei, M., He, Y., Wang, Y., Wei, H., & Tao, X. (2022). Characterization of novel exopolysaccharides from Enterococcus hirae WEHI01 and its immunomodulatory activity. *Food, 11*(21), 3538.

Juraskova, D., Ribeiro, S. C., & Silva, C. C. (2022). Exopolysaccharides produced by lactic acid bacteria: From biosynthesis to health-promoting properties. *Food, 11*(2), 156.

Khalil, E. S., Abd Manap, M. Y., Mustafa, S., Alhelli, A. M., & Shokryazdan, P. (2018). Probiotic properties of exopolysaccharide-producing Lactobacillus strains isolated from tempoyak. *Molecules, 23*(2), 398.

Khan, R., Shah, M. D., Shah, L., Lee, P. C., & Khan, I. (2022). Bacterial polysaccharides—A big source for prebiotics and therapeutics. *Frontiers in Nutrition, 9*, 1031935.

Kim, K., Lee, G., Thanh, H. D., Kim, J. H., Konkit, M., Yoon, S., et al. (2018). Exopolysaccharide from Lactobacillus plantarum LRCC5310 offers protection against rotavirus-induced diarrhea and regulates inflammatory response. *Journal of Dairy Science, 101*(7), 5702–5712.

Kim, G., Bae, J. H., Cheon, S., Lee, D. H., Kim, D. H., Lee, D., et al. (2022). Prebiotic activities of dextran from Leuconostoc mesenteroides SPCL742 analyzed in the aspect of the human gut microbial ecosystem. *Food & Function, 13*(3), 1256–1267.

Koop, H. S., de Freitas, R. A., de Souza, M. M., Savi-Jr, R., & Silveira, J. L. M. (2015). Topical curcumin-loaded hydrogels obtained using galactomannan from Schizolobium parahybae and xanthan. *Carbohydrate Polymers, 116*, 229–236.

Korcz, E., Kerényi, Z., & Varga, L. (2018). Dietary fibers, prebiotics, and exopolysaccharides produced by lactic acid bacteria: Potential health benefits with special regard to cholesterol-lowering effects. *Food & Function, 9*(6), 3057–3068.

Kšonžeková, P., Bystrický, P., Vlčková, S., Pätoprstý, V., Pulzová, L., Mudroňová, D., et al. (2016). Exopolysaccharides of Lactobacillus reuteri: Their influence on adherence of E. coli to epithelial cells and inflammatory response. *Carbohydrate Polymers, 141*, 10–19.

Kumar, R., Bansal, P., Singh, J., & Dhanda, S. (2020). Purification, partial structural characterization and health benefits of exopolysaccharides from potential probiotic Pediococcus acidilactici NCDC 252. *Process Biochemistry, 99*, 79–86.

Kwon, M., Lee, J., Park, S., Kwon, O. H., Seo, J., & Roh, S. (2020). Exopolysaccharide isolated from Lactobacillus plantarum L-14 has anti-inflammatory effects via the Toll-like receptor 4 pathway in LPS-induced RAW 264.7 cells. *International Journal of Molecular Sciences, 21*(23), 9283.

Lewis, D. E., & Blutt, S. E. (2019). *Organization of the immune system. Clinical immunology* (5th ed., p. 2). Elsevier (chapter).

Li, J., & Wang, N. (2012). The gpsX gene encoding a glycosyltransferase is important for polysaccharide production and required for full virulence in Xanthomonas citri subsp. citri. *BMC Microbiology, 12*, 1–16.

Li, S., Huang, R., Shah, N. P., Tao, X., Xiong, Y., & Wei, H. (2014). Antioxidant and antibacterial activities of exopolysaccharides from Bifidobacterium bifidum WBIN03 and Lactobacillus plantarum R315. *Journal of Dairy Science, 97*(12), 7334–7343.

Li, R. H., Zeng, T., Wu, M., Zhang, H. B., & Hu, X. Q. (2017). Effects of esterification on the structural, physicochemical, and flocculation properties of dextran. *Carbohydrate Polymers, 174*, 1129–1137.

Li, B., Chen, H., Cao, L., Hu, Y., Chen, D., & Yin, Y. (2021). *Escherichia coli* exopolysaccharides induced by ceftriaxone regulated human gut microbiota in vitro. *Frontiers in Microbiology, 12*, 634204.

Liu, C. T., Chu, F. J., Chou, C. C., & Yu, R. C. (2011). Antiproliferative and anticytotoxic effects of cell fractions and exopolysaccharides from lactobacillus casei 01. *Mutation Research/Genetic Toxicology and Environmental Mutagenesis, 721*(2), 157–162.

Liu, Z., Zhang, Z., Qiu, L., Zhang, F., Xu, X., Wei, H., & Tao, X. (2017). Characterization and bioactivities of the exopolysaccharide from a probiotic strain of Lactobacillus plantarum WLPL04. *Journal of Dairy Science, 100*(9), 6895–6905.

Mansel, B. W., Ryan, T. M., Chen, H. L., Lundin, L., & Williams, M. A. (2020). Polysaccharide conformations measured by solution state X-ray scattering. *Chemical Physics Letters, 739*, 136951.

Massier, L., Bluher, M., Kovacs, P., & Chakaroun, R. M. (2021). Impaired intestinal barrier and tissue bacteria: Pathomechanisms for metabolic diseases. *Frontiers in Endocrinology, 12*, 616506.

Mohd Nadzir, M., Nurhayati, R. W., Idris, F. N., & Nguyen, M. H. (2021). Biomedical applications of bacterial exopolysaccharides: A review. *Polymers, 13*(4), 530.

Mouro, C., Gomes, A. P., & Gouveia, I. C. (2024). Microbial exopolysaccharides: Structure, diversity, applications, and future frontiers in sustainable functional materials. *Polysaccharides, 5*(3), 241–287.

Muninathan, C., Guruchandran, S., Viswanath Kalyan, A. J., & Ganesan, N. D. (2022). Microbial exopolysaccharides: Role in functional food engineering and gut-health management. *International Journal of Food Science & Technology, 57*(1), 27–34.

Murga-Garrido, S. M., Hong, Q., Cross, T. W. L., Hutchison, E. R., Han, J., Thomas, S. P., Vivas, E. I., Denu, J., Ceschin, D. G., Tang, Z. Z., & Rey, F. E. (2021). Gut microbiome variation modulates the effects of dietary fiber on host metabolism. *Microbiome, 9*(1), 117.

Myeong, S. W., Lee, Y. J., Kim, D. H., & Kim, T. J. (2023). Improvement of inflammation, diabetes, and obesity by forest product-derived polysaccharides through the human intestinal microbiota. *Journal of the Korean Wood Science and Technology, 51*(5), 358–380.

Nishiyama, K., Yokoi, T., Sugiyama, M., Osawa, R., Mukai, T., & Okada, N. (2021). Roles of the cell surface architecture of Bacteroides and Bifidobacterium in the gut colonization. *Frontiers in Microbiology, 12*, 754819.

Oerlemans, M. M., Akkerman, R., Ferrari, M., Walvoort, M. T., & de Vos, P. (2021). Benefits of bacteria-derived exopolysaccharides on gastrointestinal microbiota, immunity and health. *Journal of Functional Foods, 76*, 104289.

Oleksy, M., & Klewicka, E. (2018). Exopolysaccharides produced by Lactobacillus sp.: Biosynthesis and applications. *Critical Reviews in Food Science and Nutrition, 58*(3), 450–462.

Osmałek, T., Froelich, A., & Tasarek, S. (2014). Application of gellan gum in pharmacy and medicine. *International Journal of Pharmaceutics, 466*(1–2), 328–340.

Paikra, S. K., Panda, J., Sahoo, G., & Mishra, M. (2022). Characterization of exopolysaccharide derived from Enterobacter ludwigii and its possible role as an emulsifier. *3 Biotech, 12*(9), 212.

Palaniraj, A., & Jayaraman, V. (2011). Production, recovery and applications of xanthan gum by Xanthomonas campestris. *Journal of Food Engineering, 106*, 1–12.

Parada Venegas, D., De la Fuente, M. K., Landskron, G., González, M. J., Quera, R., Dijkstra, G., et al. (2019). Short chain fatty acids (SCFAs)-mediated gut epithelial and immune regulation and its relevance for inflammatory bowel diseases. *Frontiers in Immunology, 10*, 277.

Patel, A. K., Michaud, P., Singhania, R. R., Soccol, C. R., & Pandey, A. (2010). Polysaccharides from probiotics: New developments as food additives. *Food Technology and Biotechnology, 48*(4), 451–463.

Patel, A., Lindström, C., Patel, A., Prajapati, J. B., & Holst, O. (2012). Probiotic properties of exopolysaccharide-producing lactic acid bacteria isolated from vegetables and traditional Indian fermented foods. *International Journal of Fermented Foods, 1*(1), 87–101.

Poole, J., Day, C. J., von Itzstein, M., Paton, J. C., & Jennings, M. P. (2018). Glycointeractions in bacterial pathogenesis. *Nature Reviews Microbiology, 16*(1), 440–452.

Prajapati, V. D., Jani, G. K., Zala, B. S., & Khutliwala, T. A. (2013). An insight into the emerging exopolysaccharide gellan gum as a novel polymer. *Carbohydrate Polymers, 93*, 670–678.

Prezotti, F. G., Cury, B. S. F., & Evangelista, R. C. (2014). Mucoadhesive beads of gellan gum/pectin intended to controlled delivery of drugs. *Carbohydrate Polymers, 113*, 286–295.

Proctor, L. M., Creasy, H. H., Fettweis, J. M., Lloyd-Price, J., Mahurkar, A., Zhou, W., et al. (2019). The integrative human microbiome project. *Nature, 569*(7758), 641–648.

Püngel, D., Treveil, A., Dalby, M. J., Caim, S., Colquhoun, I. J., Booth, C., et al. (2020). Bifidobacterium breve UCC2003 exopolysaccharide modulates the early life microbiota by acting as a potential dietary substrate. *Nutrients, 12*(4), 948–965.

Rajoka, M. S. R., Wu, Y., Mehwish, H. M., Bansal, M., & Zhao, L. (2020). Lactobacillus exopolysaccharides: New perspectives on engineering strategies, physiochemical functions, and immunomodulatory effects on host health. *Trends in Food Science & Technology, 103*, 36–48.

Rana, S., & Upadhyay, L. S. B. (2020). Microbial exopolysaccharides: Synthesis pathways, types and their commercial applications. *International Journal of Biological Macromolecules, 157*, 577–583.

Ratajczak, W., Rył, A., Mizerski, A., Walczakiewicz, K., Sipak, O., & Laszczyńska, M. (2019). Immunomodulatory potential of gut microbiome-derived short-chain fatty acids (SCFAs). *Acta Biochimica Polonica, 66*(1), 1–12.

Ray, M., Ghosh, K., Har, P. K., Ojha, D., Soren, J. P., Singh, S. N., & Mondal, K. C. (2017). Bifidobacteria and its rice fermented products on diet induced obese mice: Analysis of physical status, serum profile and gene expressions. *Journal of Beneficial Microbes, 9*(3), 441–452.

Rojo, D., M'endez-García, C., Raczkowska, B. A., Bargiela, R., Moya, A., Ferrer, M., & Barbas, C. (2017). Exploring the human microbiome from multiple perspectives: Factors altering its composition and function. *FEMS Microbiology Reviews, 41*(4), 453–478.

Sachdev, D. P., & Cameotra, S. S. (2013). Biosurfactants in agriculture. *Applied Microbiology and Biotechnology, 97*, 1005–1016.

Saldías, C., Velásquez, L., Quezada, C., & Leiva, A. (2015). Physicochemical assessment of dextran-g-poly (ε-caprolactone) micellar nanoaggregates as drug nanocarriers. *Carbohydrate Polymers, 117*, 458–467.

Santra, H. K., & Banerjee, D. (2021). Microbial exopolysaccharides: Structure and therapeutic properties. In *Microbial polymers: Applications and ecological perspectives* (pp. 375–420). Springer.

Shi, Z., Zhang, Y., Phillips, G. O., & Yang, G. (2014). Utilization of bacterial cellulose in food. *Food Hydrocolloids, 35*, 539–545.

Shi, N., Li, N., Duan, X., & Niu, H. (2017). Interaction between the gut microbiome and mucosal immune system. *Military Medical Research, 4*(1), 1–7.

Shondelmyer, K., Knight, R., Sanivarapu, A., Ogino, S., & Vanamala, J. K. P. (2018). Ancient Thali diet: Gut microbiota, immunity, and health. *Yale Journal of Biology and Medicine, 91*(2), 177–184.

Singh, H., Chopra, C., Singh, H., Malgotra, V., Wani, A. K., Dhanjal, D. S., et al. (2024). Gut-brain axis and Alzheimer's disease: Therapeutic interventions and strategies. *Journal of Functional Foods, 112*, 105915.

Sun, S., Zhang, Z., Luo, Y., Zhong, W., Xiao, M., Yi, W., et al. (2011). Exopolysaccharide production by a genetically engineered Enterobacter cloacae strain for microbial enhanced oil recovery. *Bioresource Technology, 102*(10), 6153–6158.

Sun, H., Yu, X., Li, T., & Zhu, Z. (2021). Structure and hypoglycemic activity of a novel exopolysaccharide of *Cordyceps militaris*. *International Journal of Biological Macromolecules, 166*, 496–508.

Surayot, U., Wang, J., Seesuriyachan, P., Kuntiya, A., Tabarsa, M., Lee, Y., et al. (2014). Exopolysaccharides from lactic acid bacteria: Structural analysis, molecular weight effect on immunomodulation. *International Journal of Biological Macromolecules, 68*, 233–240.

Tanes, C., Bittinger, K., Gao, Y., Friedman, E. S., Nessel, L., Paladhi, U. R., Chau, L., Panfen, E., Fischbach, M. A., Braun, J., Xavier, R. J., Clish, C. B., Li, H., Bushman, F. D., Lewis, J. D., & Wu, G. D. (2021). Role of dietary fiber in the recovery of the human gut microbiome and its metabolome. *Cell Host & Microbe, 29*(3), 394–407.E5.

Tang, W., Han, S., Zhou, J., Xu, Q., Dong, M., Fan, X., et al. (2020). Selective fermentation of Lactobacillus delbrueckii ssp. Bulgaricus SRFM-1 derived exopolysaccharide by Lactobacillus and streptococcus strains revealed prebiotic properties. *Journal of Functional Foods, 69*, 103952.

Tang, J., He, Z., Zhang, B., Cheng, J., Qiu, W., Chen, X., et al. (2024). Structural properties, bioactivities, structure-activity relationships and bio-applications of polysaccharides from Auricularia Auricula: A review. *International Journal of Biological Macromolecules, 280*, 135941.

Tian, J., Zhao, X., Tang, C., Wang, X., Zhang, X., Xiao, L., & Li, W. (2023). Protective effect of Paecilomyces cicadae TJJ11213 exopolysaccharide on intestinal mucosa and regulation of gut microbiota in immunosuppressed mice. *Food Research International, 165*, 112477.

Tiwari, B., Singh, S., Chakraborty, S., Verma, E., & Mishra, A. K. (2017). Sequential role of biosorption and biodegradation in rapid removal degradation and utilization of methyl parathion as a phosphate source by a new cyanobacterial isolate Scytonema sp. BHUS-5. *International Journal of Phytoremediation, 19*(10), 884–893.

Trabelsi, I., Slima, S. B., Chaabane, H., & Riadh, B. S. (2015). Purification and characterization of a novel exopolysaccharides produced by Lactobacillus sp. Ca6. *International Journal of Biological Macromolecules, 74*, 541–546.

Uebanso, T., Shimohata, T., Mawatari, K., & Takahashi, A. (2020). Functional roles of B-vitamins in the gut and gut microbiome. *Molecular Nutrition & Food Research, 64*(18), 2000426.

Vieira, A. T., MacIa, L., Galvão, I., Martins, F. S., Canesso, M. C. C., Amaral, F. A., et al. (2015). A role for gut microbiota and the metabolite-sensing receptor GPR43 in a murine model of gout. *Arthritis and Rheumatology, 67*(6), 1646–1656.

Wan, C., Qian, W. W., Liu, W., Pi, X., Tang, M. T., Wang, X. L., et al. (2022). Exopolysaccharide from Lactobacillus rhamnosus ZFM231 alleviates DSS-induced colitis in mice by regulating gut microbiota. *Journal of the Science of Food and Agriculture, 102*(15), 7087–7097.

Wang, J., Nie, S., Kan, L., Chen, H., Cui, S. W., Phillips, A. O., et al. (2017). Comparison of structural features and antioxidant activity of polysaccharides from natural and cultured Cordyceps sinensis. *Food Science and Biotechnology, 26*, 55–62.

Wang, J., Fang, X., Wu, T., Min, W., & Yang, Z. (2018). Exopolysaccharide producing Lactobacillus plantarum SKT109 as adjunct culture in Cheddar cheese production. *LWT, 97*, 419–426.

Wang, W., Ju, Y., Liu, N., Shi, S., & Hao, L. (2023). Structural characteristics of microbial exopolysaccharides in association with their biological activities: A review. *Chemical and Biological Technologies in Agriculture, 10*(1), 137.

Xu, X., & Zhang, X. (2015). Lentinula edodes-derived polysaccharide alters the spatial structure of gut microbiota in mice. *PLoS One, 10*(1), e0115037.

Yang, Y. J., Liu, J. Y., Tan, Y., Wang, S. H., Chen, H. M., & Zhou, A. M. (2021). Progress in understanding the structure-activity relationship and hypoglycemic mechanism of polysaccharides.

Ying, M., Yu, Q., Zheng, B., Wang, H., Wang, J., Chen, S., et al. (2020). Cultured Cordyceps sinensis polysaccharides modulate intestinal mucosal immunity and gut microbiota in cyclophosphamide-treated mice. *Carbohydrate Polymers, 235*, 115957.

Yu, Y. J., Chen, Z., Chen, P. T., & Ng, I. S. (2018). Production, characterization and antibacterial activity of exopolysaccharide from a newly isolated *Weissella cibaria* under sucrose effect. *Journal of Bioscience and Bioengineering, 126*(6), 769–777.

Zhang, G., Zhang, W., Sun, L., Sadiq, F. A., Yang, Y., Gao, J., & Sang, Y. (2019). Preparation screening, production optimization and characterization of exopolysaccharides produced by lactobacillus sanfranciscensis Ls-1001 isolated from Chinese traditional sourdough. *International Journal of Biological Macromolecules, 139*, 1295–1303.

Zhang, Y., Liu, J., Dou, P., Wu, Z., Zheng, Z., Pan, X., et al. (2021). Oral absorption characteristics and mechanisms of a pectin-type polysaccharide from Smilax China L. across the intestinal epithelium. *Carbohydrate Polymers, 270*, 118383.

Zhou, X., Zhang, D., Qi, W., Hong, T., Xiong, T., Wu, T., et al. (2021). Exopolysaccharides from Lactobacillus plantarum NCU116 facilitate intestinal homeostasis by modulating intestinal epithelial regeneration and microbiota. *Journal of Agricultural and Food Chemistry, 69*(28), 7863–7873.

Zhu, Z. Y., Guo, M. Z., Liu, F., Luo, Y., Chen, L., Meng, M., et al. (2016). Preparation and inhibition on α-d-glucosidase of low molecular weight polysaccharide from Cordyceps militaris. *International Journal of Biological Macromolecules, 93*, 27–33.

Zivkovi'c, M., Miljkovi'c, M. S., Ruas-Madiedo, P., Markeli'c, M. B., Veljovi'c, K., Tolinăcki, M., et al. (2016). EPS-SJ exopolisaccharide produced by the strain Lactobacillus paracasei subsp. paracasei BGSJ2-8 is involved in adhesion to epithelial intestinal cells and decrease on E. coli association to Caco-2 cells. *Frontiers in Microbiology, 7*(1), 1–14.

The Immune-Modulating Effects of Extracellular Polysaccharides: Mechanistic Insights and Functional Food Potential

Arindam Ganguly, Raina Kundu, Riyan Kaibarta, Suman Das, Subhajit Paramanik, Debasis Mitra, and Sourav Chattaraj

1 Introduction

Microbial extracellular polymeric substances (EPS) have also been termed as manufactured polymers or biopolymers. These biopolymers are released by microbial organisms and aid in microbial ecology and functionality. Thus, EPS can be defined as extracellular polymeric substances that originated in biological sources and contribute to microbial aggregates formation. EPS was thus used to describe a wide overlap in biological phenomena encompassing numerous types of macromolecules which includes polysaccharides, proteins, nucleic acids, phospholipids, and other polymeric substances situated within the enclosed spaces of the cells or material aggregates (Wingender, 1999). Nevertheless, in nature, the exopolysaccharides serve a variety of functions such as cell protection, adhesion, and biofilm formation but cannot be used as a carbon or energy source (Donot et al., 2012). They represent for instance high molecular weight macromolecules which exhibit varying characteristics based on the microorganism as well as environmental condition (Yang et al., 2016). EPS are capsular or slime coating enveloping the microorganisms (Bertocchi et al., 1990). EPS can be classified structurally either as capsular forms, which are located closely to the cell, or secretion, which is detached from the cell and diffused into the surroundings as EPS. EPS may be weakly or strongly attached to cell surface based on the nature of microorganism (Welman & Maddox, 2003).

A. Ganguly · R. Kundu · R. Kaibarta · S. Das · S. Paramanik
Department of Microbiology, Bankura Sammilani College, Bankura, West Bengal, India

D. Mitra
Department of Microbiology, Graphic Era (Deemed to be University), Dehradun, Uttarakhand, India

S. Chattaraj (✉)
Centre for Industrial Biotechnology Research, School of Pharmaceutical Sciences, Siksha 'O' Anusandhan Deemed to be University, Bhubaneswar, Odisha, India

T. Sarkar et al. (eds.), *Extracellular Polysaccharides*,
https://doi.org/10.1007/978-3-031-94472-7_15

The functions of extracellular polysaccharide go beyond just providing structural support, as they also safeguard microbial cells from osmotic pressure, dehydration, antibiotic and toxic chemicals, and phagocytic attack. Gram-positive bacteria often produce EPS to withstand harsh and unfavorable environmental conditions (Welman & Maddox, 2003). These macromolecules can form stable three-dimensional networks via interaction through electrostatic forces, hydrogen bonds, ionic forces, and biochemical reactions (Flemming & Wingender, 2010). EPS can be produced by submerged fermentation (SmF) as well as solid-state fermentation (SSF) methods which enhance different qualities of these polymers (Seesuriyachan et al., 2012). The synthesis and secretion of EPS does not appear to be restricted to a particular species of organism and can result from either microbial activity or from cell fracture. It has been reported for both prokaryotic and eukaryotic microorganisms in the natural and artificial environments including biofilms (Flemming & Wingender, 2001; Wingender et al., 1999). Polysaccharides being the major constituents of EPS can be seen as biological macromolecules that are obtained from plants, animals, and microorganisms. These compounds are well known as immunoactive, antioxidative, antitumorous, and prebiotic dietary fibers. Recent research highlights their significance in influencing human health through metabolization and fermentation by gut microbiota (Chen et al., 2018; Park et al., 2019). Application of biopolymers for biomedical initiatives is also attributed to their reduced toxicity and lower side effects than synthetic agents. Their signaling transduction capabilities have mostly been targeted for the development of antitumor drugs (Ying & Hao, 2023). EPS are components of microbial systems which include altering the structure of microbial communities, and interacting with the intestinal epithelial barrier and the immune system (Castro-Bravo et al., 2018; Hidalgo-Cantabrana et al., 2012). Also, they defend against environmental stresses like desiccation, phage infection, antimicrobial agents, and predation by protozoa (Ciszek-Lenda, 2011; Forde & Fitzgerald, 1999).

Microorganisms have emerged as invaluable bioresources, catalyzing innovation across diverse industries (Chattaraj et al., 2022, 2024c; Ganguly et al., 2018b). They produce a range of by-products with significant industrial potential, including enzymes (Nad et al., 2024; Ganguly et al., 2018c; Bhattacharjee et al., 2018) and bioactive compounds (Chattaraj et al., 2023b, 2024a). These microbial by-products are driving advancements in biotechnology, pharmaceuticals, food production (Ganguly et al., 2018a), agriculture (Samantaray et al., 2024), and aquaculture (Ganguly et al., 2019, 2024; Chattaraj et al., 2023a). Furthermore, the integration of biotechnology with sustainable practices is fostering sustainable economic growth (Chattaraj et al., 2024b; Ganguly & Mandal, 2024). EPS can have important applications in functional foods, especially for the purpose of immunomodulation (Poole et al., 2018). Furthermore, they are sustainable, eco-friendly, and low-cost alternatives of chemical flocculants (Wingender et al., 1999). EPS represent an invaluable resource for advancing biotechnology and healthcare. Their multifunctional nature and biomedical potential continue to position them at the forefront of scientific innovation.

2 Sources

Microorganisms are often found to be a rich source of exopolysaccharides (Table 1). Several marine and fresh water derived bacteria found to secrete EPS under certain adverse condition. Hot spring water also contains a good number of bacteria capable of synthesizing extracellular polysaccharides.

3 Structure and Composition of EPS

The main components of EPS polysaccharides are proteins (enzymes and structural proteins are included), nucleic acids (e.g., DNA), and lipids. In addition, humic substances, uronic acid, and inorganic elements are present within the EPS. As the amount of non-sugar components is lower than the polysaccharides, they are necessarily attached to sugar residues, which is the major fact responsible for distinct properties of the EPS (Decho, 1990; Sutherland, 1979). EPS may be categorized into two main types: homopolysaccharides (HoPS) and heteropolysaccharides (HePS). HoPS are made up of only a single monosaccharide type, mainly D-glucose or D-fructose. Polysaccharides based on glucose are bounded either by α-glycosidic bonds (α-glucans), or through β-glycosidic bonds (β-glucans). Fructose-based polysaccharides mostly contain β-glycosidic linkages (Dertli et al., 2018; Meng et al., 2016). Heteropolysaccharides (HePS) are made up of two or more different monosaccharides with D-glucose, D-galactose, and L-rhamnose being the main components in the EPS of the *Lactobacilli* and the *Bifidobacteria.* Besides, they might contain N-acetylated monosaccharides such as N-acetylglucosamine (GlcNAc) and N-acetyl-galactosamine (GalNAc), besides L-fucose, D-glucuronic acid, D-mannose, and glycerol (Badel et al., 2011; Hidalgo-Cantabrana et al., 2014). The synthesis of HePS is more tedious than that of HoPS because multiple genes are needed, which are located in the cps-clusters (Hidalgo-Cantabrana et al., 2014; Ferrario et al., 2016). These polysaccharide clusters consist of glycosyltransferases (GTF). The chemical composition of EPS exhibits significant diversity, with various proportions of carbohydrates, proteins, nucleic acids, lipids, and humic substances (non-secreted fractions). The majority of proteins and carbohydrates amount to 75–90% in EPS composition. It also contains lipopolysaccharides, glycoproteins, and lipoproteins (Marvasi et al., 2010; Sheng et al., 2010).

4 Carbohydrates

EPS from the outer (extracellular) part is mostly made up of the so-called carbohydrates, such as the carbon-containing neutral sugars (mostly hexoses and rarely pentoses) and uronic acids like glucuronic, galacturonic, and mannuronic acids. These

Table 1 Different sources of EPS-producing microorganisms

Sl. no	Microorganisms	Sources	Conditions	Production rate	References
1.	*Alternaria* sp. SP-32	Marine sponge	pH—6.5 Temp—28 °C Time—8 D	2.5 g/L	Chen et al. (2016)
2.	*Agrobacterium radiobacter*	Persian type culture collection (PTCC)	Temp—28 °C Time—5 D Rpm—253	1.03 g/100 mL	Bakhtiyari et al. (2015)
3.	*Brevibacillus thermoruber* strain 423	Hot springs	pH—6 Temp—55 °C Time—1 D Rpm—180	202 mg/L	Yasar Yildiz et al. (2014)
4.	*Vibrio harveyi* VB23	Different sampling sites of river	pH—7 Temp—28 °C Time—1 D Rpm—160	27.8 μg/mL	Bramhachari and Dubey (2006)
5.	*Phellinus igniarius*	Stumps of willow	pH—6.0 Temp—28 °C Time—4 D Rpm—150	1.55 g/L	He et al. (2012)
6.	*Neopestalotiopsis* sp. *SKE15*	Phenotypical and genotypical analyses (GenBank Accession No. MG649986)	Temp—30 °C Time—7 D Rpm—150	2.02 g/L	Fooladi et al. (2019)
7.	*Azotobacter AC2*	Alkaline & Acid soils	Temp—30 °C Time—4 D Rpm—200	7.5 mg/mL	Emtiazia et al. (2004)
8.	*Enterococcus faecalis DU10*	Duck intestinal homogenate	pH—6.5 Temp—35 °C Time—2 D	840 mg/L	Venkatesh et al. (2016)
9.	*Pseudomonas fluorescens BM07*	Activated sludge in a municipal wastewater treatment plant	pH—7 Temp—10 °C Time—6 D	290 mg/L	Noghabi et al. (2007)
10.	*Lactobacillus plantarum HM47*	Human breast milk	pH—5.5 Temp—37 °C (anaerobic incubator) Time—2–4 D	380 mg/L	Nambiar et al. (2018)
11.	*Morchella conica*	Sichuan Agricultural Academy of Sciences	pH—6.4 Temp—28 °C Time—7 D Rpm—50	1.06 mg/mL	Su et al. (2013)

(continued)

Table 1 (continued)

Sl. no	Microorganisms	Sources	Conditions	Production rate	References
12.	*Sphingomonas* sp. *CS101*	Soil	pH—7 Temp—30 °C Time—4 D Rpm—250	17 g/L	Seo et al. (2004)
13.	*Weissella confusa OF126*	Ogi	pH—7 Temp—30 °C Time—2 D 30 min Rpm—170	3 g/L	Adesulu-Dahunsi et al. (2018)
14.	*Pseudomonas aeruginosa (MTCC 1688)*	Microbial type culture collection (MTCC)	pH—6 Temp—32 °C Time—4 D	23.3 mg/50 mL	Chug et al. (2021)
15.	*Bacillus licheniformis WSF-1*	Sludge tank of sugar distillery plant	pH—7 Temp—30 °C Time—3 D 12 h Rpm—200	2.9 g/mL	Jenny Angel et al. (2018)

substances in addition to standard substitutes like acetate esters, pyruvate ketals, and phosphates or sulfates allow the EPS to be neutral, anionic, or cationic (Czaczyk & Myszka, 2007; Pal & Paul, 2008). HoPS (Homopolysaccharides) are polymers that are made of a single type of sugar, such as D-glucose or L-fructose, and they are generally neutral. They include dextran, curdlan, and cellulose. Dextran produced by *Leuconostoc* spp. and *Streptococcus* spp. is a polymer of glucose units with α-(1,6)-linkage and branches at α-(1,3)-, α-(1,4)-, or α-(1,2)-position. Meanwhile, curdlan, which is synthesized by *Alcaligenes faecalis* and *Agrobacterium* sp., is a linear β-(1,3)-linked polysaccharide, and cellulose, formed by numerous bacterial species, consists of β-(1,4)-linked D-glucose units (Rehm, 2010; Vu et al., 2009). HePS are among other things, the polymers that are the most studied: alginate, xanthan gum, and gellan. They consist of the repeating units of different monosaccharides. Alginate (which is found in *Pseudomonas aeruginosa* and *Azotobacter* spp.), for instance, is a linear β-(1,4)-linked polymer of mannuronic and guluronic acids. Xanthan gum, which can be obtained by *Xanthomonas campestris*, is a main chain of glucose with a trisaccharide side chain containing mannose, glucuronic acid, and acetyl groups. Gellan, a product of *Sphingomonas* spp., is made of tetrasaccharides, including glucose, glucuronic acid, and rhamnose (Czaczyk & Myszka, 2007; Flemming & Wingender, 2010). The features and characteristics of polysaccharides are vital in their physical and functional properties, especially in viewing biofilm formation and biofilm stability. For example, if alginate is not present, biofilm formation is still possible in *P. aeruginosa* but it may cause significant changes in the biofilm architecture (Rehm, 2010; Vu et al., 2009).

5 Protein

EPS consists mostly of proteins, which are enzymes and structural proteins. Enzymes are produced in EPS matrix and they are the ones that direct that degradation and utilization of the trapped organic particles. The release of these enzymes occurs in the solution of substrates, such as soluble polysaccharides, proteins, nucleic acids, and insoluble materials like cellulose, chitin, and lipids that are then acted upon by extracellular enzymes. Microorganisms can use these enzymes during the time of nutrient scarcity to digest EPS for survival, either their own or external ones (Flemming & Wingender, 2010). One such example is that *Bacillus subtilis* produces antibiotic peptide peptides that are overly structured and resist break down of protease (Marvasi et al., 2010). Structural proteins are the most responsible ones in interaction with the polysaccharide matrix formation of the floc and bacterial surfaces to the extracellular environment. The lectin-like proteins that adhere between activated sludge results floc formation (Park & Novak, 2009). Non-ribosomal peptides like polyamides, which are parallel to poly-γ-glutamate (PGA) and ε-poly-L-lysine, are important EPS components. Cells have the ability to produce PGA as slime or a capsular type; for example, *Bacillus anthracis* is the capsular PGA producer, whereas *Bacillus subtilis* is the soluble PGA producer (Rehm, 2010; Shih & Van, 2001). N-glycosylated proteins exist both in prokaryotes and eukaryotes and they are acquired as a result of a covalent bonding of sugar to proteins. This kind of protein is involved in integrity, communication, and defense of the organism (Hug & Feldman, 2011). Prokaryotic glycosylation occurs in two forms: N-glycosylation and O-glycosylation (Hitchen & Dell, 2006). The main examples of Gram-positive bacterial glycoproteins include those associated with *Bacillus amyloliquefaciens* and *Clostridium acetobutylicum*, the former; and the former species *Mycobacterium tuberculosis* (Upreti et al., 2003).

6 Extracellular DNA

E-DNA, or Extracellular DNA, is a noticeable EPS component that is derived from the various sources of the environment such as, wastewater sludge, soil, and marine sediments. This component is particularly abundant in the wastewater material, although it varies significantly between different species. E-DNA is a crucial structural element in the EPS matrix of *S. aureus*. Meanwhile, it is of less importance in *S. epidermidis* (Flemming & Wingender, 2010). Similarly, the EPS of *Bacillus cereus* relies on E-DNA for matrix formation (Marvasi et al., 2010). In the case of *Streptococcus mutans*, the appearance of E-DNA is correlated with the releasing of the competence signaling peptides, which in turn promote the horizontal gene transfer in the EPS matrix (Czaczyk & Myszka, 2007).

7 Lipids and Surfactant

The hydrophilic nature is mainly due to hydrated polysaccharides, proteins, and DNA, but there are some EPS, for example, EPS from *Rhodococcus* spp., which possess hydrophobic properties because of the polysaccharide-linked methyl and acetyl groups. Lipids and their derivatives are also constituents of the EPS matrix, where lipopolysaccharide acts as the most crucial adhesion factor, for example, *Thiobacillus ferrooxidans* to pyrite surfaces (Flemming & Wingender, 2010). Certain biosurfactants such as surfactin, viscosin, emulsan, and some others are responsible for the dispersal of hydrophobic substances and, thus, bacterial attachment or detachment from oil droplets. Biosurfactants have the ability to demonstrate antimicrobial activities. In other words, rhamnolipid, which is a glycolipid, can be seen in the EPS matrix of *Pseudomonas aeruginosa* (Flemming & Wingender, 2010; Soberón-Chávez et al., 2005), and surfactin, a lipopeptide produced by *Bacillus subtilis*, is involved in the colonization process and the competition of resources (Marvasi et al., 2010).

8 Humic Substances

EPS matrix-comprising bacterial extracellular polysaccharides-aerates a rather non-direct production of humic components though the girth of the environment as, for example, dirt or sewage sludge. These substances have significant effect on the biodegradability and the adsorption capacity of EPS (Wingender et al., 1999). Humic substances, which are natural heterogeneous ligands, account for 30–80% of the total dissolved organic matter in aquatic systems (Moura et al., 2007; Peña-Méndez et al., 2005). To be even more precise, they are three-folded in fulvic acid, humic acid, and humin. The latter two are soluble in alkalis, but the former belongs to insolubles (Peña-Méndez et al., 2005). These chemicals are synthesized during humification, wherein amino acids, lignin, pectin, and carbohydrates interact through different intermolecular forces. In other words, lignin becomes a key factor as well (Burdon, 2001; Davies & Ghabbour, 2001). In terms of origin, humic substances are hardly defined structures, having the proposed structures that show the existence of both phenolic and carboxylic groups in humic and fulvic acids (Buffle et al., 1977).

9 EPS Biosynthesis Pathway

The EPS, produced by different types of microorganisms, can be categorized into four significant groups, such as polyesters, polyamides, inorganic polyanhydrides, and polysaccharides (Tiwari et al., 2020). Microorganisms use four commons

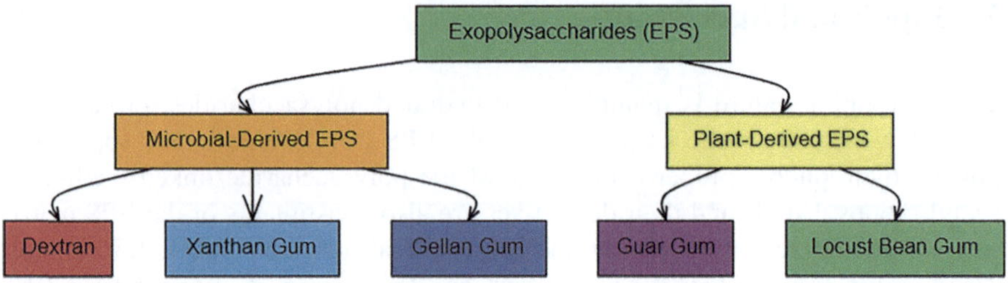

Fig. 1 Biosynthesis pathway of EPS

pathways for EPS biosynthesis, that is, extracellular, the synthase-dependent, the ATP-binding cassette transporter-dependent pathway, and the Wzx/Wzy-dependent pathway (Fig. 1). The Wzx/Wzy-dependent and extracellular synthesis pathways are common mechanisms for *Lactobacillus* EPS production (Schmid et al., 2015).

10 The Immune System: An Overview

The immune system plays a key part in the protection of the human body against various elements, in particular bacteria, viruses, fungi, and parasites, as well as other harmful environmental factors (Calder, 2013). The immune system, which is very intricate, approximately contains a trillion lymphocytes. The antibodies, which are formed and released by B lymphocytes, are indispensable for an effective immune response (McComb et al., 2019). The immune cells originate and mature in the primary lymphoid organs, such as the bone marrow and the thymus, whereas lymph nodes, spleen, Peyer's patches, the appendix, tonsils, adenoids, gut-associated lymphoid tissue (GALT), and the mucosa-related lymphoid tissue (MALT) act as secondary lymphoid organs (Calder, 2013; McComb et al., 2019). The immune system mechanism consists of the innate and the adaptive divisions (Calder, 2013). The immune system consists of the complement system, phagocytic cells (neutrophils, monocytes, and macrophages), inflammatory mediator secretor cells (basophils, mast cells, and eosinophils), and natural killer cells. The molecular basis includes complement proteins, acute phase proteins, and cytokines such as interferons (Roy & Loh, 1996; Delves & Roitt, 2000). Immuno-stimulants are often used to increase the host defense against infections. Specifically, they act by activating either innate or acquired (adaptive) immune responses and thus, they are called immunopotentiators or immune-promoting agents (Giorgi & Cochrane, 2019).

11 Mechanisms of Immune Modulation by Extracellular Polysaccharides

EPS are known to mediate both innate and adaptive immune systems. One of the most important roles of the gastrointestinal tract, where almost 80% of immune units reside, is its ability to participate in the process of regulation, in response to bacterial EPS (Lewis & Blutt, 2019; Shi et al., 2017). Negatively charged or small-sized EPS are generally responsible for the stimulation of immune cells, while neutral or larger-sized EPS are usually the ones to suppress the immune responses (Hidalgo-Cantabrana et al., 2012). EPS influence the activity of the immune system through the recruitment of mononuclear phagocytes (monocytes and macrophages), antigen-presenting cells (APCs), and other immune cells, including lymphocytes and neutrophils (Ren et al., 2021). They make their mechanism effective by binding to the receptors located on cell surfaces, triggering the activation of immune cells (Chaisuwan et al., 2020).

12 Macrophages

These cells are found throughout the whole body in connective tissues and organs, and function in both the innate and acquired immune responses. In this case, it is a type of immune cell in which it contains only one nucleus and which is able to engulf other cells (Chaisuwan et al., 2020). The cells can produce many different cytokines with the various effects. Furthermore, the major task of macrophages is to destroy the antigenic cells. Macrophages play a critical role as (Gordon, 2007; Gordon & Martinez-Pomares, 2017). The receptors called Toll-like receptors (TLRs) are involved in both innate immunity and acquired immunity, and are one of the keystone components of the immune function that also contribute to the network regulation of receptors (Li et al., 2017a, b). There is a point of exogenous polymers to activate the cells, that is, the toll-like receptor 4 (TLR4) and TLR2 on the macrophage cell surface recognize the polysaccharide and induce the cells through the nuclear factor-kappa B (NF-κB) and via PKC, MEK1, PAK, and MAPKs (Kawai & Akira, 2007; Hsu et al., 2004). The members of the NF-κB family generally belong to five highly similar proteins, which contain NF-κB1 (also known as p50), NF-κB2 (also known as p52), RelA (also known as p65), RelB, and c-Rel. The NF-κB is a group of inducible transcription factors that control the expression of many genes encoding cytokines, chemokines, proinflammatory enzymes, and adhesion molecules that are involved in the innate and adaptive immune response, inflammation, proliferation, and anti-apoptosis (Liu et al., 2017). The regulation of NF-κB signaling pathways involves the canonical and non-canonical (or alternative) ways (Sun, 2011; Vallabhapurapu & Karin, 2009). NF-kB is found in the cytoplasm as the inactive form by the interaction with the protein group called known as inhibitor proteins. The nuclear IkB family includes seven members: IkBα, IkBβ, IkBgγ, IkBε,

Bcl-3, IkBz, and IkBNS. These members are located in the nucleus and their primary role is to interact with NF-kB to regulate transcription (Kawai & Akira, 2007). EPS activate macrophages via the NF-κB signal through the canonical pathway of IκBα degradation. Initially, EPS attach to the TLR4 and TLR2 on the surface of the macrophage and recruit phosphorylated multi-subunit IκB kinase (IKK) complexes (Chaisuwan et al., 2020). This structure consists of two catalytic components, IKKα and IKKβ, which are structurally similar, and a regulatory component, NEMO (NF-κB essential modifier, also known as IKKγ) (Kawai & Akira, 2007). Furthermore, the IKK complex phosphorylates an IκBα, which binds with RelA and p50 in the cytoplasm to form a complex. Then, phosphorylated IκBα is polyubiquitinated and degraded by the 26S proteosome. The RelA-p50 complex, which is a heterodimer, is imported into the nucleus and interacts with specific DNA binding sites that regulate gene transcription (Kawai & Akira, 2007). At the end, genes are expressed for cytokines, enzymes, and adhesion molecules. EPS with immunomodulatory effects, via the TLR2–4 deficient RAW264.7 macrophages, enhanced enzymatic activity, such as the up-regulation of iNOS, COX-2, and cytokines such as TNF-α, IL1β, IL-6, IL-10, and IL-12 (Kim et al., 2011; Surayot et al., 2015). Mitogen-activated protein kinases (MAPKs) are a group of serine-threonine kinases that mediate intracellular signaling associated with many cellular activities like cell proliferation, differentiation, survival, death, and transformation (McCubrey et al., 2006; Dhillon et al., 2007). The mammalian MAPK family is made up of extracellular signal-regulated kinase (ERK), p38, and c-Jun NH2-terminal kinase (JNK; also known as stress-activated protein kinase or SAPK). Each of these kinases is multifarious: ERK1–ERK8; p38-α, -β, -γ, and -δ; and JNK1–JNK3 (Dhillon et al., 2007; Schaeffer & Weber, 1999). Macrophages bind to EPS Polysaccharides through the TLR4 or TLR2. The signals activate MAPK to phosphorylate ERK, JNK, and p38 to become pERK, p-JNK, and p-p38, respectively. The polysaccharides activate the macrophages through the regulation of the MAPK pathway (Fig. 2). The phosphorylated proteins (transcription factors) move to the nucleus and transactivate the transcription factors such as AP-1 for the gene expression regulation (Chaisuwan et al., 2020). The MAPK pathways might be very important for macrophage proliferation during the macrophage activation. Also, polysaccharides may elevate the macrophage entry into the immune response against microorganisms, which includes the macrophage phagocytosis being increased (Kim & Choi, 2010; Qi & Elion, 2005).

13 Natural Killer Cells

Also, NK cells are used for the assay of immunomodulatory activity (Chaisuwan et al., 2020). The transformation of Natural Killer (NK) cells into tumor-killing lymphocytes are also crucial for managing anti-inflammatory responses, tumor surveillance, and the regulation of autoimmune disorders (Vivier et al., 2004). NK cells do not possess any antigen-binding receptors. They are capable of direct cytotoxicity of cancer and pathogen-infected cells, which is triggered when the former are

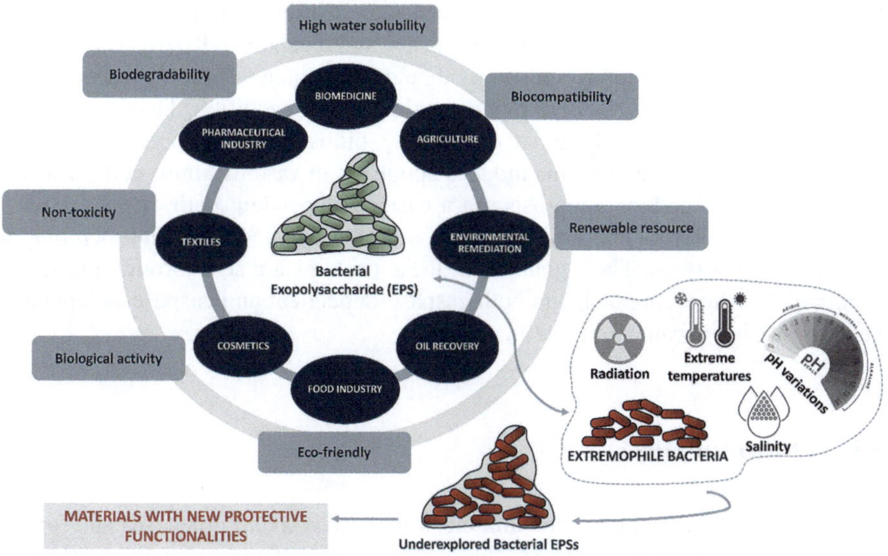

Fig. 2 The mode of action of exopolysaccharides in activating cell signaling pathways

recognized as the targets and through the balance of signals from activating and inhibitory cell surface receptors (de Saint Basile et al., 2010; Chaisuwan et al., 2020). The cells mature in the bone marrow and become active in the presence of different cytokines such as interleukins (IL), IL-2, IL-7, IL-12, IL-15, IL-18, IL-27, IL-35, and type I interferons: IFN-α and IFN-β (Waldmann & Tagaya, 1999; Abel et al., 2018). Perforins and granzymes are proteins which are synthesized in NK cells that mediate antibody-dependent cell-mediated cytotoxicity and lyse the target pathogen (Chaisuwan et al., 2020; Surayot & You, 2017). NK cells also release several tumor necrosis factors (TNF) family ligands such as FAS ligand (FASL), TNF, and TNF-related apoptosis-inducing ligands (TRAIL), which signal NK cells to activate and thus enhancing their cytolytic and antitumor functions (Cooper et al., 2001; Smyth et al., 2002; Waldmann & Tagaya, 1999). The process of NK cell development includes the expression of complement receptor type 3 (CR3) and Toll-like receptor 4 (TLR-4) both of which act as the targeted receptors for polysaccharides. The NK cell activation by the polysaccharides ranged from 87.5% to 97.0%, exhibiting significant variations (Surayot & You, 2017). Initial studies have shown that polysaccharides (e.g., *Ganoderma lucidum*, lentinan, and MD-fraction, a polysaccharide derived from *Grifola frondosa*), are able to modulate NK cells (human and mouse) activity by enhancing their cytotoxicity and cytokine expression level (Huyan et al., 2014). The EPS also promoted the cytotoxicity of NK cells due to an increase of IFN-γ and secretion of perforins and granzymes. The mechanisms for stimulation of NK cells start with the binding from polysaccharides to the receptors (CR3 and TLR4) (Chaisuwan et al., 2020). The transcription factors like the RelA-p50 complex, p-ERK, p-JNK, and p-p38 are translocated into the nucleus for control of transcription. Finally, the genes are expressed to be proteins such as cytokines (IFN-γ), FasL, granzymes, perforins, and receptor (NKp30) (Huyan et al., 2014). The NKp30 receptor binds to its ligand and is able to promote the secretion of TNF-α and IFN-γ by NK cells (Vitale et al., 2005). In addition, TRAIL and FasL, expressed by NK cells and regulated by IFN-γ, binds to their compatible receptors on a tumor cell surface, and this induces apoptosis in various tumor cells. The signals are sent and induce apoptosis via a caspase-dependent pathway (Chaisuwan et al., 2020). Perforins and granzymes are secreted by exocytosis and then bind to the target cell surface. The signals from these proteins are sent through mitochondria, which induce apoptosis via both caspase-dependent and caspase-independent pathways (Chaisuwan et al., 2020).

14 Lymphocytes

Lymphocytes are essential to the immune system to respond infections and foreign invaders (Berrington et al., 2005). These cells serve as the primary elements of the adaptive immune response (Weiss, 2014), delivering specific defense against invading microorganisms (Blum et al., 2013), providing protection from cancers like multiple myeloma (Barik, 2016), and facilitating tissue rejection during organ transplants by recognizing these tissues as foreign entities (Devine & Hoffman, 2000; Orakpoghenor et al., 2019). The activation of lymphocytes is initiated through a process called antigen presentation by antigen-presenting cells. This activation involves a complex series of signaling pathways that employ various secondary messengers and signaling molecules, including Ca2+, calcineurin (CaN), protein kinase C (PKC), and nuclear factor of activated T cells (NFAT) (Weiss, 2014). Calcium ions act as universal secondary messengers in nearly all immune cells (Clapham, 2007) and are crucial for numerous cellular activities such as proliferation, differentiation, and gene transcription (Feske, 2007). A particular polysaccharide derived from *Ganoderma atrum* known as PSG-1 has been shown to promote T lymphocyte proliferation. This homogeneous protein-bound polysaccharide not only increases intracellular levels of calcium but also enhances the activity of calcineurin along with elevating p-ERK, p-JNK, and p-p38 expression levels (Yu et al., 2013). It appears that PSG-1 may partially activate spleen lymphocytes via synergistic interactions involving both the Ca^{2+}/CaN/NFAT/IL-2 signaling pathway and the PKC/NFAT/IL-2 signaling pathway (Yu et al., 2015). Regulatory T cells (T_{regs}) are immunosuppressive and, usually, inhibit or down-regulate the induction and proliferation of effector T cells (T_C). Deficiency of T_{reg} cells may contribute to autoimmune diseases. Specific polysaccharides can stimulate B-cell proliferation and activation. Alfalfa polysaccharide APS is selective and effective in the activation of TLR4/MAPK/p38 pathway-mediated IgM production in B cells in vitro (Xie et al., 2020). In a separate experiment, a homogeneous Polysaccharide from the mycelium of marine fungus *Phoma herbarum* (YCP) was also found to affect TLR2 and TLR4 activation of p38, ERK, and JNK in cells, shifting the transcription factor NF-kB into the nucleus which finally leads to B-cell proliferation and IgM increase (Zhang et al., 2013).

15 Potential Applications in Food Industry

EPS are substances that have the ability to enhance texture, increase shelf life, and generate health benefits. Their applications are numerous, ranging from food products like dairy, bakery, meat, and fermented drinks. Thus, they offer benefits in terms of technological, sensory, and functional aspects. These are the most known applications.

15.1 Enhancing Flavor and Texture Characteristics

Bacterial and fungal EPS, for example, succinoglycan, curdlan, levan, pullulan, and EPS-605, are serving as nano-sized metal particles synthesis stabilizers, and they are also nanoreducing agents (Li et al., 2017a, b; Sathiyanarayanan et al., 2017). In food products EPS behaves as a pseudoplastic non-Newtonian fluid; thus, hydrodynamic forces lead to the loss of structural units. This, in turn, is an important feature of flavor, texture, and mouthfeel improvement (Kavita et al., 2014). Based on the study, EPS bacteria are widely applicable to achieve the optimal sensorial and textural qualities in various foods, especially in low-fat or zero-fat products (Tidona et al., 2016). Reduced fat or fat-free products are usually linked to compromised taste and other values. Incorporation of EPS will resolve these issues surrounding the preservation of key taste attributes (Tidona et al., 2016). Cheese-makers have been working on innovative solutions to upcycle cheese whey into bio-products with high economic and added value through the utilization of functional EPS. The JNULCC001 strain of *Lactiplantibacillus plantarum* has been described as a species capable of stress tolerance and production of value added biobased products. Among them are the formation of syrupy liquids, gelling, stabilizing, immobilization of methylene blue, and biosynthesis of selenium nanoparticles (Li et al., 2020). EPS from *L. plantarum* is broadly used as a textural agent and has shown potential probiotic applications (Tallon et al., 2003).

15.2 Antioxidant and Anticancer Potential

Lactobacillus plantarum JLK0142 obtained from naturally made fermented dairy tofu has probiotic potential and capacity to synthesize EPS. EPS also effectively caused apoptosis of some colon cancer cells (HT-29) showing remarkable activity in the removal of hydroxyl radicals and the immunoregulatory properties to strengthen the immune system (Wang et al., 2018). Wang et al. (2019a) investigated the application of purified EPS (as an ingredient), non-EPS-producing starter cultures, and *L. plantarum* JLK0142 (as an adjunct culture) in three different cheese production batches. The overripening period of 90 days led to a continuation of a high vital count (7.99 log CFU/g) for the flora, and the ripening was not impaired by the flora. Consequently, cheeses containing adjuncts or purified EPS showed improved sensory characteristics due to higher proteolytic activity; however, they were also harder and less cohesive than regular cheeses. In addition, water-soluble extracts from the experimental cheeses showed activated bioactivity, which included ABTS and DPPH radicals scavenging and α-amylase, angiotensin-converting enzyme, hydroxyl radicals, and the growth of HT-29 cancer cells as their inhibitors. The results pointed out that the use of *L. plantarum* JLK0142 strain, which produces EPS, in the manufacturing process of low-fat cheese, can significantly enhance the quality and bioactive features of the products (Wang et al., 2019a).

15.3 Emulsifying Properties of EPS

Recent research has pinpointed EPS as the most potent emulsifying and stabilizing agents for both oil-in-water and water-in-oil emulsions, thus providing the food industry with sustainable alternatives to stabilizers conventions (Sacco et al., 2019). EPS isolated from *Lactobacillus sakei* L3 indicated the emulsification indices of 2.61–62.30% with several types of oils, among which sunflower oil was the most reliable (Wang et al., 2019b). *Lactobacillus plantarum* S123 and *Lactobacillus paracasei* strains also showed strong emulsifying activity in toluene and benzene solvents (Saleem et al., 2021; Shankar et al., 2021). In addition to *Lactobacillus*, EPS from *Pantoea* sp. BCCS 001 GH has great antioxidant and emulsifying properties, meanwhile, strains of *Pandoraea pnomenusa*, that is, MS5 and AS3 possess emulsifying and anti-biofilm inhibiting activities (Niknezhad et al., 2018). The galactan EPS from *Weissella confusa* KR 780676 exhibited a very high emulsifying capacity and was stable with olive and groundnut oil emulsions over a wide range of temperatures and pH levels (Kavitake et al., 2020). The EPS from *Pseudomonas* ID1 acted better as emulsifying agent than the commercial gums, the EPS-BMS *from Leuconostoc citreum* (Carrión et al., 2015; Abid et al., 2021).

15.4 Rheological and Functional Roles of EPS in Food Products

Ayyash et al. (2020) researched the bioactivity of peptides that were fastened by *L. plantarum* C70, received from the lactation of camel milk. The EPS-C70 comprised 88.1% and 73.1% of a cytotoxic effect on the breast and the colon cancer cells, respectively, and it destroyed the foodborne pathogens such as *Salmonella typhimurium* and *E. coli* O157:H7, with a 2.0–3.0 log reduction of the population. Hashemi and Gholamhosseinpour (2020) showed that ultrasound treatment was an efficient way of bio-activating goat's milk, such as content and peptides that could make anti-cancer activity. However, at high amplitudes they would block EPS production. Consequently, the culture of camel milk-EPS was favored, and it was found that it could block over 70% of ACE and that it was an effective anti-proliferative compound as well (Al-Dhaheri et al., 2017). *L. lactis* strain was able to produce an antioxidative agent which has gelling properties (Almalki, 2020). *Lactiplantibacillus plantarum* (LBIO1, LBIO14, and LBIO28)-derived EPS were useful in water retention and viscosity and in low-fat dairy products (Bachtarzi et al., 2020). The *L. paracasei* obtained from Kefir has been used for the production of functional foods, which, besides immunomodulation, also exhibit stable mucus production under varying temperatures. Thus, they may be used as bio-fortifier in food industry (Bengoa et al., 2018).

15.5 EPS and Their Role as Bread Improvers

The role of EPS in improving bread quality depends significantly on factors like carbon source, sucrose concentration, and incubation time. The production of more EPS was seen in liquid sourdough (LS) made with 10% sucrose SLF of *Weissella cibaria* C43–11 (Yu et al., 2018). Similarly, a Pseudocereal-based SL with amaranth or quinoa in a 1:1 ratio with wheat flour and 3% sucrose exhibited better EPS-related properties than *L. plantarum* ITM21b which lacks EPS production in mMRS media (Valerio et al., 2020). These EPS make dough actually better and they extend its shelf life while they also reinforce its nutritional aspects not only because of the decrease of phytate components but also because of the increased availability of minerals (Milanović et al., 2020). Furthermore, incorporating pseudocereals into the composition of bakery formulations has been shown to increase the EPS production, thus adding to the improvement of both texture and flavor of the bread (Wolter et al., 2014). Nonetheless, the prime challenge of the bakery industry is to bring down the level of salt in the bread without causing any damage to its taste and structure. It has been proved that functional sourdoughs containing EPS from *Weissella cibaria* MG1 and *Lactobacillus amylovorus* DSM 19280 can allow salt reductions to as low as 0.3% and at the same time they can actually increase the volume, texture, and sensory attributes (O'Donnell et al., 2015; Belz et al., 2019). Furthermore, the use of EPS-producing LAB spp., such as *Weissella cibaria*, *Leuconostoc mesenteroides*, and *Latilactobacillus sakei* LS81, has been found to increase bread texture and specific volume as well as decrease the staling rate, consequently reducing salt content (Xu et al., 2020). In the same way, this strategy can be considered greener alternative to synthetic hydrocolloids due to the fact that it improves the physicochemical properties of both gluten-containing and gluten-free bakery products (Lynch et al., 2018).

15.6 Starter Cultures and Emerging Prebiotics

Traditionally *Bifidobacterium* has been unsuitable as a starter culture because of its poor fermentation characteristics. However, the ropy EPS-producing *Bifidobacterium longum* YS108R was found to be viable with some bioprotection capacities. This strain is capable of maintaining the microstructure of the product. Thus it minimizes the syneresis, and increases viscosity and water holding capacity of the product. The high EPS-producing strain *L. pentosus* SLC 13, which was isolated from mustard pickles, was found to possess probiotic properties. The high level of EPS production through milk fermenting could be significantly achieved by the use of *L. pentosus* SLC 13 with the addition of prebiotics (Huang et al., 2017; Wang et al., 2019a). Furthermore, yogurt pots consisting of capsules of EPS (*Streptococcus thermophilus* ASCC 285) and sticky EPS (*S. thermophilus* ASCC 1275) cultures can also boost the yogurt physical properties such as the slowdown of syneresis and the increasing of firmness (Amatayakul et al., 2006). EPS-producing cultures, YF-L903 and YC-X11 in goat milk yogurt reduce syneresis as well as enhance viscosity (Madhubasani et al., 2020). In contrast, a study included EPS-producing cultures (*S. thermophilus* LB-50, *L. delbrueckii* ssp. *bulgaricus* LB-42) and non-EPS-producing cultures (*S. thermo*philus LB-53, *L. delbrueckii* ssp. bulgaricus LB-42). The EPS-producing cultures exhibited more rigid structural reformation and aggregated milk resulting reduction of syneresis (Cartaşev & Rudik, 2017).

15.7 Moisture Retention and Enhancing Meat Juiciness

Hydrocolloids are widely used to enhance the quality of meat, especially in cooked hams. The use of in situ EPS-producing strains *Lactobacillus curvatus*, *Lactobacillus sakei*, and *Lactobacillus plantarum* has been reported (Hilbig et al., 2019). The EPS produced from *L. plantarum* S2 strain has many potential applications in the food industry by being a natural preservative and a starter culture. *L. plantarum* has shown antimicrobial efficacy against foodborne pathogen *Staphylococcus aureus* ATCC-6538, *Listeria monocytogenes* ATCC-7644, *Enterococcus faecalis* ATCC-29212, and *Pseudomonas aeruginosa* ATCC-27853 (Dincer & Kivanc, 2019). The efficiency of antimicrobial activity of this strain is heavily affected by a number of factors, for example, pH, temperature, and presence of enzymes, for example, α-chymotrypsin, α-amylase, trypsin, lysozyme, and pronase. The strain also has a positive influence on the sensory features, such as color, flavor, and aroma of meat products (Rzepkowska et al., 2017).

16 Other Applications of EPS

16.1 Edible Coatings

EPS is one of the materials among the many options for food coatings and packaging, and besides its versatility, it has additional advantages, which include antimicrobial, and antioxidant properties. Pullulan is an EPS, which is synthesized by the fungus *Aureobasidium pullulans*, which is being extensively used in food industries as it is non-toxic, heat resistant, and has oxygen barrier properties. It is made to form edible films that do not alter the shelf life, the taste, the aroma, or the appearance of food. In Japan, pullulan is used in preparation of instant beverages, sauces, and meat products, as well as being as a stabilizer and a preserver in it (Grumezescu & Holban, 2018; Cheng et al., 2011). There are some other microbial exopolysaccharides, including dextran and xanthan gum which are used in edible coatings. Addition of sorbitol to dextran improves its elasticity, thus enabling it to be used in food packaging (Sharma & Rao, 2015; Davidović et al., 2018). Microalgal EPS are used to prevent browning in fresh-cut fruit during storage besides being taste-stable. EPS from the *Porphyridium cruentum* microalgae enhances the shelf life through non-browning mechanisms (Bertrand et al., 2016). Pullulan-based films incorporated with antimicrobial agents such as essential oils and nanoparticles effectively kill pathogens like *E. coli* and *L. monocytogenes*. Also, xanthan gum that combines with cinnamic acid makes antioxidant coatings which in turn delay browning and at the same time maintain the quality of the fruit without affecting the flavor (Sharma & Rao, 2015; Duncan, 2011).

16.2 Decolorization of Dyes from Industrial Wastewater

The textile industry emits untreated effluents and dyes that are the source of water toxicity and introduce carcinogenic compounds into the aquatic ecosystems. Biosorption utilizing the microbial EPS can be an efficient solution to the problem (Zhang et al., 2009). Bacteria such as *Pseudomonas*, *Bacillus*, and *Klebsiella* have been found to attain removal efficiencies between 20% and 99.9% using EPS (Mahgoub et al., 2023). The triad of *Pseudomonas*, *Staphylococcus*, and bioflocculants can also effectively clean the dye wastewater. Furthermore, *Bacillus* species and *Bacillus megaterium* bioflocculants had dye removal rates of 82.78% and 88.14%, respectively, through adsorption bridging (via cations in effluents) (Buthelezi et al., 2012).

16.3 Elimination of Hazardous Organic Compounds

The biological decomposition of Polycyclic Aromatic Hydrocarbons (PAHs) by microbial activity is considered one of the preferable methods of pollution remediation (Zhang et al., 2011). *Zoogloea* sp. and *Aspergillus niger* have been identified as EPS-producing microbes that can degrade pyrene (Jia et al., 2011). Therefore, the enhancement of PAH biodegradation by EPS is mainly attributed to the well-ordered hydrophobic interactions that increase the solubility of these hydrophobic contaminants. For one thing, the EPS might lead to the bioavailability of phenanthrene (PHE), in which it can help its transfer from silicone oil to the water, so it can be diffused to the mass transfer limitations, which will cause the reaction to take place (Liu et al., 2001). In fact, the research has also shown that the EPS-producing bacteria might intensify the release of soil-bound phenanthrene, consequently, promoting its degradation.

16.4 Soil Restoration and Reclamation

EPS play an indispensable role in the removal of pollution caused by hydrocarbons, pesticides, and heavy metals in terrestrial and aquatic environments. The use of EPS-forming microorganisms for the pollutant immobilization and bioremediation of TPHs and PAHs has been extensively studied. The biofilm has shown the rate of diesel oil degradation up to 97% (Chandran & Das, 2011). Furthermore, EPS are also beneficial in soil stabilization and erosion control through the mechanisms of soil particles binding, gel formation, and runoff reduction. On the one hand, the usage of cyanobacteria for soil adhesion has been applied in the dry areas; on the other hand, EPS from *Rhizobium tropici* iteratively improves the process of soil cohesion and slope stabilization (Gerbersdorf et al., 2008).

17 Future Perspectives and Conclusions

Extracellular polysaccharides have a great potential for a wide variety of applications, including medicine, environmental remediation, sustainable industry, and food technology. Investigative activities are going to be focused on the improving of EPS production by resorting to the modern methods of genetic engineering and synthetic biology. These ways can result in a significant improvement in the yield and cost of production. In the field of medicine, the EPS has been found to be critical in the immune modification as well as in drug delivery and tissue engineering. Further research may be carried out on EPS with immune cells and their effects on certain diseases could actually be substantial especially in the treatment of chronic inflammation and cancer. EPS-based nanocarriers and hydrogels, have been found to be the most effective in the development of drug delivery systems and the regeneration of damaged tissues. The synergy of EPS with other biomaterials or nanotechnologies is promising to find novel medical remedies. Exopolysaccharides are multifaceted biomolecules with vast potential in environmental sustainability, healthcare, and industrial applications. EPS technologies have shown promise in bioremediation, soil stabilization, and pollutant removal. However, large-scale implementation requires further field trials to demonstrate ecological competence. EPS-based technologies have proven effective in soil degradation, water and waste management, and sustainable material development. Biodegradable EPS-based films and coatings offer a game-changing solution to plastic pollution in the packaging industry. However, cost-effective production methods and enhanced mechanical properties are needed for large-scale adoption.

Despite promising perspectives, challenges persist, including high production costs, compositional variations, and regulatory limitations. Interdisciplinary efforts should focus on promoting production technologies, exploring novel applications, and overcoming commercialization hurdles. Integrating EPS production with circular economy practices can enhance sustainability and reduce resource dependence.

Funding Not applicable. The work has not been sponsored by any funding agency.

Declaration The work is original and has not been published elsewhere, nor is it currently under consideration for publication elsewhere.

Conflict of Interest Not applicable. The authors declare that there is "no conflict of interest."

References

Abel, A. M., Yang, C., Thakar, M. S., & Malarkannan, S. (2018). Natural killer cells: Development, maturation, and clinical utilization. *Frontiers in Immunology, 9*, 1869.

Abid, Y., Azabou, S., Blecker, C., Gharsallaoui, A., Corsaro, M. M., Besbes, S., & Attia, H. (2021). Rheological and emulsifying properties of an exopolysaccharide produced by potential probiotic *Leuconostoc citreum*-BMS strain. *Carbohydrate Polymers, 256*, 117523.

Adesulu-Dahunsi, A. T., Sanni, A. I., Jeyaram, K., Ojediran, J. O., Ogunsakin, A. O., & Banwo, K. (2018). Extracellular polysaccharide from *Weissella confusa* OF126: Production, optimization, and characterization. *International Journal of Biological Macromolecules, 111*, 514–525.

Al-Dhaheri, A. S., Al-Hemeiri, R., Kizhakkayil, J., Al-Nabulsi, A., Abushelaibi, A., Shah, N. P., & Ayyash, M. (2017). Health-promoting benefits of low-fat akawi cheese made by exopolysaccharide-producing probiotic *Lactobacillus plantarum* isolated from camel milk. *Journal of Dairy Science, 100*(10), 7771–7779.

Almalki, M. A. (2020). Exopolysaccharide production by a new Lactobacillus lactis isolated from the fermented milk and its antioxidant properties. *Journal of King Saud University – Science, 32*(2), 1272–1277.

Amatayakul, T., Halmos, A. L., Sherkat, F., & Shah, N. P. (2006). Physical characteristics of yoghurts made using exopolysaccharide-producing starter cultures and varying casein to whey protein ratios. *International Dairy Journal, 16*(1), 40–51.

Ayyash, M., Abu-Jdayil, B., Itsaranuwat, P., Galiwango, E., Tamiello-Rosa, C., Abdullah, H., et al. (2020). Characterization, bioactivities, and rheological properties of exopolysaccharide produced by novel probiotic Lactobacillus plantarum C70 isolated from camel milk. *International Journal of Biological Macromolecules, 144*, 938–946.

Bachtarzi, N., Speciale, I., Kharroub, K., De Castro, C., Ruiz, L., & Ruas-Madiedo, P. (2020). Selection of exopolysaccharide-producing Lactobacillus plantarum (*Lactiplantibacillus plantarum*) isolated from algerian fermented foods for the manufacture of skim-milk fermented products. *Microorganisms, 8*(8), 1101.

Badel, S., Bernardi, T., & Michaud, P. (2011). New perspectives for Lactobacilli exopolysaccharides. *Biotechnology Advances, 29*(1), 54–66.

Bakhtiyari, M., Moosavi-Nasab, M., & Askari, H. (2015). Optimization of succinoglycan hydrocolloid production by Agrobacterium radiobacter grown in sugar beet molasses and investigation of its physicochemical characteristics. *Food Hydrocolloids, 45*, 18–29.

Barik, S. (2016). Anti-CD19 chimeric antigen receptor-modified T cells for multiple myeloma. *Journal of Blood Disorders & Transfusion, 7*(4), 342. https://doi.org/10.4172/2155-9864.1000342

Belz, M. C., Axel, C., Arendt, E. K., Lynch, K. M., Brosnan, B., Sheehan, E. M., et al. (2019). Improvement of taste and shelf life of yeasted low-salt bread containing functional sourdoughs using *Lactobacillus amylovorus* DSM 19280 and *Weisella cibaria* MG1. *International Journal of Food Microbiology, 302*, 69–79.

Bengoa, A. A., Llamas, M. G., Iraporda, C., Dueñas, M. T., Abraham, A. G., & Garrote, G. L. (2018). Impact of growth temperature on exopolysaccharide production and probiotic properties of *Lactobacillus paracasei* strains isolated from kefir grains. *Food Microbiology, 69*, 212–218.

Berrington, J. E., Barge, D., Fenton, A. C., Cant, A. J., & Spickett, G. P. (2005). Lymphocyte subsets in term and significantly preterm UK infants in the first year of life analysed by single platform flow cytometry. *Clinical & Experimental Immunology, 140*(2), 289–292.

Bertocchi, C., Navarini, L., Cesàro, A., & Anastasio, M. (1990). Polysaccharides from cyanobacteria. *Carbohydrate Polymers, 12*(2), 127–153.

Bertrand, C., Raposo, M. F., Morais, R. M., & Morais, M. M. (2016). Effects of different edible coatings on polyphenol oxidase activity and colour of fresh-cut apple during cold storage. *International Journal of Postharvest Technology and Innovation, 5*(2), 91–104.

Bhattacharjee, M., Banerjee, M., Mitra, P., & Ganguly, A. (2018). Investigation of alkaline phosphatase enzyme of a novel *Bacillus* species isolated from rhizospheric soil of potato field. *Research Journal of Life Sciences, Bioinformatics, Pharmaceutical and Chemical Sciences, 4*(2), 129–144.

Blum, J. S., Wearsch, P. A., & Cresswell, P. (2013). Pathways of antigen processing. *Annual Review of Immunology, 31*(1), 443–473.

Bramhachari, P. V., & Dubey, S. K. (2006). Isolation and characterization of exopolysaccharide produced by Vibrio harveyi strain VB23. *Letters in Applied Microbiology, 43*(5), 571–577.

Buffle, J., Greter, F. L., & Haerdi, W. (1977). Measurement of complexation properties of humic and fulvic acids in natural waters with lead and copper ion-selective electrodes. *Analytical Chemistry, 49*(2), 216–222.

Burdon, J. (2001). Are the traditional concepts of the structures of humic substances realistic? *Soil Science, 166*(11), 752–769.

Buthelezi, S. P., Olaniran, A. O., & Pillay, B. (2012). Textile dye removal from wastewater effluents using bioflocculants produced by indigenous bacterial isolates. *Molecules, 17*(12), 14260–14274. https://doi.org/10.3390/molecules171214260

Calder, P. C. (2013). Feeding the immune system. *Proceedings of the Nutrition Society, 72*(3), 299–309.

Carrión, O., Delgado, L., & Mercade, E. (2015). New emulsifying and cryoprotective exopolysaccharide from Antarctic Pseudomonas sp.ID1. *Carbohydrate Polymers, 117*, 1028–1034.

Cartaşev, A., & Rudik, V. (2017). The effect of starter culture producing exopolysaccharide on physicochemical properties of yoghurt. *Chemistry Journal of Moldova, 12*(2), 7–12.

Castro-Bravo, N., Wells, J. M., Margolles, A., & Ruas-Madiedo, P. (2018). Interactions of surface exopolysaccharides from Bifidobacterium and Lactobacillus within the intestinal environment. *Frontiers in Microbiology, 9*, 2426.

Chaisuwan, W., Jantanasakulwong, K., Wangtueai, S., Phimolsiripol, Y., Chaiyaso, T., Techapun, C., et al. (2020). Microbial exopolysaccharides for immune enhancement: Fermentation, modifications and bioactivities. *Food Bioscience, 35*, 100564.

Chandran, P., & Das, N. (2011). Degradation of diesel oil by immobilized *Candida tropicalis* and biofilm formed on gravels. *Biodegradation, 22*(6), 1181–1189. https://doi.org/10.1007/s10532-011-9473-1

Chattaraj, S., Ganguly, A., Mandal, A., & Das Mohapatra, P. K. (2022). A review of the role of probiotics for the control of viral diseases in aquaculture. *Aquaculture International, 30*(5), 2513–2539. https://doi.org/10.1007/s10499-022-00915-6

Chattaraj, S., Ganguly, A., Mitra, D., Mitra, D., & Das Mohapatra, P. K. (2023a). Study of intestinal bacteria of *Cirrhinus reba* and characterization of a new probiotic bacteria: An initiative to save the threatened species of *Cirrhinus*. *Applied Biochemistry and Microbiology, 59*(6), 833–849. https://doi.org/10.1134/S0003683823060248

Chattaraj, S., Mitra, D., Chattaraj, A., Chattaraj, M., Kundu, M., Ganguly, A., & Mohapatra, P. K. D. (2023b). Antioncogenic potential of probiotics: Challenges and future prospective. *Indian Journal of Microbiology Research, 10*(1), 1–10. https://doi.org/10.18231/j.ijmr.2023.001

Chattaraj, S., Mitra, D., Ganguly, A., Thatoi, H., & Mohapatra, P. K. D. (2024a). A critical review on the biotechnological potential of Brewers' waste: Challenges and future alternatives. *Current Research in Microbial Sciences, 6*, 100228. https://doi.org/10.1016/j.crmicr.2024.100228

Chattaraj, S., Mitra, D., Chattaraj, M., Ganguly, A., Thatoi, H., & Das Mohapatra, P. K. (2024b). Brewers' spent grain as fish feed ingredient: Evaluation of bio-safety and analysis of its impact on gut bacteria of *Cirrhinus reba* by 16S metagenomic sequencing. *Current Research in Microbial Sciences, 7*, 100286. https://doi.org/10.1016/j.crmicr.2024.100286

Chattaraj, S., Das, S., Ganguly, A., Thatoi, H., & Das Mohapatra, P. K. (2024c). Valorization of wastewater by the application of probiotic, nitrifying, and denitrifying bacterial consortium. In *Application of microbial technology in wastewater treatment and bioenergy recovery* (pp. 133–156). Springer Nature. https://doi.org/10.1007/978-981-97-3458-0_6

Chen, Y., Mao, W. J., Yan, M. X., Liu, X., Wang, S. Y., Xia, Z., et al. (2016). Purification, chemical characterization, and bioactivity of an extracellular polysaccharide produced by the marine sponge endogenous fungus *Alternaria* sp. SP-32. *Marine Biotechnology, 18*, 301–313.

Chen, Q., Qi, C., Peng, G., Liu, Y., Zhang, X., & Meng, Z. (2018). Immune-enhancing effects of a polysaccharide PRG1-1 from Russula griseocarnosa on RAW264.7 macrophage cells via the MAPK and NF-κB signalling pathways. *Food and Agricultural Immunology, 29*(1), 833–844.

Cheng, K. C., Demirci, A., & Catchmark, J. M. (2011). Pullulan: Biosynthesis, production, and applications. *Applied Microbiology and Biotechnology, 92*(1), 29–44. https://doi.org/10.1007/s00253-011-3477-y

Chug, R., Mathur, S., Kothari, S. L., & Gour, V. S. (2021). Maximizing EPS production from Pseudomonas aeruginosa and its application in Cr and Ni sequestration. *Biochemistry and Biophysics Reports, 26*, 100972.

Ciszek-Lenda, M. (2011). Review paper Biological functions of exopolysaccharides from probiotic bacteria. *Central European Journal of Immunology, 36*(1), 51–55.

Clapham, D. E. (2007). Calcium signaling. *Cell, 131*(6), 1047–1058.

Cooper, M. A., Fehniger, T. A., & Caligiuri, M. A. (2001). The biology of human natural killer-cell subsets. *Trends in Immunology, 22*(11), 633–640.

Czaczyk, K., & Myszka, K. (2007). Biosynthesis of extracellular polymeric substances (EPS) and its role in microbial biofilm formation. *Polish Journal of Environmental Studies, 16*(6), 799–806.

Davidović, S., Miljković, M., Tomić, M., Gordić, M., Nešić, A., & Dimitrijević, S. (2018). Response surface methodology for optimisation of edible coatings based on dextran from Leuconostoc mesenteroides T3. *Carbohydrate Polymers, 184*, 207–213.

Davies, G., & Ghabbour, E. A. (2001). Humic acids: Marvelous products of soil chemistry. *Journal of Chemical Education, 78*(12), 1609–1614. https://doi.org/10.1021/ed078p1609

de Saint Basile, G., Ménasché, G., & Fischer, A. (2010). Molecular mechanisms of biogenesis and exocytosis of cytotoxic granules. *Nature Reviews Immunology, 10*(8), 568–579.

Decho, A. W. (1990). Microbial exopolymer secretions in ocean environments: Their role(s) in food webs and marine processes. *Oceanography and Marine Biology: An Annual Review, 28*(7), 73–153.

Delves, P. J., & Roitt, I. M. (2000). The immune system. *New England Journal of Medicine, 343*(1), 37–49.

Dertli, E., Colquhoun, I. J., Côté, G. L., Le Gall, G., & Narbad, A. (2018). Structural analysis of the α-D-glucan produced by the sourdough isolate Lactobacillus brevis E25. *Food Chemistry, 242*, 45–52.

Devine, S. M., & Hoffman, R. (2000). Role of mesenchymal stem cells in hematopoietic stem cell transplantation. *Current Opinion in Hematology, 7*(6), 358–363.

Dhillon, A. S., Hagan, S., Rath, O., & Kolch, W. (2007). MAP kinase signalling pathways in cancer. *Oncogene, 26*(22), 3279–3290.

Dincer, E., & Kivanc, M. (2019). Characterization of Lactobacillus plantarum strains isolated from Turkish pastırma and possibility to use of food industry. *Food Science and Technology, 40*, 498–507.

Donot, F., Fontana, A., Baccou, J. C., & Schorr-Galindo, S. (2012). Microbial exopolysaccharides: Main examples of synthesis, excretion, genetics and extraction. *Carbohydrate Polymers, 87*(2), 951–962.

Duncan, T. V. (2011). Applications of nanotechnology in food packaging and food safety: Barrier materials, antimicrobials and sensors. *Journal of Colloid and Interface Science, 363*(1), 1–24.

Emtiazia, G., Ethemadifara, Z., & Habibib, M. H. (2004). Production of extra-cellular polymer in Azotobacter and biosorption of metal by exopolymer. *African Journal of Biotechnology, 3*(6), 330–333.

Ferrario, C., Milani, C., Mancabelli, L., Lugli, G. A., Duranti, S., Mangifesta, M., et al. (2016). Modulation of the eps-ome transcription of bifidobacteria through simulation of human intestinal environment. *FEMS Microbiology Ecology, 92*(4), fiw056.

Feske, S. (2007). Calcium signalling in lymphocyte activation and disease. *Nature Reviews Immunology, 7*(9), 690–702.

Flemming, H. C., & Wingender, J. (2001). Relevance of microbial extracellular polymeric substances (EPSs)-Part I: Structural and ecological aspects. *Water Science and Technology, 43*(6), 1–8.

Flemming, H., & Wingender, J. (2010). The biofilm matrix. *Nature Reviews Microbiology, 8*, 623–633.

Fooladi, T., Soudi, M. R., Alimadadi, N., Savedoroudi, P., & Heravi, M. M. (2019). Bioactive exopolysaccharide from Neopestalotiopsis sp. strain SKE15: Production, characterization and optimization. *International Journal of Biological Macromolecules, 129*, 127–139.

Forde, A., & Fitzgerald, G. F. (1999). Bacteriophage defence systems in lactic acid bacteria. *Antonie Van Leeuwenhoek, 76*, 89–113.

Ganguly, A., & Mandal, A. (2024). Production and optimization of xylanase enzyme from *Bacillus cereus* BSA1 by submerged fermentation. *International Research Journal of Basic and Applied Science, 9*(1), 13–20.

Ganguly, A., Banerjee, A., Mandal, A., & Das Mohapatra, P. K. (2018a). Optimization of effective dose of a newly isolated probiotic bacteria for growth and disease resistance of *Carias batrachus* (Linn.). *Research Journal of Life Sciences, Bioinformatics, Pharmaceutical and Chemical Sciences, 4*(4), 344–354.

Ganguly, A., Banerjee, A., Mandal, A., & Das Mohapatra, P. K. (2018b). Probiotic-based cultivation of *Clarias batrachus*: Importance and future perspective. *Acta Biologica Szegediensis, 62*(2), 158–168. https://doi.org/10.14232/abs.2018.2.158-168

Ganguly, A., Banerjee, A., Mandal, A., & Das Mohapatra, P. K. (2018c). Study of indigenous freshwater fish diversity of Bankura (West Bengal), India with special reference to *Clarias batrachus*. *Journal of Applied and Natural Science, 10*, 1162–1172. https://doi.org/10.31018/jans.v10i4.1892

Ganguly, A., Banerjee, A., Mandal, A., Khan, M. A., & Das Mohapatra, P. K. (2019). Isolation and characterization of bacteria from the intestine of Clarias batrachus for probiotic organism. *Proceedings of the Zoological Society, 72*(4), 411–419. https://doi.org/10.1007/s12595-018-0283-x

Ganguly, A., Chattaraj, S., Ganguly, M., Chattaraj, M., Banerjee, A., Mandal, A., Ali Khan, M., Sen, S. K., & Das Mohapatra, P. K. (2024). Effect of three probiotic *Bacillus* strains supplemented feeds on growth, carcass composition and blood parameters of *Clarias magur* (Hamilton). *Journal of Applied Aquaculture, 37*, 19–44. https://doi.org/10.1080/10454438.2024.2330499

Gerbersdorf, S. U., Jancke, T., Westrich, B., & Paterson, D. M. (2008). Microbial stabilization of riverine sediments by extracellular polymeric substances. *FEMS Microbiology Ecology, 6*(1), 57–69. https://doi.org/10.1111/j.1472-4669.2007.00120.x

Giorgi, A., & Cochrane, Z. (2019). About immunosuppressant drugs. *Healthline*. https://www.healthline.com/health/immunosuppressant-drugs. Last viewed: 01.07.2021.

Gordon, S. (2007). The macrophage: Past, present and future. *European Journal of Immunology, 37*(S1), S9–S17.

Gordon, S., & Martinez-Pomares, L. (2017). Physiological roles of macrophages. *Pflügers Archiv: European Journal of Physiology, 469*(3), 365–374.

Grumezescu, A. M., & Holban, A. M. (Eds.). (2018). *Biopolymers for food design* (Vol. 20). Academic Press.

Hashemi, S. M. B., & Gholamhosseinpour, A. (2020). Effect of ultrasonication treatment and fermentation by probiotic Lactobacillus plantarum strains on goat milk bioactivities. *International Journal of Food Science & Technology, 55*(6), 2642–2649.

He, P., Geng, L., Wang, J., Wang, Z., Mao, D., & Xu, C. (2012). Purification, characterization and bioactivity of an extracellular polysaccharide produced from Phellinus igniarius. *Annals of Microbiology, 62*, 1697–1707.

Hidalgo-Cantabrana, C., López, P., Gueimonde, M., de Los Reyes-Gavilán, C. G., Suárez, A., Margolles, A., & Ruas-Madiedo, P. (2012). Immune modulation capability of exopolysaccharides synthesised by lactic acid bacteria and bifidobacteria. *Probiotics and Antimicrobial Proteins, 4*, 227–237.

Hidalgo-Cantabrana, C., Sánchez, B., Milani, C., Ventura, M., Margolles, A., & Ruas-Madiedo, P. (2014). Genomic overview and biological functions of exopolysaccharide biosynthesis in Bifidobacterium spp. *Applied and Environmental Microbiology, 80*(1), 9–18.

Hilbig, J., Loeffler, M., Herrmann, K., & Weiss, J. (2019). Application of exopolysaccharide-forming lactic acid bacteria in cooked ham model systems. *Food Research International, 119*, 761–768.

Hitchen, P. G., & Dell, A. (2006). Bacterial glycoproteomics. *Microbiology, 152*(6), 1575–1580. https://doi.org/10.1099/mic.0.28650-0

Hsu, H. Y., Hua, K. F., Lin, C. C., Lin, C. H., Hsu, J., & Wong, C. H. (2004). Extract of reishi polysaccharides induces cytokine expression via TLR4-modulated protein kinase signaling pathways. *Journal of Immunology, 173*(10), 5989–5999. https://doi.org/10.4049/jimmunol.173.10.5989

Huang, J. Y., Kao, C. Y., Liu, W. S., & Fang, T. J. (2017). Characterization of high exopolysaccharide-producing *Lactobacillus* strains isolated from mustard pickles for potential probiotic applications. *International Microbiology, 20*(2), 75–84.

Hug, I., & Feldman, M. F. (2011). Analogies and homologies in lipopolysaccharide and glycoprotein biosynthesis in bacteria. *Glycobiology, 21*(2), 138–151.

Huyan, T., Li, Q., Yang, H., Jin, M. L., Zhang, M. J., Ye, L. J., et al. (2014). Protective effect of polysaccharides on simulated microgravity-induced functional inhibition of human NK cells. *Carbohydrate Polymers, 101*, 819–827.

Jenny Angel, S., Vidyadharani, G., Santhosh, S., & Dhandapani, R. (2018). Optimization and characterisation of thermo stable exopolysaccharide produced from Bacillus licheniformis WSF-1 strain. *Journal of Polymers and the Environment, 26*, 3824–3833.

Jia, C., Li, P., Li, X., Tai, P., Liu, W., & Gong, Z. (2011). Degradation of pyrene in soils by extracellular polymeric substances (EPS) extracted from liquid cultures. *Process Biochemistry, 46*(8), 1627–1631. https://doi.org/10.1016/j.procbio.2011.05.005

Kavita, K., Singh, V. K., Mishra, A., & Jha, B. (2014). Characterisation and anti-biofilm activity of extracellular polymeric substances from *Oceanobacillus iheyensis*. *Carbohydrate Polymers, 101*, 29–35.

Kavitake, D., Balyan, S., Devi, P. B., & Shetty, P. H. (2020). Evaluation of oil-in-water (O/W) emulsifying properties of galactan exopolysaccharide from *Weissella confusa* KR780676. *Journal of Food Science and Technology, 57*, 1579–1585.

Kawai, T., & Akira, S. (2007). Signaling to NF-κB by Toll-like receptors. *Trends in Molecular Medicine, 13*(11), 460–469.

Kim, E. K., & Choi, E. J. (2010). Pathological roles of MAPK signaling pathways in human diseases. *Biochimica et Biophysica Acta (BBA) – Molecular Basis of Disease, 1802*(4), 396–405.

Kim, J. K., Cho, M. L., Karnjanapratum, S., Shin, I. S., & You, S. G. (2011). In vitro and in vivo immunomodulatory activity of sulfated polysaccharides from Enteromorpha prolifera. *International Journal of Biological Macromolecules, 49*(5), 1051–1058.

Lewis, D. E., & Blutt, S. E. (2019). Organization of the immune system. In *Clinical immunology* (pp. 19–38). Elsevier.

Li, W. J., Tang, X. F., Shuai, X. X., Jiang, C. J., Liu, X., Wang, L. F., et al. (2017a). Mannose receptor mediates the immune response to Ganoderma atrum polysaccharides in macrophages. *Journal of Agricultural and Food Chemistry, 65*(2), 348–357.

Li, C., Zhou, L., Yang, H., Lv, R., Tian, P., Li, X., et al. (2017b). Self-assembled exopolysaccharide nanoparticles for bioremediation and green synthesis of noble metal nanoparticles. *ACS Applied Materials & Interfaces, 9*(27), 22808–22818.

Li, C., Ding, J., Chen, D., Shi, Z., & Wang, L. (2020). Bioconversion of cheese whey into a hetero-exopolysaccharide via a one-step bioprocess and its applications. *Biochemical Engineering Journal, 161*, 107701.

Liu, A., Ahn, I. S., Mansfield, C., Lion, L. W., Shuler, M. L., & Ghiorse, W. C. (2001). Phenanthrene desorption from soil in the presence of bacterial extracellular polymer: Observations and model predictions of dynamic behavior. *Water Research, 35*(3), 835–843. https://doi.org/10.1016/S0043-1354(00)00324-9

Liu, T., Zhang, L., Joo, D., & Sun, S. C. (2017). NF-κB signaling in inflammation. *Signal Transduction and Targeted Therapy, 2*(1), 1–9.

Lynch, K. M., Coffey, A., & Arendt, E. K. (2018). Exopolysaccharide producing lactic acid bacteria: Their techno-functional role and potential application in gluten-free bread products. *Food Research International, 110*, 52–61.

Madhubasani, G. B. L., Prasanna, P. H. P., Chandrasekara, A., Gunasekara, D. C. S., Senadeera, P., Chandramali, D. V. P., & Vidanarachchi, J. K. (2020). Exopolysaccharide producing starter cultures positively influence on microbiological, physicochemical, and sensory properties of probiotic goats' milk set-yoghurt. *Journal of Food Processing and Preservation, 44*(3), e14361.

Mahgoub, S. A., Qattan, S. Y. A., Salem, S. S., Abdelbasit, H. M., Raafat, M., Ashkan, M. F., Al-Quwaie, D. A., Motwali, E. A., Alqahtani, F. S., & Abd El-Fattah, H. I. (2023). Characterization and biodegradation of phenol by *Pseudomonas aeruginosa* and *Klebsiella variicola* strains isolated from sewage sludge and their effect on soybean seeds germination. *Molecules, 28*(3), 1203. https://doi.org/10.3390/molecules28031203

Marvasi, M., Visscher, P. T., & Casillas Martinez, L. (2010). Exopolymeric substances (EPS) from Bacillus subtilis: Polymers and genes encoding their synthesis. *FEMS Microbiology Letters, 313*(1), 1–9.

McComb, S., Thiriot, A., Akache, B., Krishnan, L., & Stark, F. (2019). Introduction to the immune system. In *Immunoproteomics: Methods and protocols* (pp. 1–24). Springer.

McCubrey, J. A., LaHair, M. M., & Franklin, R. A. (2006). Reactive oxygen species-induced activation of the MAP kinase signaling pathways. *Antioxidants & Redox Signaling, 8*(9–10), 1775–1789.

Meng, X., Pijning, T., Dobruchowska, J. M., Yin, H., Gerwig, G. J., & Dijkhuizen, L. (2016). Structural determinants of alternating (α1→ 4) and (α1→ 6) linkage specificity in reuteransucrase of Lactobacillus reuteri. *Scientific Reports, 6*(1), 35261.

Milanović, V., Osimani, A., Garofalo, C., Belleggia, L., Maoloni, A., Cardinali, F., et al. (2020). Selection of cereal-sourced lactic acid bacteria as candidate starters for the baking industry. *PLoS One, 15*(7), e0236190.

Moura, M. N., Martín, M. J., & Burguillo, F. J. (2007). A comparative study of the adsorption of humic acid, fulvic acid and phenol onto *Bacillus subtilis* and activated sludge. *Journal of Hazardous Materials, 149*(1), 42–48.

Nad, S., Konar, U., Chattaraj, S., & Ganguly, A. (2024). Valorization of feather waste by microbial enzymatic activity: Bioconversion, production, and application. In *Agro-waste to microbe assisted value added product: Challenges and future prospects* (pp. 337–363). Springer Nature. https://doi.org/10.1007/978-3-031-58025-3_15

Nambiar, R. B., Sellamuthu, P. S., Perumal, A. B., Sadiku, E. R., Phiri, G., & Jayaramudu, J. (2018). Characterization of an exopolysaccharide produced by Lactobacillus plantarum HM47 isolated from human breast milk. *Process Biochemistry, 73*, 15–22.

Niknezhad, S. V., Najafpour-Darzi, G., Morowvat, M. H., & Ghasemi, Y. (2018). Eexopolysaccharide production of Pantoea sp. BCCS 001 GH: Physical characterizations, emulsification, and antioxidant activities. *International Journal of Biological Macromolecules, 118*, 1103–1111.

Noghabi, K. A., Zahiri, H. S., & Yoon, S. C. (2007). The production of a cold-induced extracellular biopolymer by Pseudomonas fluorescens BM07 under various growth conditions and its role in heavy metals absorption. *Process Biochemistry, 42*(5), 847–855.

O'Donnell, M., Mente, A., & Yusuf, S. (2015). Sodium intake and cardiovascular health. *Circulation Research, 116*(6), 1046–1057.

Orakpoghenor, O., Avazi, D. O., Markus, T. P., & Olaolu, O. S. (2019). Lymphocytes: A brief review. *SciRes Literature, 3*(1), 5–8.

Pal, A., & Paul, A. (2008). Microbial extracellular polymeric substances: Central elements in heavy metal bioremediation. *Indian Journal of Microbiology, 48*, 49–64.

Park, C., & Novak, J. T. (2009). Characterization of lectins and bacterial adhesins in activated sludge flocs. *Water Environment Research, 81*(8), 755–764.

Park, Y. M., Lee, H. Y., Kang, Y. G., Park, S. H., Lee, B. G., Park, Y. J., et al. (2019). Immune-enhancing effects of Portulaca oleracea L.–based complex extract in cyclophosphamide-induced splenocytes and immunosuppressed rats. *Food and Agricultural Immunology, 30*(1), 13–24.

Peña-Méndez, E. M., Havel, J., & Patočka, J. (2005). Humic substances-compounds of still unknown structure: Applications in agriculture, industry, environment, and biomedicine. *Journal of Applied Biomedicine, 3*(1), 13–24.

Poole, J., Day, C. J., von Itzstein, M., Paton, J. C., & Jennings, M. P. (2018). Glycointeractions in bacterial pathogenesis. *Nature Reviews Microbiology, 16*(1), 440–452. https://doi.org/10.1038/s41579-018-0007-2

Qi, M., & Elion, E. A. (2005). MAP kinase pathways. *Journal of Cell Science, 118*(16), 3569–3572.

Rehm, B. H. (2010). Bacterial polymers: Biosynthesis, modifications and applications. *Nature Reviews Microbiology, 8*(8), 578–592.

Ren, L., Zhang, J., & Zhang, T. (2021). Immunomodulatory activities of polysaccharides from ganoderma on immune effector cells. *Food Chemistry, 340*, 127933. https://doi.org/10.1016/j.foodchem.2020.127933

Roy, S., & Loh, H. H. (1996). Effects of opioids on the immune system. *Neurochemical Research, 21*, 1375–1386.

Rzepkowska, A., Zielińska, D., Ołdak, A., & Kołożyn-Krajewska, D. (2017). Organic whey as a source of Lactobacillus strains with selected technological and antimicrobial properties. *International Journal of Food Science & Technology, 52*(9), 1983–1994.

Sacco, L. P., Castellane, T. C. L., Polachini, T. C., de Macedo Lemos, E. G., & Alves, L. M. C. (2019). Exopolysaccharides produced by Pandoraea shows emulsifying and anti-biofilm activities. *Journal of Polymer Research, 26*, 1–11.

Saleem, M., Malik, S., Mehwish, H. M., Ali, M. W., Hussain, N., Khurshid, M., et al. (2021). Isolation and functional characterization of exopolysaccharide produced by *Lactobacillus plantarum* S123 isolated from traditional Chinese cheese. *Archives of Microbiology, 203*, 3061–3070.

Samantaray, A., Chattaraj, S., Mitra, D., Ganguly, A., Kumar, R., Gaur, A., Das Mohapatra, P. K., Santos-Villalobos, S. L., Rani, A., & Thatoi, H. (2024). Advances in microbial based bio-inoculum for amelioration of soil health and sustainable crop production. *Current Research in Microbial Sciences, 7*, 100251. https://doi.org/10.1016/j.crmicr.2024.100251

Sathiyanarayanan, G., Dineshkumar, K., & Yang, Y. H. (2017). Microbial exopolysaccharide-mediated synthesis and stabilization of metal nanoparticles. *Critical Reviews in Microbiology, 43*(6), 731–752.

Schaeffer, H. J., & Weber, M. J. (1999). Mitogen-activated protein kinases: Specific messages from ubiquitous messengers. *Molecular and Cellular Biology, 19*, 2435.

Schmid, J., Sieber, V., & Rehm, B. (2015). Bacterial exopolysaccharides: Biosynthesis pathways and engineering strategies. *Frontiers in Microbiology, 6*, 496.

Seesuriyachan, P., Kuntiya, A., Hanmoungjai, P., Techapun, C., Chaiyaso, T., & Leksawasdi, N. (2012). Optimization of exopolysaccharide overproduction by Lactobacillus confusus in solid state fermentation under high salinity stress. *Bioscience, Biotechnology, and Biochemistry, 76*(5), 912–917.

Seo, E. J., Yoo, S. H., Oh, K. W., Cha, J., Lee, H. G., & Park, C. S. (2004). Isolation of an exopolysaccharide-producing bacterium, Sphingomonas sp. CS101, which forms an unusual type of sphingan. *Bioscience, Biotechnology, and Biochemistry, 68*(5), 1146–1148.

Shankar, T., Palpperumal, S., Kathiresan, D., Sankaralingam, S., Balachandran, C., Baskar, K., et al. (2021). Biomedical and therapeutic potential of exopolysaccharides by Lactobacillus paracasei isolated from sauerkraut: Screening and characterization. *Saudi Journal of Biological Sciences, 28*(5), 2943–2950.

Sharma, S., & Rao, T. R. (2015). Xanthan gum based edible coating enriched with cinnamic acid prevents browning and extends the shelf-life of fresh-cut pears. *LWT-Food Science and Technology, 62*(1), 791–800.

Sheng, G. P., Yu, H. Q., & Li, X. Y. (2010). Extracellular polymeric substances (EPS) of microbial aggregates in biological wastewater treatment systems: A review. *Biotechnology Advances, 28*(6), 882–894.

Shi, N., Li, N., Duan, X., & Niu, H. (2017). Interaction between the gut microbiome and mucosal immune system. *Military Medical Research, 4*, 1–7.

Shih, L., & Van, Y. T. (2001). The production of poly-(γ-glutamic acid) from microorganisms and its various applications. *Bioresource Technology, 79*(3), 207–225.

Smyth, M. J., Hayakawa, Y., Takeda, K., & Yagita, H. (2002). New aspects of natural-killer-cell surveillance and therapy of cancer. *Nature Reviews Cancer, 2*(11), 850–861.

Soberón-Chávez, G., Lépine, F., & Déziel, E. (2005). Production of rhamnolipids by Pseudomonas aeruginosa. *Applied Microbiology and Biotechnology, 68*, 718–725.

Su, C. A., Xu, X. Y., Liu, D. Y., Wu, M., Zeng, F. Q., Zeng, M. Y., et al. (2013). Isolation and characterization of exopolysaccharide with immunomodulatory activity from fermentation broth of *Morchella conica. DARU Journal of Pharmaceutical Sciences, 21*, 1–6.

Sun, S. C. (2011). Non-canonical NF-κB signaling pathway. *Cell Research, 21*(1), 71–85.

Surayot, U., & You, S. (2017). Structural effects of sulfated polysaccharides from Codium fragile on NK cell activation and cytotoxicity. *International Journal of Biological Macromolecules, 98*, 117–124.

Surayot, U., Wang, J., Lee, J. H., Kanongnuch, C., Peerapornpisal, Y., & You, S. (2015). Characterization and immunomodulatory activities of polysaccharides from Spirogyra neglecta (Hassall) Kützing. *Bioscience, Biotechnology, and Biochemistry, 79*(10), 1644–1653.

Sutherland, I. W. (1979). Microbial exopolysaccharides. Control of synthesis and acylation. In *Microbial polysaccharides and polysaccharases*. Academic Press.

Tallon, R., Bressollier, P., & Urdaci, M. C. (2003). Isolation and characterization of two exopolysaccharides produced by Lactobacillus plantarum EP56. *Research in Microbiology, 154*(10), 705–712.

Tidona, F., Francolino, S., Zhang, H., Contarini, G., Cui, S. W., Giraffa, G., & Carminati, D. (2016). Design of a starter culture to produce a reduced-fat soft cheese with added bio-value. *Journal of Food and Nutrition Research, 55*(1), 33–39.

Tiwari, O. N., Sasmal, S., Kataria, A. K., & Devi, I. (2020). Application of microbial extracellular carbohydrate polymeric substances in food and allied industries. *3 Biotech, 10*, 1–17.

Upreti, R. K., Kumar, M., & Shankar, V. (2003). Bacterial glycoproteins: Functions, biosynthesis and applications. *Proteomics, 3*(4), 363–379.

Valerio, F., Bavaro, A. R., Di Biase, M., Lonigro, S. L., Logrieco, A. F., & Lavermicocca, P. (2020). Effect of amaranth and quinoa flours on exopolysaccharide production and protein profile of liquid sourdough fermented by Weissella cibaria and Lactobacillus plantarum. *Frontiers in Microbiology, 11*, 967.

Vallabhapurapu, S., & Karin, M. (2009). Regulation and function of NF-κB transcription factors in the immune system. *Annual Review of Immunology, 27*(1), 693–733.

Venkatesh, P., Balraj, M., Ayyanna, R., Ankaiah, D., & Arul, V. (2016). Physicochemical and biosorption properties of novel exopolysaccharide produced by Enterococcus faecalis. *LWT-Food Science and Technology, 68*, 606–614.

Vitale, M., Chiesa, M. D., Carlomagno, S., Pende, D., Arico, M., Moretta, L., & Moretta, A. (2005). NK-dependent DC maturation is mediated by TNFα and IFNγ released upon engagement of the NKp30 triggering receptor. *Blood, 106*(2), 566–571.

Vivier, E., Nunès, J. A., & Vély, F. (2004). Natural killer cell signaling pathways. *Science, 306*(5701), 1517–1519.

Vu, B., Chen, M., Crawford, R. J., & Ivanova, E. P. (2009). Bacterial extracellular polysaccharides involved in biofilm formation. *Molecules, 14*(7), 2535–2554.

Waldmann, T. A., & Tagaya, Y. (1999). The multifaceted regulation of interleukin-15 expression and the role of this cytokine in NK cell differentiation and host response to intracellular pathogens. *Annual Review of Immunology, 17*(1), 19–49.

Wang, J., Wu, T., Fang, X., Min, W., & Yang, Z. (2018). Characterization and immunomodulatory activity of an exopolysaccharide produced by *Lactobacillus plantarum* JLK0142 isolated from fermented dairy tofu. *International Journal of Biological Macromolecules, 115*, 985–993.

Wang, J., Wu, T., Fang, X., & Yang, Z. (2019a). Manufacture of low-fat Cheddar cheese by exopolysaccharide-producing *Lactobacillus plantarum* JLK0142 and its functional properties. *Journal of Dairy Science, 102*(5), 3825–3838.

Wang, B., Song, Q., Zhao, F., Han, Y., & Zhou, Z. (2019b). Production optimization, partial characterization and properties of an exopolysaccharide from *Lactobacillus sakei* L3. *International Journal of Biological Macromolecules, 141*, 21–28.

Weiss, N. (2014). *Meeting the needs of student users in academic libraries: Reaching across the great divide*. By Michael J. Crump and LeiLani S. Freund. Chandos Publishing, 2012. xviii + 204 pages. ISBN 9781843346845.

Welman, A. D., & Maddox, I. S. (2003). Exopolysaccharides from lactic acid bacteria: Perspectives and challenges. *Trends in Biotechnology, 21*(6), 269–274.

Wingender, J. (1999). What are bacterial extracellular polymeric substances. In *Microbial extracellular polymeric substances: Characterization, structure, and function* (pp. 1–20). Springer.

Wingender, J., Neu, T. R., & Flemming, H. C. (1999). What are bacterial extracellular polymeric substances? In J. Wingender, T. R. Neu, & H. C. Flemming (Eds.), *Microbial extracellular polymeric substances*. Springer. https://doi.org/10.1007/978-3-642-60147-7_1

Wolter, A., Hager, A. S., Zannini, E., Galle, S., Gänzle, M. G., Waters, D. M., & Arendt, E. K. (2014). Evaluation of exopolysaccharide producing Weissella cibaria MG1 strain for the production of sourdough from various flours. *Food Microbiology, 37*, 44–50.

Xie, Y., Wang, L., Sun, H., Shang, Q., Wang, Y., Zhang, G., et al. (2020). A polysaccharide extracted from alfalfa activates splenic B cells by TLR4 and acts primarily via the MAPK/p38 pathway. *Food & Function, 11*(10), 9035–9047.

Xu, D., Hu, Y., Wu, F., Jin, Y., Xu, X., & Gänzle, M. G. (2020). Comparison of the functionality of exopolysaccharides produced by sourdough lactic acid bacteria in bread and steamed bread. *Journal of Agricultural and Food Chemistry, 68*(33), 8907–8914.

Yang, Y., Wikieł, A. J., Dall'Agnol, L. T., Eloy, P., Genet, M. J., Moura, J. J., et al. (2016). Proteins dominate in the surface layers formed on materials exposed to extracellular polymeric substances from bacterial cultures. *Biofouling, 32*(1), 95–108.

Yasar Yildiz, S., Anzelmo, G., Ozer, T., Radchenkova, N., Genç, S., Di Donato, P., et al. (2014). Brevibacillus themoruber: A promising microbial cell factory for exopolysaccharide production. *Journal of Applied Microbiology, 116*(2), 314–324.

Ying, Y., & Hao, W. (2023). Immunomodulatory function and anti-tumor mechanism of natural polysaccharides: A review. *Frontiers in Immunology, 14*, 1147641.

Yu, Q., Nie, S. P., Li, W. J., Zheng, W. Y., Yin, P. F., Gong, D. M., & Xie, M. Y. (2013). Macrophage immunomodulatory activity of a purified polysaccharide isolated from Ganoderma atrum. *Phytotherapy Research, 27*(2), 186–191.

Yu, Q., Nie, S. P., Wang, J. Q., Huang, D. F., Li, W. J., & Xie, M. Y. (2015). Signaling pathway involved in the immunomodulatory effect of Ganoderma atrum polysaccharide in spleen lymphocytes. *Journal of Agricultural and Food Chemistry, 63*(10), 2734–2740.

Yu, Y. J., Chen, Z., Chen, P. T., & Ng, I. S. (2018). Production, characterization and antibacterial activity of exopolysaccharide from a newly isolated *Weissella cibaria* under sucrose effect. *Journal of Bioscience and Bioengineering, 126*(6), 769–777.

Zhang, Z., Xia, S., Wang, X., Yang, A., Xu, B., Chen, L., Zhu, Z., Zhao, J., Jaffrezic-Renault, N., & Leonard, D. (2009). A novel biosorbent for dye removal: Extracellular polymeric substance (EPS) of *Proteus mirabilis* TJ-1. *Journal of Hazardous Materials, 163*(1), 279–284. https://doi.org/10.1016/j.jhazmat.2008.06.096

Zhang, Y., Wang, F., Yang, X., et al. (2011). Extracellular polymeric substances enhanced mass transfer of polycyclic aromatic hydrocarbons in the two-liquid-phase system for biodegradation. *Applied Microbiology and Biotechnology, 90*(3), 1063–1071. https://doi.org/10.1007/s00253-011-3134-5

Zhang, X., Ding, R., Zhou, Y., Zhu, R., Liu, W., Jin, L., et al. (2013). Toll-like receptor 2 and Toll-like receptor 4-dependent activation of B cells by a polysaccharide from marine fungus *Phoma herbarum* YS4108. *PLoS One, 8*(3), e60781.

EPS in Baked Goods Assessing the Impact of EPS on the Quality of Baked Products: Exploring Gluten-Free and Healthier Alternatives

Susmita Ranjan, Pradeep Kumar, and Simran Kirti

1 Introduction

EPS, or exopolysaccharides, are extruded by various microorganisms, subdivided into high molecular weight polymers and include bacteria, fungi, and yeasts. These polymers are biochemically synthesized to respond to certain environmental conditions and serve as cellular protectants as well as structural components for the microorganisms that produce them. EPS displays a varied composition, which includes differences in its structure such as branching patterns, types of linkages, and monosaccharide building blocks (Xiao & Zheng, 2016). These features, when combined, give EPS astonishing functional diversity which can be effectively utilized in modern baking in the food industry. In addition, the growing need for sustainable and health-oriented food products has, yet again, increased interest in EPS as functional ingredients to improve product quality as well as appeal to consumers seeking clean label formulations (Nabeshima et al., 2024).

EPS gained traction because of the growing worries about effects of environmental pollution due to synthetic food additives. EPS are harvested and biogenically manufactured and thus are renewable in nature and offer beneficial aid. In these times when food security and nutrition sustainability is on the rise, Minervini et al. (2022) predicted that EPS could enhance health benefits in polymers while maintaining high levels of quality in processed foods. Manzoor and others (2020) outline

S. Ranjan (✉)
Food Technology, Warner College of Dairy Technology, SHUATS,
Prayagraj, Uttar Pradesh, India

P. Kumar (✉)
ICAR-Indian Grassland and Fodder Research Institute, Jhansi, Uttar Pradesh, India

S. Kirti
Department of Agricultural Biotechnology and Molecular Biology, Dr. Rajendra Prasad
Central Agricultural University, Pusa, Samastipur, Bihar, India

T. Sarkar et al. (eds.), *Extracellular Polysaccharides*,
https://doi.org/10.1007/978-3-031-94472-7_16

these biopolymers as the ones that not only add value to the structure and texture but also have antioxidant and anti-inflammatory properties which extensively support the health claims, making them a prime ingredient in the functional food system.

The baking business has one important component, EPS, known for its capability of altering dough characteristics, enhancing crumb quality, and increasing shelf life. Microbial EPS like xanthan gum, dextran, and gellan gum are indeed the most used microbial EPS because of their ability to stabilize gas cells during fermentation, improve elasticity, and enhance the quality of baked goods (Pourjafar et al., 2023). These polymers work with proteins and starches in the dough via the "zip-lock" mechanism to form structures that improve the textural and stability properties of the final product. Similarly, guar gum and locust bean gum, which are also plant origin EPS, are widely used in bakery formulations because of their excellent water absorption capacity and enhancement of dough rheology (Barak & Mudgil, 2014). EPS are multifunctional which allows for their incorporation in virtually all bakery products, including but not limited to bread, cakes, and even gluten-free products.

One of the areas in which EPS offers a game-changing solution is in gluten-free baked products. The key protein of grains, gluten, gives elasticity and structural support to the wheat-based products. In gluten-free formulations, its absence is linked to textural and shelf-life challenges such as crumbly texture and poor product. EPS serve as functional replacements within numerous systems by providing viscoelasticity to the doughs and stabling the matrices, thus, improving the quality of the end-products (Korcz & Varga, 2021). Examples are the inclusion of xanthan gum and guar gum in gluten-free formulations with the aim of enhancing the dough stability and retention of moisture. Research evidence shows that EPS not only enhances the physical characteristics of gluten-free baked products but also improves the sensory characteristics of food that is used by many consumers who have gluten intolerance or celiac disease (Gustafson, 2016). EPS helps to improve the most critical issues of gluten-free baking by lessening the amount of crumbly texture and staling, thereby providing the much-needed product development assistance.

The EPS is not only charged with enhancing the quality of products but also proves to be beneficial for health in an assortment of ways. EPS such as beta-glucans are placed under indigestible dietary fibers rather classified. Such fibers are essential for the epidemiology of the intestines because they are useful in the proliferation of beneficial microbiota and improving digestive activity. In addition, some types of EPS can possibly be useful in benefiting our health owing to their prebiotic characteristics, which improve gut flora metabolic activity. Recent studies have found EPS as beneficial for lowering cholesterol and better glycemic control, making it possible for the development of functional foods for health-conscious markets (Aryee & Boye, 2014). The health benefits of EPS will always complement the international inclination toward low-calorie, high-fiber, and nutrient-dense food products.

A different and very important area of EPS is where its clean-label production intersects with sustainability. EPS-producing microbial fermentation processes are green biotechnologies that utilize renewable substrates and produce little waste.

Lactobacillus and *Leuconostoc* spp. strains, for instance, are very effective at producing high-value EPS and are therefore amenable to industrial-scale use. The fact that EPS is biodegradable and renewable renders supports green EPS technology principles. Additionally, this characteristic facilitates their incorporation into sustainable food systems (Rachwał & Gustaw, 2024). Moreover, their ability to work with vegan, free-from, and ultra-processed products meets the increased need for ethical and healthy foods.

Manufacturing EPS on an industrial scale has its merits; however, it poses a few challenges. One of the major obstacles is the expense involved in the EPS manufacturing process, particularly in the production of microbial EPS. Fermentation processes are capital intensive due to the need to control pH levels, temperature, and substrate availability (Singh et al., 2017). Additionally, the source and the method of production of EPS determine its functional properties, such as its gelling, thickening, or emulsifying capabilities. This challenge in bakery products formulation and application is always accompanied by the concern for quality and consistency. However, there is hope with engineering innovation in EPS production, as well as bioprocess technology improvements, that have helped in solving these issues by improving the yield and properties of EPS (Kiran et al., 2024).

The effective incorporation of EPS in bakery formulations must consider dosage, pH, as well as temperature. If EPS is overused it can produce undesired textures, while high temperature or low pH conditions that are dealt with insufficient stability can affect the product quality. More efficient and sustainable EPS production is made possible by technological advancements such as co-cultured fermentation, as well as bioengineering. These innovations, in addition to enhancing the economic viability of EPS, make it possible to broaden its use in different types of food products (More et al., 2014).

Furthermore, these EPS characteristics can also be useful in improving the sensory properties and shelf life of different baked products. It is because of EPS' hydrophilic properties that moisture remains inside the dough which slows down the staling process and keeps products fresh for longer. For commercial baked goods, where long shelf status is necessary for the products' success, this property is very useful (Galić et al., 2009). On the other hand, EPS helps to improve the crumb structure and makes the products soft, which is essential for gaining acceptance from consumers. These attributes are why EPS is a modern era baking innovation and aids in the manufacture of efficient, preservative free, and quality products (Anumudu et al., 2024).

In this chapter, the diverse range of uses of EPS in baking is discussed along with their impact on product quality, health, and sustainability. It also discusses the hurdles faced and the prospects for improved EPS production and incorporation. EPS can significantly change the landscape of the baking industry due to the increasing need for healthier and more functional food.

2 Overview of Exopolysaccharides (EPS) in Food Systems

Exopolysaccharides capture interest due to their high demands in the food industry and EPS is a particular type of Exopolysaccharides that are primarily polysaccharides with a high molecular weight produced by many microorganisms. EPSs are used extensively in the food industry, especially by different species of bacteria, fungi, and yeasts, which are beneficial for improving food texture, consistency, and stability. Nowadays it is commonly used in yogurt, kefir, and sauerkraut and even on bakery goods which improves the overall feeling one gets with consumables such as bread or cake and even EPS is used on for increasing other baked products (Asqardokht-Aliabadi et al., 2025).

EPS can be divided into classes according to their origin and structural features as shown in Fig. 1. Two categories of EPS are microbial-derived and plant-derived. Microbial derived encompass EPS that can be produced by microorganisms like *Lactobacillus*, *Leuconostoc*, and *Pediococcus* (Ponzio et al., 2024). Some examples are dextrin, xanthan gum, and gellan gum. Often dextran is used for enhancing viscosity and stabilizing the dough system, while xanthan gum is largely employed as a thickening and stabilizing agent as it also largely increases the dough flexibility. Gellan has moderate gel strength, but high cooking stability which makes it useful in strong baking recipes. These EPS serve several purposes in the baking industry including emulsification, stabilization, and even moisture retention (Zannini et al., 2016).

Guar gum and locust bean gum are examples of plant-derived EPS sourced from plants. Guar gum is most often used as a thickener and stabilizer to improve dough handling and ensure a smooth texture in baked goods. Other gum hydrocolloids, such as locust bean, are also used because of their synergistic effect; improved gel strength and solution viscosity are observed in the bakery formulation (Singh et al., 2022). The structural variability in EPS concerning the composition of monomers, degree of polymerization, and branching shows diversity in functionality in baked goods. Such diversity makes it possible to use EPS as a thickener, gelling agent, or stabilizer, thereby increasing the overall quality of baked products (Alam et al., 2024).

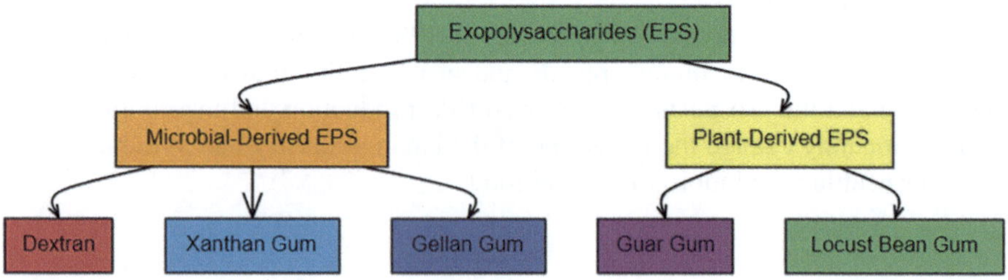

Fig. 1 Classification of exopolysaccharides (EPS)

Microorganisms like *Lactobacillus* spp., *Leuconostoc* spp., and *Pediococcus* spp. are heavily known for their efficient methods for EPS production in food. Among these microorganisms, *Lactobacillus* strains stand out because they produce EPS with good emulsifying, gelling, and stabilizing properties suitable for improving texture and shelf-life of fermented and baked products. The recent advances in microbial fermentation techniques have paved the way for the use of engineered strains with high EPS productivity which further improves the commercial applicability in the food sector (Mahmoud et al., 2021).

3 Contribution of EPS to Baked Goods

The introduction of EPS, especially in dough and final product quality, has brought about significant transformation in baking. Enhancing dough quality is possibly the most useful and impactful of those factors. According to Sun et al. (2023), in addition to the EPS elasticity, gums help dough proteins and starch from the matrix to form macromolecular chains. The synergy between EPS and water is required for optimal leavening and increasing the loaf volume. With EPS, optimal gas retention in the dough is achieved by stabilizing the dough structure during fermentation. EPS also promotes uniform development of the dough crumb, handles the deformation from processing, and greatly enhances uniformity during baking (Bender & Schönlechner, 2020).

Another benefit of EPS in bakery is its ability to retain moisture. EPS capture water molecules within networks and, due to their hydrophilic tendencies, help retain moisture during storage and baking sessions. This feature mitigates the common problem of staleness in baked products (Table 1). While overcoming this challenge, EPS add further value by providing prominence to the baked goods arms ensuring they stay light and fresh for longer (Islam & Islam, 2024).

EPS blend well with the dough starch and thus create a gel-like structure which serves as the framework for the dough. This relationship augments the dough stability for fermentation while increasing the gas retention, resulting in a less dense and more fluffy final crumb structure of the baked product. Besides, when EPS are added into the dough, they can also inhibit the moisture migration responsible for stalling, which allows the bread to remain soft and fresh for a longer time (Mitelut et al., 2021). This quality is very important for delivered and stored products in low-humidity environments or regions where the transport and storage times are prolonged.

Tabibloghmany and Ehsandoost (2014) argues that this EPS expands the durability of baked goods and is still a justified argument for economic success. EPS keeps structure while resisting staleness: It guarantees longer marketability. The multifunctional characteristic of EPS is a rather important reason for their impossibility of nonuse in today's baking, enabling the bakers to obtain improved structure and uniformity of quality of baked products with better shelf life (Bhise & Kaur, 2013).

Table 1 Comparison of moisture retention in EPS-enhanced vs. non-EPS baked products (Roşca et al., 2023)

Parameter	EPS-enhanced baked products	Non-EPS baked products
Moisture retention	High (due to EPS' hydrophilic nature)	Moderate (without EPS)
Crumb freshness	Retains softness for longer periods	Tends to dry out faster
Shelf life	Extended (due to moisture retention)	Shorter shelf life due to moisture loss
Texture	Soft, uniform, less crumbliness	More crumbly and dry over time
Storage conditions	Stable even under varying humidity	Prone to stalling and texture degradation in different storage environments

4 Application of EPS in Gluten-Free Baking

The lack of gluten in gluten-free baking results in dry and crumbly products which make formulation complex. EPS can enhance texture and retain moisture in gluten-free products which is assisted by EPS mimicking some properties of gluten (Table 2). Guar gum and xanthan gum are EPS that aid in the formation of the dough's matrix structure which decreases the hardness of gluten-free products and increases the product's texture (Naqash et al., 2017).

The lack of gluten in gluten-free baking results in dry and crumbly products which make formulation complex. EPS can enhance texture and retain moisture in gluten-free products which is assisted by EPS mimicking some properties of gluten. Guar gum and xanthan gum are EPS that aid in the formation of the dough's matrix structure which decreases the hardness of gluten-free products and increases the product's texture (Naqash et al., 2017).

4.1 Common EPS Used in Gluten-Free Baking

The lack of gluten in gluten-free baking results in dry and crumbly products which make formulation complex. EPS can enhance texture and retain moisture in gluten-free products which is assisted by EPS mimicking some properties of gluten. Guar gum and xanthan gum are EPS that aid in the formation of the dough's matrix structure which decreases the hardness of gluten-free products and increases the product's texture (Naqash et al., 2017).

Table 2 Comparison of EPS in gluten-free and gluten-based baked products (Ikram et al., 2023)

Parameter	Traditional gluten-based products	EPS-enhanced gluten-free products
Structure	Elastic gluten network	Mimics gluten structure using EPS
Moisture retention	Moderate	High (due to EPS hydrophilic properties)
Texture	Uniform and cohesive	Improved uniformity and reduced crumbliness with EPS
Shelf life	Moderate	Extended due to moisture retention and stability from EPS
Elasticity	Naturally present	Achieved through EPS (e.g., xanthan, guar gum)
Suitability for consumers	Limited for gluten-sensitive individuals	Suitable for gluten-sensitive individuals with enhanced sensory appeal
Common applications	Bread, cakes, cookies	Bread, cakes, cookies, pizza crust

5 EPS in Sugar-Free and Low-Sugar Baked Goods

EPS, because of their replicative functional abilities with sugar, have become a more popular choice in sugar-free or less sugar containing baked products. Inulin, pectin, and xanthan gum are diet sugar substitutes that not only give sweetness but also improve texture, hydration, and overall mouthfeel in low-sugar formulation Kenzie & Lee, 2022). In sugar-free baked products, the incorporation of EPS attempts to mitigate the absence of sugar related textural issues such as hardness and crumbliness. EPS are also known to increase the water holding capacity of dough, which aids in avoiding the dryness of the final product, thereby increasing its taste appetites (Gyawali & Ibrahim, 2016).

Inulin and pectin type EPS can help recover some of the texture characteristics which are diminished due to sugar modification or elimination. These polysaccharides facilitate the formation of gels in the dough increasing the products elasticity and cohesive nature and hence, providing resistance to shredding (Ahmed & Rashid, 2019). Alongside the smoothness, EPS may also improve the sweetness of baked goods that are sugar-free. Part of why EPS enhances consumer acceptance of sugar-free offerings is that they improve moistness and provide a more enjoyable experience in addition to constructing stable matrices within the product (Sayas-Barberá et al., 2023).

6 EPS and Health Benefits

EPS is believed to have potential health benefits such as being a dietary fiber. Many EPS have been shown to be non-digestible by humans, thus adding to the fiber intake which helps with digestion. This is very helpful since most people are now interested in low calories, low sugar, and high fiber foods which means EPS can be distributed in baked products and marketed as a healthier alternative since they are highly fibrous without adding extra calories or sugar (Fardet, 2018). A specific type of EPS called beta-glucans is noted for its cholesterol reducing properties and can be added to baked products such as loaves of bread and muffins to reduce low-density lipoprotein (LDL) cholesterol. Another added benefit is EPS like inulin have been studied to improve blood sugar regulation which helps diabetic consumers who want to indulge in baked goods but are conscious of their glycemic response (Wee & Henry, 2020).

The addition of EPS such as beta-glucans and mannans on baked products can enhance cholesterol-lowering effects and improve gut health. These types of formulations aid in the nutritional enhancement of products while safeguarding the sensory parameters (Dong & Karboune, 2021).

7 Mechanisms Behind EPS-Enhanced Rheology and Texture

Exopolysaccharides (EPS) increase the elasticity, viscosity, and stability of dough, hence improving dough rheology (Fig. 2). Preceding studies attributed EPS's interactions with starches and flour proteins to their large molecular size and hydrophilic nature because it alters their degree of hydration and increases water retention (Mir et al., 2016). Moreover, these properties of EPS assist to modify and control the dough consistency and texture to limit overhydration which makes the dough difficult to work with. The unique properties of EPS also aid in the enhanced handling of dough in gluten-free formulations, which tend to be dry and crumbly due to lack of gluten. EPS have been shown to mimic gluten functionalities and in the process, improve the structural integrity of the dough while also providing the much-needed high viscosity and elasticity to the dough which improves fermentation and leavening. All these lead to improved texture with lighter and more uniform crumb structure.

The gelation characteristics of EPS aid in the formation of an air-holding stable dough matrix that facilitates gas retention and availability during fermentation processes which make baked products light and fluffy. This is critical for gluten-free products because their lack of gluten leads to inferior gas retention and density problems. EPS serves as a functional ingredient that alleviates such problems (Nguyen et al., 2024).

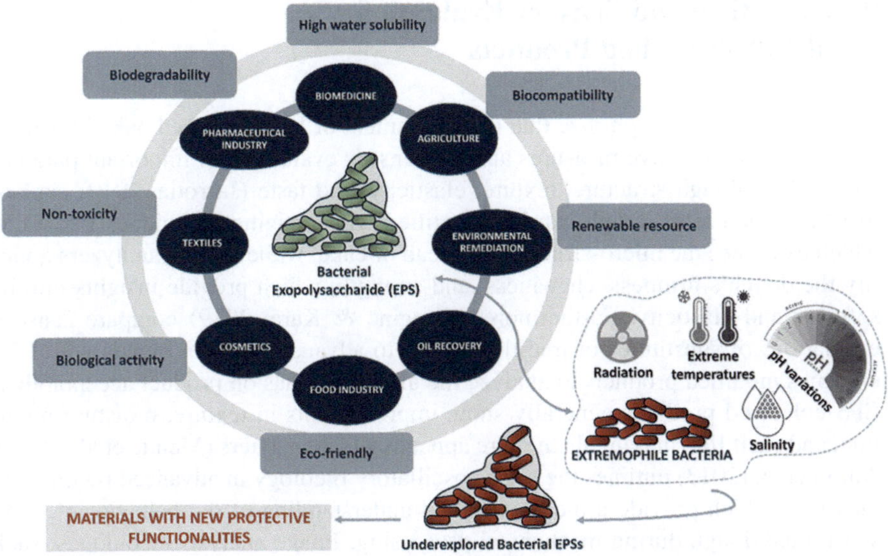

Fig. 2 EPS structure and its functional role in dough rheology (Daba et al., 2021)

8 Innovations in EPS Processing and Integration

With baked formulations, the appropriate type and dosage of EPS are paramount. Different EPS have different thickening, gelling, or emulsion effects which means their dosage needs to be altered for usage. If EPS is overdosed, then it can worsen the texture and disrupt fermentation, which needs precision control adjustments (Sharafi et al., 2024). EPS can be rendered inactive and may fall apart at elevated baking especially within glucose-rich formulations. To enhance the EPSs' programs, stabilizers such as calcium salts or manipulations of the baking parameters like decrease oven temperatures or elongate bake times may be required (Arepally et al., 2022).

Mixture of EPS requires changing the temperature, mixing time, and even the water volume so that it does not sacrifice quality during incorporation into the baking process. EPS might disintegrate at abnormal conditions, in other words the great temperature or high acid or base concentration, which affects the texture of the product and its expiration date (De Corato, 2020).

9 Analytical and Sensory Evaluation of EPS-Enriched Products

Roşca et al. (2023) emphasize that the assessment of EPS-modified baked goods is based on both objective measures and the sensory evaluation of important parameters such as dough structure, texture, elasticity, and taste (Jasrotia, 2010) explain that the structure of dough can be quantified with imaging techniques like SEM which examines the micro-structure of bread or cake, while texture analyzers quantify the dough's firmness, chewiness, and strength, which provide insights into its softness and elasticity. Furthermore, (Gezginc & Kara, 2019) compare sensory evaluations of substitutes of traditional EPS to advanced EPS in blends with EPS enzymes modified products to analyze the affect EPS has on product acceptability. EPS enhanced products generally show improvements in texture, moisture retention, and shelf life, making them more appealing to consumers (Mende et al., 2016). Amjid et al. (2013) outlines the use of oscillatory rheology in advanced rheological measures which provide a more enhanced understanding of the behavior of EPS-substituted dough during mechanical processing. Image analysis techniques, such as confocal laser scanning microscopy (CLSM), may provide valuable insights into the micro-structure of EPS-modified baked goods that is strengthened by the dough components interacts across the microscope minor (Lynch et al., 2018).

10 EPS and Sustainability in Baking

Due to their ability to improve the texture, stability, and shelf life of many baked items, exopolysaccharides (EPS) had garnered interest among modern bakers. About the texture of breads, cakes, cookies, and even gluten-free products, EPS are anticipated to be useful due to their water binding capacity which not only aids in improving mouthfeel but also enhances moisture retention (Muninathan et al., 2022). For example, EPS can be used in bread making to extend the period of softness and freshness of the bread. EPS such as xanthan and guar gum, build a gel-like structure that keeps the moisture of the dough and prevents the baked product from drying or hardening within a short period. The longer shelf life becomes vital especially in industrial production of food, where shelf stability of baked products is of utmost importance (Nhouchi et al., 2018).

EPS severely impacts GFB because of the lack of gluten, which traditionally provides structure to the dough, provides EPS with value-adding characteristics within the process (Wang et al., 2021). Traditionally, gluten-free formulations are of concern due to poor textural quality and crumbling which result in lack of cohesiveness. EPS makes it possible for gluten-free products to mimic the texture and elastic properties of those that contain gluten such as bread and cake. EPS simulates how gluten re-arranges itself, giving additional structure and stiffness which is usually absent during gluten-free bread making (Xu et al., 2019). Often gluten-free items such as bread are dense in nature and have a crumbly texture which EPS heavily aids in improving. Moreover, EPS as a hydrophilic polymer increases the water retention of the dough during the kneading process, resulting in a softer bakeable dough and more appealing product (O'Shea et al., 2014).

In addition to texture and structure, EPS also aids in prolonging the freshness and life of specialized baked goods. EPS comprises high water activity which enhances the freshness of the product and hinders staling which is essential for products like bread and cakes in remaining soft and edible for elongated periods of time. This is especially useful on commercial scale baking where the products endure long distribution periods. Furthermore, EPS may lessen the quantity of synthetic preservatives that are avoided by health-conscious people. Therefore, EPS is not beneficial in baking for the quality of the product but is also advantageous in the context of growing need for clean label ingredients in the food industry (Hamid et al., 2024).

10.1 Clean-Label and Functional Applications of EPS

There has been a growing interest in exopolysaccharides (EPS) as clean-label substitutes to EPS centuries of synthetic additives and stabilizers. This is part of the trend where consumers prefer food products without additives or with minimally processed ingredients. Synthetic preservatives and emulsifiers are considered poor due to their potential health effects and negative perceptions around their use (Lakshmayya et al., 2023). This has made food companies shift towards EPS as clean label functional ingredients for EPS their foods such as baked products, beverages, and dairy alternatives. EPS can replace synthetic ingredients because they are of high molecular weight and possess varying functional characteristics which, do not alter the quality of the final product. Consumers consider the ingredients and additives for which documentation is hard to obtain, so this trend also aligns with the demand for transparency on food labels (Meijer et al., 2021).

Xanthan gum and guar gum have a wide range in vegan and gluten-free recipes, serving more than a mere vegetarian alternative to animal-based additives, but also taking on the role of structure forming agents in unique ways. These exopolysaccharides (EPS) are in high demand within many modern recipes as they function in binding or thickening water. Anticipated by many, EPS are virtually necessary in accommodating vegan and plant-based consumers. For example, they are needed in enhancing the viscosity of batters and doughs to ensure the product is well structurally modified without using milk or eggs. Himashree et al., in 2022 demonstrated the efficiency of xanthan gum works in increasing viscosity.

For gluten-free items, EPS serve crucial functional characteristics, particularly in texture and structure as well as dough elasticity. Since gluten has a very significant importance in the texture of most baked products, its absence in gluten-free products causes the food to be dense and crumbly. EPS such as gellan gum and pullulan exhibit some of the structural properties of gluten and thus, assist in gluten-free doughs (Bangar et al., 2022). In addition, these EPS stabilize the dough so that it is not overworked during mixing or baking. More importantly, these EPS increase the texture of the food, ensuring a more pleasant mouthfeel. Furthermore, EPS also helps in retaining moisture to reduce the staling process which has a positive impact on the shelf life of gluten-free products. As markets are continuously looking for clean-label gluten-free and vegan options, EPS will have a greater role in meeting these requirements with few or no compromises on functionality (Pierre et al., 2019).

10.2 Healthier and Sustainable Alternatives: The Future of EPS in Baking

Exopolysaccharides (EPS) have a bright future in the baking industry owing to humanity's pursuit of self-nutritional health protection goals. The presence of any EPS like inulin, which is known to be a prebiotic, offers profound benefits towards the digestive system and health. It is a well-established fact that inulin is a potent prebiotic, which enhances the growth of probiotics in the gut, resulting in improved microbiome balance. Importantly, inulin is compatible with the baking process and can be added to formulations without detracting from the expected sensory attributes of the product. With the increasing consumer awareness related to gut health, the addition of EPS and prebiotics like inulin into the formulation of baked products can form a novel class of functional foods that boost gastrointestinal health without adversely affecting the taste and texture of the food (Birch & Bonwick, 2019).

As the demand for healthy substitutes continues to grow, EPS has proved to be useful in the creation of low-calorie and low-fat pastries, which is always positive. The sector of EPS is quite broad, covering inulin, for example, which can substitute oils and fats that are typically needed in baking without affecting the texture or flavor of the finished product (Warrand, 2006). These water-soluble multifunctional EPS enhance moisture and elasticity of the food which therefore enhances the experience of the individual, making them believe that the food which they are consuming is of a higher quality. In cakes, cookies, and bread, fats can be replaced by inulin without compromising the quality and lowering the caloric content of the product. This enhances the value of such products, since the consumers will find ways to lower the impact of fat and calorie intake from their diet along Chandon and Wansink (2012) making it easier to manage their weight and have healthy living.

Moreover, the ability to produce EPS sustainably is an important aspect that makes them a green option within the baking sector. EPS, like most synthetic additives, are not made from single use resources. Instead, they are made during fermentation processes that utilize agricultural waste or plant derived sugars. This implies that EPS can be manufactured in an eco-friendlier manner. Furthermore, there is neither significant waste produced during the fermentation of EPS nor excessive amounts of energy required, thus it is a much cleaner method of production than the advertisement of conventional synthetic food additives EPS. As society shifts towards greater demand for more sustainable products, EPS presents a powerful opportunity to provide healthy products whilst simultaneously achieving environmental health targets through the supply of clean-label ingredients (Pua et al., 2022).

11 Consumer Perception and Market Trends in EPS-Enhanced Bakery Products

As consumers prioritize healthier, functional, and clean-label products, the food industry has started using exopolysaccharides (EPS) to improve the nutritional and sensory properties of baked goods. In recent years, consumers have started favoring products free from synthetic additives and preservatives. Natural alternatives like EPS which come from microbially fermented processes are in much higher demand as people around the world have become incredibly informed about the ingredients used in the food they consume (Prete et al., 2021). Research claims that consumers are noticeably valuing the benefits EPS provide, especially when it comes to the texture, moisture retention, and shelf life of baked goods. This increased awareness around EPS attributes has allowed strong EPS incorporation into bakery formulations, replacing traditional additives like gums and emulsifiers (Ali et al., 2024).

The next decade will bring potentially remarkable growth in the market for bakery products with EPS components, especially due to various factors in recent years. The current primary driver of this growth is the market's increased concern with health matters related to sugar, gluten, and interest in consuming plant-based food products. These enquiries are met with functional EPS such as xanthan gum, gellan gum, and inulin, which have versatile structural properties (Domżał-Kędzia et al., 2023). For example, EPS are utilized in gluten-free bakery product formulations to perform the gluten's structural textural forming functions that are necessary for soft-stable bread, cake, and cookie production. Likewise, increased interest in meatless and vegetarian diets has led to increased use of EPS as a clean label, plant-based substitute for eggs and milk in baked goods. These tendencies track greater movement towards the use of functional, natural ingredients in the food industry, thus creating better conditions for the bakery industry using EPS (Yemenicioğlu et al., 2020).

With the increasing attention related to functional foods, such as those fortified with EPS, there is a unique window for EPS to emerge as a prominent ingredient in the baking sector. The capabilities of EPS provide functional properties including rheological improvement of dough, increasing moisture retention, and even possessing prebiotic properties, these features enhance their value in catering to a large range of baked products. As the demand for clean labels rises, more bakeries and food producers seem to try out EPS to cater to the consumers who prefer ingredients that are less processed and more transparent (Lehto, 2024). Not only that, but since EPS are sourced from renewable and more sustainable sources, their use in bakery products compliments the shifting consumer preferences towards eco-conscious and sustainable diets. Hence, as the consumers become increasingly concerned about health, sustainability, and eco-friendly products, the use of EPS in the baking sector is anticipated to continue growing (Matías et al., 2024).

12 Future Challenges and Directions

Even with its transformative capabilities, the implementation of exopolysaccharides (EPS) in the banking sector surmounts to a great deal of challenges and opportunities for further development. One important aspect is the essence of microbial EPS production technology due to its cost-ineffectiveness on a business scale. Moving forward, available technology must be linked to the creation of profitable sustainable production procedures such as the use of renewable feedstock or the improvement of microbial fermentation technologies. EPS in Exopolysaccharide (EPS) fortified baked products is still a major knowledge gap especially in the areas of taste and texture for gluten-free and low sugar items. Business studies and more advanced analytical approaches are essential to EPS for optimal functional and sensory criteria.

There is also the need to investigate the interaction of EPS with other baking ingredients under varied processing conditions for stable quality and performance. The creation of new EPS with specialized properties like higher thermal stability or additional health properties can also broaden the range of applications they can be used for. Overcoming these obstacles will enhance the usability of EPS in commercial baking, while simultaneously meeting the increasing consumer preferences for eco-friendly, clean label, and nutritious food products.

The development of exopolysaccharides (EPS) for edible uses, for instance, in baking, fosters sustainability. The classic techniques of EPS production include chemical techniques that are often energy intensive and wasteful. Nevertheless, the EPS's microbial synthesis from renewable resources like agro wastes is made possible by recent biotechnological innovations and in turn, lessens the ecological burden. Also, biodegradable, non-polluting, microbial polysaccharides respond to the rising need for green food additives. Further work in EPS bioprocessing, especially optimizing fermentation parameters and discovering new EPS producing microbes, could make polysaccharide production even more environmentally acceptable for the food industry.

13 Conclusion

In modern consumption trends, exopolysaccharides (EPS) can serve as a groundbreaking innovation within the baking industry due to the unique health functions it offers. EPS introduction to the baking industry makes it important to note their multifunctional capabilities such as providing favorable improvements to shelf life, dough composition, and texture as well as moisture retention. EPS plays a structural role of Gluten, especially in gluten-free formulations. Even though EPS has its functional advantages, it also improves gut health as it can be beneficial to prebiotic fibers. Even though there are issues such as sensory integration, optimization, and production costs, the developments in technology and research make it possible to deal with these exploitable challenges. EPS within the baking industry has its potential to create new possibilities which could result in accomplishing the nutritional and sensory needs of the modern consumer. The ability to serve varying demographics of consumers while striving for sustainability within the baking industry makes it truly revolutionary.

References

Ahmed, W., & Rashid, S. (2019). Functional and therapeutic potential of inulin: A comprehensive review. *Critical Reviews in Food Science and Nutrition, 59*(1), 1–13.

Alam, M., Dar, B. N., & Nanda, V. (2024). Hydrocolloid-based fruit fillings: A comprehensive review on formulation, techno-functional properties, synergistic mechanisms, and applications. *Journal of Texture Studies, 55*(4), e12861.

Ali, S. S., Alsharbaty, M. H., Al-Tohamy, R., Naji, G. A., Elsamahy, T., Mahmoud, Y. A. G., et al. (2024). A review of the fungal polysaccharides as natural biopolymers: Current applications and future perspective. *International Journal of Biological Macromolecules, 273*, 132986.

Amjid, M. R., Shehzad, A., Hussain, S., Shabbir, M. A., Khan, M. R., & Shoaib, M. (2013). A comprehensive review on wheat flour dough rheology. *Pakistan Journal of Food Sciences, 23*, 105–123.

Anumudu, C. K., Miri, T., & Onyeaka, H. (2024). Multifunctional applications of lactic acid bacteria: Enhancing safety, quality, and nutritional value in foods and fermented beverages. *Foods, 13*(23), 3714.

Arepally, D., Reddy, R. S., Goswami, T. K., & Coorey, R. (2022). A review on probiotic microencapsulation and recent advances of their application in bakery products. *Food and Bioprocess Technology, 15*(8), 1677–1699.

Aryee, A. N., & Boye, J. I. (2014). Current and emerging trends in the formulation and manufacture of nutraceuticals and functional food products. In *Nutraceutical and functional food processing technology* (pp. 1–63). Wiley.

Asqardokht-Aliabadi, A., Sarabi-Aghdam, V., Homayouni-Rad, A., & Hosseinzadeh, N. (2025). Postbiotics in the bakery products: Applications and nutritional values. *Probiotics and Antimicrobial Proteins, 17*, 292–314.

Bangar, S. P., Whiteside, W. S., Singh, A., Özogul, F., Gupta, A., & Gahlawat, S. K. (2022). Properties, preparation methods, and application of sour starches in the food. *Trends in Food Science & Technology, 121*, 44–58.

Barak, S., & Mudgil, D. (2014). Locust bean gum: Processing, properties and food applications—A review. *International Journal of Biological Macromolecules, 66*, 74–80.

Bender, D., & Schönlechner, R. (2020). Innovative approaches towards improved gluten-free bread properties. *Journal of Cereal Science, 91*, 102904.

Bhise, S., & Kaur, A. (2013). Polyols to improve quality and shelf life of baked products: A review. *International Journal of Advanced Scientific and Technical Research, 1*(3), 262–272.

Birch, C. S., & Bonwick, G. A. (2019). Ensuring the future of functional foods. *International Journal of Food Science & Technology, 54*(5), 1467–1485.

Chandon, P., & Wansink, B. (2012). Does food marketing need to make us fat? A review and solutions. *Nutrition Reviews, 70*(10), 571–593.

Daba, G. M., Elnahas, M. O., & Elkhateeb, W. A. (2021). Contributions of exopolysaccharides from lactic acid bacteria as biotechnological tools in food, pharmaceutical, and medical applications. *International Journal of Biological Macromolecules, 173*, 79–89.

De Corato, U. (2020). Improving the shelf-life and quality of fresh and minimally-processed fruits and vegetables for a modern food industry: A comprehensive critical review from the traditional technologies into the most promising advancements. *Critical Reviews in Food Science and Nutrition, 60*(6), 940–975.

Domżał-Kędzia, M., Ostrowska, M., Lewińska, A., & Łukaszewicz, M. (2023). Recent developments and applications of microbial levan, a versatile polysaccharide-based biopolymer. *Molecules, 28*(14), 5407.

Dong, Y., & Karboune, S. (2021). A review of bread qualities and current strategies for bread bioprotection: Flavor, sensory, rheological, and textural attributes. *Comprehensive Reviews in Food Science and Food Safety, 20*(2), 1937–1981.

Fardet, A. (2018). Characterization of the degree of food processing in relation with its health potential and effects. *Advances in Food and Nutrition Research, 85*, 79–129.

Galić, K., Ćurić, D., & Gabrić, D. (2009). Shelf life of packaged bakery goods—A review. *Critical Reviews in Food Science and Nutrition, 49*(5), 405–426.

Gezginc, Y., & Kara, Ü. (2019). The effect of exopolysaccharide producing Lactobacillus plantarum strain addition on sourdough and wheat bread quality. Quality assurance and safety of crops & foods, 11(1), 95-106.

Gustafson, K. L. (2016). *Impact of ingredients on quality and sensory characteristics of gluten-free baked goods*. Kansas State University.

Gyawali, R., & Ibrahim, S. A. (2016). Effects of hydrocolloids and processing conditions on acid whey production with reference to Greek yogurt. *Trends in Food Science & Technology, 56*, 61–76.

Hamid, M. S., Khalid, D., Noman, I., Ahmed, D. F., Solangi, N. H., Mubarak, N. M., & Jatoi, A. S. (2024). Challenges & perspectives of bio-based additives of food industry. In *Application of bio-additives for the food industry* (pp. 219–239). Springer.

Himashree, P., Sengar, A. S., & Sunil, C. K. (2022). Food thickening agents: Sources, chemistry, properties and applications – A review. *International Journal of Gastronomy and Food Science, 27*, 100468.

Ikram, A., Saeed, F., Noor, R. A., Imran, A., Afzaal, M., Rasheed, A., et al. (2023). A comprehensive review on biochemical and technological properties of rye (Secale cereale L.). *International Journal of Food Properties, 26*(1), 2212–2228.

Islam, M. A., & Islam, S. (2024). Sourdough bread quality: Facts and factors. *Foods, 13*(13), 2132.

Jasrotia, A. K. S. (2010). The Effects of Functional Ingredients and Frozen Storage on Wheat Dough and Bread Quality Studied Using Time-Domain Nuclear Magnetic Resonance, Dynamic Rheology, and Environmental Scanning Electron Microscopy. North Carolina State University.

Kiran, N. S., Yashaswini, C., Singh, S., & Prajapati, B. G. (2024). Revisiting microbial exopolysaccharides: A biocompatible and sustainable polymeric material for multifaceted biomedical applications. *3 Biotech, 14*(4), 95.

Korcz, E., & Varga, L. (2021). Exopolysaccharides from lactic acid bacteria: Techno-functional application in the food industry. *Trends in Food Science & Technology, 110*, 375–384.

Lakshmayya, N. S. V., Mishra, A. K., Mohanta, Y. K., Panda, J., Naik, B., Mishra, B., & Varma, R. S. (2023). Essential oils-based nano-emulsion system for food safety and preservation: Current status and future prospects. *Biocatalysis and Agricultural Biotechnology, 53*, 102897.

Lehto, A. (2024). *Exploring circularity approaches for a bakery ingredient wholesaler in the context of ESRS E5* (Bachelor's thesis).

Lynch, K. M., Coffey, A., & Arendt, E. K. (2018). Exopolysaccharide producing lactic acid bacteria: Their techno-functional role and potential application in gluten-free bread products. *Food Research International, 110*, 52–61.

Mahmoud, Y. A. G., El-Naggar, M. E., Abdel-Megeed, A., & El-Newehy, M. (2021). Recent advancements in microbial polysaccharides: Synthesis and applications. *Polymers, 13*(23), 4136.

Manzoor, M., Singh, J., Bandral, J. D., Gani, A., & Shams, R. (2020). Food hydrocolloids: Functional, nutraceutical and novel applications for delivery of bioactive compounds. *International Journal of Biological Macromolecules, 165*, 554–567.

Matías, J., Rodríguez, M. J., Carrillo-Vico, A., Casals, J., Fondevilla, S., Haros, C. M., et al. (2024). From "Farm to Fork": Exploring the potential of nutrient-rich and stress-resilient emergent crops for sustainable and healthy food in the Mediterranean region in the face of climate change challenges. *Plants, 13*(14), 1914.

McKenzie, E., & Lee, S. Y. (2022). Sugar reduction methods and their application in confections: A review. *Food Science and Biotechnology, 31*(4), 387–398.

Meijer, G. W., Detzel, P., Grunert, K. G., Robert, M. C., & Stancu, V. (2021). Towards effective labelling of foods. An international perspective on safety and nutrition. *Trends in Food Science & Technology, 118*, 45–56.

Mende, S., Rohm, H., & Jaros, D. (2016). Influence of exopolysaccharides on the structure, texture, stability and sensory properties of yoghurt and related products. *International Dairy Journal, 52*, 57–71.

Minervini, F., Comitini, F., De Boni, A., Fiorino, G. M., Rodrigues, F., Tlais, A. Z. A., et al. (2022). Sustainable and health-protecting food ingredients from bioprocessed food by-products and wastes. *Sustainability, 14*(22), 15283.

Mir, S. A., Shah, M. A., Naik, H. R., & Zargar, I. A. (2016). Influence of hydrocolloids on dough handling and technological properties of gluten-free breads. *Trends in Food Science & Technology, 51*, 49–57.

Mitelut, A. C., Popa, E. E., Popescu, P. A., & Popa, M. E. (2021). Trends of innovation in bread and bakery production. In *Trends in wheat and bread making* (pp. 199–226). Elsevier.

More, T. T., Yadav, J. S. S., Yan, S., Tyagi, R. D., & Surampalli, R. Y. (2014). Extracellular polymeric substances of bacteria and their potential environmental applications. *Journal of Environmental Management, 144*, 1–25.

Muninathan, C., Guruchandran, S., Viswanath Kalyan, A. J., & Ganesan, N. D. (2022). Microbial exopolysaccharides: Role in functional food engineering and gut-health management. *International Journal of Food Science & Technology, 57*(1), 27–34.

Nabeshima, E. H., Tavares, P. E. D. R., Lemos, A. L. D. S. C., & Moura, S. C. S. R. D. (2024). Emerging ingredients for clean label products and food safety. *Brazilian Journal of Food Technology, 27*, e2023160.

Naqash, F., Gani, A., Gani, A., & Masoodi, F. A. (2017). Gluten-free baking: Combating the challenges – A review. *Trends in Food Science & Technology, 66*, 98–107.

Nguyen, H. T., Pham, T. T., Nguyen, P. T., Le-Buanec, H., Rabetafika, H. N., & Razafindralambo, H. L. (2024). Advances in microbial exopolysaccharides: Present and future applications. *Biomolecules, 14*(9), 1162.

Nhouchi, Z., Botosoa, E. P., & Karoui, R. (2018). Critical assessment of formulation, processing and storage conditions on the quality of alveolar baked products determined by different analytical techniques: A review. *Trends in Food Science & Technology, 81*, 159–171.

O'Shea, N., Arendt, E., & Gallagher, E. (2014). State of the art in gluten-free research. *Journal of Food Science, 79*(6), R1067–R1076.

Pierre, G., Delattre, C., Dubessay, P., Jubeau, S., Vialleix, C., Cadoret, J. P., et al. (2019). What is in store for EPS microalgae in the next decade? *Molecules, 24*(23), 4296.

Ponzio, A., Rebecchi, A., Zivoli, R., & Morelli, L. (2024). Reuterin, phenyllactic acid, and exopolysaccharides as main antifungal molecules produced by lactic acid bacteria: A scoping review. *Foods, 13*(5), 752.

Pourjafar, H., Ansari, F., Sadeghi, A., Samakkhah, S. A., & Jafari, S. M. (2023). Functional and health-promoting properties of probiotics' exopolysaccharides; isolation, characterization, and applications in the food industry. *Critical Reviews in Food Science and Nutrition, 63*(26), 8194–8225.

Prete, R., Alam, M. K., Perpetuini, G., Perla, C., Pittia, P., & Corsetti, A. (2021). Lactic acid bacteria exopolysaccharides producers: A sustainable tool for functional foods. *Foods, 10*(7), 1653.
Pua, A., Tang, V. C. Y., Goh, R. M. V., Sun, J., Lassabliere, B., & Liu, S. Q. (2022). Ingredients, processing, and fermentation: Addressing the organoleptic boundaries of plant-based dairy analogues. *Foods, 11*(6), 875.
Rachwał, K., & Gustaw, K. (2024). Lactic acid bacteria in sustainable food production. *Sustainability, 16*(8), 3362.
Roşca, M. F., Păucean, A., Man, S. M., Chiş, M. S., Pop, C. R., Pop, A., & Fărcaş, A. C. (2023). Leuconostoc citreum: A promising sourdough fermenting starter for low-sugar-content baked goods. *Foods, 13*(1), 96.
Sayas-Barberá, E., Paredes, C., Salgado-Ramos, M., Pallarés, N., Ferrer, E., Navarro-Rodríguez de Vera, C., & Pérez-Álvarez, J. Á. (2023). Approaches to enhance sugar content in foods: Is the date palm fruit a natural alternative to sweeteners? *Foods, 13*(1), 129.
Sharafi, H., Divsalar, E., Rezaei, Z., Liu, S. Q., & Moradi, M. (2024). The potential of postbiotics as a novel approach in food packaging and biopreservation: A systematic review of the latest developments. *Critical Reviews in Food Science and Nutrition, 64*(33), 12524–12554.
Singh, V., Haque, S., Niwas, R., Srivastava, A., Pasupuleti, M., & Tripathi, C. (2017). Strategies for fermentation medium optimization: An in-depth review. *Frontiers in Microbiology, 7*, 2087.
Singh, A. K., Malviya, R., & Rao, G. S. N. K. (2022). Locust bean gum: Processing, properties and food applications. *Recent Advances in Food Nutrition & Agriculture, 13*(2), 93–102.
Sun, X., Wu, S., Li, W., Koksel, F., Du, Y., Sun, L., et al. (2023). The effects of cooperative fermentation by yeast and lactic acid bacteria on the dough rheology, retention and stabilization of gas cells in a whole wheat flour dough system – A review. *Food Hydrocolloids, 135*, 108212.
Tabibloghmany, F. S., & Ehsandoost, E. (2014). An overview of healthy and functionality of exopolysaccharides produced by lactic acid bacteria in the dairy industry. *European Journal of Nutrition & Food Safety, 4*(2), 63–86.
Wang, Y., Maina, N. H., Coda, R., & Katina, K. (2021). Challenges and opportunities for wheat alternative grains in breadmaking: Ex-situ-versus in-situ-produced dextran. *Trends in Food Science & Technology, 113*, 232–244.
Warrand, J. (2006). Healthy polysaccharides. *Food Technology and Biotechnology, 44*(3), 355–370.
Wee, M. S., & Henry, C. J. (2020). Reducing the glycemic impact of carbohydrates on foods and meals: Strategies for the food industry and consumers with special focus on Asia. *Comprehensive Reviews in Food Science and Food Safety, 19*(2), 670–702.
Xiao, R., & Zheng, Y. (2016). Overview of microalgal extracellular polymeric substances (EPS) and their applications. *Biotechnology Advances, 34*(7), 1225–1244.
Xu, Y., Coda, R., Holopainen-Mantila, U., Laitila, A., Katina, K., & Tenkanen, M. (2019). Impact of in situ produced exopolysaccharides on rheology and texture of fava bean protein concentrate. *Food Research International, 115*, 191–199.
Yemenicioğlu, A., Farris, S., Turkyilmaz, M., & Gulec, S. (2020). A review of current and future food applications of natural hydrocolloids. *International Journal of Food Science & Technology, 55*(4), 1389–1406.
Zannini, E., Waters, D. M., Coffey, A., & Arendt, E. K. (2016). Production, properties, and industrial food application of lactic acid bacteria-derived exopolysaccharides. *Applied Microbiology and Biotechnology, 100*, 1121–1135.

Exploring Bacterial Exopolysaccharide as Prebiotics and Their Role in Gut Health and Immunomodulation

Parama Das Gupta, Upal Das Ghosh, and Sudipta Chakraborty

1 Introduction

Bacterial exopolysaccharides (EPS) have recently gained much attention due to their distinct structure and characteristics. They emerge as a suitable candidate for application in different fields of medicine and therapeutics, food, and other industries. The term "exopolysaccharide" was first coined by Sutherland in 1972 (Netrusov et al., 2023). They are high molecular weight extracellular carbohydrate polymer, substituted with proteins, DNA, phospholipids, etc. (Angelin et al., 2020) and may also contain various functional groups like hydroxyl, carboxyl, phosphate, carbonyl, and acetate (Netrusov et al., 2023; Angelin et al., 2020). Generally EPS is divided into two groups: homopolysaccharides that are composed of a single type of monosaccharide like glucose or fructose, joined through glycosidic linkage and heteropolysaccharides that contain two or more types of monosaccharide units like glucose, fructose, galactose, mannose, and N-acetyl glucosamine. On the basis of their localization, EPS is divided into capsular polysaccharides that are closely associated with bacterial cell surface and free slime polysaccharides loosely attached or may be completely secreted outside the cell (Netrusov et al., 2023). Bacterial exopolysaccharides (EPSs) are extreme outer layer of bacterial cells, either loosely attached to form slime layers, where many bacterial cells are embedded in a common matrix or the comparatively more stable structure called "capsule." A large

Parama Das Gupta and Upal Das Ghosh contributed equally with all other contributors.

P. D. Gupta · U. D. Ghosh
Department of Microbiology, Bidhannagar College, Kolkata, West Bengal, India

S. Chakraborty (✉)
Department of Microbiology, Govt General Degree College, Narayangarh, West Midnapore, India

T. Sarkar et al. (eds.), *Extracellular Polysaccharides*,
https://doi.org/10.1007/978-3-031-94472-7_17

number of bacteria consisting of both Gram-positive and Gram-negative ones can produce EPS during their metabolic processes (Netrusov et al., 2023). EPS are produced both intra and extracellularly. Extracellular production involves secretion of extracellular enzymes that catalyze the polymerization of monosaccharide units. On the contrary during intracellular production, some exopolysaccharides are produced by intracellular process and then exported outside the cell (Jurášková et al., 2022).

Bacterial EPS play significant role in virulence and pathogenicity of several bacteria. They act as antiphagocytic agent, prevent opsonization, help in binding to epithelial cells, and also protect bacteria from various biotic and abiotic stresses. They have also some commercial utilization. Bacterial EPS have a vast application in various industries as well in therapeutics (Angelin et al., 2020; Tiwari et al., 2020). Due to their antiviral, antimicrobial, anticancer activities they are used in medical fields. Besides this, they have prominent immunomodulatory activity, can reduce cholesterol, have antidiabetic properties, and act as a carrier for drug delivery system. They also possess antibiofilm activity so that they can be used in prevention of infections caused by biofilm producing pathogenic bacteria (Jurášková et al., 2022). The high water holding capacity and rheological properties make bacterial EPS a prominent candidate as adhesives, stabilizer, gelling, suspending, thickening agent in food industries (Tiwari et al., 2020). Recently bacterial EPS has been emerged as a potential candidate having prebiotic activities. Prebiotics are generally carbohydrate polymers that are resistant to digestion at the upper intestinal tract and stimulate the growth of specific gut microflora by providing energy source for them (Oerlemans et al., 2021). In this chapter, we try to provide a thorough insight about the prebiotic potential of EPS, role of EPS in gut health and current strategies and future prospects of using EPS as prebiotics.

2 EPS Producing Bacteria

Both Gram-positive and Gram-negative bacteria are able to synthesize various kinds of EPS, both homopolysaccharide and heteropolysaccharide. Table 1 shows various kinds of EPS and their producer organisms.

3 Biosynthesis of Exopolysaccharides

Bacteria synthesize exopolysaccharides either intracellularly or extracellularly. One of the four mechanisms are generally involved in the production of bacterial EPS. These are (i) the Wzx/Wzy-dependent pathway; (ii) the ATP-binding cassette (ABC) transporter-dependent pathway; (iii) the synthase-dependent pathway; and (iv) the extracellular synthesis by use of a single sucrase protein (Schmid et al., 2015).

The Wzx/Wzy-dependent pathway generally involves biosynthesis of heteropolymers (Schmid et al., 2015). The proteins involved in this pathway are glycosyltransferases (GT), flippase (Wzx) and polymerase (Wzy). Glycosyltransferases catalyzes the subsequent addition of monosaccharide units to an undecaprenol

Table 1 List of various kinds of EPS and their producer organisms

EPS name	Type	Monosaccharide units	Producer organism	Gram character	Reference
Xanthan	Heteropolysaccharide	Glucose, mannose, glucuronic acid	*Xanthomonas campestris*	Gram negative	Netrusov et al. (2023), Khan et al. (2022)
Dextran	Homopolysaccharide	Glucose	*Lactobacillus*, *Leuconostoc*, *Streptococcus*	Gram positive	Netrusov et al. (2023)
Alginate	Heteropolysaccharide	Glucuronic acid, mannuronic acid	*Azotobacter vinelandii*, *Pseudomonas*	Gram negative	Netrusov et al. (2023), Khan et al. (2022)
Curdlan	Homopolysaccharide	Glucose	*Alcaligenes faecalis*, *Rhizobium radiobacter*, *Cellulomonas* spp., *Agrobacterium* sp.	Gram negative	Netrusov et al. (2023), Khan et al. (2022)
Gellan	Heteropolysaccharide	Glucose, rhamnose, glucuronic acid	*Sphingomonas elodea*, *Pseudomonas* sp.	Gram negative	Netrusov et al. (2023), Khan et al. (2022)
Hyaluronan	Heteropolysaccharide	Glucuronic acid and N-acetyl D glucosamine	*Streptococcus*, *Pasteurella multocida*	Gram positive	Netrusov et al. (2023), Khan et al. (2022), Guan et al. (2020)
Levan	Homopolysaccharide	Fructose	*Bacillus subtilis*, *Streptococcus mutans*, *Streptococcus salivarius*, *Gluconacetobacter xylinus*, *Zymomonas mobilis*	Gram positive	Khan et al. (2022), Domżał-Kędzia et al. (2023)
Fucogel	Heteropolysaccharide	Fucose, galactose, 4 O acetyl galacturonic acid	*Klebsiella pneumoniae*	Gram negative	Netrusov et al. (2023), Xiao et al. (2022)
Clavan	Heteropolysaccharide	Glucose, galactose, fucose	*Clavibactermichiganensis*	Gram positive	Netrusov et al. (2023), Roca et al. (2015)
Fucopol	Heteropolysaccharide	Fucose, galactose, glucose, glucuronic acid and acyl groups	*EnterobacterA47*	Gram negative	Netrusov et al. (2023), Baptista et al. (2022)
Kefiran	Heteropolysaccharide	Glucose, galactose	*Lactobacillus kefiranofaciens*, *Lacidophilus*, *L. delbruckii*, *L. casei*, *L. rhamnosus*, *L. kefiri*	Gram positive	Netrusov et al. (2023), Prado et al. (2015)
Mauran	Heteropolysaccharide	High uronic acid content	*Halomonasmaura*	Gram negative	Bouchotroch et al. (2001), Arias et al. (2003)

diphosphate residue attached to inner membrane and Wzx or flippase catalyzes the translocation of it to the outer membrane. Then polymerization continues at the periplasmic space, catalyzed by Wzy protein. Polymers are then transported to the cell surface by the aid of two additional proteins; polysaccharide co-polymerase and outer membrane polysaccharide export protein (Cuthbertson et al., 2009; Morona et al., 2009; Islam & Lam, 2014).

The ABC transporter-dependent pathway mainly involves biosynthesis of capsular polysaccharides. Like the Wzx/Wzy pathway, this pathway also involves glycosyltransferases for the assembly of monosaccharide units at the cytoplasmic face of the plasma membrane. However the transport across the inner membrane and transportation to the cell surface is carried out by a protein complex composed of ABC transporters spanning the inner membrane and periplasmic proteins of PCP and OPX family (Whitney & Howell, 2013) (Cuthbertson et al., 2009; Morona et al., 2009).

The synthase-dependent pathway is generally involved in the synthesis of curdlan, bacterial cellulose, hyaluronan, etc. A single synthase enzyme is sufficient for polymerization and translocation process. This pathway is independent of flippase (Rehm, 2015).

The fourth process involves extracellular synthesis of polysaccharides like dextran or levan. This process is catalyzed by glycosyltransferases which are secreted and covalently attached to the cell surface (Schmid et al., 2015).

4 Role of EPS in Maintaining Gut Health

EPS are able to restore and retain the biological environment of guts by stimulating the growth of normal flora, by inhibiting the adhesion of pathogens to the Intestinal epithelium, by increasing the integrity of intestinal mucosal barrier and/or by influencing immune system interacting with Toll like receptors. EPS can act as a potential source of carbohydrate for gut microflora. Besides this, it can also improve the gut health by providing energy to the gut epithelium, nourish the growth of villi, and induce crypt development and mucin production. This mucin glycoprotein in turn improves the integrity of the gut barrier (Yue et al., 2022). For acting as a prebiotic, EPS should be utilized by the gut microflora as their substrate (Gibson et al., 2017) and hence the components must remain undigested within the upper gastrointestinal tract. The extent of degradation/metabolization of EPS is dependent on the structure of the constituent monosaccharides. Presence of highly charged or uncharged long side chains minimize the degree of digestion, whereas, presence of simple side chain aggravates the same. Homopolysaccharides like fructan, β glucan, and heteropolysaccharides like alginate, Xanthan gum, and hyaluronic acid have been well established in their prebiotic role. Exo homopolysaccharides produced by *Lactobacillus sanfranciscensis* has been reported to be utilized by Bifidobacteria and Lactic acid bacteria of gut (Korakli et al., 2002). In another investigation, utilization of *Streptococcus thermophilus* derived heteropolysaccharide by normal gut microflora has been established (Ruijssenaars et al., 2000). In an in vitro fermentation model, it has been established that the EPS produced by *Lactobacillus rhamnosus* has been converted into short chain fatty acids propionic acid, butyric acid, acetic acid, and formic acid by the human gut microflora (Zhu et al., 2021). These fatty acids are responsible for reducing inflammation, onset of cancer, anabolism of glucose, and many other beneficial actions in intestine. On the other hand, these gut commensals are also well known for their production of EPS, which in turn shows anti-inflammatory and anti-microbial activity within the gut environment (Kaur & Dey, 2023). EPS can also upregulate the production of tight junction proteins, activity of enzymes involved in carbohydrate metabolism, EPS produced by the normal microflora of gut also helps them to attach in the intestinal mucosa thereby inhibiting the colonization of the pathogens. Intestinal health is also known to be nourished by the improved blood flow and enhanced immune activity mediated by the EPS produced from the normal gut microflora (Nadzir et al., 2021). For the nourishment of the intestinal normal microflora as well as the intestinal mucosal cells, probiotic bacteria that produce exopolysaccharides are commonly used as dietary supplement. These bacteria can withstand and survive in the low pH environment of gastric juices and act as a source of exopolysaccharide required for the maintenance of intestinal health (Angelin & Kavitha, 2020). Bacterial EPS has been evident to have regressive activity over obesity, dyslipidemia, diabetes, colon cancer, inflammatory bowel disease, diarrhea, and hepatic diseases (Xue et al., 2024). So, to boost the immune system and restore the normal microflora of intestine after prolonged treatment with antibiotic, probiotic exopolysaccharides are being prescribed by the medical practitioners as a continuation of the treatment.

5 Role of EPS in Host Immunomodulation

Immunomodulatory capacity of EPSs depends on several physicochemical properties like monosaccharide composition, average molecular weight, water solubility, electric charges and their stereochemistry. Several studies have confirmed that small, negative charge EPSs of low molecular weight are strong immunomodulators and neutral and/or larger polymers act as immunosuppressors (Lopez et al., 2012; Hidalgo-Cantabrana et al., 2014). Some charged, zwitterionic EPS present in *Bacillus fragile* play a promising role in the activation of the host immune response in the gastrointestinal tract (Round et al., 2011). Most of the evidences of immunomodulatory activities of EPS are the outcome of in vitro observations conducting experiments on cultured animal cells like macrophage, dendritic cells, peripheral mononuclear cells (PBMCs) by the assessment of the pattern of cytokine released (Górska et al., 2016) or in vivo experiments in mice (Kamiya et al., 2018). The gut mucosal immune system is the largest immune sites of the body, constitutes 80% of the total immune cells (Lewis & Blutt, 2019). Prebiotics like EPS is recognized by receptors of immune cells, and have been shown to modulate the innate and adaptive immunity of gut environment and considered to have a great impact on overall human health.

Determination of the process of interaction of different EPSs with the various receptors of immune cells and the mechanisms of modulation of immunity in the gastrointestinal tract would definitely enrich the existing knowledge of using EPS to treat different diseases and to promote human health.

Different bacterial EPS have been reported to modulate both innate and adaptive immunity. Their immunomodulatory activities are mainly confined to the stimulation of the proliferation of T/B lymphocytes, natural killer (NK) c ells, increment of their tumoricidal activity, enhancement in the phagocytic capacity of macrophages, increased mitogenic activity of lymphocytes, and most importantly the induction of cytokines and collective enhancement of the host immune defense against common pathogens (Shao et al., 2014). However, the precise mechanisms of immunomodulation by different kinds of bacterial EPSs are still under investigations. EPS of probiotic bacteria are recognized as MAMPs (Microbe Associated Molecular Patterns) mainly by TLR (toll like receptor)-4, 2 and C-type lectins (such as Dectin 1) present in intestinal epithelium (Castro-Bravo et al., 2019; Castro-Bravo et al., 2018; Paulraj et al., 2018). Heteropolysaccharide EPSs produced by *Bifidobacteria* spp. or *Lactobacilli* spp. mainly acts as a ligand for TLR-4 (Delgado et al., 2020). The process of immunomodulation follows some steps.

(a) *Recognition bacterial EPS by immune system*: Bacterial EPSs are recognized by several pattern-recognition receptors (PRRs) present on immune cells. They mainly interact with Toll-like receptors (TLRs), C-type lectin receptors (CLRs), and nucleotide oligomerization domain-like receptors (NODs) present on variety of lymphocytes and antigen presenting cells (Hill & Diehl, 2018).

(b) *Interactions of EPS with Toll-like receptors and C-type lectin receptors A*: Toll-like receptors (TLRs) are considered to be one of the important pathogen-associated molecular patterns (PAMPs) in host immune system. Bacterial EPSs promote protection against several enteric pathogens in a TLR-dependent manner. EPS-producing *Bacillus subtilis* can reduce inflammation induced by the enteric mouse pathogen *Citrobacter rodentium* after interaction with TLR4 receptors are present in mucosal cells and immune cells (Jones et al., 2014). There are growing number of reports that demonstrate both activation and inhibition immune responses by various EPS through their interactions with TLRs either directly or indirectly (Janeway Jr & Medzhitov, 2002). C-type lectin receptors A (CLR) are sugar binding protein expressed on the surface of dendritic cells, monocytes, and macrophages (McGreal et al., 2005; Castro-Bravo et al., 2018). Several reports are there which deals with the recognition and interaction of bacterial EPSs. It occurs mainly by three types of CLRs like Dectin-1, Dectin-2, and DC-SIGN. Surface polysaccharides of *Bifidobacterium bifidum* strain *PRI1* promotes both immunosuppressive and immunostimulatory effects in a structure-dependent manner interacting with Dectin-1 on dendritic cells (Speciale et al., 2019). Dectin-2 interacts with mannosylated O-antigens of Gram-negative bacteria LPS, mannose-capped lipoarabinomannan from *Mycobacterium tuberculosis* and α-mannan in fungal cell walls. DC-SIGN interacts with mannose and fucose residues present on bacterial cell wall including *M. tuberculosis* and *H. pylori* and *Pseudomonas aeruginosa* biofilms (Gringhuis et al., 2009), modulates T cell polarization, and regulates the proliferation and maturation of dendritic cells (Sahasrabudhe et al., 2016).

(c) *Immunomodulation by EPS in gut environment*: Several studies have documented the immunomodulatory activities of various bacterial EPSs which encompass the triggering of pro-inflammatory and anti-inflammatory cytokines released from repertoire of immune cells in gut. Several research publications have reported the stimulation of cell mediated immune response by EPS from *Lactobacillus* and *Bifidobacterium* strains which elevate the recruitment of mucosal immune cell populations, antigen presentation, and immune activation and elevation IgA and serum IgG (Fanning et al., 2012; Laiño et al., 2016). It has been also documented that purified EPS along with other (MAMPs) such as lipoteichoic acid, or lipoprotein that are recognized by TLR2/1 and TLR2/6 receptors increase the activation markers and cytokine production of lymphocytes (Angelin et al., 2020; Górska et al., 2016). Polysaccharide A (PSA) produced by *B. fragilis* is found to be engulfed by dendritic cells. The antigen is later processed by the endocytic pathway and presented to the CD4 $^{+}$ T cells carrying αβ-T-cell receptors via MHC class II. These activated T cells then secrete anti-inflammatory cytokines like IL-12, TNF-α, and IFN-γ which promotes immunosuppressive effect (Round et al., 2011; Wang et al., 2006; Cobb et al., 2004). Different in vivo studies also report anti-inflammatory effects like IL-10 secretions leading to proliferation of Foxp3 $^{+}$ T-regulatory cells that allow the protection of *B. fragilis* in the gastrointestinal tract (Kalka-Moll et al., 2002). EPSs of *Lactobacillus reuteri* consisting of galactose have been shown to have anti-inflammatory effects on macrophages by enhancing the anti-inflammatory cytokines (Chen et al., 2019). Anti-inflammatory effect has been also displayed by EPS synthesized by *F. prausnitzii HTF-F* by inducing the production of anti-inflammatory cytokines like IL-12 in antigen presenting cells in a TLR-2-dependent manner (Rossi et al., 2015), now considered as an anti-inflammatory next-generation probiotic. It has been also shown to enhance pro-inflammatory cytokine like IL-12p70 and of IL-10 in in vivo mouse model in response to the co-administration of *L. plantarum* EPS (Rossi et al., 2015). EPS of *B. Subtilis* protect gut mucosa conferring protection against *Citrobacter rodentium* (Jones et al., 2014). EPS isolated from *L. casei* WXD030 can trigger nitric oxide, TNF-α, IL-1β and IL-6 production, elevate antibodies, and enhanced T-cell proliferation in vitro (Xiu et al., 2018). Recently, some bacterial EPSs have shown anti-inflammatory effects in some animal experimental models of diseases like IBD (Inflammatory Bowel Diseases) and DSS-induced colitis by increasing T regulatory (Treg) cells in mucosal epithelium of gastrointestinal tract and have also shown to modulate the Treg/Th17 axis (Hidalgo-Cantabrana et al., 2016; Yu et al., 2019). In some studies EPS has been shown to activate Th1, Th2, and Treg cells to influence anti-inflammatory response and suppress pro-inflammatory response (Fig. 1). EPS produced by *Leuconostoc mesenteroides* strain NTM048 and Kefiran was isolated from *L. kefiranofaciens* has been found to activate humoral immune response by activating the produc-

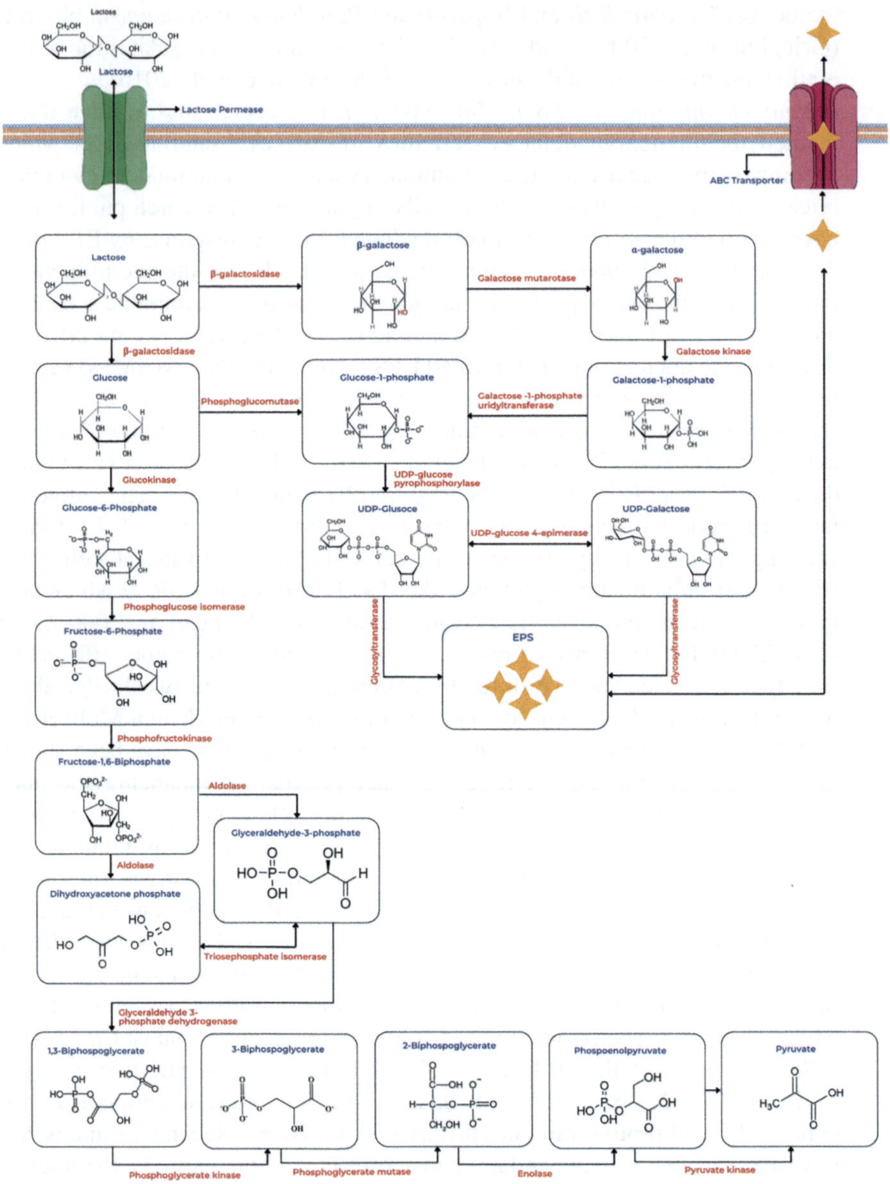

Fig. 1 Different bacterial EPS are recognized by specific PARs (Pattern Recognition Receptors) present on mucosal epithelial cells to present the antigen via professional APCs like dendritic cells to activate Th1, Th2, and Treg cells and influence anti-inflammatory response and suppress pro-inflammatory response (Oerlemans et al., 2021)

tion of IgA, upregulation of TGF-β genes, and modulate Th1 and Th2 cell-mediated responses in splenocytes (Kaur et al., 2023; Matsuzaki et al., 2015). EPS isolated from the biofilm of *Lactobacillus plantarum*, bifidobacteria, and capsular polysaccharide from *Salmonella typhi* has been reported to reduce pro-inflammatory cytokines like IL-6, IL-8, and MCP-1 after their interaction with TLR5 and TLR4 receptors indirectly (Murofushi et al., 2015; Raffatellu et al., 2006). In a similar study, it has been found that EPS isolated from the biofilm of *Thermus aquaticus* YT-1 interacts directly with TLR2 and induces the secretion of TNF-α and IL-6 in macrophages (Lin et al., 2011). Different EPSs can also induce immunosuppressive effects. *Streptococcus suis* were recognized by human TLR2 present on monocytes, triggering the release of pro-inflammatory cytokines like IL-1, IL-6, IL-8, and MCP-1 (Graveline et al., 2007). It has been documented that EPS interacts with TLR-4/5 on LPS activated macrophage cell surface and activates NF-κB and MAPK signal transduction pathways. Binding of EPS with TLR-4 leads to the phosphorylation of MAP kinase and NF-κB inhibitor (Iκ-B). Activated MAP kinase then recruits different activated transcription factors like ERK, p38, JNK, API. Consequently, the secretion of pro-inflammatory mediators like IL-6 and IL-12 decrease anti-inflammatory mediators like IL-10 increase from the activated macrophage cells. qRT-PCR and Western blot are used to confirm the expression of relevant mRNAs and proteins in macrophages co-treated with EPS and LPS. Activation of MAPK pathway results in enhanced proliferation of macrophages (Fig. 2). EPS provide important health benefits in human body by immunomodulation mechanisms. EPS producers *Leuconostoc citreum* and *Bifidobacterium longum* have been shown to reduce the symptoms of allergy by reducing the recruitment eosinophils, Th2 cells, and IgE (Domingos-Lopes et al., 2017; Schiavi et al., 2018). EPS from *L. rhamnosus* has also been shown to reduce several inflammatory diseases (Nowak et al., 2020). Other studies show the role of EPS in balancing the immune system at the time of infection or inflammation and secretion of IgA against potential mucosal pathogens (Matsuzaki et al., 2015; Ruas-Madiedo, 2014). Bacterial EPS also take important role in the protection of IBD (Inflammatory Bowel Diseases) and ulcerative colitis (Thomson et al., 2018; Yu et al., 2019). Extensive research work is required to elucidate the role of EPS as a prebiotic to define their effects on the host immune system and overall gut health.

6 Current Strategies and Future Prospect of Using EPS as Prebiotic

Microbial exopolysaccharides have wide range of commercial uses in food, pharmaceutical, and medical industry. In food industry, EPS is being used to maintain texture, structure and flavor of the food. Though the use of such exopolysaccharides

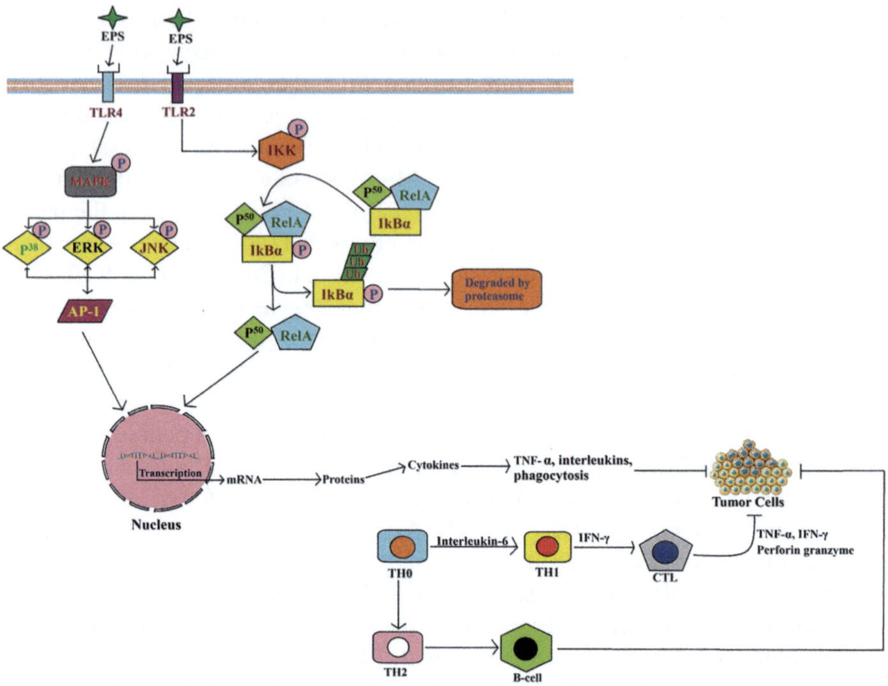

Fig. 2 Bacterial EPS interacts with TLR-4/5 on LPS activated macrophage cell surface and activates NF-κB and MAPK signal transduction pathways to induce the proliferation of macrophages, positively regulates anti-inflammatory responses, and negatively regulates pro-inflammatory response (Kaur et al., 2023; Khan et al., 2022; Oerlemans et al., 2021; Angelin et al., 2020; Jurášková et al., 2022; Delgado et al., 2020; Castro-Bravo et al., 2018)

have been initially started with an aim to increase the standard of the food products, but additionally the added exopolysaccharides, after being consumed can exhibit their prebiotic potential. Commercialization of microbial exopolysaccharide has been started from the advent of industrial production of bacterial cellulose, a more advanced and better alternative of plant cellulose. Different strains of Gram-negative *Glucanobacter*, *Achromobacter*, *Agrobacterium*, *Rhizobium*, *Pseudomonas*, and *Salmonella* are reported for their extracellular cellulose production. Manipulation of strains and culture condition for better production of such extracellular cellulose with commercial perspective is still under review (Netrusov et al., 2023). Besides cellulose, intestinal microflora like *Bifidobacterium* also produce around 11 types of fermentable EPS component (Salazar et al., 2008) which can be commercialized as prebiotic product after proper downstream processing. EPS obtained from lactic acid bacteria are widely being used to prepare culture milk product like cheese, yogurt, and cream (Mende et al., 2016). Glucan produced by such *Lactobacillus rhamnosus* exerts its prebiotic effect while added in chocolate pudding to reduce syneresis (İspirli et al., 2018). Microbial homopolysaccharide dextran is used in bakery industry to increase the softness of the product. It is also being used in ice cream and confectionary industries as cryoprotectant and protection against oxidative damage of the food product. Other commercialized microbial exopolysaccharides like Xanthan, Gellan, Alginate are also being used in bakery, dairy, and beverage industry to increase the stability of the food (Suryawanshi et al., 2022). A wide variety of EPS obtained from fungi are being used culinary, pharmaceutical, and medical industry (Rusinova-Videva et al., 2023).

EPS produced by different probiotic bacteria have added advantage of being used as prebiotic in the gut environment. EPS produced by different *Lactobacillus* species have been identified with their antitumor, antioxidant, antidiabetic, and wound healing property (Angelin & Kavitha, 2020). Probiotic bacteria derived prebiotic exopolysaccharide are well established in traditional Bulgarian Yogurt, Tibetan Kefir, Chinese Paocai, Camel Milk, and Tofu.

Besides being used as dietary supplement, biomedical application of bacterial exopolysaccharide as prebiotic is also well known. Use of bacterial EPS in biomedical sphere was started since the application of dextran as plasma expander. EPS secreted by different extremophiles are well accepted in pharmaceutical industry. EPS is commercially widely used as drug encapsulating agent, antibacterial surfactant, and also as substitute of plasma (Nadzir et al., 2021). EPS secreted by *Leuconostoc* sp. is being used as nanogels for intracellular drug delivery (Curcio et al., 2017). Commercially available prebiotic supplement made up off fungal EPS obtained from *Ophiocordyceps dipterigena* increase the probiotic consortium in chicken feces when mixed with regular diet (Prathumpai et al., 2015). Alginate extracted from seaweeds is used in the preparation of antacid stomach protector tablets and microspheric drug delivery. Further experiments in favor of EPS as useful drug delivery system are still going on using multiple in vitro and in vivo model.

References

Angelin, J., & Kavitha, M. (2020). Exopolysaccharides from probiotic bacteria and their health potential. *International Journal of Biological Macromolecule, 162*, 853–865.

Arias, S., del Moral, A., Ferrer, M. R., Tallon, R., Quesada, E., & Béjar, V. (2003). Mauran, an exopolysaccharide produced by the halophilic bacterium *Halomonas maura*, with a novel composition and interesting properties for biotechnology. *Extremophiles, 7*(4), 319–326. https://doi.org/10.1007/s00792-003-0325-8. Epub 2003 May 27.

Baptista, S., Torres, C. A. V., Sevrin, C., Grandfils, C., Reis, M. A. M., & Freitas, F. (2022). Extraction of the bacterial extracellular polysaccharide FucoPol by membrane-based methods: Efficiency and impact on biopolymer properties. *Polymers, 14*(3), 390.

Bouchotroch, S., Quesada, E., Del Moral, A., Llamas, I., & Béjar, V. (2001). Halomonas maura sp. nov., a novel moderately halophilic, exopolysaccharide-producing bacterium. *International Journal of Systemic and Evolutionary Microbiology, 51*(Pt 5), 1625–1632.

Castro-Bravo, N., Wells, J. M., Margolles, A., & Ruas-Madiedo, P. (2018). Interactions of surface exopolysaccharides from Bifidobacterium and Lactobacillus within the intestinal environment. *Frontiers in Microbiology, 9*, 2426.

Castro-Bravo, N., Margolles, A., Wells, J. M., & Ruas-Madiedo, P. (2019). Exopolysaccharides synthesized by *Bifidobacterium animalis* subsp. lactis interact with TLR4 in intestinal epithelial cells. *Anaerobe, 56*, 98–101.

Chen, Y. C., Wu, Y. J., & Hu, C. Y. (2019). Monosaccharide composition influence and immunomodulatory effects of probiotic exopolysaccharides. *International Journal of Biological Macromolecules, 133*, 575–582.

Cobb, B. A., Wang, Q., Tzianabos, A. O., & Kasper, D. L. (2004). Polysaccharide processing and presentation by the MHCII pathway. *Cell, 117*(5), 677–687.

Curcio, M., Diaz-Gomez, L., Cirillo, G., Concheiro, A., Iemma, F., & Alvarez-Lorenzo, C. (2017). pH/redox dual-sensitive dextran nanogels for enhanced intracellular drug delivery. *European Journal of Pharmaceutics and Biopharmaceutics, 117*, 324–332.

Cuthbertson, L., Mainprize, I. L., Naismith, J. H., & Whitfield, C. (2009). Pivotal roles of the outer membrane polysaccharide export and polysaccharide copolymerase protein families in export of extracellular polysaccharides in gram-negative bacteria. *Microbiology and Molecular Biology Reviews, 73*, 155–177.

Delgado, S., Sánchez, B., Margolles, A., Ruas-Madiedo, P., & Ruiz, L. (2020). Molecules produced by probiotics and intestinal microorganisms with immunomodulatory activity. *Nutrients, 12*(2), 391.

Domingos-Lopes, M. F. P., Nagy, A., Stanton, C., Ross, P. R., Gelencsér, E., & Silva, C. C. G. (2017). Immunomodulatory activity of exopolysaccharide producing *Leuconostoc citreum* strain isolated from Pico cheese. *Journal of Functional Foods, 33*, 235–243.

Domżał-Kędzia, M., Ostrowska, M., Lewińska, A., & Łukaszewicz, M. (2023). Recent developments and applications of microbial levan, a versatile polysaccharide-based biopolymer. *Molecules, 28*(14), 5407.

Fanning, S., Hall, L. J., Cronin, M., Zomer, A., MacSharry, J., Goulding, D., Motherway, M. O.'. C., et al. (2012). Bifidobacterial surface-exopolysaccharide facilitates commensal-host interaction through immune modulation and pathogen protection. *Proceedings of the National Academy of Sciences, 109*(6), 2108–2113.

Gibson, G. R., Hutkins, R., Sanders, M. E., Prescott, S. L., Reimer, R. A., Salminen, S. J., et al. (2017). Expert consensus document: The International Scientific Association for Probiotics and Prebiotics (ISAPP) consensus statement on the definition and scope of prebiotics. *Nature Reviews Gastroenterology and Hepatology, 14*(8), 491–502.

Górska, S., Sandstrőm, C., Wojas-Turek, J., Rossowska, J., Pajtasz-Piasecka, E., Brzozowska, E., & Gamian, A. (2016). Structural and immunomodulatory differences among lactobacilli exopolysaccharides isolated from intestines of mice with experimentally induced inflammatory bowel disease. *Scientific Reports, 6*(1), 37613.

Graveline, R., Segura, M., Radzioch, D., & Gottschalk, M. (2007). TLR2-dependent recognition of *Streptococcus suis* is modulated by the presence of capsular polysaccharide which modifies macrophage responsiveness. *International Immunology, 19*(4), 375–389.

Gringhuis, S. I., Den Dunnen, J., Litjens, M., Van Der Vlist, M., & Geijtenbeek, T. B. (2009). Carbohydrate-specific signaling through the DC-SIGN signalosome tailors immunity to *Mycobacterium tuberculosis*, HIV-1 and *Helicobacter pylori*. *Nature Immunology, 10*(10), 1081–1088.

Guan, L., Zhang, L., Xue, Y., Yang, J., & Zhao, Z. (2020). Molecular pathogenesis of the hyaluronic acid capsule of *Pasteurella multicide*. *Microbial Pathogenesis, 149*, 104380.

Hidalgo-Cantabrana, C., Sánchez, B., Milani, C., Ventura, M., Margolles, A., & Ruas-Madiedo, P. (2014). Genomic overview and biological functions of exopolysaccharide biosynthesis in Bifidobacterium spp. *Applied and Environmental Microbiology, 80*(1), 9–18.

Hidalgo-Cantabrana, C., Francesca, A., Rodriguez-Nogalesn, A., Vezza, T., Martinez-Camblor, P., Margolles, A., Ruas-Madiedo, P., & Galvez, J. (2016). Effect of a ropy exopolysaccharide-producing Bifidobacterium animalis subsp. lactis strain orally administered on DSS-induced colitis mice model. *Frontiers in Microbiology, 7*, 868.

Hill, A., & Diehl, G. (2018). Identifying the patterns of pattern recognition receptors. *Immunity, 49*(3), 389–391.

Islam, T., & Lam, S. (2014). Synthesis of bacterial polysaccharides via the Wzx/Wzy-dependent pathway. *Canadian Journal of Microbiology, 60*, 697–716.

İspirli, H., Demirbaş, F., & Dertli, E. (2018). Glucan type exopolysaccharide (EPS) shows prebiotic effect and reduces syneresis in chocolate pudding. *Journal of Food Science and Technology, 55*(9), 3821–3826.

Janeway, A., Jr., & Medzhitov, R. (2002). Innate immune recognition. *Annual Review of Immunology, 20*(1), 197–216.

Jones, S., Paynich, M., Kearns, D., & Knight, K. (2014). Protection from intestinal inflammation by bacterial exopolysaccharides. *The Journal of Immunology, 192*(10), 4813–4820.

Jurášková, D., Ribeiro, S. C., & Silva, C. C. (2022). Exopolysaccharides produced by lactic acid bacteria: From biosynthesis to health-promoting properties. *Foods, 11*(2), 156.

Kalka-Moll, W. M., Tzianabos, A. O., Bryant, P. W., Niemeyer, M., Ploegh, H. L., & Kasper, D. L. (2002). Zwitterionic polysaccharides stimulate T cells by MHC class II-dependent interactions. *The Journal of Immunology, 169*(11), 6149–6153.

Kamiya, T., Tang, C., Kadoki, M., Oshima, K., Masahira, H., Saijo, S., Adachi, Y., Ohno, N., & Iwakura, Y. (2018). β-Glucans in food modify colonic microflora by inducing antimicrobial protein, calprotectin, in a Dectin-1-induced-IL-17F-dependent manner. *Mucosal Immunology, 11*(3), 763–773.

Kaur, N., & Dey, P. (2023). Bacterial exopolysaccharides as emerging bioactive macromolecules: From fundamentals to applications. *Research in Microbiology, 174*(4), 104024.

Khan, R., Shah, M. D., Shah, L., Lee, P. C., & Khan, I. (2022). Bacterial polysaccharides—A big source for prebiotics and therapeutics. *Frontiers in Nutrition, 9*, 1031935.

Korakli, M., Gänzle, M., & Vogel, R. (2002). Metabolism by bifidobacteria and lactic acid bacteria of polysaccharides from wheat and rye, and exopolysaccharides produced by *Lactobacillus sanfranciscensis*. *Journal of Applied Microbiology, 92*(5), 958–965.

Laiño, J., Villena, J., Kanmani, P., & Kitazawa, H. (2016). Immunoregulatory effects triggered by lactic acid bacteria exopolysaccharides: New insights into molecular interactions with host cells. *Microorganisms, 4*(3), 27.

Lewis, D., & Blutt, S. (2019). Organization of the immune system. In *Clinical immunology* (pp. 19–38). Elsevier.

Lin, M., Yang, Y., Chen, Y., Hua, K., Lu, C., Sheu, F., & Wu, S. (2011). A novel exopolysaccharide from the biofilm of *Thermus aquaticus* YT-1 induces the immune response through Toll-like receptor 2. *Journal of Biological Chemistry, 286*(20), 17736–17745.

Lopez, P., Monteserin, D. C., Gueimonde, M., de los Reyes-Gavilan, C. G., Margolles, A., Suarez, A., & Ruas-Madiedo, P. (2012). Exopolysaccharide-producing Bifidobacterium strains elicit different in vitro responses upon interaction with human cells. *Food Research International, 46*(1), 99–107.

Matsuzaki, C., Hayakawa, A., Matsumoto, K., Katoh, T., Yamamoto, K., & Hisa, K. (2015). Exopolysaccharides produced by *Leuconostoc mesenteroides* strain NTM048 as an immunostimulant to enhance the mucosal barrier and influence the systemic immune response. *Journal of Agricultural and Food Chemistry, 63*(31), 7009–7015.

McGreal, E., Miller, J., & Gordon, S. (2005). Ligand recognition by antigen-presenting cell C-type lectin receptors. *Current Opinion in Immunology, 17*(1), 18–24.

Mende, S., Rohm, H., & Jaros, D. (2016). Influence of exopolysaccharides on the structure, texture, stability and sensory properties of yoghurt and related products. *International Dairy Journal, 52*(1), 57–71.

Morona, R., Purins, L., Tocilj, A., Matte, A., & Cygler, M. (2009). Sequence-structure relationships in polysaccharide co-polymerase (PCP) proteins. *Trends in Biochemical Sciences, 34*, 78–84.

Murofushi, Y., Julio, V., Kyoko, M., Paulraj, K., Masanori, T., Tomoyuki, S., & Hisashi, A. (2015). The toll-like receptor family protein RP105/MD1 complex is involved in the immunoregulatory effect of exopolysaccharides from *Lactobacillus plantarum* N14. *Molecular Immunology, 64*(1), 63–75.

Nadzir, M., Nurhayati, R., Idris, F., & Nguyen, M. (2021). Biomedical applications of bacterial exopolysaccharides: A review. *Polymers (Basel), 3*(4), 530–553.

Netrusov, A., Liyaskina, E., Kurgaeva, I., Liyaskina, A., Yang, G., & Revin, V. (2023). Exopolysaccharides producing bacteria: A review. *Microorganisms, 11*(6), 1541.

Nowak, B., Śróttek, M., Ciszek-Lenda, M., Skałkowska, A., Gamian, A., Górska, S., & Marcinkiewicz, J. (2020). Exopolysaccharide from *Lactobacillus rhamnosus* KL37 inhibits T cell-dependent immune response in mice. *Archivum Immunologiae et Therapiae Experimentalis, 68*(3), 17.

Oerlemans, M. M., Akkerman, R., Ferrari, M., Walvoort, M. T., & de Vos, P. (2021). Benefits of bacteria-derived exopolysaccharides on gastrointestinal microbiota, immunity and health. *Journal of Functional Foods, 76*, 104289.

Paulraj, K., Albarracin, L., Kobayashi, H., Iida, H., Komatsu, R., Kober, A. K. M. H., & Ikeda-Ohtsubo, W. (2018). Exopolysaccharides from *Lactobacillus delbrueckii* OLL1073R-1 modulate innate antiviral immune response in porcine intestinal epithelial cells. *Molecular Immunology, 93*, 253–265.

Prado, M. R., Blandón, L. M., Vandenberghe, L. P., Rodrigues, C., Castro, G. R., Thomaz-Soccol, V., & Soccol, C. R. (2015). Milk kefir: Composition, microbial cultures, biological activities, and related products. *Frontiers in Microbiology, 6*, 1177.

Prathumpai, W., Rachtawee, P., & Khajeeram, S. (2015). Potential of fungal exopolysaccharide as novel source for prebiotic supplement to broiler chicken diet. *The Indian Journal of Animal Sciences, 85*(12), 1362–1369.

Raffatellu, M., Chessa, D., Wilson, R. P., Tükel, C., Akçelik, M., & Bäumler, A. J. (2006). Capsule-mediated immune evasion: A new hypothesis explaining aspects of typhoid fever pathogenesis. *Infection and Immunity, 74*(1), 19–27.

Rehm, B. H. (2015). Synthetic biology towards the synthesis of custom-made polysaccharides. *Microbiology and Biotechnology, 8*, 19–20.

Roca, C., Alves, V., Freitas, F., & Reis, M. A. M. (2015). Exopolysaccharides enriched in rare sugars: Bacterial sources, production, and applications. *Frontiers in Microbiology, 6*, 288.

Rossi, O., Tanweer Khan, M., Schwarzer, M., Hudcovic, T., Srutkova, D., Duncan, S. H., Stolte, E. H., et al. (2015). *Faecalibacterium prausnitzii* strain HTF-F and its extracellular polymeric matrix attenuate clinical parameters in DSS-induced colitis. *PLoS One, 10*(4), e0123013.

Round, J., Lee, S., Li, J., Tran, G., Jabri, B., Chatila, T. A., & Mazmanian, S. (2011). The Toll-like receptor 2 pathway establishes colonization by a commensal of the human microbiota. *Science, 332*(6032), 974–977.

Ruas-Madiedo, P. (2014). Biosynthesis and bioactivity of exopolysaccharides produced by probiotic bacteria. In *Food oligosaccharides: Production, analysis and bioactivity* (pp. 118–133). Wiley.

Ruijssenaars, H. J., Stingele, F., & Hartmans, S. (2000). Biodegradability of food-associated extracellular polysaccharides. *Current Microbiology, 40*(3), 194–199.

Rusinova-Videva, S., Ognyanov, M., Georgiev, Y., Petrova, A., Dimitrova, P., & Kambourova, M. (2023). Chemical characterization and biological effect of exopolysaccharides synthesized by Antarctic yeasts *Cystobasidium ongulense* AL101 and *Leucosporidium yakuticum* AL102 on murine innate immune cells. *World Journal of Microbiology and Biotechnology, 39*(2), 39.

Sahasrabudhe, N. M., Schols, H. A., Faas, M. M., & de Vos, P. (2016). Arabinoxylan activates Dectin-1 and modulates particulate β-glucan-induced Dectin-1 activation. *Molecular Nutrition & Food Research, 60*(2), 458–467.

Salazar, N., Gueimonde, M., Hernández-Barranco, A. M., Ruas-Madiedo, P., & de los Reyes-Gavilán, C. G. (2008). Exopolysaccharides produced by intestinal Bifidobacterium strains act as fermentable substrates for human intestinal bacteria. *Applied and Environmental Microbiology, 74*(15), 4737–4745.

Schiavi, E., Plattner, S., Rodriguez-Perez, N., Barcik, W., Frei, R., Ferstl, R., & Kurnik-Lucka, M. (2018). Exopolysaccharide from *Bifidobacterium longum* subsp. longum 35624™ modulates murine allergic airway responses. *Beneficial Microbes, 9*(5), 761–773.

Schmid, J., Sieber, V., & Rehm, B. (2015). Bacterial exopolysaccharides: Biosynthesis pathways and engineering strategies. *Frontiers in Microbiology, 6*, 2015.

Shao, L., Wu, Z., Zhang, H., Chen, W., Ai, L., & Guo, B. (2014). Partial characterization and immunostimulatory activity of exopolysaccharides from *Lactobacillus rhamnosus* KF5. *Carbohydrate Polymers, 107*, 51–56.

Speciale, I., Verma, R., Di Lorenzo, F., Molinaro, A., Im, S. H., & De Castro, C. (2019). *Bifidobacterium bifidum* presents on the cell surface a complex mixture of glucans and galactans with different immunological properties. *Carbohydrate Polymers, 218*, 269–278.

Suryawanshi, N., Naik, S., & Jujjawarapu, S. (2022). Exopolysaccharides and their applications in food processing industries. *Food Science and Applied Biotechnology, 5*, 22.

Thomson, P., Medina, D. A., Ortúzar, V., Gotteland, M., & Garrido, D. (2018). Anti-inflammatory effect of microbial consortia during the utilization of dietary polysaccharides. *Food Research International, 109*, 14–23.

Tiwari, O. N., Sasmal, S., Kataria, A. K., & Devi, I. (2020). Application of microbial extracellular carbohydrate polymeric substances in food and allied industries. *3 Biotech, 10*(5), 221.

Wang, Q., McLoughlin, R. M., Cobb, B. A., Charrel-Dennis, M., Zaleski, K. J., Golenbock, D., & Kasper, D. L. (2006). A bacterial carbohydrate links innate and adaptive responses through Toll-like receptor 2. *The Journal of Experimental Medicine, 203*(13), 2853–2863.

Whitney, J., & Howell, P. (2013). Synthase-dependent exopolysaccharide secretion in Gram-negative bacteria. *Trends in Microbiology, 21*, 63–72.

Xiao, M., Ren, X., Yu, Y., Gao, W., Zhu, C., Sun, H., Kong, Q., Fu, X., & Mou, H. (2022). Fucose-containing bacterial exopolysaccharides: Sources, biological activities, and food applications. *Food Chemistry X, 13*, 100233.

Xiu, L., Zhang, H., Hu, Z., Liang, Y., Guo, S., Yang, M., & Wang, X. (2018). Immunostimulatory activity of exopolysaccharides from probiotic *Lactobacillus casei* WXD030 strain as a novel adjuvant in vitro and in vivo. *Food and Agricultural Immunology, 29*(1), 1086–1105.

Xue, H., Liang, B., Wang, Y., Gao, H., Fang, S., Xie, K., & Tan, J. (2024). The regulatory effect of polysaccharides on the gut microbiota and their effect on human health: A review. *International Journal of Biological Macromolecules, 270*(2), 132170.

Yu, R., Zuo, F., Ma, H., & Chen, S. (2019). Exopolysaccharide-producing *Bifidobacterium adolescentis* strains with similar adhesion property induce differential regulation of inflammatory immune response in Treg/Th17 axis of DSS-colitis mice. *Nutrients, 11*(4), 782.

Yue, L., Jiaqi, W., & Changxin, W. (2022). Modulation of gut microbiota and immune system by probiotics, pre-biotics, and post-biotics. *Frontiers in Nutrition, 8*, 634897.

Zhu, Y., Zhou, J. M., Liu, W., Pi, X., Zhou, Q., Li, P., Zhou, T., & Gu, Q. (2021). Effects of exopolysaccharide from *Lactobacillus rhamnosus* on human gut microbiota in in vitro fermentation model. *LWT – Food Science and Technology, 139*, 110524.

Sustainable Biopolymers: Economic Feasibility of Microbial Exopolysaccharides Production

Dhritiksha M. Baria, Shivani M. Yagnik, Rakeshkumar R. Panchal, Kiransinh N. Rajput, and Vikram H. Raval

1 Introduction

Technologies in the microbial synthesis of biomolecules such as polymers, enzymes, antibiotics, and metabolites have advanced significantly since the turn of the twentieth century. Today, the industrial manufacturing of a broad range of goods, including biopharmaceuticals and medicines in the healthcare industry, biopolymers and biofuels in the energy and environmental sectors, and pesticides, fertilizers, and feed additives in the agrochemical sector, employs microorganisms. By 2016, the global market for microbial goods is predicted to reach over 250 billion US dollars due to rising end-use sector needs and increased environmental concerns, per current market reports (Souza et al., 2022).

The majority of cells are covered in polysaccharides, which are naturally occurring, non-toxic, and biodegradable polymers that are crucial to many biological processes, including immune response, adhesion, infection, and signal transmission (Hosseinkhani & Domb, 2019; Peng et al., 2021). Studies on the alternative therapies used by various countries throughout history have shown that the plants and fungi used were rich in bioactive polysaccharides that have been shown to exhibit immunomodulatory activity and possess health-promoting properties when used to treat cancer and inflammatory diseases. As a result, a lot of studies have been done

D. M. Baria · R. R. Panchal · K. N. Rajput · V. H. Raval (✉)
Department of Microbiology and Biotechnology, University School of Sciences, Gujarat University, Ahmedabad, Gujarat, India
e-mail: vhraval@gujaratuniversity.ac.in

S. M. Yagnik
Department of Microbiology and Biotechnology, University School of Sciences, Gujarat University, Ahmedabad, Gujarat, India

Department of Biotechnology, M.G. Science Institute (Autonomous), Navrangpura, Ahmedabad, Gujarat, India

T. Sarkar et al. (eds.), *Extracellular Polysaccharides*,
https://doi.org/10.1007/978-3-031-94472-7_18

to use structure-function analysis to clarify the biological activity mechanism of these polysaccharides (Zhou et al., 2019). In addition to their use in the biotechnology and health sectors, polysaccharides are employed as emulsifiers, biosorbents, and bio flocculants in the environmental sector (Baria et al., 2022) and as thickeners, stabilizers, probiotics, and gelling agents in the food and cosmetic industries (Basiri, 2021; Meng et al., 2023).

The production of natural polymers that are beneficial for a range of industries and applications has garnered more attention throughout the last 10 years (Baria et al., 2020; Rana & Upadhyay, 2020). According to Zang et al. (2023), polysaccharides are primarily utilized as food additives and are derived from plants. However, exopolysaccharides (EPS) originating from microorganisms are now the focus of investigation. Sugar residues are polymerized to synthesize EPS. The molecular weights range from 10 to 1000 kDa and the chain lengths, of the EPS molecules are extremely varied. These are either heteropolysaccharides, which are made up of several monosaccharides, or homopolysaccharides, which are made up of just one kind of monosaccharide (Bhatt et al., 2022).

Polysaccharides exhibit an extensive amount of variation in their structure and content. Based on the monomeric nature of these molecules, they are typically categorized as either homopolysaccharides or heteropolysaccharides (Nabot et al., 2022). A single type of monosaccharide repeating unit, known as a homopolysaccharide, is made up of sugar monomers bonded together to form either ramified chains (like dextran) or linear chains (like pullulan, levan, curdlan, or bacterial cellulose). Typically found as numerous copies of oligosaccharides with three to eight residues (gellan or xanthan), heteropolysaccharides are made up of two or more types of monosaccharides (Rana & Upadhyay, 2020; Nabot et al., 2022).

Bacteria make up the majority of the microorganisms used in industry to produce EPS. The species of *Xanthomonas*, *Leuconostoc*, *Sphingomonas*, and *Alcaligenes* that produce xanthan, dextran, gellan, and curdlan are the most well known and often used (Table 1). The highly regulated carbon and energy-intensive process of EPS synthesis causes a range of environmental and dietary requirements for the strains that synthesize EPS (Rana & Upadhyay, 2020).

For novel polymers, such as hydrocolloids, to become commodities in large markets, production costs must be reduced. Utilizing less costly substrates, optimizing fermentation conditions to increase product output, investigating more productive strains (e.g., by mutagenesis or genetic editing), and simplifying downstream processing are all ways to lower production costs (Yemenicioğlu et al., 2020).

2 Properties and Significance of Microbial Exopolysaccharides in Food Industry

Industry prefers microbial extracellular polymers (EPS) because of their unique functions, repeatable physicochemical characteristics, consistent price, and plentiful supply (Birch et al., 2019; Hamidi et al., 2022). The structure and composition

Table 1 List of EPS-producing microorganisms using various low-cost substrates

Microorganisms	EPS	Monomers	Charge	Low-cost substrate	Yield	References
Leuconostoc mesenteroides V-2317D	Dextran	Glucose	Neutral	Sugar beet molasses	50 g/L	Zhang et al. (2021)
Sphingomonas paucimobilis ATCC-31461	Gellan	Glucose Rhamnose Glucuronic acid	Anionic	Sugarcane molasses	13.81 g/L	Liu et al. (2023)
Paenibacillus polymyxa NRRL B-18475	Levan	Fructose	Neutral	Sugarcane syrup	19.6 g/L	Liyaskina et al. (2021)
Aureobasidium pullulans NRRLY-6220	Pullulan	Glucose	Neutral	Grape pomace	22.3 g/L	Singh et al. (2019)
Sclerotium rolfsii MTCC 2156	Scleroglucan	Glucose	Neutral	Sugarcane juice	23.87 g/L	Valdez et al. (2019)
Xanthomonas campestris	Xanthan	Glucose Mannose Glucuronic acid	Anionic	Carob extracts	0.126 g/L	Zakeri et al. (2017)
Xanthomonas campestris EBK-4	Xanthan	Glucose Mannose Glucuronic acid	Anionic	Ram horn hydrolysate	25.6 g/L	Dey and Chatterji (2023)
Botryosphaeria rhodina	β-Glucan	Glucose	Anionic	Olive mill wastewater	17.2 g/L	Zerva et al. (2017)

of the polysaccharides, which in turn depend on the microorganism from which they have been extracted, influence the rheological properties of microbial EPS solutions, which are crucial for managing and comprehending their industrial applications (Ayyash et al., 2020). Moreover, due to their diversity, these polysaccharides may have a range of applications across several industries. The distinctive structure and special rheological qualities, microbial polysaccharides can significantly change the physical characteristics of food products, making them useful in the food industry for stabilizing, thickening, and gelling applications as well as preventing the formation of ice crystals in frozen foods (Lovegrove et al., 2017). Instant meals, salad dressings, sauces, toppings, processed cheese, and dairy products all include polysaccharides with the aforementioned characteristics (Thakur et al., 2022). Microorganisms that produce EPS are commonly employed in the dairy industry to stabilize dairy products and improve their texture and flavor by binding water (Korcz & Varga, 2021). In bakery goods, dextran is used to increase viscosity, retain moisture, and stop sugar crystallization (Rana & Upadhyay, 2020). Furthermore, the food sector may use EPS-stabilizing emulsions between water and hydrophobic substances as natural emulsifiers (Zammouri et al., 2024).

Emulsifiers are frequently subjected to variations in temperature, pH, and salinity during industrial processes (McClements & Jafari, 2018). Extremophile-derived bioemulsifiers exhibit excellent emulsion stability throughout a broad pH, salinity, and temperature range (Mouro et al., 2024). In a variety of food formulations, including sauce, dairy, cake, salad dressing, syrup, and pudding, high pseudoplasticity is a desirable rheological property (Himashree et al., 2022). Good sensory attributes including tongue feel and flavor release are produced by EPS's pseudoplastic characteristic. In food processing, it is also helpful for pumping, pouring, and mixing at various operative shear rates (Yalmanci et al., 2023).

2.1 Rheological Properties

Dextran can be added to baked goods to enhance their rheological characteristics. Sourdough produces dextran on-site, which improves the bread's softness, volume, and airiness. Additionally, it enhances the bread's texture and mouthfeel (Wang et al., 2020a, b). Sourdough dextran should have a high molecular weight and fewer branching bonds. It is said that dextran produced in sourdough is more efficient than adding it externally (Zhang et al., 2024). Ayyash et al. (2020) recently assessed the rheological and bioactive potential of EPS produced by *Leuconostoc plantarum* C70, a novel probiotic. Camel milk was considered a potential source for EPS-C70 isolation in this investigation. Furthermore, EPSC70 exhibits a potent bactericidal effect against *Salmonella typhimurium*, *Listeria monocytogenes*, *E. coli* O157:H7, and *Staphylococcus aureus* ATCC are among the harmful foodborne pathogens. This EPS reduces the initial growing populations of these pathogenic microorganisms by 2.0–3.0 log (CFU/mL) (Bibi et al., 2021). An overview of EPS's technological and functional characteristics in the food industry is represented in Fig. 1.

2.2 Flavor and Texture-Enhancing Properties

A well-known, lactose-rich by-product, Whey initially formed during the coagulation of milk in the cheese-making industry (Guo & Wang, 2019). A β-galactosidase enzyme is used in a few more stages to transform the lactose in whey. Additionally, this mechanism catalyzes lactose's hydrolysis and transgalactosylation, which are further processed to produce useful by-products (Xavier et al., 2018). Numerous bacterial and fungal extracellular polymers (EPSs), including succinoglycan, curdlan, levan, pullulan, and EPS-605, are currently employed as stabilizing and reducing agents for the development of metal-based nanoparticles (Li et al., 2020). Since EPS structural units are essential for generating an appropriate mouthfeel and flavor-enhancing qualities (Bibi et al., 2021), they may be lost in food items as a result of shear thinning properties (pseudoplastic non-Newtonian fluids), primarily due to hydrodynamic forces.

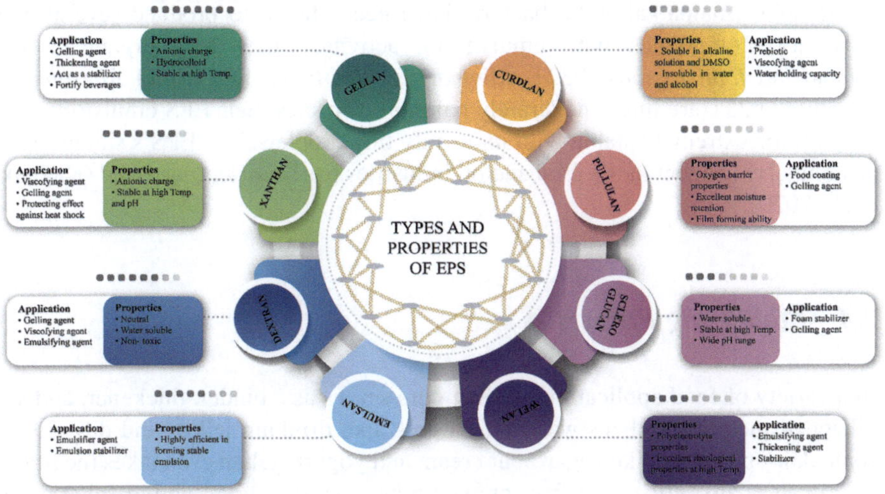

Fig. 1 Overview of types, properties, and applications of exopolysaccharides

These days, the possibility of turning cheese whey into beneficial products is drawing attention to this research. Therefore, *L. plantarum*-produced EPS is typically utilized as a textural agent, while it may also be used as a probiotic (Silva et al., 2019). Because of its capacity to biosorb methylene blue dye, a particular strain of *L. plantarum* (JNULCC001) may turn cheese whey into a valuable EPS product. It also contributes to the thickening, gelling, stabilizing, and synthesizing of selenium nanoparticles (Li et al., 2020).

2.3 *Emulsifying Property*

Numerous novel EPSs have surfaced as promising stabilizing and emulsifying agents in recent years. Both water-in-oil and oil-in-water emulsions could be developed and stabilized with the help of these. This EPS activity could offer the food industry a good substitute for emulsifiers and stabilizers to improve organoleptic properties (Sacco et al., 2019). A recent study examined the emulsifying activity of EPS derived from *Lactobacillus sakei* L3 concerning several technological characteristics. The *L. sakei* L3 EPS's emulsification indices with various vegetable oils, such as soybean, peanut, sunflower, and mustard oils, ranged from 2.61% to 62.30%. Sunflower oil at a concentration of 2.0 mg/mL L3 EPS produced the most stable and effective emulsion (Wang et al., 2019).

More recently, EPS obtained from *L. plantarum* S123 has shown strong emulsifying action because of its very porous and spongy shape. Food products may benefit from the use of this EPS as a stabilizer. Compared to coconut oil, another study found that the *Lactobacillus paracasei* EPS emulsification index was significantly higher in toluene (79.20) and benzene (78.867) boosted media (Shankar et al., 2021).

Halophilic/haloalkaliphilic bacteria have been shown to produce several exopolymeric compounds with emulsifying activity. Many microorganisms like *Idiomarina fontislapidosi* F32T, *Idiomarina ramblicola* R22T, and *Alteromonas hispanica* F23T are three moderately halophilic species their EPS emulsions were found to be extremely stable with uniformly tiny droplets. The EPS's suitability as an emulsifying agent in the food and oil industries has been demonstrated (Kavitake et al., 2020).

2.4 Stabilizing Properties

In a variety of food applications, gellan gum serves as a binder, thickener, and stabilizer. It mostly stabilizes water-based gels, like drinking jellies and desserts. In some dairy products, like vegan sour cream and yogurt, gellan even takes the role of gelatin. It is also utilized in fruit preparations, yogurt, sauces, nonfat salad dressings, low-calorie (sugar-free) jams where pectin is ineffective, and films. The simplest way to shorten the setting time of starch confections is to add gellan, which also keeps candies from adhering together in a heated environment (Kane et al., 2024).

2.5 Gelling Properties

The food industry seems to use biopolymer gellan gum as a thickening and gelling agent when alternative polymers are not suitable (Himashree et al., 2022). The degree of acylation of gellan gums determines the characteristics of the gels they produce. In other words, when low-acyl gellan gum is present, the gels are brittle, stiff, and thermally irreversible; but, when high-acyl gellans are present, the gels are flexible, elastic, and thermoreversible (Fig. 1). The viscose and spreadable paste that gellan gums make is different from the delicate pourable gel (Kia et al., 2024).

The ionic strength of the solution and the concentration of the polymer are found to be the primary determinants of gel appearance. The strongest gels are made in acidic conditions, and it is important to note that increasing the ionic strength will result in higher gel turbidity mostly because of increased intermolecular aggregation. Gellan gums are also suited to fortify beverages as a suspending agent for protein, minerals, vitamins, etc. They are utilized in the production of dairy goods such as ice cream, milkshakes, and yogurt, as well as confections, jams, jellies, and manufactured meals (Kurniawan et al., 2024). Gellan gums are suitable for use in the production of vegetarian and religiously restricted foods because they are sourced from non-animal sources. The food industry frequently uses gellan gum, a natural, water-soluble, anionic EPS (*Sphingomonas elodea*), as an ingredient (stabilizer, thickening agent, and gelling agent). However, the food industry can also use gellan gum for membranes and barrier coatings, including breading and batters for potatoes, fish, poultry, cheese, and vegetables (Dev et al., 2022).

2.6 Thickening Properties

Pullulan is used in food as a coating agent, thickening (0.2–3%), and binder. In Japan, pullulan edible packaging sheets are widely used. Pullulan is frequently found in pastries, confections, soy sauces, icings, creams, quick drinks, and other sauces. Pullulan is a low-calorie food additive that is a biodegradable polymer. In coatings, it takes the place of gelatin (Singh et al., 2019). The first industrial polysaccharide made by LAB, such as *Leuconostoc mesenteroides*, is dextran. In 1880, it was determined that dextran was the cause of the thickening and gelation of sugar cane or beet syrups (Halima et al., 2020). Some dextrans are water soluble, whereas others are insoluble due to structural variations. In confections, dextran can be added to increase viscosity, retain moisture, and prevent sugar crystallization. The exopolysaccharide produced by the bacteria during the fermentation process gives fermented milk products their thick, gel-like texture (Tian et al., 2024). In Japan, curdlan is most frequently used to enhance the texture of a variety of foods, such as sausage, noodles, jams, jellies, and so on. Curdlan gives cooked meals more thickness and solidity while preserving their shape (Shang et al., 2024).

2.7 Water-Holding Capacity and Juiciness of Meat Products

Only a few studies introducing EPS-producing LAB spp. to meat products are carried out to assess the sustainability and vulnerability, even though EPS-producing bacteria are currently employed to increase the water-holding capacity and juiciness of meat products (Dertli et al., 2016). Hydrocolloid applications enhance the quality of meat products, such as cooked hams. Except for isolated products, the in situ use of EPS-producing *Lactobacillus* species, such as *L. curvatus, L. sakei,* and *L. plantarum*, in meat matrices enhances product qualities and facilitates the tumbling process during the manufacturing of ham (Hilbig et al., 2019).

In addition to serving as a natural preservative and an antimicrobial agent against four pathogenic strains—*Staphylococcus aureus* ATCC-6538, *Listeria monocytogenes* ATCC-7644, *Enterococcus feacalis* ATCC-29212, and *Pseudomonas aeroginosa* ATCC-27853—EPS of the *L. plantarum* S2 strain was demonstrated to be a promising candidate for use in the food industry (Dincer & Kivanc, 2019). They also suggested that several enzymes, including α-chymotrypsin, α-amylase, trypsin, lysozyme, and pronase, as well as pH and temperature, are necessary for antimicrobial activity. This strain enhanced the food's color, flavor, aroma, and appearance for extended use in the consumer market in addition to seemingly serving as a starting culture (Rzepkowska et al., 2017).

3 Low-Cost EPS Production Resources

3.1 Sugar Beet Pulp

The fibrous substance that remains after the sugar is removed from sugar beets is known as sugar beet pulp (SBP), a by-product of the sugar beet industry. It is mostly made up of cellulose, hemicellulose, and pectin. In nations with robust cattle-rearing sectors, beet pulp is utilized as animal feed. It is disposed of in landfills in other countries. Nonetheless, SBP has the potential to be a valuable renewable resource, and its bioconversion seems to have significant biotechnological value (Marzo et al., 2020).

The pretreatments used for SBP have been the subject of multiple reports. It has been observed that autoclaving SBP for 1 h at 122 °C and 136 °C alters its composition and physicochemical characteristics, increasing swelling and improving the solubility of pectins and arabinans. It was discovered that the cellulose component's hydrolysis efficiency was significantly increased by the ammonia pressurization depressurization (APD) procedure, in which SBP is exploded by the abrupt evaporation of ammonia (Abbès et al., 2020).

3.2 Syrups and Molasses

The fermentative production of commercial polysaccharides like pullulan (Aquinas et al., 2024), xanthan (da Silva et al., 2017), dextran (Díaz-Montes, 2021), scleroglucan (Kozlowicz et al., 2020), levan (de Siqueira & Öner, 2023), and gellan (Dev et al., 2022) are essential. Low pullulan and scleroglucan yields from using crude molasses highlighted the necessity of pretreatments. Moreover, their many benefits, including high sucrose and other nutrient contents, low cost and ready availability, and ease of storage, syrups, and molasses have long been used as substrates for fermentative production. Sulfuric acid pretreatment of sugar beet molasses has been studied for the production of pullulan and levan (Gojgic-Cvijovic et al., 2019). However, when acid treatment was coupled with activated carbon treatment, notable increases in pullulan and levan yields were achieved, most likely as a result of the removal of colored substances and heavy metals. The effectiveness of activated carbon in eliminating heavy metal contaminants is especially well known. *L. mesenteroides* bacterial cultures can also produce dextran from crude beet molasses, and the yields were comparable with those of medium containing pure sucrose (Esmaeilnejad-Moghadam et al., 2019).

3.3 Cheese Whey

The main by-product that is produced when dairy products like cheese are made is whey. Whey's nutritional makeup is determined by the nutritional makeup of the milk it comes from, which is influenced by several variables, including the milk's processing method. The main ingredient, lactose, makes up around 70% of the whey's total solids. A pool of nutrients and growth factors found in whey can also promote the growth of microorganisms; nevertheless, the usefulness of whey for EPS formation is mostly dependent on the microorganism's capacity to use lactose. Cheese whey has been utilized as a source of carbon and nitrogen for the manufacture of gellan (Wang et al., 2020a, b) and xanthan (Cancella et al., 2024). Additionally, two distinct *X. campestris* strains have been utilized to produce xanthan from mozzarella cheese whey; while both strains produced equivalent yields, the polymers' chemical properties were found different (Abdel-Salam et al., 2023). The *X. campestris* strains inefficient lactose utilization has been attributed to the low yields. Conversely, Gamal et al. (2018) assessed the *S. paucimobilis* ATCC 31461 strain's ability to produce gellan gum in a medium that contained lactose, glucose, and sweet cheese whey as substrates. The strain was known to thrive on lactose and to convert lactose directly into an extremely viscous gellan. Three cycles of heat treatment at 80 °C for 30 min were used to neutralize and disinfect sweet cheese whey that was purchased from the industry. After 100 h of fermentation, the flask cultures could provide a maximum gellan yield of 7.9 g/L (Wang et al., 2020a, b). Additionally, cheese whey has been studied as a possible substrate for *L. mesenteroides* NRRL B512 cultures to produce dextran. This was accomplished by precipitating proteins out of whey using autoclaving and centrifugation. When carob extract was added to the medium, 7.23 g/L of dextran could be generated, although lactose in the supernatant was reported to suppress the dextransucrase activity (Esmaeilnejad-Moghadam et al., 2019).

3.4 Waste Loquat Kernel

The high protein and carbohydrate content of waste loquat kernels makes them another viable biomass resource for the synthesis of EPS. After being dried and ground, waste kernels are autoclaved to undergo acid hydrolysis with 2 M HCl. After being neutralized and detoxified with Ca $(OH)_2$, the hydrolysates were utilized for the synthesis of EPS by *Morchella esculenta* (Guo et al., 2024) and scleroglucan by *S. rolfsii* (Zeng et al., 2021).

3.5 Olive Mill Wastewater

OMW, a dark-colored juice made from a combination of olive water, cooling water from machinery, fruit washings, and leftover fruit, is produced in vast quantities during the production of olive oil. OMW typically contains 15% organic material, which is made up of lipids, proteins, and carbohydrates in addition to a variety of other organic compounds, such as monoaromatic and polyaromatic molecules (Diamantopoulou et al., 2021). Its extremely high organic load and the presence of recalcitrant organic compounds, like polyphenols, which have potent antimicrobial qualities, are the main causes of its toxic effects. Therefore, in Mediterranean nations, the valuation of OMW generated by the olive oil business has long been a source of environmental concern. Biovalorization of OMW to value-added compounds is thought to be the most economical and environmentally friendly alternative, in addition to the many traditional technological treatment techniques used. It has been suggested that it can be used as a good substrate for the synthesis of microbial polymers because of its high carbon-to-nitrogen ratio, and it has been used to make xanthan gum and pullulan (Foti et al., 2021).

3.6 Pomace

Grape pomace, a highly significant waste product in the agriculture industry today, has only been the subject of a small number of published studies. In the wine industry, grape pomace is the residue that remains after grapes are pressed to obtain their juice. The annual production of grape pomace (seeds, skin, and stem) is approximately ten million tons worldwide. More than 250 million kg of this by-product is utilized annually in Spain alone, either as low-nutritional animal feed or to produce ethanol through fermentation and distillation, which has little utility. Due to its underutilization and the fact that the majority of it is typically disposed of in open spaces, this material causes major environmental issues (Kachrimanidou et al., 2023). Akdeniz Oktay et al. (2022) utilized hot water at 65–70 °C to extract the sugars from the grape pulp. Then after, they purified the solution and used it to produce pullulan by *Aureobasidium pullulans* NRRLY-6220 (Özcan & Öner, 2015).

The volume and molecular weight of the polymer formed were extremely close to those of the pullulan produced in the prescribed medium since the grape pomace extract was primarily composed of sugars and very little protein. Additionally, after 7 days of fermentation, the pullulan yields were substantial, reaching 22.3 g/L. To produce xanthan using solid-state fermentation (SSF) with *X. campestris*, various substrates such as grape pomace, tangerine peels, and apple pomace can be also utilized (Özcan & Öner, 2015). After neutralizing the organic acids with an alkaline solution, these substrates were introduced to the fermentation media. These feedstocks' performances were also assessed with and without spent malt grains since apple pomace and inert support were shown to be superior substrates that produced significant amounts of xanthan in both scenarios. Due to the low sugar content and the solid material's poor absorption capacity, grape pomace produced low xanthan yields.

3.7 Lignocellulosic Biomass

A cheap and plentiful substitute for microbial biopolymer production is lignocellulosic biomass, particularly for microbial systems that can hydrolyze using endoglucanases or cellobiose. Otherwise, pretreatments are necessary because it is only used to a limited degree during fermentation. The filamentous fungi *S. rolfsii* and other medicinal mushrooms (Basidiomycetes) are particularly well suited for producing EPS from lignocellulosic substrates because they can naturally metabolize various five-carbon sugars, such as xylose and arabinose. *Schizophyllum commune* strain ATCC 38548 cells cultivated on alkaline H_2O_2-pretreated corn fiber as a sole carbon source has been shown to produce schizophyllan and arabinoxylan simultaneously (Debnath et al., 2022).

The same strain was also utilized to produce schizophyllan from rice hull hydrolysate detoxified by activated charcoal (Singh et al., 2017). This study methodically examined the effects of single or combination inhibitors as well as the significance of the detoxification phase in the synthesis of EPS. *Lentinus squarrosulus* MBFBL 201, a high temperature-tolerant white rot fungus, has been shown to break down cornstalks quickly under SSF conditions, and it has been possible to collect up to 5 g/L of EPS from the fermentation environment (Das et al., 2021).

3.8 Ram Horn Hydrolysates

In addition to their high mineral and amino acid content, ram horn hydrolysates were also found to be an appropriate stimulant for *X. campestris* EBK-4 xanthan synthesis. Ram horns, which are often discarded as waste in slaughterhouses, were acid hydrolyzed with sulfuric acid and then heated to 130 °C to produce the hydrolysates. The hydrolysates were added to the fermentation medium after being neutralized and filtered to make them clear (Ozdal & Başaran Kurbanoglu, 2019).

4 Economic Evaluation of EPS Production Cost

4.1 Cost-Benefit Analysis of Incorporating EPS into Food Production

Using microbes to synthesize value-added biochemicals from biomass is a viable substitute for harsh chemical synthesis methods that use costly, dangerous, and non-renewable raw materials. A beneficial biopolymer cannot find its right place in the polymer market unless it can be made inexpensively, even though its functional features determine its market potential. The change in feedstock consumption in microbial polysaccharide production necessitates extensive research efforts to implement novel ideas on a broad scale. These ideas may include fermentation and downstream processing methods, as well as new resources and pretreatments. The metabolic requirements of the microbe for producing EPS play a major role in determining the feedstock's suitability. Syrups and molasses of different origins are recognized alternatives if the production process depends on glucose or sucrose (de Siqueira & Öner, 2023). To extract the precursors for polysaccharide synthesis, however, additional biomass resources that are high in pectin, sucrose, or glucose, such as pomace, ram horn, and food industry pulps, could also be utilized in conjunction with a suitable pretreatment. Similarly, the producer strain's lactose metabolism directly affects the cheese whey's potential for gellan production as opposed to xanthan and dextran production. Although cellulose-rich biomass is a well-established source of fungal bioactive EPSs, further research is required to isolate new cellulolytic strains that produce EPSs, develop pretreatment techniques, and optimize fermentation conditions to produce bacterial EPSs from lignocellulosic biomass resources. The fact that the chemical and physical characteristics of the polymers are mostly dictated by the growth conditions is another crucial problem in the synthesis of microbial EPS. This fluctuation in polymer properties is particularly noticeable when biomass residues are utilized as feedstock. When the product is introduced to the market, this issue becomes even more crucial. To find appropriate biomass supplies, it is crucial to accurately characterize the EPS and the mechanism underlying its biosynthesis. Consistency in both product quality and yield is more critical for biopolymers with high-value applications than production yields. This can be achieved to a large degree by limiting the carryover of contaminants and metabolic by-products. In these situations, complex media with variable composition are typically not as suitable for cultivation as chemically defined medium conditions. Cell-free microalgal culture media may be a viable source for the recovery of bioactive polysaccharides since microalgae make it possible to employ CO_2 as an inexpensive, easy, and plentiful carbon source. An increasing amount of research is now concentrated on improving the fermentation conditions for the formation of microalgal polysaccharides due to rising knowledge and awareness of their biological significance (Zhen et al., 2023). Additional research on the creation of multi-stage process strategies may be able to address issues related to the variety of culture conditions for the synthesis of lipid-rich biomass and EPS. The long-term viability of the entire process depends heavily on this research on the integration of value-added chemical synthesis with microalgal biofuel generation using a biorefinery technique (Korcz & Varga, 2021).

4.2 Evaluating Economic Sustainability for Producers and Consumers

Numerous studies on the discovery of novel microbial exopolysaccharides (EPSs) with multifunctional properties have recently been published. All of these microbial EPSs are quite attractive, but using them in a variety of sectors is more difficult than anticipated. One of the likely causes is that, while investigating microbial EPSs with potential for industrial use, an economic perspective is typically absent or even disregarded. Due to a lack of accurate information, conducting a proper cost-benefit analysis (CBA) at the start of such a study is fairly difficult or impossible. When compared to synthetic polymers, microbial EPS, which can be made from renewable resources, has no harmful impact on the environment and is produced more quickly than plant-based EPS (Gu, 2021; Baria et al., 2022).

Additionally, they typically exhibit a wide range of chemical structures, stability in a variety of environmental circumstances, and outstanding rheological qualities, all of which give them distinct physicochemical properties (Miao et al., 2022). Significant biological activities (such as anticancer, antibacterial, antioxidant, anti-inflammatory, immunomodulatory, or prebiotic activity) that are absent from synthetic polymers are typically displayed by microbial extracellular polymers (Qi et al., 2022). As a result, some of them might be included in high-end goods like medications, cosmetics, medical equipment, or functional meals (Niknezhad et al., 2022). Despite all of these benefits, the application of microbial EPS is restricted because of the high cost of production. The quest for novel compounds with improved rheological properties and the creation of more economical and efficient manufacturing processes have been spurred in recent years by the growing demand for biopolymers. The goal is to make these molecules more attractive to the industry (Barcelos et al., 2020). The total cost of producing microbial EPS is significantly influenced by the cost of the culture medium. Therefore, it is essential to use alternative low-cost substrates to lower the end product's price. So, many wastes and by-products from agro-industrial processes have been evaluated as low-cost substrates for different microorganisms to produce EPS (Antunes et al., 2017; Niknezhad et al., 2022).

5 Market Potential of EPS and Challenges

To give dairy products viscosity, stability, and water-binding properties that improve the mouthfeel, texture, and flavor perception of fermented dairy products, EPS-producing bacteria are frequently utilized in dairy products. One of the most significant commercial uses of EPS is in the production of yogurt, lactic acid bacteria, such as *S. thermophilus*, are utilized in dairy products with authority (Ağagündüz et al., 2022). Unique EPS obtained from *Lactobacillus helveticus*, biscuit production process water absorption, dough development time, and dough stability are all increased, extending biscuit shelf life (Finamore et al., 2019). Pullulan and other exopolysaccharides are utilized in baked products and pastas as a partial starch substitute. Pullulan solutions are employed as low-viscosity fillers in sauces and beverages because of their comparatively low viscosity, which is similar to gum Arabic. Pullulan is used to hold nuts to cookies, as well as a binder and stabilizer in food pastes and as an adhesive for dentures. Drying a pullulan solution on a suitable flat surface produces pullulan film. This characteristic is utilized to prolong the shelf life of fruits like apples. The films are clear, highly oxygen-impermeable, and possess outstanding mechanical qualities. Their thickness ranges from 5 to 60 μm, and they keep fruits fresh for extended periods while shielding them from deterioration during transit. These films have no negative consequences, unlike the wax coatings that are typically applied to fruits to keep them fresh because they dissolve easily in water and melt in the mouth like edible food coatings. Specialty films could have scents or colors. Pullulan chips are used for culinary purposes and are decorative. As an alternative, pullulan can be brushed on food directly as a glazing to protect it (Chen et al., 2020). Fucose, galactose, and glucose combined with acyl groups (pyruvate, succinate, and acetate) form a novel high molecular weight heteropolysaccharide that has great potential for use in the food industry as a thickener, stabilizer, emulsifier, or binding agent, such as in icings, jams, jelly, pastries, cheese, sauces, pet food, and salad dressing (Muthukumar et al., 2021). Recently, microbial cellulose has also gained a lot of traction in the food industry. *Acetobacter xylinum* produces cellulose which was introduced as nata de coco, a popular snack in the Philippines and other countries, and demonstrated the ability to lower cholesterol (Vadanan, 2020). It was also introduced in Japan's diet drinks as Manchurian tea for health purposes.

6 Conclusion

Exopolysaccharides are employed in the food industry due to their unique properties such as thickening agents, stabilizers, and emulsifiers. The need for novel and practical additives is growing as food items become more varied and complicated. Various polysaccharides recently developed to alter the texture and viscosity of food. As non-fat alternatives, microbial exopolysaccharides are often used in small amounts and have positive health effects on users. Even though several microbial exopolysaccharides may be used in the food industry, only xanthan, with smaller amounts of dextran and gellan, are synthesized industrially and are the most widely available. Eliminating artificial ingredients and additives from the food industry is now imperative. Customers have demanded for safe and natural ingredients to be added to food goods. This sector makes extensive use of microbial EPS that is specifically produced in-situ due to its unique rheological and techno-functional properties. Different kinds of exopolysaccharides have unique and developing capabilities depending on their chain structure and features, as evidenced by the numerous ways in which food's viscoelastic qualities are changing. Examples of EPS usage in edible films and coatings are numerous. They possess numerous uses, including food preservation, improving nutritional value, preventing moisture loss, improving texture, mouthfeel, plasticizers, and properties of mechanical strength and flexibility.

Acknowledgments Author DMB acknowledges SHODH (ScHeme Of Developing High-Quality Research) fellowship, Education Department, Government of Gujarat, India, with (Reference No. 201901380189) and DMB also acknowledges CSIR-Direct SRF (Senior Research Fellowship) with (Application No. 112-2158-501/2k23/1) during her doctoral research at Gujarat University. VHR heartfully acknowledges the funding provided by the UGC-BSR Start-up Research Grant (BSR) (grant No.F.30-521/2020) and financial support provided by Gujarat State Biotechnology Mission (GSBTM) (GSBTM/JD(R&D)/662/2022-23/00292647). All authors acknowledge the infrastructural support and facilities provided by the Department of Microbiology and Biotechnology, Gujarat University, Ahmedabad.

References

Abbès, B., Lacoste, C., Bliard, C., Maalouf, C., Simescu-Lazar, F., Bogard, F., & Polidori, G. (2020). Novel extruded starch-beet pulp composites for packaging foams. *Materials, 13*(7), 1571.

Abdel-Salam, A. B., Saad, M., Ahmed, R. F., & Soliman, N. S. (2023). Investigation of the influence of xanthan on mozzarella cheese characteristics focusing on its antimicrobial effect. *Journal of Advanced Veterinary and Animal Research, 10*(3), 469.

Ağagündüz, D., Şahin, T. Ö., Ayten, Ş., Yılmaz, B., Güneşliol, B. E., Russo, P., et al. (2022). Lactic acid bacteria as pro-technological, bioprotective, and health-promoting cultures in the dairy food industry. *Food Bioscience, 47*, 101617.

Akdeniz Oktay, B., Bozdemir, M. T., & Özbaş, Z. Y. (2022). Evaluation of some agro-industrial wastes as fermentation medium for pullulan production by *Aureobasidium pullulans* AZ-6. *Current Microbiology, 79*(3), 93.

Antunes, S., Freitas, F., Sevrin, C., Grandfils, C., & Reis, M. A. (2017). Production of FucoPol by *Enterobacter* A47 using waste tomato paste by-product as sole carbon source. *Bioresource Technology, 227*, 66–73.

Aquinas, N., Chithra, C. H., & Bhat, M. R. (2024). Progress in bioproduction, characterization, and applications of pullulan: A review. *Polymer Bulletin, 81*, 1–36.

Ayyash, M., Abu-Jdayil, B., Olaimat, A., Esposito, G., Itsaranuwat, P., Osaili, T., et al. (2020). Physicochemical, bioactive, and rheological properties of an exopolysaccharide produced by a probiotic *Pediococcus pentosaceus* M41. *Carbohydrate Polymers, 229*, 115462.

Barcelos, M. C., Vespermann, K. A., Pelissari, F. M., & Molina, G. (2020). Current status of biotechnological production and applications of microbial exopolysaccharides. *Critical Reviews in Food Science and Nutrition, 60*(9), 1475–1495.

Baria, D. M., Patel, N. Y., Arya, P. S., Rajput, K. N., Panchal, R. R., & Raval, V. H. (2020). Exploration of marine bacteria for their exopolysaccharide (EPS) producing ability along various regions of Gujarat coast. *Bioscience Biotechnology Research Communications, 13*(1), 78–84.

Baria, D. M., Patel, N. Y., Yagnik, S. M., Panchal, R. R., Rajput, K. N., & Raval, V. H. (2022). Exopolysaccharides from marine microbes with prowess for environment cleanup. *Environmental Science and Pollution Research, 29*(51), 76611–76625.

Basiri, S. (2021). Applications of microbial exopolysaccharides in the food industry. *Avicenna Journal of Medical Biochemistry, 9*(2), 107–120.

Bhatt, H. B., Baria, D. M., Raval, V. H., & Singh, S. P. (2022). Multifunctional properties of polysaccharides produced by halophilic bacteria and their new applications in biotechnology. In *Microbial syntrophy-mediated eco-enterprising* (pp. 41–70). Academic Press.

Bibi, A., Xiong, Y., Rajoka, M. S. R., Mehwish, H. M., Radicetti, E., Umair, M., et al. (2021). Recent advances in the production of exopolysaccharide (EPS) from Lactobacillus spp. and its application in the food industry: A review. *Sustainability, 13*(22), 12429.

Birch, J., Van Calsteren, M. R., Pérez, S., & Svensson, B. (2019). The exopolysaccharide properties and structures database: EPS-DB. Application to bacterial exopolysaccharides. *Carbohydrate Polymers, 205*, 565–570.

Cancella, M. J., Cerqueira, A. F. L. W., da Costa Teodoro, L., Pereira, J. R., da Costa Ludwig, Z. M., de Carvalho Anjos, V., et al. (2024). Xanthan gum produced from milk permeate and deproteinized cheese whey: A comparative analysis with commercial xanthan gums. *Biocatalysis and Agricultural Biotechnology, 56*, 103053.

Chen, L., McClements, D. J., Zhang, Z., Zhang, R., Bian, X., Jin, Z., & Tian, Y. (2020). Effect of pullulan on oil absorption and structural organization of native maize starch during frying. *Food Chemistry, 309*, 125681.

da Silva, L. C. C., Targino, B. N., Furtado, M. M., de Oliveira Pinto, M. A., Rodarte, M. P., & Hungaro, H. M. (2017). Xanthan: Biotechnological production and applications. In *Microbial production of food ingredients and additives* (pp. 385–422). Academic Press.

Das, S., Bisoyi, S. K., Anmoldeep, P. D., & Srivastava, S. (2021). Role of cultivated mushrooms in bioremediation: A review. *Biological Forum – An International Journal, 13*(1), 160–168.

de Siqueira, E. C., & Öner, E. T. (2023). Co-production of levan with other high-value bioproducts: A review. *International Journal of Biological Macromolecules, 235*, 123800.

Debnath, S., Hore, S., Das, P., & Saha, A. K. (2022). The evaluation of selected bioactivities of fungal mycelium of a wild *Schizophyllum commune* Fr. from submerged cultures. *Journal of Microbiology, Biotechnology, and Food Sciences, 12*(2), e3250–e3250.

Dertli, E., Yilmaz, M. T., Tatlisu, N. B., Toker, O. S., Cankurt, H., & Sagdic, O. (2016). Effects of in situ exopolysaccharide production and fermentation conditions on physicochemical, microbiological, textural, and microstructural properties of Turkish-type fermented sausage (sucuk). *Meat Science, 121*, 156–165.

Dev, M. J., Warke, R. G., Warke, G. M., Mahajan, G. B., Patil, T. A., & Singhal, R. S. (2022). Advances in fermentative production, purification, characterization, and applications of gellan gum. *Bioresource Technology, 359*, 127498.

Dey, R., & Chatterji, B. P. (2023). Sources and methods of manufacturing xanthan by fermentation of various carbon sources. *Biotechnology Progress, 39*(6), e3379.

Diamantopoulou, P., Gardeli, C., & Papanikolaou, S. (2021). Impact of olive mill wastewaters on the physiological behavior of a wild-type new *Ganoderma resinaceum* isolate. *Environmental Science and Pollution Research, 28*, 20570–20585.

Díaz-Montes, E. (2021). Dextran: Sources, structures, and properties. *Polysaccharides, 2*(3), 554–565.

Dincer, E., & Kivanc, M. (2019). Characterization of *Lactobacillus plantarum* strains isolated from Turkish pastırma and possibility to use of food industry. *Food Science and Technology, 40*, 498–507.

Esmaeilnejad-Moghadam, B., Mokarram, R. R., Hejazi, M. A., Khiabani, M. S., & Keivaninahr, F. (2019). Low molecular weight dextran production by *Leuconostoc mesenteroides* strains: Optimization of a new culture medium and the rheological assessments. *Bioactive Carbohydrates and Dietary Fibre, 18*, 100181.

Finamore, A., Roselli, M., Donini, L., Brasili, E., Rami, R., Carnevali, P., et al. (2019). Supplementation with *Bifidobacterium longum* Bar33 and *Lactobacillus helveticus* Bar13 mixture improves immunity in elderly humans (over 75 years) and aged mice. *Nutrition, 63*, 184–192.

Foti, P., Romeo, F. V., Russo, N., Pino, A., Vaccalluzzo, A., Caggia, C., & Randazzo, C. L. (2021). Olive mill wastewater as renewable raw materials to generate high added-value ingredients for agro-food industries. *Applied Sciences, 11*(16), 7511.

Gamal, F., Abdel-hady, M., Amin, A., & Ali, K. (2018). Production of gellan gum by *Sphingomonas paucimobilis* on crude sweet whey using different bioreactor feeding strategies. *South Asian Journal of Research in Microbiology, 1*(4), 1–12.

Gojgic-Cvijovic, G. D., Jakovljevic, D. M., Loncarevic, B. D., Todorovic, N. M., Pergal, M. V., Ciric, J., et al. (2019). Production of levan by *Bacillus licheniformis* NS032 in sugar beet molasses-based medium. *International Journal of Biological Macromolecules, 121*, 142–151.

Gu, J. D. (2021). Biodegradability of plastics: The issues, recent advances, and future perspectives. *Environmental Science and Pollution Research, 28*(2), 1278–1282.

Guo, M., & Wang, G. (2019). History of whey production and whey protein manufacturing. In *Whey protein production, chemistry, functionality, and applications* (pp. 1–12). https://doi.org/10.1002/9781119256052.ch1

Guo, W., Yun, J., Wang, B., Xu, S., Ye, C., Wang, X., et al. (2024). Comparative study on physicochemical properties and hypoglycemic activities of intracellular and extracellular polysaccharides from submerged fermentation of Morchella esculenta. *International Journal of Biological Macromolecules, 278*, 134759.

Halima, B. A., Alkali, Z. D., & Shafiu, N. A. (2020). Lactic acid bacteria: A review. *International Journal of Advanced Academic Research | Sciences, Technology and Engineering, 6*(3), 21 36.

Hamidi, M., Okoro, O. V., Milan, P. B., Khalili, M. R., Samadian, H., Nie, L., & Shavandi, A. (2022). Fungal exopolysaccharides: Properties, sources, modifications, and biomedical applications. *Carbohydrate Polymers, 284*, 119152.

Hilbig, J., Loeffler, M., Herrmann, K., & Weiss, J. (2019). Application of exopolysaccharide-forming lactic acid bacteria in cooked ham model systems. *Food Research International, 119*, 761–768.

Himashree, P., Sengar, A. S., & Sunil, C. K. (2022). Food thickening agents: Sources, chemistry, properties and applications-A review. *International Journal of Gastronomy and Food Science, 27*, 100468.

Hosseinkhani, H., & Domb, A. J. (2019). Biodegradable polymers in gene-silencing technology. *Polymers for Advanced Technologies, 30*(10), 2647–2655.

Kachrimanidou, V., Alexandri, M., Papapostolou, H., Papadaki, A., & Kopsahelis, N. (2023). Valorization of grape pomace for *Trametes versicolor* mycelial mass and polysaccharides production. *Sustainability, 15*(20), 15080.

Kane, A., Diallo, Y., Tamba, A., Diakhaté, P. A., Sagna, C. G., Mbodji, H., et al. (2024). Frequency and diversity of stabilizers, thickeners, and gelling agents used as food additives in food products sold on Dakar markets. *Open Access Library Journal, 11*(1), 1–13.

Kavitake, D., Balyan, S., Devi, P. B., & Shetty, P. H. (2020). Evaluation of oil-in-water (O/W) emulsifying properties of galactan exopolysaccharide from *Weissella confusa* KR780676. *Journal of Food Science and Technology, 57*, 1579–1585.

Kia, E. M., Rezaei, M., Ezzati, S., & Ghasempour, Z. (2024). Impact of environmental conditions on the preparation and functionality of gellan gum. In *Application of gellan gum as a biomedical polymer* (pp. 21–38). Academic Press.

Korcz, E., & Varga, L. (2021). Exopolysaccharides from lactic acid bacteria: Techno-functional application in the food industry. *Trends in Food Science & Technology, 110*, 375–384.

Kozlowicz, B., Jensen, T., Khambete, M., Kadhum, M. J., Garshol, F. K., Vik, E. A., et al. (2020). Scleroglucan polymer stability: Thermal, chemical, and microbial. In *SPE improved oil recovery conference?* (p. D021S019R003). SPE.

Kurniawan, A. W., Tawali, A. B., & Fitri, F. (2024). The effect of gelatin and xanthan gum concentration on diversification of dark chocolate praline jelly coffee products. In *BIO web of conferences* (Vol. 96, p. 01029). EDP Sciences.

Li, C., Ding, J., Chen, D., Shi, Z., & Wang, L. (2020). Bioconversion of cheese whey into a hetero-exopolysaccharide via a one-step bioprocess and its applications. *Biochemical Engineering Journal, 161*, 107701.

Liu, J., Li, H., Zhang, X., Yue, L., Lu, W., Ma, S., et al. (2023). Cost-efficient production of the sphingan WL gum by *Sphingomonas* sp. WG using molasses and sucrose as the carbon sources. *Marine Biotechnology, 25*(1), 192–203.

Liyaskina, E. V., Rakova, N. A., Kitykina, A. A., Rusyaeva, V. V., Toukach, P. V., Fomenkov, A., et al. (2021). Production and characterization of the exopolysaccharide from strain *Paenibacillus polymyxa* 2020. *PLoS One, 16*(7), e0253482.

Lovegrove, A., Edwards, C. H., De Noni, I., Patel, H., El, S. N., Grassby, T., et al. (2017). Role of polysaccharides in food, digestion, and health. *Critical Reviews in Food Science and Nutrition, 57*(2), 237–253.

Marzo, C., Díaz, A. B., Caro, I., & Blandino, A. (2020). Conversion of exhausted sugar beet pulp into fermentable sugars from a biorefinery approach. *Food, 9*(10), 1351.

McClements, D. J., & Jafari, S. M. (2018). Improving emulsion formation, stability and performance using mixed emulsifiers: A review. *Advances in Colloid and Interface Science, 251*, 55–79.

Meng, Q., Wang, Q., Chen, L., Li, J., Fan, L., Gu, Z., et al. (2023). Rheological properties and thickening effect of high molecular weight exopolysaccharide and intracellular polysaccharide from *Schizophyllum commune*. *Food Hydrocolloids, 144*, 108920.

Miao, S. J., Zhou, J., Qi, G. N., Liu, J. F., Yang, S. Z., Gang, H., & Mu, B. Z. (2022). Rheological properties of a new microbial exopolysaccharide produced by *Sphingomonas* sp. HS and its potential in enhanced oil recovery. *Energy & Fuels, 36*(4), 1792–1798.

Mouro, C., Gomes, A. P., & Gouveia, I. C. (2024). Microbial exopolysaccharides: Structure, diversity, applications, and future frontiers in sustainable functional materials. *Polysaccharides, 5*(3), 241–287.

Muthukumar, J., Chidambaram, R., & Sukumaran, S. (2021). Sulfated polysaccharides and its commercial applications in food industries—A review. *Journal of Food Science and Technology, 58*, 2453–2466.

Nabot, M., Guérin, M., Sivakumar, D., Remize, F., & Garcia, C. (2022). Variability of bacterial homopolysaccharide production and properties during food processing. *Biology, 11*(2), 171.

Niknezhad, S. V., Kianpour, S., Jafarzadeh, S., Alishahi, M., Najafpour Darzi, G., Morowvat, M. H., et al. (2022). Biosynthesis of exopolysaccharide from waste molasses using *Pantoea* sp. BCCS 001 GH: A kinetic and optimization study. *Scientific Reports, 12*(1), 10128.

Özcan, E., & Öner, E. T. (2015). Microbial production of extracellular polysaccharides from biomass sources. In *Polysaccharides* (pp. 161–184). Springer.

Ozdal, M., & Başaran Kurbanoglu, E. (2019). Use of chicken feather peptone and sugar beet molasses as low-cost substrates for xanthan production by *Xanthomonas campestris* MO-03. *Fermentation, 5*(1), 9.

Peng, X., Dong, K., Wu, Z., Wang, J., & Wang, Z. L. (2021). A review on emerging biodegradable polymers for environmentally benign transient electronic skins. *Journal of Materials Science, 56*, 16765–16789.

Qi, M., Zheng, C., Wu, W., Yu, G., & Wang, P. (2022). Exopolysaccharides from marine microbes: Source, structure, and application. *Marine Drugs, 20*(8), 512.

Rana, S., & Upadhyay, L. S. B. (2020). Microbial exopolysaccharides: Synthesis pathways, types, and their commercial applications. *International Journal of Biological Macromolecules, 157*, 577–583.

Rzepkowska, A., Zielińska, D., Ołdak, A., & Kołożyn-Krajewska, D. (2017). Organic whey as a source of Lactobacillus strains with selected technological and antimicrobial properties. *International Journal of Food Science & Technology, 52*(9), 1983–1994.

Sacco, L. P., Castellane, T. C. L., Polachini, T. C., de Macedo Lemos, E. G., & Alves, L. M. C. (2019). Exopolysaccharides produced by *Pandoraea* show emulsifying and anti-biofilm activities. *Journal of Polymer Research, 26*, 1–11.

Shang, J., Liu, W., Xing, Y., Liu, S., Chang, G., Yu, Y., et al. (2024). Influence of the addition of Curdlan gel on the textural properties of beef emulsified sausage. *International Journal of Agriculture and Food Sciences Research, 1*(1), 103–114.

Shankar, T., Palpperumal, S., Kathiresan, D., Sankaralingam, S., Balachandran, C., Baskar, K., et al. (2021). Biomedical and therapeutic potential of exopolysaccharides by *Lactobacillus paracasei* isolated from sauerkraut: Screening and characterization. *Saudi Journal of Biological Sciences, 28*(5), 2943–2950.

Silva, L. A., Lopes Neto, J. H. P., & Cardarelli, H. R. (2019). Exopolysaccharides produced by *Lactobacillus plantarum*: Technological properties, biological activity, and potential application in the food industry. *Annals of Microbiology, 69*, 321–328.

Singh, M. K., Kumar, M., & Thakur, I. S. (2017). Proteomic characterization and schizophyllan production by *Schizophyllum commune* ISTL04 cultured on Leucaena leucocephala wood under submerged fermentation. *Bioresource Technology, 236*, 29–36.

Singh, R. S., Kaur, N., & Kennedy, J. F. (2019). Pullulan production from agro-industrial waste and its applications in food industry: A review. *Carbohydrate Polymers, 217*, 46–57.

Souza, M. A. D., Vilas-Boas, I. T., Leite-da-Silva, J. M., Abrahão, P. D. N., Teixeira-Costa, B. E., & Veiga-Junior, V. F. (2022). Polysaccharides in agro-industrial biomass residues. *Polysaccharides, 3*(1), 95–120.

Thakur, K., Singh, D., & Rajput, R. (2022). Effects of food additives and preservatives and shelf life of the processed foods. *Journal of Current Research in Food Science, 3*, 11–22.

Tian, H., Gu, Y., Lv, Z., & Wang, L. (2024). The exopolysaccharides produced by *Leuconostoc mesenteroides* XR1 and its effect on silk drawing phenomenon of yoghurt. *International Journal of Biological Macromolecules, 262*, 129952.

Vadanan, S. V. (2020). *Bacterial cellulose: From culturing to the development of functional materials*. Nanyang Technological University.

Valdez, A. L., Babot, J. D., Schmid, J., Delgado, O. D., & Fariña, J. I. (2019). Scleroglucan production by *Sclerotium rolfsii* ATCC 201126 from amylaceous and sugarcane molasses-based media: Promising insights for sustainable and eco-friendly scaling-up. *Journal of Polymers and the Environment, 27*, 2804–2818.

Wang, B., Song, Q., Zhao, F., Han, Y., & Zhou, Z. (2019). Production optimization, partial characterization, and properties of an exopolysaccharide from *Lactobacillus sakei L3. International Journal of Biological Macromolecules, 141*, 21–28.

Wang, D., Kim, H., Lee, S., Kim, D. H., & Joe, M. H. (2020a). Improved gellan gum production by a newly isolated *Sphingomonas azotifigens* GL-1 in a cheese whey and molasses-based medium. *Process Biochemistry, 95*, 269–278.

Wang, Y., Trani, A., Knaapila, A., Hietala, S., Coda, R., Katina, K., & Maina, N. H. (2020b). The effect of in situ produced dextran on flavor and texture perception of wholegrain sorghum bread. *Food Hydrocolloids, 106*, 105913.

Xavier, J. R., Ramana, K. V., & Sharma, R. K. (2018). β-galactosidase: Biotechnological applications in food processing. *Journal of Food Biochemistry, 42*(5), e12564.

Yalmanci, D., Dertli, E., Tekin-Cakmak, Z. H., & Karasu, S. (2023). Utilization of exopolysaccharide produced by *Leuconostoc lactis* GW-6 as an emulsifier for low-fat mayonnaise production. *International Journal of Biological Macromolecules, 226*, 772–779.

Yemenicioğlu, A., Farris, S., Turkyilmaz, M., & Gulec, S. (2020). A review of current and future food applications of natural hydrocolloids. *International Journal of Food Science & Technology, 55*(4), 1389–1406.

Zakeri, A., Pazouki, M., & Vosoughi, M. (2017). Use of response surface methodology analysis for xanthan biopolymer production by *Xanthomonas campestris*: Focus on agitation rate, carbon source, and temperature. *Iranian journal of chemistry and chemical engineering, 36*(1), 173–181.

Zammouri, A., Ziadi, M., Gharsallaoui, A., Fguiri, I., Sbissi, I., Hammadi, M., & Khorchani, T. (2024). Characterization of novel exopolysaccharides from *Weissella cibaria* and *Lactococcus lactis* strains and their potential application as bio-hydrocolloid agents in emulsion stability. *Fermentation, 10*(10), 532.

Zang, E., Jiang, L., Cui, H., Li, X., Yan, Y., Liu, Q., et al. (2023). Only plant-based food additives: An overview on application, safety, and key challenges in the food industry. *Food Reviews International, 39*(8), 5132–5163.

Zeng, W., Wang, J., Shan, X., Yu, S., & Zhou, J. (2021). Efficient production of scleroglucan by *Sclerotium rolfsii* and insights into molecular weight modification by high-pressure homogenization. *Frontiers in Bioengineering and Biotechnology, 9*, 748213.

Zerva, A., Papaspyridi, L. M., Christakopoulos, P., & Topakas, E. (2017). Valorization of olive mill wastewater for the production of β-glucans from selected basidiomycetes. *Waste and Biomass Valorization, 8*, 1721–1731.

Zhang, S., Wang, J., & Jiang, H. (2021). Microbial production of value-added bioproducts and enzymes from molasses, a by-product of sugar industry. *Food Chemistry, 346*, 128860.

Zhang, Y., Wang, D., Zhang, Z., Guan, H., Zhang, Y., Xu, D., et al. (2024). Improvement on wheat bread quality by in situ produced dextran—A comprehensive review from the viewpoint of starch and gluten. *Comprehensive Reviews in Food Science and Food Safety, 23*(3), e13353.

Zhen, H., Yuan, K., Qiao, Y., Li, J., Waqas, M. A., Tian, G., et al. (2023). Eco-compensation quantification of sustainable food waste management alternatives based on economic and environmental life cycle cost-benefit assessment. *Journal of Cleaner Production, 382*, 135289.

Zhou, Y., Cui, Y., & Qu, X. (2019). Exopolysaccharides of lactic acid bacteria: Structure, bioactivity, and associations: A review. *Carbohydrate Polymers, 207*, 317–332.

Development and Futuristic Applications of Genetically Modified Extracellular Polymeric Substances

Smritikana Paul, Ritojo Basu, Issheka Biswas, Sayani Datta, Pijush Basak, and Arup Kumar Mitra

1 Introduction

Research on microbial adhesion to surfaces using submerged slides started in the 1930s with ZoBell's investigations on bacteria attaching to glass surfaces. He found that sessile bacteria secrete a cementing substance, forming a film that retains extracellular enzymes and aids in nutrient acquisition. By demonstrating the significance of the matrix in biofilms, Costerton et al. demonstrated how bacteria are adhered to a network of fibers known as the "glycocalyx." It was found that bacteria anchor themselves to substrates by producing polysaccharide fibers. This matrix also aids in nutrient acquisition and enzyme retention, contributing to the structure and function of biofilms (Costerton et al., 1978). Sutherland introduced the term "extracellular polymeric substances" (EPSs) to describe the hydrated gel network of fibers produced by microbial species in biofilms (Cerning, 1995).

Bacteria and other various microbes produce extracellular polymeric substances, or EPS. They vary in sizes, composition, and their chemical properties. EPS helps in cell adaptability, resiliency, and have functional roles in environment. They mainly consist of DNA, polysaccharides, and proteins. Production of EPS is energetically expensive and is primarily induced by environmental signals (Costa et al., 2018). The arrangement of structural unit changes for different kind of EPS. The polysaccharides, being the most studied component, vary vastly in their structure (Roca et al., 2015). The EPS matrix has crucial role in terms of maintaining cell proximity, capturing of resources, mechanical strength including greater tolerance to antimicrobials, and other stressful conditions. Because of the complex and highly

S. Paul (✉) · R. Basu · I. Biswas · S. Datta · A. K. Mitra
Postgraduate and Research Department of Microbiology, St. Xavier's College (Autonomous), Kolkata, India, West Bengal

P. Basak
Department of Biotechnology, JBNSTS, Kolkata, West Bengal, India

T. Sarkar et al. (eds.), *Extracellular Polysaccharides*,
https://doi.org/10.1007/978-3-031-94472-7_19

dynamic nature of the EPS, biofilms are the most prevalent type of life (Flemming et al., 2016). The highly hydrated EPS forms a matrix which holds the biofilm cells together and helps in water retention. It has sorption capabilities which help to sequestrate particulate and dissolved chemicals from the environment (Flemming et al., 2007).

Research on EPS has evolved over the years, with a focus on isolation procedures to obtain accurate components of EPS and their role in biofilm formation. The term "EPS" has become standard in biofilm research, encompassing the diverse polymers that make up the matrix in microbial communities (Wingender et al., 1999). The complex interactions between different polymers within the EPS matrix contribute to the structural integrity and function of biofilms. EPS is poly-anionic in nature and therefore through negative staining or customized stains it can be viewed under light microscope. While the presence of EPS is determined by scanning electron microscopy, the transmission electron microscopy provides information about the surface structures (Tiwari et al., 2020).

The EPS enveloping microbiological cells serve as an innate defense mechanism, isolating the cell from its surroundings. As a result, they exhibit high levels of adhesion, self-defense, source of carbon, and environmental energy production (Sijia et al., 2023). Bacterial EPS play a various role such as adhesion, cohesion, and aggregation of soil particles. In the rhizospheric regions, they retain water molecules and serve as potential barriers (Maitra et al., 2024). It enhances the synthesis of plant-available nutrients (Flemming & Wingender, 2010). The hydrophilic polysaccharides and proteins help in the retaining of water therefore desiccation tolerant in water-scarce environment. Protection against antimicrobials is also due the polysaccharides and proteins. All the components of EPS are essentially nutrient source for the biofilm community (Flemming et al., 2016). From a study on EPS, on its growing influence in ocean systems, it was concluded that the secretion of EPS is an adaptive response employed by the microbes to increase their survivability and metabolic efficiency (Decho & Gutierrez, 2017). EPS can act both as an electron donor or an electron acceptor and helps in electron transportation (Flemming et al., 2016). EPS production improves quality of soil, increase soil aggregation, fertility, and its presence in soil and roots expand the uptake of water and nutrients for the plant and microorganisms both, thus furthering the environment as a whole. EPS-producing bacteria EPS are employed in wastewater treatment and environmental effluent remediation processes. The possible function of EPS is eliminating heavy metals from wastewater because of their chelating ability in a solution containing metal ions (Tiwari et al., 2020).

EPS has valuable properties and therefore various genetic engineering techniques have been developed to increase the production of EPS, or to change its substituents generating novel polysaccharides. Various novel strategies adopted to achieve this include interruption of other processes hampering EPS formation, overexpression of genes responsible for its formation, assembly, and proteins. The primary advantage of microbial polysaccharides lies in their adjustability in chemical composition, alongside high yields, diverse sources, and easy separation and

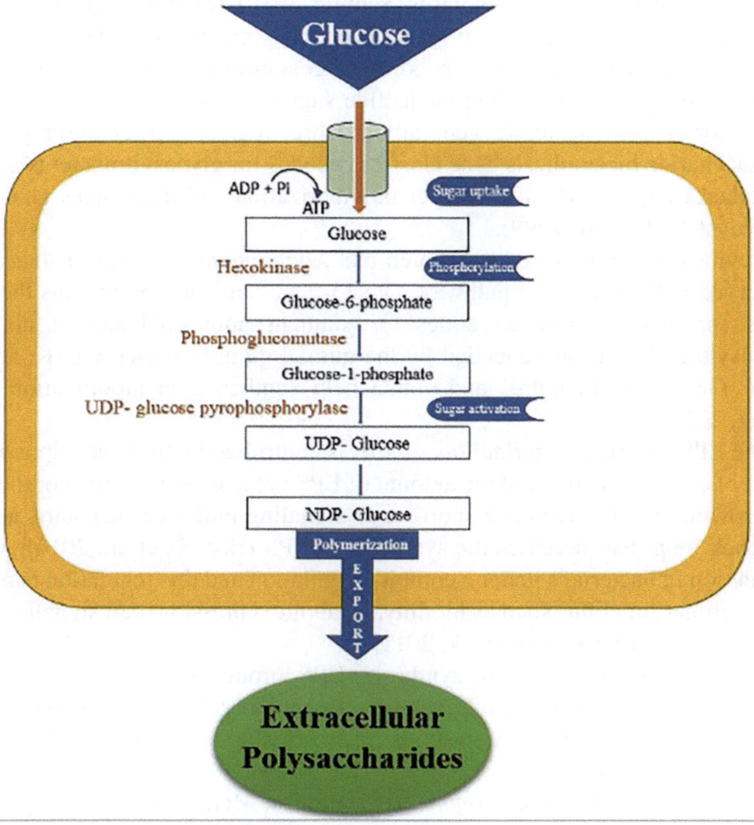

Fig. 1 Extracellular polymeric substances (EPSs) biosynthesis pathway

extraction (Sun & Zhang, 2021). Figure 1 shows extracellular polymeric substances (EPSs) biosynthesis pathway.

2 Genes Involved in EPS Production

The genes involved and the immediate environment, where matrix develops, decides the biochemical arrangement of EPS (Sutherland et al., 2001). Heteropolysaccharides and homopolysaccharides are the two kinds of polysaccharides found in bacteria which are composed of repeated units of xanthan and dextran, respectively (Kumar et al., 2018). Three diverse pathways are employed in EPS production, i.e., the ABC transporter-dependent pathway, synthase-dependent pathway, and Wzx-wzy-dependent pathway. Genes code for various types of glycosyl transferases, enzymes modifying sugar moiety, polymerases (Schmid, 2018). In spite of extensive research, the genetic activity of majority of genes is not fully understood. The genes are classified into three categories: one encoding glycosyltransferases, one for polysaccharide processing and one coding nucleotide sugars (Reeves et al., 1996). The gene coding for the polysaccharide generally has three regions. The central region of the polysaccharide biosynthesis gene clusters encodes for glycosyltransferases and this is flanked by genes which codes for polymerization and determines chain length (De Vuyst & Degeest, 1999).

Previous research has demonstrated that *Xanthomonas campestris* has a highly conserved EPS production pathway. The 13-gene xanthan operon plus the xanAB genes (precursor synthesis) codes for xanthan gum (Wibberg et al., 2015). Glycosyltransferases are encoded by the gumD, gumM, gumH, gumK, and gumI genes. Genes gumL, gumF, and gumG help xanthan gum modification (Sun & Zhang, 2021).

The EPS production in *Bacillus subtilis* is controlled by the transcriptional regulation of epsA-O operon and the amount of EPS present at that time. For the cells in the growing biofilm, EPS is a short-range signaling molecule that starts a positive feedback loop that increases the synthesis of EPS (Elsholz et al., 2014). The EPS production in bacteria is under a complex regulation and this regulation is related to other cellular functions such as motility, catabolic repression, cell signaling that are critical for its survival (Janczarek, 2011).

Like other bacteria, rhizobia synthesize EPS through a structured procedure that needs the coordinated action of several enzymes. The genes involved in this process are typically found on mega plasmids or rhizobial chromosomes in huge clusters (Finan et al., 2001).

In a study conducted on *Streptococcus mutans*, MarR transcription factor EpsR negatively regulated the gtfB gene which codes for EPS production. The regulation of its expression may lead to the prevention of cariogenic biofilm favoring diseases (Chen et al., 2021). Figure 2 shows extracellular polymeric substances (EPSs) gene cluster.

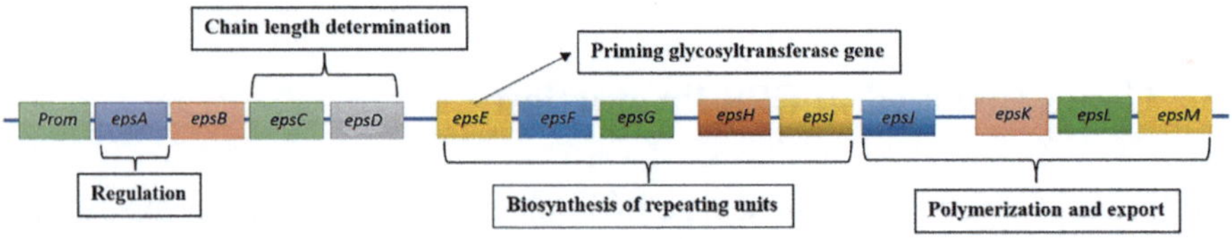

Fig. 2 Extracellular polymeric substances (EPSs) gene cluster

3 Isolation Techniques for Extracellular Polymeric Substances (EPSs) in Biofilm

EPS can be extracted using various techniques. The majority of researchers employ a variety of physical and chemical techniques to enhance the quantity of EPS released by bacteria, although centrifugation alone can be used to extract them. Alkaline reagents, EDTA, and aldehydic solutions are frequently used in chemical extractions, while sonication, ultracentrifugation, cation exchange resin, and heating are examples of physical extractions.

3.1 Physical Extraction Methods for EPS

3.1.1 Centrifugation

Biofilm samples were first collected with sterile techniques and washed with phosphate-buffered saline (PBS) to eliminate any loosely bound cells. When necessary, samples were homogenized to ensure even consistency. They were then centrifuged at 3000–5000 × g for 10–20 minutes to settle larger particles, after which the supernatant was carefully decanted. To further refine the separation, a second centrifugation was carried out at 10,000–15,000 × g for 10–15 minutes, concentrating the smaller particles and leaving a purified supernatant. Finally, the extracellular polymeric substance (EPS) solution was stored at 4 °C to maintain its integrity (Felz et al., 2016).

3.1.2 Ultrasonication

Biofilm samples were first collected and homogenized to ensure uniform consistency. The homogenized samples were then placed in an ultrasonic bath or sonicator containing a suitable liquid medium, such as phosphate-buffered saline (PBS). The frequency was set between 20 and 40 kHz, and the exposure time was adjusted to span 1–30 minutes, with careful monitoring of the temperature to prevent overheating. After sonication, the sample was centrifuged at 3500 × g for 15 minutes at 4 °C to separate the solids, and the EPS-rich supernatant was subsequently collected. Finally, the extracellular polymeric substance (EPS) solution was stored at 4 °C to maintain its stability (Felz et al., 2016).

3.1.3 Steaming Extraction and Heat Treatment

Biofilm samples were collected and thoroughly rinsed. The rinsed biofilms were then placed in a heat-resistant container with a small amount of buffer. To extract extracellular polymeric substances (EPSs), the samples were either steamed for 10–30 minutes or heated in a water bath at 70–100 °C, with temperature carefully monitored to prevent EPS degradation. Once the sample was heated, it was cooled to room temperature to preserve the integrity of the EPS. The samples were then centrifuged at 3500 × g to separate the solids, and the supernatant containing EPS was collected. The final EPS solution was stored at 4 °C for preservation (Pan et al., 2016).

3.2 Chemical Extraction Methods for EPS

3.2.1 EDTA (Ethylene Diamine Tetraacetic Acid)

Biofilm samples were collected using sterile techniques and rinsed with a buffer to remove unattached cells. A 0.1–0.5 M EDTA solution was added to the biofilm, followed by incubation at either room temperature or 4 °C for 30 minutes to several hours, with gentle agitation to facilitate EPS release. After incubation, the samples were then centrifuged at 3500 × g for 15 minutes, and the EPS rich supernatant was collected. The resulting EPS solution was stored at 4 °C to maintain stability (Felz et al., 2016).

3.2.2 Cation Exchange Resins (CERs)

Biofilm samples were collected and rinsed thoroughly. The resin, such as Amberlite IR120, was conditioned in a sodium chloride solution to prepare it for use. The conditioned resin was then added to the biofilm sample at a 1:10 ratio, followed by gentle agitation and incubation for several hours to promote EPS release. After incubation, the mixture was centrifuged at 3500 × g, and the EPS-rich supernatant was collected. The final EPS solution was stored at 4 °C to preserve its properties (Pan et al., 2016).

3.2.3 Formaldehyde–NaOH Extraction

Biofilm granules were transferred into a glass bottle and mixed with demineralized water, followed by the addition of formaldehyde (0.3 mL). The bottle was gently shaken and stored at 4 °C for 1 hour. Afterward, 20 mL of 1 M NaOH was added to the mixture, and it was again gently shaken before being refrigerated at 4 °C for 3 hours. The suspension was then evenly distributed into two centrifuge tubes and centrifuged at 35,000 × g for 15 minutes. Finally, the supernatant containing the extracellular polymeric substances (EPSs) was collected (Felz et al., 2016).

3.2.4 Formamide–Sodium Hydroxide (NaOH) Extraction

Biofilm granules were mixed with demineralized water in a glass bottle, followed by the addition of 0.3 mL of formamide. The mixture was gently agitated and incubated at 4 °C for 1 hour. Subsequently, 20 mL of 1 M NaOH was added, and the solution was shaken and kept at 4 °C for 3 hours. The sample was aliquoted into centrifuge tubes and centrifuged at 3500 × g for 15 minutes to separate the components. The resulting supernatant was carefully collected (Felz et al., 2019).

3.3 Combined Physical and Chemical Methods for EPS Extraction

3.3.1 Ultrasound-NaOH Combination

Biofilm samples were collected and rinsed thoroughly. A 0.1–1.0 M NaOH solution was prepared, with the concentration adjusted based on the characteristics of the biofilm. The samples were immersed in the NaOH solution in an ultrasonic bath or sonicator. The sonication parameters were set with a frequency of 20–40 kHz and a duration of 5–30 minutes, with temperature carefully monitored to prevent overheating. Pulsed sonication was applied to enhance the extraction efficiency. Following sonication, the samples were centrifuged at 3500 × g, and the supernatant containing the extracellular polymeric substances (EPSs) was carefully collected. The EPS solution was stored at 4 °C for preservation (Adav & Lee, 2007).

3.3.2 Heat-NaOH Combination

Biofilm samples were collected and rinsed thoroughly. A 0.1–1.0 M NaOH solution was prepared, and the samples were immersed in this solution. The samples were heated to 70–100 °C, either using steam or a water bath, for 10–20 minutes to aid in the extraction of extracellular polymeric substances (EPSs). Afterward, the samples were allowed to cool to room temperature. The mixture was then centrifuged at 3000–3500 × g, and the supernatant, which contained the EPS, was collected. Finally, the EPS solution was stored at 4 °C for preservation (Adav & Lee, 2007).

4 Characterization of Extracted Extracellular Polymeric Substances (EPSs) from Biofilms

The EPS characterizations show qualitative and quantitative differences depending on the method used.

4.1 Chemical Techniques for Characterizing Extracellular Polymeric Substances (EPSs)

4.1.1 Ultraviolet-Visible (UV-Vis) Spectrophotometry

EPS solution was diluted to ensure the absorbance falls within the linear range (0.1–1.0 absorbance units), using a blank for baseline correction. The sample was transferred to a cuvette, and the spectrophotometer was adjusted to the suitable wavelength, such as 280 nm, for protein analysis. Absorbance was measured, and standard curves were used to quantify the concentration of specific EPS components based on the obtained readings (Pan et al., 2016).

4.1.2 Chromatography

4.1.2.1 HPLC or High-Performance Liquid Chromatography

The EPS polysaccharides were hydrolyzed into monosaccharides, and the sample was then filtered to remove any residual particles. The filtered sample was injected into the chromatographic system, with a mobile phase, such as an acetonitrile gradient, used to elute the monosaccharides through a column. Elution was monitored using a refractive index or UV detector, and the peak areas were compared with those of standards to identify and quantify the specific monosaccharide components.

4.1.2.2 Gas Chromatography-Mass Spectrometry (GC-MS)

The hydrolyzed monosaccharides were treated with a silylation reagent, like trimethylsilyl chloride, to improve their volatility. The derivatized sample was subsequently injected into the GC system. The temperature of the GC oven was progressively raised to facilitate the separation of components on the basis of their volatility as they moved through a capillary column with a non-polar stationary phase. As the components eluted, they were directed into the mass spectrometer for ionization and fragmentation. The resulting mass spectrum was analyzed by comparing the fragmentation patterns to those of known standards to identify the individual monosaccharide components.

4.1.3 Colorimetric Methods for Polysaccharide Analysis

4.1.3.1 Anthrone-Sulfuric Acid Method

A 0.5 mL aliquot of the extracted EPS solution was transferred into a test tube, followed by the careful addition of 5 mL of anthrone reagent (0.2% in concentrated sulfuric acid). The test tube was placed in a boiling water bath and heated for 10 minutes. After heating, the sample was quickly cooled to room temperature. The absorbance was then measured at 620 nm using a UV-Vis spectrophotometer, with a blank consisting of water and anthrone reagent to calibrate the instrument to zero. A standard curve was created using known concentrations of glucose, and the polysaccharide concentration of the EPS sample was interpolated from the curve (Pan et al., 2016).

4.1.3.2 Phenol-Sulfuric Acid Method

1 mL of the EPS sample was transferred into a test tube, followed by the addition of 1 mL of 5% phenol solution and 5 mL of concentrated sulfuric acid. The mixture was allowed to sit at room temperature for 10 minutes to develop color, with gentle mixing to ensure a complete reaction and then absorbance was recorded at 490 nm using a UV-Vis spectrophotometer, with a blank used for baseline correction. The concentration of polysaccharides was determined by comparing absorbance to a glucose standard curve (S. Felz et al., 2019).

4.1.4 Protein Quantification Techniques

4.1.4.1 Lowry Foline-Phenol Method

The EPS sample was diluted to the appropriate concentration range and added to a test tube. Lowry reagent (alkaline copper tartrate solution) was first added, followed by Folin-Ciocalteu reagent after 10 minutes. The mixture was left to incubate at room temperature for 30 minutes to facilitate the reaction. The absorbance was then recorded at 750 nm using a UV-Vis spectrophotometer. Bovine serum albumin (BSA) was used as a protein standard to quantify the concentration of the EPS sample (Felz et al., 2019).

4.1.4.2 Bradford Coomassie Brilliant Blue Dye Method

The EPS sample was diluted to the desired concentration, and Bradford reagent was added to the sample in a test tube. The solution was mixed well and left to stand for 5–10 minutes to allow the color to develop. The absorbance was then recorded at 595 nm using a spectrophotometer. Bovine serum albumin (BSA) was used as a standard to quantify the protein concentration in the EPS sample (Teasdale et al., 2018).

4.1.5 Spectroscopy

4.1.5.1 3D-EEM or Three-Dimensional Excitation–Emission Matrix Spectroscopy

The EPS sample was diluted as needed to achieve the appropriate concentration. The spectrofluorometer was set to scan across multiple excitation wavelengths, typically ranging from 200 to 500 nm, while measuring emission over a wide range (e.g., 300–600 nm). The resulting excitation-emission matrix was recorded. Fluorescence peaks were analyzed to identify proteins and humic substances, with the fluorescence intensities used as a quantitative measure for their concentration in the sample (Pan et al., 2016).

4.1.5.2 Fourier Transform Infrared (FTIR) Spectroscopy

The EPS sample was dehydrated to eliminate water, as water strongly absorbs infrared radiation. The sample was then placed on an ATR (attenuated total reflection) crystal or prepared as a KBr pellet. The FTIR spectrometer was run to record the spectrum in the mid-IR range, typically from 4000 to 500 cm^{-1}. Functional groups, such as hydroxyl and carboxyl, were identified by analyzing the peak positions in the spectrum and comparing them with known reference spectra (Teasdale et al., 2018).

4.1.5.3 Nuclear Magnetic Resonance (NMR) Spectroscopy in EPS Analysis

Biofilms are kept hydrated to preserve their natural structure during analysis, with solid-state NMR used to examine them in situ on substrates. Samples are exposed to a strong magnetic field and radiofrequency pulses, causing certain nuclei (like ^13C, ^1H, or ^15N) to shift between energy levels and emit resonance signals. These signals are unique to their chemical environments, allowing for the identification of functional groups such as hydroxyl (-OH), amide groups or carboxyl (-COOH). The resulting NMR spectra show chemical shifts that correspond to specific molecular structures, enabling the identification of EPS components like polysaccharides and proteins. This method provides detailed information on the biofilm's chemical composition without altering its native state (Pan et al., 2016).

4.2 *Microscopy*

4.2.1 Scanning Electron Microscopy (SEM)

Biofilms are chemically fixed by immersing them in aldehyde solutions, such as glutaraldehyde, to preserve their structural integrity. The samples are then dehydrated by gradually immersing them in increasing concentrations of acetone or ethanol, which removes the water content. Once dried, the biofilm is sputter-coated with a thin layer of conductive metal, such as platinum or gold which enhances electron conductivity during Scanning Electron Microscopy (SEM) imaging. The samples are then placed in a vacuum chamber and scanned using an electron beam. The interaction between the electron beam and the sample surface generates high-resolution 2D images, providing detailed information about the biofilm's morphology (Pan et al., 2016).

4.2.2 Confocal Laser Scanning Microscopy (CLSM)

Fluorescent dyes are used to stain biofilms, selectively binding to various components of the EPS, such as lectins for polysaccharides or SYTO 9 for nucleic acids. Multiple fluorochromes can be used simultaneously for multichannel imaging, enabling the visualization of distinct biofilm components in different channels.

A laser beam scans the stained biofilm sample in thin optical sections, capturing 2D images at various depths. These 2D images are then processed and reconstructed into 3D models, allowing for a detailed spatial analysis of the biofilm structure and the distribution of EPS within the biofilm matrix (Adav & Lee, 2007).

4.2.3 Atomic Force Microscopy (AFM)

The biofilm sample is first prepared for AFM analysis, and the AFM probe is positioned in the appropriate mode based on the desired resolution and sample characteristics. In contact mode, the probe maintains direct contact with the biofilm surface, scanning in a systematic pattern to capture high-resolution data. In tapping mode, the probe intermittently touches the surface, preventing damage to the biofilm by minimizing prolonged contact, and in non-contact mode, the probe is kept just above the surface, relying on proximity forces to gather data without physical contact. The probe scans the surface, and the deflection is measured to create high-resolution 3D topographic images of the biofilm structure. After imaging, force-distance curves are measured to quantify the mechanical properties of the EPS matrix and assess the interactions between biofilm cells and the surface (Pan et al., 2016).

4.2.4 Raman Microscopy (RM)

A laser beam is directed onto the biofilm sample, typically using a monochromatic laser source, which interacts with the molecules within the biofilm. As the laser light interacts with the molecules, it induces vibrational modes in the chemical bonds, causing the light to scatter in different directions. The scattered light, including both elastically scattered (Rayleigh scattering) and inelastically scattered light (Raman scattering), is collected by a detector. The inelastically scattered light, which undergoes a shift in energy corresponding to the vibrational modes of the molecules, is then analyzed to generate a Raman spectrum. This spectrum displays characteristic peaks that correspond to specific molecular vibrations of functional groups within the biofilm (Pan et al., 2016).

4.2.5 Fourier Transform Infrared (FTIR) and Attenuated Total Reflection (ATR) Microscopy

The biofilm sample is placed on an ATR (attenuated total reflection) crystal, and an infrared beam is directed through the sample. As the infrared light reaches the surface, it undergoes total internal reflection within the ATR crystal, generating an evanescent wave that penetrates the sample to a depth of approximately 1.0 μm. This wave interacts with the surface molecules, collecting spectral data from the biofilm's surface. The resulting infrared spectra provide information on the functional groups present, such as hydroxyl, carboxyl, and amide groups, which are characteristic of the components within the extracellular polymeric substances (EPSs) (Pan et al., 2016).

5 Comparison of Different EPS Extraction and Analytical Methods

The methods for extracting extracellular polymeric substances (EPSs) are broadly categorized into physical, chemical, and combined techniques, each with distinct advantages and limitations that affect the quality and quantity of extracted EPS. Physical methods such as centrifugation, ultrasonication, and steaming rely on mechanical or thermal forces to disrupt the biofilm matrix. Centrifugation is effective for processing large volumes but may not efficiently extract tightly bound EPS, often leading to incomplete extraction. Ultrasonication offers higher extraction yields and is particularly useful for tightly bound EPS, though excessive sonication can degrade the EPS matrix, especially with improper frequency and temperature control. Steaming is beneficial for dense biofilms but can denature heat-sensitive EPS components, compromising their functional properties.

Chemical extraction methods, including EDTA, cation exchange resins (CER), and alkaline agents like NaOH, provide more complete extraction by breaking down extracellular matrix components and disrupting ionic bonds. EDTA is highly effective in chelating divalent cations, facilitating EPS release, but it may alter biofilm integrity. Cation exchange resins are gentler but less efficient, while NaOH can enhance EPS yield by breaking down cell wall components, though it may cause cell lysis, affecting the purity of the extracted EPS.

Combined physical-chemical methods, such as ultrasonication with NaOH or heat treatment with NaOH, can improve extraction efficiency by synergistically disrupting the biofilm matrix. However, these methods can also degrade certain EPS components, requiring optimization based on the biofilm type and research objectives. Choosing the appropriate extraction method depends on the biofilm's characteristics, the desired EPS yield and quality, and the possible effect of extraction conditions on the biofilm's structural integrity.

In chemical analysis, several methods offer distinct advantages and limitations in EPS characterization. UV-Visible (UV-Vis) Spectrophotometry is effective for quantifying proteins and specific EPS components but has limited selectivity, complicating readings in complex samples. Chromatography techniques like HPLC and GC-MS provide detailed separation and quantification of monosaccharides, crucial for understanding polysaccharide composition, but require extensive sample preparation that can alter sensitive EPS components. Colorimetric methods, including the Phenol-Sulfuric Acid and Anthrone-Sulfuric Acid and assays, offer high sensitivity for polysaccharide quantification but can degrade EPS under acidic conditions, affecting the accuracy of the results. For protein quantification, the Lowry and Bradford methods are widely used, though they may be influenced by other EPS constituents, leading to inaccuracies.

Spectroscopic techniques like 3D-Excitation Emission Matrix (EEM) Spectroscopy, Nuclear Magnetic Resonance (NMR) Spectroscopy, and Fourier Transform Infrared (FTIR) Spectroscopy, provide in-depth insights into EPS composition, offering valuable functional and structural information. However, these methods can be complex and require careful sample preparation to prevent the degradation of sensitive EPS components. Figure 3 is the Overview of extracellular polymeric substances (EPSs) extraction and characterization methods.

6 Different Genetic Engineering Processes Involved in EPS Production

Extracellular polymeric substances (EPSs) are high-molecular weight polysaccharides secreted by microorganisms into their surrounding environment. They play important roles in biofilm formation, and protection against microorganisms to enhance or modify exopolysaccharide (EPS) production involving several

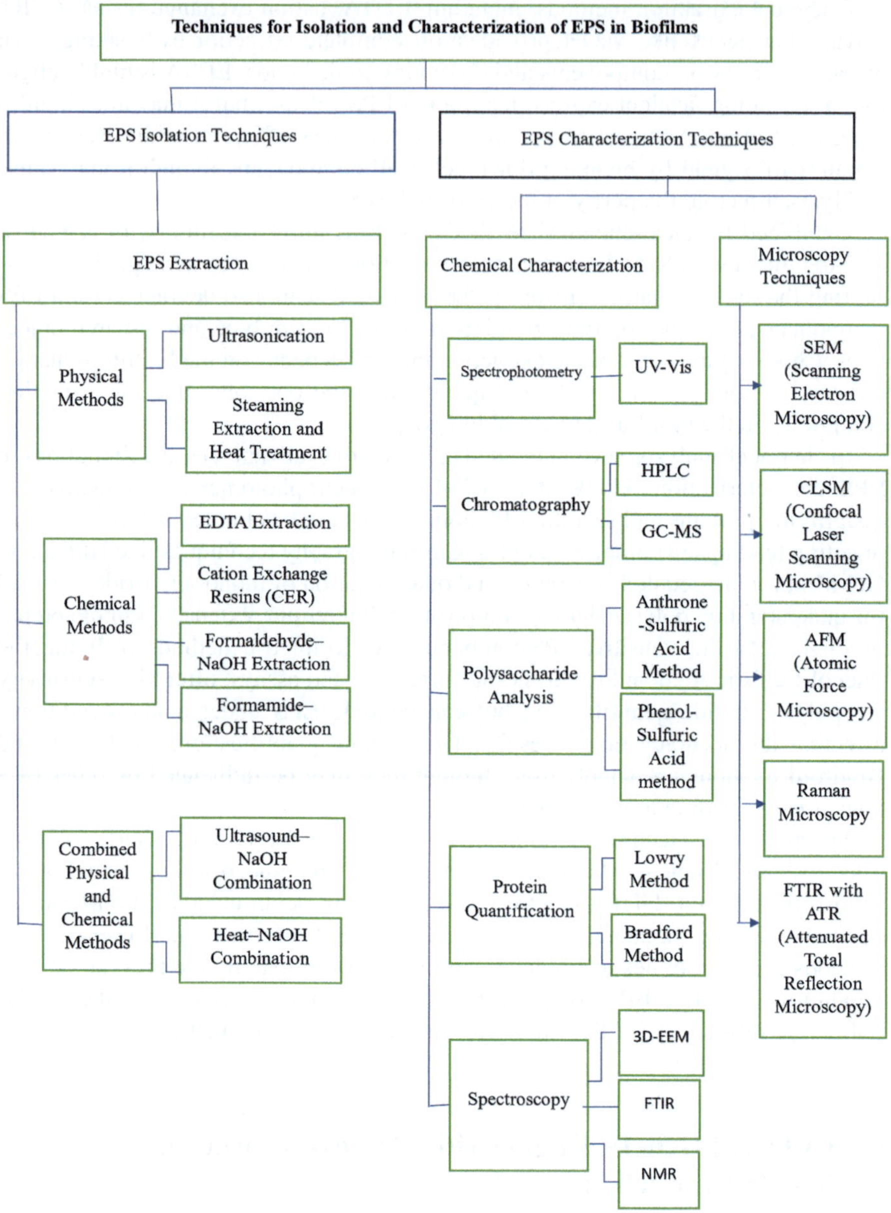

Fig. 3 Overview of extracellular polymeric substances (EPSs) extraction and characterization methods

processes, including gene manipulation and metabolic pathway optimization (Marvasi et al., 2010).

Here is an overview of the key genetic engineering processes involved in extracellular polymeric substances (EPSs).

1. **Gene Identification and Characterization**

   A. **Key enzymes-** Multiple genes and proteins are involved in the process and the biosynthetic pathway, making the genetic overview of EPS production by LAB an auxiliary area of interest. Among them **Glycosyltranferase** (gtfs) and **Polymerases** are the key enzymes involved in EPS production or biosynthesis. Glycosyltranferases catalyze the transfer of sugar moieties. Polymerases assemble sugar monomers into long chains (Soumya & Madhavan Nampoothiri, 2021).

   B. **Genomic Analysis**

   An effective method for comprehending the set of genes a microbe uses to produce biopolymers in genome sequencing and annotation. The discovery of the EPS biosynthesis pathway could open up new avenues for designing overproduction and custom EPSs with desired characteristics through metabolic and genomic engineering. This includes genes coding for precursor synthesis, polymerization, and export machinery (Wang et al., 2019).

2. **Pathway Engineering**

   A. **Precursor Synthesis**

   The synthesis of sugar nucleotide precursors, such as **UDP-glucose** and **UDP-galactose**, is the first step in the creation of EPS. To increase EPS generation, it is imperative to improve the pathways that produce these precursors using methods like **gene overexpression** and **deletion techniques.** For example, **UDP-glucose** availability can be raised by overexpressing the genes (**UGP2, PGM1, GALK1, HK2, GPI**) producing **glucose-1-phosphate uridylyltransferase** (Degeest & De Vuyst, 2000).

   For example**, uridine diphosphate glucose (UDPG) pyrophosphorylase** and **glycosyltransferase** activity, for instance, uridine diphosphate (UDP) glucose is the direct precursor of cellulose synthesis by ***Acetobacter xylinum*** and pullulan manufacture by ***Aurebasidium pullulans***.

   B. **Metabolic engineering—*Lactobacillus casei*** **LC2W** is a probiotic-producing strain that produces exopolysaccharides (EPS). The primary barriers to EPS biosynthesis use in the food business are its low efficiency and hazy regulatory system. This strain was subjected to metabolic engineering in order to better understand the primary rate-limiting elements of EPS biosynthesis and increase its production. Eight relevant genes related to **glycosyltransferase, sugar-nucleotide supply, central metabolism,** and cofactor engineering were cloned and overexpressed. The results showed that the genes **galT, pfk, rfbB,** and nox made the largest contributions to EPS biosynthesis in the study. By increasing EPS biosynthesis, the EPS output increased by 46.0,20,17.4, and 19.6%, respectively. The aerobic circumstances that the recombinant strain **LC nox** notable produced the highest amount of EPS (263.7 mg) were not conductive to the growth of ***Lactobacillus casei*** (Degeest & De Vuyst, 2000).

3. **Gene Expression**
   The genes encoding enzymes involved in EPS biosynthesis are often overexpressed to increase yield. The three eps genes (gtf27, gtf28, and orf29) that confer *Lactobacillus brevis* KB290 are found on the 42.4-kb plasmid Pkb290-1. The membrane-bound glycosyltransferase family includes the predicted products of GTF27 and GTF28, while the ABC transporter is homologous to the orf29 gene product. For instance, overexpressing **glycosyltransferase genes** enhances the assembly of the sugar monomers into EPS. Promoters can be modified for strong and constitutive expression of these genes (Fukao et al., 2019). Another strategy is a **heterologous expression** of EPS biosynthesis genes from one microorganism into another more industrially relevant host, such as ***E.coli*** or **yeast** (Zhang & Poh, 2018).
4. **Regulatory Element Modification**

   A. **Promoter Engineering**
   One of the most significant probiotic lactic acid bacteria (LAB), ***Streptococcus thermophilus,*** is utilized extensively in the dairy sector and is a hot topic in synthetic biology and metabolic engineering. Nevertheless, the well-characterized constitutive promoters that are now available are largely restricted to modifying ***S. thermophilus gene expression***. Using the reporter red fluorescent protein, a pool of constitutive promoters was characterized in *S. thermophilus*, *Lactobacillus casei*, and *Escherichia coli* after being screened by RNA-seq analysis. Six constitutive promoters were chosen for superoxide dismutase expression and markedly increased enzyme activity in the three bacteria mentioned above in order to evaluate their potential for use. Moreover, two strong constitutive promoters were used to construct a dual-expression vector for overexpressing **eps and epsE**, key proteins of exopolysaccharide (EPS) biosynthesis, resulting in the change of molecular weight and the titer of EPS (Kong et al., 2019).

   B. **Transcriptional Activators or Repressors**
   Genetic circuits that modulate the activity of transcription factors can also be introduced to increase EPS biosynthesis under specific environmental conditions (Kamat, 2021).

5. **Heterologous Pathway Introduction**
   It is possible to produce novel EPS in organisms that are simpler to cultivate or scale up industrially by introducing complete EPS biosynthesis gene clusters from EPS producing bacteria into non-EPS producing hosts. For instance, a novel microbial cell factory can be produced by introducing ***Lactobacillus* sp.** Genes that produce EPS into ***E.coli*** (Han et al., 2024).

6. **Transport and Export System Engineering**
   Clarifying the relationships between metabolic pathways and EPS biosynthesis mechanism is necessary to control and thereby increase its microbial productivity, as the need to overcome economic obstacles and the growing importance of microbial EPSs in industrial and medical biotechnology demand. Furthermore, improving the quality and characteristics of products as well as designing new strains requires a deeper comprehension of the biosynthetic mechanism. Consequently, a system-based approach improves the metabolic performance of EPS biosynthesis for industrial use and is a crucial step in comprehending the interaction between metabolism and biosynthesis (Ates, 2015).
7. **Stain Optimization Through Adaptive Evolution**
   Halophilic bacteria that produce microbial exopolysaccharides (EPS) have drawn a lot of attention as a result of growing awareness of the applications of EPSs in a variety of disciplines. The recently discovered ***Kocuria rosea*** **ZJUQH CCTCC M2016754** was successfully adapted to an environment with 10% NaCl, indicating that it is a moderate halobacterium. The best mix of fermentation medium compositions for the formation of EPS was investigated. A fractional factorial design was used in this study to examine the important variables influencing the formation of EPS. It was discovered that the KCl and MgSO4 variables significantly affected the generation of EPS (Gu et al., 2017).
8. **Use of Synthetic Biology Tools:**

   A. **CRISPR-Cas9** and other gene editing tools can be employed for precise modification of EPS related genes. **CRISPR interference (CRISPRi)** can be used to regulate gene expression levels (Kong et al., 2019).
   B. **Metabolic flux analysis and synthetic circuits:** They are used to optimize the balance between growth and EPS production (Kim et al., 2012).

9. **Optimizing yield and composition of EPS:** The type and composition of EPS can be engineered by manipulating the enzymes that determine the monosaccharide composition and their linkages. By modifying glycosyltransferases and altering the types of sugar precursors fed into the pathway, the physical properties (e.g., viscosity, charge) and biological activities of EPS can be tailored for specific applications. Whenever *Lactobacillus fermentum* Lf2 is used as a food ingredient in dairy matrices, it produces large amounts of exopolysaccharides (EPS) (~1 g/L). These EPS have characteristics that make their composition with other similar molecules from lactic acid bacteria (LAB) intriguing.

Those characteristics encouraged us to optimize the production of EPS (Ale et al., 2019).

7 Conclusion

As we went over in the introduction, genetic engineering is becoming more and more relevant to enhancing extracellular polysaccharides for an abundance of applications. We have talked about and detailed a number of significant research findings that were anticipated by top researchers. Various genetic engineering methods have been employed to enhance and modify the qualities and quantities of extracellular polysaccharides. Sugar nucleotide precursor production improvement, metabolic engineering, Gene expression, promoter engineering, transcriptional activators or repressors, transport and export system engineering, CRISPR-Cas9, and, metabolic flux analysis leading to genetically modify the products. According to recent research references, a wide variety of genetic modifications have been made. Together with the creation of biomaterials using genetic engineering technologies, the enhancement of chemical and physical characteristics that increase the applicable capacity are highlighted. It was reported that genetically modified extracellular polysaccharide (EPS) would have a major impact on intestinal health, immune function, wastewater treatment, and the removal of heavy metals from wastewater. Genetically modified EPS with enhanced physical and chemical properties can be utilized in biomedical treatment and environmental effluent remediation processes, expanding their range of applications.

8 Future Prospects

Research on genetically engineered extracellular polysaccharides (EPS) is still ongoing because of its enormous potential for use in biological, medical, and chemical applications as well as in solving global issues. Flexible genome editing techniques include promoter engineering, metabolic engineering, transport and export system engineering, and CRISPR-Cas9 RNA-based gene targeting. As explained, these modified EPS can be created and applied to environmental remediation and biomedical applications. In order to successfully exploit bacterial extracellular polysaccharides (EPS), it is crucial to strategically design the EPS through chemical modification or genetically modified bacterial strains. This calls for a deeper comprehension of the biosynthesis pathways of extracellular polymeric substances and how they relate to cellular metabolism. In the near future, industrial clinical settings will be able to use these genetically modified products. The development of future EPS will be significantly influenced by genetic engineering techniques, which are pertinent to these discussions.

References

Adav, S. S., & Lee, D. (2007). Extraction of extracellular polymeric substances from aerobic granule with compact interior structure. *Journal of Hazardous Materials, 154*(1–3), 1120–1126. https://doi.org/10.1016/j.jhazmat.2007.11.058

Ale, E. C., Batistela, V. A., Olivar, G. C., Ferrado, J. B., Sadiq, S., Ahmed, H. I., Reinheimer, J. A., Candioti, L. V., Laws, A. P., & Binetti, A. G. (2019). Statistical optimisation of the exopolysaccharide production by *Lactobacillus fermentum* Lf2 and analysis of its chemical composition. *International Journal of Dairy Technology, 73*(1), 1–4. https://doi.org/10.1111/1471-0307.12639

Ates, O. (2015). System biology of microbial exopolysaccharide production. *Frontiers in Bioengineering and Biotechnology, 3*. https://doi.org/10.3389/fbioe.2015.00200

Cerning, J. (1995). Production of exopolysaccharides by lactic acid bacteria and dairy propionibacteria. *Le Lalit, 75*(4–5), 463–472.

Chen, J., Zhang, A., Xiang, Z., Lu, M., Huang, P., Gong, T., Pan, Y., Lin, Y., Zhou, X., & Li, Y. (2021). EpsR negatively regulates Streptococcus mutans exopolysaccharide synthesis. *Journal of Dental Research, 100*(9), 968–976. https://doi.org/10.1177/00220345211000668. Epub 2021 Mar 20. PMID: 33749354.

Costa, O. Y. A., Raaijmakers, J. M., & Kuramae, E. E. (2018). Microbial extracellular polymeric substances: Ecological function and impact on soil aggregation. *Frontiers in Microbiology, 9*, 1636. https://doi.org/10.3389/fmicb.2018.01636

Costerton, J. W., Geesey, G. G., & Cheng, K. J. (1978). How bacteria stick. *Scientific American, 238*, 85–95.

De Vuyst, L., & Degeest, B. (1999). Heteropolysaccharides from lactic acid bacteria. *FEMS Microbiology Reviews, 23*(2), 153–177.

Decho, A. W., & Gutierrez, T. (2017). Microbial extracellular polymeric substances (EPSs) in ocean systems. *Frontiers in Microbiology, 8*, 922. https://doi.org/10.3389/fmicb.2017.00922

Degeest, B., & De Vuyst, L. (2000). Correlation of activities of the enzymes α-phosphoglucomutase, UDP-galactose 4-epimerase, and UDP-glucose pyrophosphorylase with exopolysaccharide biosynthesis by *Streptococcus thermophilus* LY03. *Applied and Environmental Microbiology, 66*(8), 3519–3520.

Elsholz, A. K., Wacker, S. A., & Losick, R. (2014). Self-regulation of exopolysaccharide production in Bacillus subtilis by a tyrosine kinase. https://doi.org/10.1101/gad.246397.114. PMID: 25085422; PMCID: PMC4117945

Felz, S., Al-Zuhairy, S., Aarstad, O. A., Van Loosdrecht, M. C., & Lin, Y. M. (2016). Extraction of structural extracellular polymeric substances from aerobic granular sludge. *Journal of Visualized Experiments, 115*. https://doi.org/10.3791/54534

Felz, S., Vermeulen, P., Van Loosdrecht, M. C., & Lin, Y. M. (2019). Chemical characterization methods for the analysis of structural extracellular polymeric substances (EPS). *Water Research, 157*, 201–208. https://doi.org/10.1016/j.watres.2019.03.068

Finan, T. M., Weidner, S., Womg, K., Buhrmester, J., Chain, P., Vorhölter, F. J., Hernandez-Lucas, I., Becker, A., Cowie, A., Gouzy, J., et al. (2001). The complete sequence of the 1683-kb pSymB megaplasmid from the N2-fixing endosymbiont *Sinorhizobium meliloti*. *Proceedings of the National Academy of Sciences of the United States of America, 98*, 9889–9894.

Flemming, H. C., & Wingender, J. (2010). The biofilm matrix. *Nature Reviews. Microbiology, 8*(9), 623–633. https://doi.org/10.1038/nrmicro2415. Epub 2010 Aug 2. PMID: 20676145.

Flemming, H. C., Neu, T. R., & Wozniak, D. J. (2007). The EPS matrix: The "house of biofilm cells". *Journal of Bacteriology, 189*(22), 7945–7947. https://doi.org/10.1128/JB.00858-07. Epub 2007 Aug 3. PMID: 17675377; PMCID: PMC2168682.

Flemming, H. C., Wingender, J., Szewzyk, U., et al. (2016). Biofilms: An emergent form of bacterial life. *Nature Reviews. Microbiology, 14*, 563–575. https://doi.org/10.1038/nrmicro.2016.94

Fukao, M., Zendo, T., Inoue, T., Nakayama, J., Suzuki, S., Fukaya, T., Yajima, N., & Sonomoto, K. (2019). Plasmid-encoded glycosyltransferase operon is responsible for exopolysaccharide production, cell aggregation, and bile resistance in a probiotic strain, Lactobacillus brevis KB290. *Journal of Bioscience and Bioengineering, 128*(4), 391–397. https://doi.org/10.1016/j.jbiosc.2019.04.008

Gu, D., Jiao, Y., Jianan, W., Liu, Z., & Chen, Q. (2017). Optimization of EPS production and characterization by Halophilic Bacterium, *Kocuria rosea* ZJUQH from Chaka salt lake with response surface methodology. *Molecules, 22*, 3519–3520. https://doi.org/10.3390/molecules22050814

Han, N.-N., Yang, J. H., Fan, N. S., & Jin, R. C. (2024). Mechanistic insight into microbial interaction and metabolic pattern of anammox consortia on surface modified biofilm carrier with extracellular polymeric substances. *Bioresource Technology, 407*, 131092.

Janczarek, M. (2011). Environmental signals and regulatory pathways that influence exopolysaccharide production in rhizobia. *International Journal of Molecular Sciences, 12*, 7898–7933. https://doi.org/10.3390/ijms12117898

Kamat, S. (2021). Molecular basis and genetic regulation of EPS. *Microbial Exopolysaccharides as Novel and Significant Biomaterials* (pp. 45–83). Springer Nature. https://doi.org/10.1007/978-3-030-75289-7_3

Kim, H. U., Kim, T. Y., & Lee, S. Y. (2012). Metabolic flux analysis and metabolic engineering of microorganisms. *Molecular BioSystems, 4*, 1–3.

Kong, L.-H., Xiong, Z.-Q., Song, X., Xia, Y.-J., Zhang, N., & Ali, L.-Z. (2019). Characterization of a panel of strong constitutive promoters from *Streptococcus thermophilus* for fine-tuning gene expression. *ACS Synthetic Biology, 8*(6), 1469–1471. https://doi.org/10.1021/acssynbio.9b00045

Kumar, A., Rao, K. M., & Han, S. S. (2018). Application of xanthan gum as polysaccharide in tissue engineering: A review. *Carbohydrate Polymers, 180*, 128–144.

Maitra, D., Roy, B., Dan, R., Sarkar, S., Choudhury, S., & Mitra, A. (2024). A study on the application and efficiency of novel biofertilizer on paddy: A small-scale study. *Journal of Mycopathological Research, 62*, 2583–6315. https://doi.org/10.57023/JMycR.62.2.2024.275

Marvasi, M., Visscher, P. T., & Martinez, L. C. (2010). Exopolymeric substances (EPS) from *Bacillus subtilis:* Polymers and genes encoding their synthesis. *FEMS Microbiology Letters, 313*(1), 1–9. https://doi.org/10.1111/j.1574-6968.2010.02085.x

Pan, M., Zhu, L., Chen, L., Qiu, Y., & Wang, J. (2016). Detection techniques for extracellular polymeric substances in biofilms: A review. *BioResources, 11*(3), 8092–8115.

Reeves, P. R., Hobbs, M., Valvano, M. A., Skurnik, M., Whitfield, C., Coplin, D., Kido, N., Klena, J., Maskell, D., Raetz, C. R., & Rick, P. D. (1996). Bacterial polysaccharide synthesis and gene nomenclature. *Trends in Microbiology, 4*(12), 495–503.

Roca, C., Alves, V. D., Freitas, F., & Reis, M. A. M. (2015). Exopolysaccharides enriched in rare sugars: Bacterial sources, production, and applications. *Frontiers in Microbiology, 6*, 288. https://doi.org/10.3389/fmicb.2015.00288

Schmid, J. (2018). Recent insights in microbial exopolysaccharide biosynthesis and engineering strategies. *Current Opinion in Biotechnology, 53*, 130–136.

Sijia, W., Huo, H., Shi, Y., Zhang, F., Tingting, G., & Li, Z. (2023). Chapter three – Extraction and application of extracellular polymeric substances from fungi. In G. M. Gadd & S. Sariaslani (Eds.), *Advances in applied microbiology* (Vol. 125, pp. 79–106). Academic Press. https://doi.org/10.1016/bs.aambs.2023.08.001. ISSN 0065-2164, ISBN 9780443192760.

Soumya, M. P., & Madhavan Nampoothiri, K. (2021). An overview of functional genomics and relevance of glycosyltransferases in exopolysaccharide production by lactic acid bacteria. *International Journal of Biological Macromolecules, 184*, 1015–1017. https://doi.org/10.1016/j.ijbiomac.2021.06.131

Sun, X., & Zhang, J. (2021). Bacterial exopolysaccharides: Chemical structures, gene clusters and genetic engineering. *International Journal of Biological Macromolecules, 173*, 481–490. https://doi.org/10.1016/j.ijbiomac.2021.01.139. ISSN 0141-8130.

Sutherland, I . W. (2001). Microbial polysaccharides from Gram-negative bacteria. *International Dairy Journal, 11*, 663–674. https://doi.org/10.1016/S0958-6946(01)00112-1

Teasdale, S. L., Wu, X., Ye, L., & Hackney, C. R. (2018). The identification and extraction of extracellular polymeric substances (EPS) from sediments. In *Geomorphological techniques* (pp. 1–14). British Society for Geomorphology.

Tiwari, O. N., Sasmal, S., Kataria, A. K., & Devi, I. (2020). Application of microbial extracellular carbohydrate polymeric substances in food and allied industries. *3 Biotech, 10*(5), 221. https://doi.org/10.1007/s13205-020-02200-w. Epub 2020 Apr 28. PMID: 32355595; PMCID: PMC7188750.

Wang, J., Goh, K. M., Salem, D. R., & Sani, R. K. (2019). Genome analysis of a thermophilic exopolysaccharide producing bacterium–*Geobacillus sp.* WSUCF1. *Scientific Reports, 9*(1), 1–9. https://doi.org/10.1038/s41598-018-36983-z

Wibberg, D., Alkhateeb, R. S., Winkler, A., Albersmeier, A., Schatschneider, S., Albaum, S., Niehaus, K., Hublik, G., Pühler, A., & Vorhölter, F.-J. (2015). Draft genome of the xanthan producer Xanthomonas campestris NRRL B-1459 (ATCC 13951). *Journal of Biotechnology, 204*, 45–46.

Wingender, J., Neu, T., & Flemming, H.-C. (1999). *Microbial extracellular polymer substances*. Springer.

Zhang, J., & Poh, C. L. (2018). Regulating exopolysaccharide gene wcaF allows control of Escherichia coli biofilm formation. *Scientific Reports, 8*(1), 13127. https://doi.org/10.1038/s41598-018-31161-7

Index

T. Sarkar et al. (eds.), *Extracellular Polysaccharides*,
https://doi.org/10.1007/978-3-031-94472-7

GPSR Compliance

The European Union's (EU) General Product Safety Regulation (GPSR) is a set of rules that requires consumer products to be safe and our obligations to ensure this.

If you have any concerns about our products, you can contact us on ProductSafety@springernature.com

In case Publisher is established outside the EU, the EU authorized representative is:

Springer Nature Customer Service Center GmbH
Europaplatz 3
69115 Heidelberg, Germany

Batch number: 10755329

Printed by Printforce, the Netherlands